D0218436

MATHEMATICAL PHYSICS

BY

DONALD H. MENZEL
Director of Harvard College Observatory
Paine Professor of Practical Astronomy
and Professor of Astrophysics
Harvard University

DOVER PUBLICATIONS, INC.
NEW YORK

Published in Canada by General Publishing Company, Ltd., 30 Lesmill Road, Don Mills, Toronto, Ontario.
Published in the United Kingdom by Constable and Company, Ltd., 10 Orange Street, London WC 2.

This Dover edition, first published in 1961, is an unabridged and corrected republication of the second (1953) edition of the work first published by Prentice-Hall, Inc., in 1947 under the title *Theoretical Physics*.

Standard Book Number: 486-60056-4

Manufactured in the United States of America
Dover Publications, Inc.
180 Varick Street
New York, N.Y. 10014

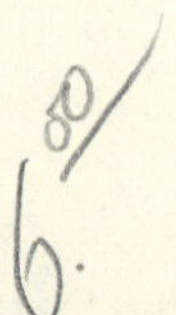

Preface to Dover Edition

Mathematical Physics appears to fulfill the purpose for which it was originally intended: to guide the student through the maze of mathematical analysis necessary to understand theoretical physics. It is particularly adapted for self-study, in that the derivations have few major gaps in logical development of the formulas. Although the book serves as an introduction to classical physics, it carries the reader into such modern applications as matrix algebra and operator calculus, especially that leading into quantum mechanics. Major changes from the earlier edition include correction of typographical errors and revision of the tabulated values of the physical constants, to agree with those given in the latest edition of *Fundamental Formulas of Physics*, Dover Edition S595 and S596.

DONALD H. MENZEL

Cambridge, Mass.
January, 1961

Preface to 1947 Edition

In discussing the external physical world, we may start with Newton's laws of motion, Hamilton's principle, or similar basic postulates and develop therefrom the essential whole of macroscopic classical physics. Or we can adopt Schrödinger's wave equation and study the microscopic properties of atoms. Modern physicists, with their natural inclination to emphasize the newer phases of the subject, sometimes skip over or omit the classical problems entirely.

This volume, although primarily concerned with classical physics, specially emphasizes those topics closely related to modern phases, either in technique or subject matter. Several topics seldom presented, e.g., electron theory or general relativity, appear in considerable detail, because an understanding of them is increasingly vital to the student of atomic physics.

The book has been designed for use in junior, senior, or graduate courses in mathematical physics. The student is expected to possess a good working knowledge of differential and integral calculus. Some prior experience with differential equations is desirable, though not absolutely necessary.

All other phases of mathematics, function theory, vectors, matrices, dyadics, tensors, partial differential equations, etc., flow directly from the physics during the presentation of the various subjects. Such topics as existence theorems, convergence, and high degree of mathematical rigor do not receive special emphasis because they chiefly concern the pure mathematician. Rigor in physics is an important consideration, but a feeling for the mathematics as part of a physical process is even more important.

To develop the student's abilities and to assist his understanding of new principles, the methods followed in the book are not always the shortest or the most elegant in the mathematical sense. Some deliberate duplication occurs to explain a topic more fully or to demonstrate the superiority of a new technique. Also, as an aid to both student and teacher, the intervening steps and auxiliary argument in the development of a formula appear in greater detail than usual. The book generally avoids phrases like "after a little algebra, we get" or "it is obvious that," which commonly imply that the reader must spend some time in laborious if not recondite verification of the intermediate steps. As a consequence, the book will also be suited to the independent reader.

A general simplification of the derivations permits the treatment of topics often omitted. To compensate for the fact that selected topics and problems appear in considerably greater detail than usual, other less essential topics have been omitted or shortened. As far as possible, assumptions and postulates have been distinguished from conclusions that follow directly from the analysis.

In the references to sections, equations, etc., the roman numeral signifies the major Part (I to V), the numeral following indicates the section, with a decimal to mark the equation. Thus II-20.18 refers to equation 18, section 20, Part II. Omission of the roman numeral signifies that the given equation falls in the same Part under immediate consideration. Omission of the numerical part before the decimal indicates that the equation falls within the same section wherein the reference occurs.

In Part IV, Electromagnetic Theory, the basic formulae appear in both Gaussian and rationalized MKS notation. The reader can choose for himself which one he prefers. Each system has certain advantages and disadvantages, but both are here to stay and a scientist must understand something of both systems if he desires to read the literature.

I wish to acknowledge especially the generous help and assistance given to me by George H. Shortley, Carl Anderson, Max Krook, and Cecilia Payne-Gaposchkin, who read the manuscript and made many helpful suggestions. I am indebted to Edward U. Condon, former editor of this Prentice-Hall Physics Series, for his encouragement. Everett Dulitt gave particularly helpful advice on MKS units and on Part IV in general. Mrs. Stanley P. Wyatt, Jr. and Mrs. Richard M. Adams aided me in the preparation of the manuscript in its successive draft forms. And to my students, who have used the original notes in various mimeographed editions, go my special thanks for helpful advice and criticism.

DONALD H. MENZEL

Cambridge, Mass.

Contents

MATHEMATICAL PHYSICS

PART I

Physical Dimensions and Fundamental Units

Units and Dimensions in Physics

1. The significance of an observation. Any direct observation of a physical nature ordinarily results in a number expressing the magnitude of the measured quantity. The simpler measures are those of lengths, of masses, or of times. More complicated ones may be of velocities, of energies, or of angular momenta. The number, by itself, does not indicate what is being observed; its magnitude depends upon the type of measuring scale employed. We may express lengths in centimeters, miles, or light years; we may define masses in grams, tons, or in units of solar mass. All basic physical measurements are, therefore, ratios.

We see that a physical observation has a dual significance: (a) the reading itself, expressed in some chosen system of units and (b) the type of quantity that is being measured. If we refer to the velocity of the earth in its orbit, we do not say simply 29.8, but 29.8 kilometers per second. When the magnitude of the numerical result depends on the units chosen we say that the quantity has dimensions. For example, the physical dimensions of a velocity are those of a length, L, divided by the time, T, or $[LT^{-1}]$.

In Part I, we shall define and discuss briefly many types of physical parameters, many of which we shall encounter again in later sections. Most readers should be satisfied to give §§ 6–13 only a brief glance at the first reading. The material of these sections is presented for completeness and ready reference, with respect to later chapters.

2. Conversion factors. We frequently find it necessary to convert measures in units of one system to those of another, a process ordinarily effected through multiplication* by a conversion factor. The application of conversion factors is greatly simplified because the number of fundamental physical quantities is limited. Disregarding for the moment quan-

*Occasionally an additive constant is also involved, as in the conversion of Fahrenheit temperatures to the centigrade scale. Absolute Fahrenheit and centigrade conform, however, to the regular conversion rules.

tities of an electromagnetic nature, we have the following fundamental physical quantities: length, mass, and time, indicated respectively by the letters L, M, and T. To these we may add the temperature, Θ, expressed on an absolute scale. We choose these quantities as the fundamental physical dimensions, and express measurements of other quantities in terms of them. Area has physical dimensions $[L^2]$ and volume $[L^3]$. No new parameters appropriate only to area and volume are required. We have seen that a velocity has dimensions $[LT^{-1}]$. Similarly the dimensions of acceleration are $[LT^{-2}]$, of force $[MLT^{-2}]$, of energy $[ML^2T^{-2}]$, etc.

The factor necessary to change a measure in one system into units of another is simply determined. We substitute the conversion factor for each of the fundamental units into the dimensional formula for the measured quantity. The result gives the conversion factor. As an example, one astronomical unit* is 1.495×10^8 km and one year is 3.16×10^7 sec. Hence, the factor to convert velocities expressed in astronomical units per year to kilometers per second is 1.495×10^8 divided by 3.16×10^7, or 4.73. In other words, one astronomical unit per year is equal to 4.73 km per sec. This application of the theory of dimensions is well known and should require no further discussion.

In this book we shall employ the centimeter-gram-second (cgs) system of units as far as possible. The International Angstrom Unit (I.A.) is useful for measures of wavelength. For certain astronomical measures, the radius of the earth's orbit (astronomical unit), or the parsec (the distance at which an astronomical unit subtends one second of arc), are more convenient as units of length. Likewise the mass of the sun, and the tropical year may be used in place of the gram and the second. The appropriate conversion factors appear in Table I.

TABLE I

1 I.A.	$= 1 \times 10^{-8}$ cm	1 cm	$= 10^8$ I.A.
1 solar radius	$= 6.953 \times 10^{10}$ cm	1 cm	$= 1.438 \times 10^{-11}$ sol radius
1 ast unit	$= 1.495 \times 10^{13}$ cm	1 cm	$= 6.691 \times 10^{-14}$ ast unit
1 parsec	$= 3.084 \times 10^{18}$ cm	1 cm	$= 3.242 \times 10^{-19}$ parsec
1 solar mass	$= 1.983 \times 10^{33}$ gram	1 gram	$= 5.043 \times 10^{-34}$ solar mass
1 trop yr	$= 3.1557 \times 10^7$ sec*	1 sec	$= 3.1689 \times 10^{-8}$ year

*One sidereal year equals 1.000038773 tropical years.

3. Dimensional constants. Many of the quantities of physics to which names have been given, such as energy, action, force, are not readily visualized. They have found a place as the result of theoretical developments interpreting the data of observation. As we shall see later on, these quantities occur naturally in the mathematical equations that form the

*The semi-major axis of the earth's orbit.

basis of physical theory. Many of these equations are differential in character.

The various "constants" of physics also have physical dimensions, since their values depend on the system in which they are expressed. As an example we have the so-called "constant" G defined by Newton's expression for the law of gravitation, that the attractive force between two bodies is directly proportional to the product of their masses and inversely proportional to the square of the distance between them:

$$F = Gm_1m_2/r^2, \tag{1}$$

where G is the "constant" of proportionality.

Physical equations, like the above, must be true, independent of the units in which they are expressed. If we write in the dimensions, which are known for all but the constant G, we have

$$[\mathrm{MLT^{-2}}] = G[\mathrm{M^2L^{-2}}], \tag{2}$$

where the purely dimensional parts of the equation have, in accord with custom, been enclosed in square brackets. Equation (1), therefore, will be independent of the units only if G has dimensions defined by

$$[G] = [\mathrm{M^{-1}L^3T^{-2}}]. \tag{3}$$

Such a constant is "dimensional" because its magnitude depends on the units used to express the other quantities in the equation. Most constants of nature are of this variety.

4. Electromagnetic quantities. The units for electric charge and magnetic pole strength remain to be fixed. The formula for the force between two charged spheres, containing charges q_1 and q_2 (Coulomb's law), is analogous to that for gravitation 3.1:

$$F = C_1q_1q_2/\kappa r^2. \tag{1}$$

Similarly the force between two magnetic poles of strengths p_1 and p_2 is

$$F = C_2p_1p_2/\mu r^2. \tag{2}$$

The quantities κ and μ are characteristic of the media between the attractive bodies. Electric and magnetic differ from gravitational forces in that they depend on whether the intervening medium is conductive or insulating, magnetically permeable or not. A copper screen, for example, interposed between two bodies, will alter the electric forces, whereas the gravitational action passes through unimpeded. We term the respective quantities κ and μ the "dielectric constant" and the "magnetic permeability."

Several alternative choices of units present themselves. We may select some arbitrary units of electric charge and pole strength, adopt some convenient values for κ and μ in a perfect vacuum, and determine the constants C_1 and C_2 to agree with the measured forces in cgs or other basic units. We defined our units of mass in such an arbitrary manner. This procedure has some advantages, as we shall note later on.

For the moment, we take both C_1 and κ as dimensionless, with $C_1 = 1$ and $\kappa = 1$ for a vacuum, select our unit of electric charge to satisfy equation (1), and write

$$F = q_1 q_2 / \kappa r^2. \tag{3}$$

Units defined in this way are said to be on the "electrostatic system," since the formula depends upon the concept of only stationary electric charges. This definition determines the physical dimensions of electric charge, q, on the electrostatic system.

$$[\mathrm{MLT}^{-2}] = \varepsilon^2[\mathrm{L}^2\kappa^{-1}], \tag{4}$$

or

$$[q] = [\mathrm{M}^{1/2}\mathrm{L}^{3/2}\mathrm{T}^{-1}\kappa^{1/2}] = [\mathrm{M}^{1/2}\mathrm{L}^{3/2}\mathrm{T}^{-1}], \tag{5}$$

since we have assumed that κ is dimensionless.

We proceed to make a similar definition for magnetic poles. Setting $C_2 = 1$, we define the pole strength in terms of dynamical units, thus:

$$F = p_1 p_2 / \mu r^2. \tag{6}$$

the so-called "electromagnetic system." We set μ equal to unity for a vacuum.

The two systems of units are not independent, since we may also define an electric current in terms of the magnetic field produced. There are thus two possible expressions for the electric charge, one in the electrostatic (es) and one in the electromagnetic (em) system. The values are not the same, and we must reconcile the two systems of units. From (6) we see that p has dimensions

$$[p] = [\mathrm{M}^{1/2}\mathrm{L}^{3/2}\mathrm{T}^{-1}\mu^{1/2}]. \tag{7}$$

The force that a magnetic field exerts upon a pole is $F = Hp$. Hence the dimensions of H are

$$[H] = [F/p] = [\mathrm{ML}^{-1/2}\mathrm{T}^{-1}\mu^{-1/2}]. \tag{8}$$

In the electromagnetic system we define the current, i, in terms of the resultant field H. For example, a current flowing in a circle of radius r gives

$$H = 2\pi i/r, \tag{9}$$

so that current has dimensions

$$[i]_{\mathrm{em}} = [\mathrm{M}^{1/2}\mathrm{L}^{1/2}\mathrm{T}^{-1}\mu^{-1/2}]. \tag{10}$$

Now, on the electrostatic system, current has dimensions of charge divided by time, or

$$[i]_{\mathrm{es}} = [\mathrm{M}^{1/2}\mathrm{L}^{3/2}\mathrm{T}^{-2}\kappa^{1/2}]. \tag{11}$$

In one sense, the definition of current on the electrostatic system seems far less artificial than in the electromagnetic. Even so, we measure current in the laboratory with an ammeter, whose operation depends on the magnetic field produced.

We must, of course, define the current so that its physical dimensions are independent of the system. The ratio of our two sets of units requires that

$$[\kappa\mu]^{-1/2} = [\mathrm{LT}^{-1}], \tag{12}$$

the physical dimensions of a velocity. Maxwell pointed out that this quantity is c, the velocity of light in the medium.

Hence, if we use the electrostatic system, we shall reconcile the quantities defined electromagnetically, if we take $\mu = c^2$ for free space. Conversely, if we take the electromagnetic system as fundamental, we shall have to set $\kappa = c^2$ for free space.

The so-called "Gaussian" system provides an alternative to the foregoing. Here we adopt the electrostatic system as basic and take the constant C_2 in equation (2) equal to $1/c$. This procedure enables one to keep $\kappa = \mu = 1$ for free space.

The other widely used system of units is the so-called meter-kilogram-second system. Where the Gaussian system employs only three fundamental quantities [MLT], the MKS system employs four, the extra one being electric charge. This choice makes μ and κ also dimensional, so that we cannot omit them from our equations, even for free space. The Gaussian system is "unrationalized," by which term we mean that factors like 4π continually crop up in unlikely spots. By an arbitrary choice of our unit of electric charge, we can make factors involving π disappear in problems that involve rectangular symmetry and reappear in problems that possess axial or spherical symmetry.

For MKS, Coulomb's law takes the form

$$F = q_1 q_2/4\pi\epsilon r^2, \tag{13}$$

where q is the charge in coulombs and r the distance in meters. The factor 4π in the denominator achieves the aforementioned "rationalization." Since the unit of charge is arbitrary, and we must measure F in "newtons," i.e. kilogram meter/second^2, we must define the quantity ϵ, a characteristic of the medium related to κ in the foregoing, so as to satisfy the dynamical relation (3). For a vacuum, $\epsilon = \epsilon_0 = (10^{-9}/36\pi)$ farad/meter.

Equation (2), for the force between magnetic poles, has some limitations, arising from the fact that magnetic poles, unlike electric charges, are fictitious quantities. If poles did occur singly or if they could be even reasonably isolated we should find (2) a satisfactory representation of the force field. However, we ordinarily operate only with "induced" rather than "real" poles. Hence we introduce an induced pole strength or "magnetic charge," $\bar{p}$, such that

$$\bar{p} = \mu p. \tag{14}$$

Thus we write, instead of (2)

$$F = C_2 \mu \bar{p}_1 \bar{p}_2 / r^2 = \bar{p}_2 B, \tag{15}$$

where

$$B = \mu H, \tag{16}$$

the magnetic induction as defined by Maxwell's equations. Now, for MKS, we set $C_2 = 1/4\pi$, and

$$F = \mu \bar{p}_1 \bar{p}_2 / 4\pi r^2. \tag{17}$$

We must take $\mu = \mu_0 = 4\pi \times 10^{-7}$ henry/meter. Then

$$(1/\epsilon_0 \mu_0)^{1/2} = 3 \times 10^8 \text{ meters/sec} = c, \tag{18}$$

the velocity of light, as required by (12). We shall find, later, that

$$\mathbf{F} = \mathbf{i} \times \mathbf{B}. \tag{19}$$

Hence the physical dimensions of $\bar{p}$ must be the same as that of current, or $[QLT^{-1}]$.

Maxwell, himself, overlooked the fact that **B** is more fundamental than **H**. Many of the older texts fail to point out the difference. Indeed some inconsistencies occasionally enter as a result. Some writers have unjustly attributed the resulting confusion to a supposed defect of Gaussian units. However, the MKS system avoids the difficulty because of the necessity of keeping the constants ϵ and μ in the equations.

In the MKS system, we employ such practical quantities as coulomb, volt, ampere, ohm, watt, farad, etc., along with meter, kilogram, and second. The fact that we can employ such units is the chief argument for the MKS system.

Although engineers have specially favored MKS, many physicists have felt that it is unduly cumbersome, especially when both electrical and dynamical quantities appear in the same equation. Despite the heavy campaign for general adoption of MKS, the Gaussian system will undoubtedly persist for some time to come. Even if MKS ultimately takes over completely, necessity for reading and understanding the older books and journals would seem to require a knowledge of both systems.

Here we hold no special brief for one over the other. In Part IV, which deals with electricity and magnetism, we shall employ both systems concurrently.

5. Definitions and dimensions of physical quantities. One should remember that the definitions of the various physical quantities depend upon the way in which they enter into the mathematical equations of physics. Although there is no a priori reason why quantities of dimensions like

$M^{27}L^{-61}T^{13}$ should not appear in physics, nature fortunately seems to prefer simplicity.

We introduce at this time definitions of various physical quantities and their dimensions in terms of basic units. The list is somewhat more complete than necessary for the purposes of this volume. The letters or combinations of letters that will conventionally be used to denote the given quantity are given in parentheses. Occasional duplication should cause no confusion. Some, but not all, of the vector quantities (i.e., quantities having direction as well as magnitude) have been distinguished by bold-face letters.

Note that a dot above the quantity indicates the first derivative with respect to the time, two dots the second derivative, etc. The physical dimensions appear, as usual, in brackets. The name or designation of the unit is given at the end; cgs units are employed when available. For additional formulae, see the *Smithsonian Physical Tables,* which is the basis of the following compilation.

Dimensions of Various Physical Quantities

6. Fundamental units.

LENGTH. $(l, x, y, z, r, s, ds, d, \mathbf{r}, d\mathbf{s}$, etc.) [L] cm. (1)

MASS. The quantity of matter a body contains. (m, M) [M] g (gram). (2)

TIME. (t) [T] sec (second). (3)

TEMPERATURE. (T) [Θ] deg (absolute or K). (4)

DIELECTRIC CONSTANT. (κ) $[\kappa]$. (5)

MAGNETIC PERMEABILITY. (μ) $[\mu]$. (6)

7. Definitions from geometry and mechanics.

AREA is expressed in terms of a unit square. (A, S, dS) $[L^2]$ cm^2. (1)

VOLUME is expressed in terms of a unit cube. $(V, dV, d\tau)$ $[L^3]$ cm^3. (2)

DENSITY is mass per unit volume. $(\rho = M/V)$ $[ML^{-3}]$ $g\ cm^{-3}$. (3)

FREQUENCY is the number of times per unit of time that a given physical quantity assumes the same value. (ν) $[T^{-1}]$ sec^{-1}. (4)

LINEAR VELOCITY is the rate of change of the distance of an object from a point.

$$(v = \dot{l} = dl/dt, \mathbf{v}) \quad [LT^{-1}] \quad cm\ sec^{-1}. \tag{5}$$

LINEAR ACCELERATION is the rate of change of velocity.

$$(a, \alpha, \ddot{l}, d^2l/dt^2, \dot{\mathbf{v}}) \quad [LT^{-2}] \quad cm\ sec^{-2}. \tag{6}$$

ANGLE is measured by the ratio of the length of an arc to its radius. Since the two lengths are mutually perpendicular, the true dimensional formula may be written $[L_xL_y^{-1}]$, although it is frequently given as unity, (i.e., dimensionless). When L_x and L_y are measured in similar units, the angles are expressed in radians. The possibility of using different measures for the two linear dimensions provides for the use of other units of angular measure, such as degrees (θ). (7)

ANGULAR VELOCITY is the angle described in unit time.

$$(\omega = \dot{\theta} = d\theta/dt, \boldsymbol{\omega}).$$

$$[L_xL_y^{-1}T^{-1}] \quad \text{or} \quad [T^{-1}] \quad \text{sec}^{-1}. \tag{8}$$

ANGULAR ACCELERATION is the rate of change of angular velocity.

$$(\dot{\omega} = \ddot{\theta} = d^2\theta/dt^2) \quad [L_xL_y^{-1}T^{-2}] \quad \text{or} \quad [T^{-2}] \quad \text{sec}^{-2}. \tag{9}$$

AREAL VELOCITY is the rate of transcription of area by a radius vector.

$$(\tfrac{1}{2}l^2\, d\theta/dt) \quad [L_xL_yT^{-1}] \quad \text{or} \quad [L^2T^{-1}]. \tag{10}$$

MOMENTUM is the product of the mass of a moving body by its velocity.

$$(p = m\, dl/dt = m\dot{l}, \mathbf{p}) \quad [MLT^{-1}] \quad \text{g cm sec}^{-1}. \tag{11}$$

MOMENT of an infinitesimal body with respect to a plane is the product of its mass by its perpendicular distance from the plane. [ML] g cm. For an extended body, of density ρ, the moment with respect to the yz plane is $\int \rho x\, d\tau$. With respect to a point, the moment (a vector) is $\int \rho \mathbf{r}\, d\tau$; $d\tau$ is an element of volume. (12)

MOMENT OF INERTIA of a body about an axis is given by the integral $I = \int \rho r^2\, d\tau$, where ρ is the density at any given point, r its distance from the axis, and $d\tau$ an element of volume.

$$[ML^2] \quad \text{g cm}^2. \tag{13}$$

ANGULAR MOMENTUM is the product of the moment of inertia of a body by its angular velocity.

$$\left(L = p_\theta = \int \rho r^2 (d\theta/dt)\, d\tau, \mathbf{L}\right)$$

$$[ML_x^3L_y^{-1}T^{-1}] \quad \text{or} \quad [ML^2T^{-1}] \quad \text{g cm}^2\,\text{sec}^{-1}. \tag{14}$$

MOMENT OF MOMENTUM of an infinitesimal body with respect to a a point is the product of the momentum of the body by its distance from the point.

$$(M(dl_x/dt)l_y, \quad p_xl_y)$$

$$[ML_xL_yT^{-1}] \quad \text{or} \quad [ML^2T^{-1}] \quad \text{g cm}^2\,\text{sec}^{-1}. \tag{15}$$

For an extended body, we may best express the moment of momentum as a vector product, $\int \rho \mathbf{v} \times \mathbf{r}\, d\tau$. (In most problems the angular momentum and moment of momentum are equivalent.)

FORCE is the rate of change of momentum. In non-relativistic formulae, where the mass does not depend on the velocity, it may be defined as mass times the acceleration.

$$\left(F = \frac{d}{dt} mv \sim m \frac{d^2 l}{dt^2}, \mathbf{F}\right) \quad [\mathrm{MLT}^{-2}] \quad \text{g cm sec}^{-2} \text{ or dyne} \tag{16}$$

FORCE INTENSITY or force at a point. To calculate the value of the force, we must know the mass of the accelerated body. In many instances the value of the mass is of no great importance. We may investigate the intensity of the force field surrounding any body in terms of the force acting per unit mass. The magnitude of this quantity is termed the force intensity. Its physical dimensions are those of an acceleration.

$$(F, \mathbf{F}) \quad [F/M = \mathrm{LT}^{-2}]. \tag{17}$$

We shall occasionally refer to the force intensity as the force vector.

WORK is done when a body, acted upon by a force, moves in the direction of the force. Work is the product of the component of the force in the direction of motion, by the distance moved through.

$$\left(W = \int F\, dl\right)$$

$$[\mathrm{ML^2T^{-2}}] \quad \text{g cm}^2 \text{ sec}^{-2} \text{ or dyne cm or erg.} \tag{18}$$

ENERGY results from work done on a body. The work produces a change of shape, of position, of velocity. In the first two instances there is a change of potential energy, in the latter a change of kinetic.

$$T = \text{KINETIC ENERGY} = \tfrac{1}{2}Mv^2.$$

When energy is conserved, i.e., when the processes are completely reversible,

$$V = \text{POTENTIAL ENERGY} = \int F\, dl.$$

Under these conditions, we may derive the force from the potential energy, thus: $F = -\partial V/\partial l$. When the system is non-conservative, part of the work goes into heat, etc.

TOTAL ENERGY. (E, W, H). H refers specifically to the sum of the kinetic and potential energies; $H = T + V$. H is the Hamiltonian function. The Lagrangian function, L, will also be used for certain problems; $L = T - V$. The dimensional formula for energy is the same as that for work.

$$[\mathrm{ML^2T^{-2}}] \quad \text{g cm}^2 \text{ sec}^{-2} \text{ or erg.} \tag{19}$$

POTENTIAL. On occasion we shall find it convenient to define a function that bears the same relation to the potential energy that the force intensity bears to the true force. In other words, this new function is the potential energy that a unit mass would possess at the specified point. When the potential, V, is given as a function of the coordinates, the negative of its partial derivative with respect to a coordinate defines the force intensity in the direction of the coordinate.

$$F = -\partial V/\partial l. \quad [L^2T^{-2}] \quad \text{cm}^2\,\text{sec}^{-2}. \tag{20}$$

TORQUE, moment of force, or centrifugal couple, is the product of a force by a length. It is sometimes defined as the rate of change of angular momentum.

$$(T,\ Fl,\ dL/dt,\ \mathbf{T}) \quad [ML^2T^{-2}] \quad \text{dyne cm}. \tag{21}$$

PRESSURE is the force per unit area.

$$(P,\ p) \quad [ML^{-1}T^{-2}] \quad \text{dyne cm}^{-2}. \tag{22}$$

SOLID ANGLE. To measure the solid angle at a point, P, subtended by a surface, first connect P, by an envelope of straight lines, to the extremities of the surface and then draw, about P as a center, a sphere of radius R. Let S be the area of the spherical surface included in the solid angle. Then

$$\omega = \text{solid angle} = S/R^2 \quad (\omega,\ d\omega,\ \Omega)$$

$$\text{Dimensions: [unity] or } [L_x^2 L_y^{-2}] \quad \text{steradian}. \tag{23}$$

8. Definitions in radiation theory.

ENERGY DENSITY, the energy per unit volume.

$$(\rho = E/V) \quad [ML^2T^{-2}/L^3] = [ML^{-1}T^{-2}] \quad \text{erg cm}^{-3}. \tag{1}$$

ENERGY DENSITY OF ν-RADIATION, the energy per unit volume per unit frequency.

$$(\rho_\nu) \quad \rho = \int_0^\infty \rho_\nu\, d\nu \quad [ML^{-1}T^{-1}] \quad \text{erg cm}^{-3}\,\text{sec}. \tag{2}$$

FLUX OR RADIANT FLUX, the energy flowing per unit time through a given element of surface.

$$(F) \quad [MT^{-3}] \quad \text{erg cm}^{-2}\,\text{sec}^{-1} \quad \text{or} \quad \text{g sec}^{-3}. \tag{3}$$

FLUX OF ν-RADIATION, energy per unit frequency flowing per unit time through a given element of surface. (F_ν).

$$F = \int_0^\infty F_\nu\, d\nu \quad [MT^{-2}] \quad \text{erg cm}^{-2}. \tag{4}$$

FLUX OF λ-RADIATION, energy per unit wavelength flowing per unit time through a given element of surface.

$$(F_\lambda) \quad F = \int_0^\infty F_\lambda \, d\lambda \quad [\mathrm{ML^{-1}T^{-3}}] \quad \mathrm{erg\,cm^{-3}\,sec^{-1}}. \tag{5}$$

SPECIFIC INTENSITY is the flux per unit solid angle (I), (I_ν), or (I_λ). The dimensions are the same as for flux, with which intensity is all too often confused. Further and more precise definitions will appear in the sections on radiation. (6)

ATOMIC ABSORPTION COEFFICIENT represents the area presented by an atom that absorbs energy from an incident beam of radiation.

$$(\alpha, \alpha_\nu) \quad [\mathrm{L^2}] \quad \mathrm{cm^2}. \tag{7}$$

MASS ABSORPTION COEFFICIENT represents the total absorbing area presented by all the atoms in a unit mass of material.

$$(k, k_\nu) \quad [\mathrm{L^2M^{-1}}] \quad \mathrm{cm^2\,g^{-1}}. \tag{8}$$

9. Heat units.

TEMPERATURE. (T) $[\Theta]$. Temperature ordinarily refers to the reading of a thermometer scale. T occurs most frequently in physical equations when multiplied by a dimensional constant. Under such conditions the product has the dimensions of energy. We shall indicate the presence of this multiplying constant by the dimensional symbol $[\Theta']$. The expression

$$[\Theta'] = \mathrm{ML^2T^{-2}} \tag{1}$$

is symbolic of the equivalence of heat and energy. Ordinary heat measurements are based on the energy necessary to raise a unit mass of water at a given temperature, by one degree. In thermal units, quantity of heat has dimensions $[\mathrm{M}\Theta]$. The cgs unit for heat or thermal energy is the calorie; that for ordinary energy is the erg. The factor necessary to convert calories into ergs is called the mechanical equivalent of heat.

SPECIFIC HEAT of a substance is the ratio of the amount of heat necessary to raise the temperature of a given mass by one degree C to the amount required to raise the temperature of the same mass of water by one degree. Being a ratio, the specific heat is dimensionless. (2)

TEMPERATURE GRADIENT is the rate of change of temperature with distance.

$$(dT/dl) \quad [\Theta\mathrm{L^{-1}}]. \tag{3}$$

THERMAL CONDUCTIVITY is the quantity of heat transmitted per unit time per unit area per unit temperature gradient.

$$\begin{aligned} &[\mathrm{ML^{-1}T^{-1}}] \quad \text{in thermal units, or} \\ &[\mathrm{MLT^{-3}}\Theta^{-1}] \quad \text{in dynamical units.} \end{aligned} \tag{4}$$

10. Electrostatic units.

QUANTITY OF ELECTRICITY has already been defined.

$$(Q,\ q,\ Z\varepsilon) \quad [\mathrm{M}^{1/2}\mathrm{L}^{3/2}\mathrm{T}^{-1}\kappa^{1/2}] \text{ esu.} \tag{1}$$

ELECTRIC INTENSITY is the ratio of the force on a quantity of electricity to the quantity of electricity; compare argument of (7.17).

$$(F/Q) \quad [\mathrm{M}^{1/2}\mathrm{L}^{-1/2}\mathrm{T}^{-1}\kappa^{-1/2}]. \tag{2}$$

ELECTRIC POTENTIAL or electromotive force is the ratio of the work done in an electric circuit per unit of electricity operative.

$$(W/Q) \quad [\mathrm{M}^{1/2}\mathrm{L}^{1/2}\mathrm{T}^{-1}\kappa^{-1/2}]. \tag{3}$$

ELECTRIC MOMENT with respect to a plane is the product of the charge by its distance from the plane.

$$(P,\ Ql) \quad [\mathrm{M}^{1/2}\mathrm{L}^{5/2}\mathrm{T}^{-1}\kappa^{1/2}]. \tag{4}$$

In general, the electric moment with respect to a point is a vector, $\mathbf{P} = \int \rho \mathbf{r}\, d\tau$, where ρ is the density of electric charge.

CAPACITY is proportional to the ratio of a charge to the potential of the charge.

$$C = Q/\text{potential} \quad [\mathrm{L}\kappa]. \tag{5}$$

Capacity is expressed in centimeters in es units.

ELECTRIC CURRENT is the rate at which electricity flows past a given point.

$$i = \dot{Q} = dQ/dt \quad [\mathrm{M}^{1/2}\mathrm{L}^{3/2}\mathrm{T}^{-2}\kappa^{1/2}]. \tag{6}$$

POTENTIAL GRADIENT is the rate of change of electric potential with distance.

$$[\mathrm{M}^{1/2}\mathrm{L}^{-1/2}\mathrm{T}^{-1}\kappa^{-1/2}]. \tag{7}$$

ELECTRIC CONDUCTIVITY is the quantity of electricity transmitted per unit time per unit area per unit potential gradient.

$$[T^{-1}\kappa]. \tag{8}$$

CONDUCTANCE is the ratio of a current flowing through a conductor to the difference of potential between its ends.

$$[\mathrm{L}\mathrm{T}^{-1}\kappa]. \tag{9}$$

RESISTANCE is the reciprocal of the conductance.

$$[\mathrm{L}^{-1}\mathrm{T}\kappa^{-1}]. \tag{10}$$

11. Electromagnetic units. The definitions for many of the various electromagnetic quantities are analogous to those for electrostatic quantities and need not be repeated in detail.

MAGNETIC POLE STRENGTH is defined by equations (4.2), analogous to (4.1) for the electrostatic system.

$$(p) \quad [M^{1/2}L^{3/2}T^{-1}\mu^{1/2}]. \tag{1}$$

MAGNETIC FIELD INTENSITY. $[M^{1/2}L^{-1/2}T^{-1}\mu^{-1/2}]$. (2)

MAGNETIC POTENTIAL. $[M^{1/2}L^{1/2}T^{-1}\mu^{-1/2}]$. (3)

MAGNETIC MOMENT. $[M^{1/2}L^{5/2}T^{-1}\mu^{1/2}]$. (4)

CURRENT. $[M^{1/2}L^{1/2}T^{-1}\mu^{-1/2}]$. (5)

QUANTITY OF ELECTRICITY is the product of the current by the time.

$$[M^{1/2}L^{1/2}\mu^{-1/2}]. \tag{6}$$

ELECTRIC POTENTIAL. $[M^{1/2}L^{3/2}T^{-2}\mu^{1/2}]$. (7)

CAPACITANCE. $[L^{-1}T^{2}\mu^{-1}]$. (8)

CONDUCTANCE. $[L^{-1}T\mu^{-1}]$. (9)

RESISTANCE. $[LT^{-1}\mu]$. (10)

SELF INDUCTANCE is the emf (electromotive force) produced in a circuit, per unit rate of variation of the current through the circuit.

$$\text{emf}/(di/dt) \quad [L\mu]. \tag{11}$$

12. The physical constants. We have already studied one example of the way in which certain dimensional constants enter naturally into the equations of physics, that of the constant of gravitation, G, which appeared in equation (3.1). Many such constants exist. They appear sometimes singly and sometimes in combination with one another. Their values are never fixed by a priori reasoning.* The equations are intended to be quantitatively representative of nature; therefore the values of the constants must be determined by experiment. Since successive experiments always yield the same results, within the limits of experimental error, the quantities are said to be "constants of nature." The experimental methods of determining the values of various physical constants do not fall within the scope of this book. Birge has given an admirable summary.†

A number of choices for the fundamental constants are presented, from which we may derive the others. The constants ordinarily chosen are not necessarily the simplest in the dimensional sense, but the ones most easily and accurately determined from experiment. The constants fall into two categories, the one relative to fundamental physical relations and the other dependent on reconciling various systems of units employed. Table (12.1)

*Various investigators have proposed recondite semi-metaphysical theories intended to fix the values of certain dimensionless ratios, but no definite and entirely satisfactory proofs of this reasoning have appeared.

†Cf. *Reviews of Modern Physics*, 1, 1, 1929; or *Smithsonian Physical Tables*, 8th rev. ed., 1934, pp. 73 ff.

gives the fundamental and derived constants, as calculated by DuMond and Cohen.

LEAST-SQUARES ADJUSTED VALUES OF PRIMITIVE UNKNOWNS AND FUNCTIONS THEREOF

(See end of table for explanation of Γ coefficients)

N Avogadro's number:
$(6.02486 \pm 0.00016) \times 10^{23}$ gm mol^{-1} (phys) (-1.241Γ)

c Velocity of light:
(299793.0 ± 0.3) km sec^{-1} (0.023Γ)

ε Electronic charge:
$(4.80286 \pm 0.00009) \times 10^{-10}$ esu (1.291Γ)

$\varepsilon' = \varepsilon c^{-1}$ Electronic charge:
$(1.60206 \pm 0.00003) \times 10^{-20}$ emu (1.268Γ)

m Electron rest mass:
$(9.1083 \pm 0.0003) \times 10^{-28}$ g (1.077Γ)

h Planck's constant:
$(6.62517 \pm 0.00023) \times 10^{-27}$ erg sec (2.073Γ)

$\hbar = h/2\pi$ (h-"bar"):
$(1.05443 \pm 0.00004) \times 10^{-27}$ erg sec (2.073Γ)

λ_g/λ_s Conversion factor from Siegbahn x-units to milliangstroms:
(1.002039 ± 0.000014) (0.349Γ)

$F = N\varepsilon$ Faraday constant:
$(2.89366 \pm 0.00003) \times 10^{14}$ esu g mol^{-1} (phys) (0.050Γ)

$F' = N\varepsilon'$ Faraday constant:
(9652.19 ± 0.11) emu g mol^{-1} (phys) (0.027Γ)

h/ε $(1.37942 \pm 0.00002) \times 10^{-17}$ erg sec esu^{-1} (0.782Γ)

ε/m Specific charge of the electron:
$(5.27305 \pm 0.00007) \times 10^{17}$ esu g^{-1} (0.214Γ)

ε'/m Specific charge of the electron:
$(1.75890 \pm 0.00002) \times 10^{7}$ emu g^{-1} (0.191Γ)

h/m (7.27377 ± 0.00006) cm^2 sec^{-1} or erg sec g^{-1} (0.966Γ)

$\alpha = 2\pi\varepsilon^2 h^{-1} c^{-1}$ Fine structure constant:
$(7.29729 \pm 0.00003) \times 10^{-3}$ (0.486Γ)

α^{-1} 137.0373 ± 0.0006 (-0.486Γ)

α^2 $(5.32504 \pm 0.00005) \times 10^{-5}$ (0.972Γ)

$\alpha/2\pi$ $(1.161398 \pm 0.000005)10^{-3}$ (0.486Γ)

$(1 - \alpha^2)^{1/2}$ $1 - (0.266252 \pm 0.000002) \times 10^{-4}$ $(-2.6 \times 10^5\Gamma)$

$\lambda_{ce} = hm^{-1}c^{-1}$ Compton wavelength of the electron:
$(2.42626 \pm 0.00002) \times 10^{-10}$ cm (0.973Γ)

$\lambda_{ce} = hm^{-1}c^{-1}(2\pi)^{-1}$ Compton radian length of the electron:
$(3.86151 \pm 0.00004) \times 10^{-11}$ cm (0.973Γ)

Nm Atomic wt (electron):
$(5.48763 \pm 0.00006) \times 10^{-4}$ (phys) (−0.164Γ)

$(4\pi c)^{-1}(\varepsilon'/m)$ Zeeman displ. per gauss:
$(4.66885 \pm 0.00006) \times 10^{-5}$ cm^{-1} gauss^{-1} (0.168Γ)

$a_0 = h^2(4\pi^2 m\varepsilon^2)^{-1}$ First Bohr radius:
$(5.29172 \pm 0.00002) \times 10^{-9}$ cm (0.487Γ)

$a_0' = a_0(1 - \alpha^2)^{1/2}$ Separation of electron and proton in ground state: of H^1
$(5.29158 \pm 0.00002) \times 10^{-9}$ cm (0.487Γ)

$a_0'' = a_0' R_\infty / R_H$ Radius of electron orbit referred to center of mass for normal H^1:
$(5.29446 \pm 0.00002) \times 10^{-9}$ cm (0.487Γ)

$r_0 = \varepsilon^2 m^{-1} c^{-2}$ Classical radius of the electron:
$(2.81785 \pm 0.00004) \times 10^{-13}$ cm (1.459Γ)

r_0^2 $(7.94030 \pm 0.00021) \times 10^{-26}$ cm^2 (2.918Γ)

H Atomic wt hydrogen:
1.008142 ± 0.000003 (phys)

$H^+ = H - Nm$ Atomic wt proton:
1.007593 ± 0.000003 (phys) (8.94×10^{-5}Γ)

$H/H^+ = R_\infty/R_H$ $1.000544613 \pm 0.000000006$ (-8.94×10^{-5}Γ)

$H^+/H = R_H/R_\infty$ $0.999455683 \pm 0.000000006$ (8.94×10^{-5}Γ)

H^+/Nm Ratio proton mass to electron mass:
1836.12 ± 0.02 (0.164Γ)

Nm/H^+ Ratio electron mass to proton mass:
$(5.44627 \pm 0.00005) \times 10^{-4}$ (−0.164Γ)

R_∞ Rydberg for infinite mass:
$(109737.309 \pm 0.012$ cm$^{-1})$

$R_H = (1 - Nm/H)R_\infty$ Rydberg for hydrogen:
109677.576 ± 0.012 cm^{-1} (8.94×10^{-5}Γ)

$\mu = mH^+/H$ Reduced mass of electron in hydrogen atom:
$(9.1034 \pm 0.0003) \times 10^{-28}$ g (1.077Γ)

$(1/16)R_H\alpha^2$ Fine structure doublet separation in hydrogen:
$(0.3650234 \pm 0.0000035)$ cm^{-1} (0.972Γ)

$\sigma = 2\pi^5 R_0^4/15c^2h^3N^4$ Stefan-Boltzmann constant:
$(5.6687 \pm 0.00010) \times 10^{-5}$ erg cm^{-2} deg^{-4} sec^{-1} (−1.301Γ)

$c_1 = 8\pi hc$ First radiation constant:
$(4.9918 \pm 0.0002) \times 10^{-15}$ erg cm (2.096Γ)

$c_2 = hcNR_0^{-1}$ Second radiation constant:
(1.43880 ± 0.00007) cm deg (0.855Γ)

c_2/c Atomic specific heat constant:
$(4.79931 \pm 0.00023) \times 10^{-11}$ sec deg (0.832Γ)

$k = R_0 N^{-1}$ Boltzmann's constant:
$(1.38044 \pm 0.00007) \times 10^{-16}$ erg deg^{-1} (1.241Γ)

$\lambda_{max} T = chN/4.965114R_0 = 0.2014052c_2$ Wien's displacement law constant:
(0.289782 ± 0.000013) cm deg (0.855Γ)

$\mu_0 = h\varepsilon'/(4\pi m)$ Bohr magneton:
$(0.92731 \pm 0.00002) \times 10^{-20}$ erg gauss^{-1} (2.264Γ)

$\mu_0(1 + \alpha/2\pi - 3\alpha^2/\pi^2)$ Theor. magnetic moment of the electron:
$(0.92837 \pm 0.00002) \times 10^{-20}$ erg gauss^{-1} (2.269Γ)

$(3k/N)^{1/2} = (3R_0/N^2)^{1/2}$ Multiplier of (Curie const.)$^{1/2}$ to give magnetic moment per molecule:
$(2.62178 \pm 0.00010) \times 10^{-20}$ (erg mole deg^{-1})$^{1/2}$ (phys) (1.241Γ)

$8\pi^2 mh^{-2}$ Schrödinger constant for a fixed nucleus:
$(1.63836 \pm 0.00007) \times 10^{27}$ erg^{-1} cm^{-2} (−3.069Γ)

$8\pi^2 \mu h^{-2}$ Schrödinger constant for the hydrogen atom:
$(1.63748 \pm 0.00007) \times 10^{27}$ erg^{-1} cm^{-2} (−3.069Γ)

$E_0 = c^2 F'^{-1} 10^{-14}$ Conversion factor from atomic mass units to Mev:
(931.141 ± 0.010) Mev (amu)$^{-1}$ (phys) (0.019Γ)

$E_g = c^2 \epsilon'^{-1} 10^{-14}$ Conversion factor from grams to Mev:
$(5.61000 \pm 0.00011) \times 10^{26}$ Mev g^{-1} (−1.222Γ)

$E_e = c^2(\epsilon'/m)^{-1} 10^{-14}$ Energy equiv. of electron mass in Mev:
(0.510976 ± 0.000007) Mev electron-mass^{-1} (−0.145Γ)

$E_p = E_e \mathrm{H}^+/Nm$ Energy equiv. of proton mass in Mev:
(938.211 ± 0.010) Mev proton-mass^{-1} (0.019Γ)

$\lambda_{cp} = \lambda_{ce} Nm/\mathrm{H}^+$ Compton wavelength of the proton:
$(1.32141 \pm 0.00002) \times 10^{-13}$ cm (0.809Γ)

$\bar{\lambda}_{cp} = \bar{\lambda}_{ce} Nm/\mathrm{H}^+$ Compton radian-length of the proton:
$(2.10308 \pm 0.00003) \times 10^{-14}$ cm (0.809Γ)

$\lambda_0 = (hc^2\varepsilon^{-1})10^{-8}$ Wavelength associated with 1 ev:
$(12397.67 \pm 0.22) \times 10^{-8}$ cm (0.828Γ)

$\nu_0 = 10^8 \varepsilon h^{-1} c^{-1}$ Frequency associated with 1 ev:
$(2.41814 \pm 0.00004) \times 10^{14}$ sec^{-1} (−0.805Γ)

$\tilde{\nu}_0 = 10^8\, \varepsilon h^{-1} c^{-2}$ Wave number associated with 1 ev:
(8066.03 ± 0.14) cm^{-1} (−0.828Γ)

$\varepsilon 10^8/c$ Energy associated with 1 ev:
$(1.60206 \pm 0.00003) \times 10^{-12}$ erg (1.268Γ)

$E_1 = hc$ Energy associated with unit wave number:
$(1.98618 \pm 0.00007) \times 10^{-16}$ erg (2.096Γ)

$v_0 = [2 \cdot 10^8(e'/m)^{1/2}$ Speed of 1 ev electron:
$(5.93284 \pm 0.00009) \times 10^7$ cm sec^{-1} (0.0905Γ)

$(R_0/F')10^{-8}$ Energy associated with 1° Kelvin:
$(8.6167 \pm 0.0004) \times 10^{-5}$ ev (-0.027Γ)

$(F'/R_0)10^{8}$ "Temperature" associated with 1 ev:
(11605.4 ± 0.5) deg Kelvin (0.027Γ)

$n_0 = N/V_0$ Loschmidt's number:
$(2.68719 \pm 0.00010) \times 10^{19}$ cm^{-3} (-1.241Γ)

$$S_0/R_0 = \ln \left\{ \begin{matrix} (2\pi k)^{3/2}(2.71828)^{5/2}h^{-3}N^{-5/2} \\ \text{Sakur-Tetrode constant} \end{matrix} \right\}$$

(-5.57324 ± 0.00007) (0.225Γ)

S_0 $(-4.63524 \pm 0.00020) \times 10^{8}$ erg mole^{-1} deg^{-1} (0.225Γ)

In the preceding table, the uncertainties of the most poorly determined value limit the accuracy of each expression. Among the atomic quantities, the one least well known is a factor related to the hyperfine structure shift of $^2S_{1/2}$ of hydrogen. Theory proposes the relationship:

$$\Delta\nu_H = \alpha^2(\mu_p/\mu_0)R_\infty G^{-2}.$$

The quantity G^2, whose nature we need not record here, is imperfectly known and is subject to future changes as we revise it from refined experimental determinations. Designate our best present guess as $\mathcal{G}^2$, and define a quantity Γ, such that

$$\Gamma = 10^6(G^2/\mathcal{G}^2 - 1).$$

The quantities set in parentheses to follow each value of the above table indicate the change, in parts per million, needed to correct the tabulated value for any error in the estimated G, if later studies show that $\mathcal{G}$ is not the correct value.

For example, under N we find the quantity (-1.241Γ). If we should later find that $(G^2/\mathcal{G}^2 - 1) = -30.6 \times 10^{-6}$, $(-1.241\Gamma = 38)$. Consequently we should have to increase the tabulated value of N by 38×10^{-6} or 38 parts per million.

13. The checking of physical equations. As was mentioned in the previous section, we are at liberty to choose any of the constants we wish as fundamental. The constants appearing most frequently in physical equations are the following:

gravitation $G = 6.670 \times 10^{-8}$ dyne cm^2 g^{-2} $[M^{-1}L^3T^{-2}]$ (1)

velocity of light $c = 2.99793 \times 10^{10}$ cm sec^{-1} $[LT^{-1}]$ (2)

electronic charge $\varepsilon = 4.80286 \times 10^{-10}$ es units $[M^{1/2}L^{3/2}T^{-1}]$ (3)

electronic mass $m = 9.1083 \times 10^{-28}$ g $[M]$ (4)

Planck's constant $h = 6.62517 \times 10^{-27}$ erg sec $[ML^2T^{-1}]$ (5)

Boltzmann's constant $k = 1.38044 \times 10^{-16}$ erg deg^{-1} $[ML^2T^{-2}\Theta^{-1}]$ (6)

Loschmidt number $n_0 = 2.68719 \times 10^{19}$ cm^{-3} $[L^{-3}]$ (7)

The constants usually occur in combination with the physical variables, time, velocity, energy, etc., which represent the parameters of the particular physical systems under investigation. An equation is a statement of the functional relationship between the parameters. Newton's equation, for example, expresses the dependence of the gravitational force upon the masses of and the distance between two bodies. Assignment of numerical values to the physical constants and to the independent parameters leads to a numerical value of the force.

The two sides of the equation must agree dimensionally as well as numerically. If, after carrying through a laborious analysis, we substitute the dimensions of the various quantities into the equation and find that the left-hand side represents a force $[MLT^{-2}]$ and the right-hand side an acceleration $[LT^{-2}]$, we must have inadvertently omitted or lost a factor of dimensions $[M]$ in the calculation. Dimensions, accordingly, form a rapid means of checking the accuracy of algebraic computation. If physical constants and parameters occur in a logarithm or an exponent, their combination must ordinarily be dimensionless. Otherwise the value of the quantity would depend on the units employed.

In checking the dimensions of differential equations, the student should note that derivatives also possess dimensions. The following examples are self-explanatory and will be a guide to more complicated forms.

$$\dot{l} = dl/dt, \quad [LT^{-1}]; \qquad \dot{l}^2 = (dl/dt)^2, \quad [L^2T^{-2}];$$

$$\ddot{l} = d^2l/dt^2, \quad [LT^{-2}]; \qquad d^3l/dt^3, \quad [LT^{-3}].$$

Dimensional Analysis

14. Derivation of equations. The requirement that equations must balance dimensionally, plus the fact that the number of fundamental dimensions and physical constants is limited, places a powerful tool in our hands. If we suspect that certain parameters may enter into a certain physical relationship, we can set up a general equation between the constants and variables. The condition of dimensional equality requires that the variables combine with one another so that the exponents for each dimension are identical on each side of the equation. Thus we may often determine the functional form of an equation except for the numerical factors of zero dimension. We call this procedure *dimensional analysis.*

To illustrate the method we shall first of all investigate the classical example of the relationship between the parameters of the pendulum and the period of its swing, under gravitational action. The constant of gravitation G may be expected to enter, but none of the others. Atomic constants could play no part in the process. The constants k and n_0 will not come in, if we suppose the pendulum to be placed in an evacuated chamber at constant temperature. The parameters characteristic of the pendulum are its length, l, and the mass, m, of the bob. Let t be the period of its swing. We should expect the swing to depend on the acceleration of gravity, g, which is given by

$$g = \frac{GM_E}{R_E^2}, \qquad [\mathrm{LT}^{-2}] \tag{1}$$

where M_E is the mass and R_E the radius of the earth. Then we may write

$$t = g^\alpha m^\beta l^\gamma, \tag{2}$$

where α, β, and γ are numerical exponents to be determined. The dimensional analogue of this equation is

$$[\mathrm{T} = \mathrm{M}^\beta \mathrm{L}^{\gamma+\alpha} \mathrm{T}^{-2\alpha}], \tag{3}$$

by which we see, equating exponents on both sides of the equation, that

$$\beta = 0 \qquad \gamma + \alpha = 0 \qquad -2\alpha = 1$$

$$\alpha = -1/2 \qquad \beta = 0 \qquad \gamma = 1/2.$$

Thus

$$t = 2\pi\sqrt{l/g}. \tag{4}$$

The numerical factor 2π, which dimensional analysis leaves undetermined, has been inserted for reference. The period of swing proves to be independent of the mass, a result that experiment and further theoretical analysis substantiate. In this development we have omitted consideration of θ, the angle of swing, which actually does enter into the problem for large amplitudes. Here dimensional analysis gives no information, except that we may expect a dimensionless factor $f(\theta)$, to multiply the right-hand side of (4).

The method finds immediate application to many problems of physics and astronomy. It has been suggested that certain stars, the Cepheid variables, whose light and radial velocities undergo periodic variations, are actually pulsating. We suspect that the period, t, may depend in some way upon the star's radius, R, and mass, M. We also expect that G will enter because the restoring force that tends to limit the pulsations is gravitational. Hence we write

$$t = G^\alpha R^\beta M^\gamma, \tag{5}$$

$$[\mathrm{T} = \mathrm{M}^{-\alpha+\gamma} \mathrm{L}^{3\alpha+\beta} \mathrm{T}^{-2\alpha}], \tag{6}$$

$$-2\alpha = 1 \qquad 3\alpha + \beta = 0 \qquad -\alpha + \gamma = 0$$

$$\alpha = -1/2 \qquad \beta = 3/2 \qquad \gamma = -1/2.$$

Thus
$$t = (2)\left(\frac{R^3}{GM}\right)^{1/2} = \left(\frac{3}{\pi}\right)^{1/2}(G\rho)^{-1/2}, \tag{7}$$

where ρ is the mean stellar density. Sterne gives the expression

as
$$t = (6\pi\beta)^{1/2}(G\rho)^{-1/2}, \tag{8}$$

where β is a dimensionless parameter depending on the mode of vibration of the star and the ratio of the specific heats of stellar material. Equation (8) fits the observational data very closely, within the uncertainty of our knowledge of the parameter β.

Let us investigate the problem presented by Kepler's third law of planetary motion, the relation between R, the planet's mean distance from the sun, and t, its periodic time. When we introduce the parameters G and M, the solar mass, the resulting equations are identical with equations (5)–(7), except for an undetermined factor π. We find, on introducing the factor π, and squaring (7), that

$$t^2 = (4\pi^2)R^3/GM. \tag{9}$$

The squares of the periodic times are proportional to the cubes of the mean distances. A more rigorous analysis shows that we should replace M by the sum of the masses of sun and planet. This difference becomes important for the problem of the spectroscopic binary, a double star whose components possess comparable masses.

We may determine the functional form of many important physical laws in similar manner. Take Stefan's law, for example, which expresses the relation between energy radiated per cm^2 per second (flux) by a hot, perfectly emitting body, and the temperature of the body. We select the physical constants, c, h, and k, as relative to the problem, in addition to the absolute temperature, T and the flux F.

$$F = c^\alpha h^\beta k^\gamma T^\delta. \tag{10}$$

$$[\mathrm{MT}^{-3} = \mathrm{M}^{\beta+\gamma}\mathrm{L}^{\alpha+2\beta+2\gamma}\mathrm{T}^{-\alpha-\beta-2\gamma}\Theta^{-\gamma+\delta}]. \tag{11}$$

(In the dimensional equation, T represents time and Θ the temperature). We have four equations and four unknowns. The solution is

$$\alpha = -2, \qquad \beta = -3, \qquad \gamma = \delta = 4.$$

Note that the temperature must always be accompanied by the factor k, or a parameter of similar dimensions. One may save some time in checking equations by immediately setting kT equal to an energy. Stefan's law becomes

$$F = \left(\frac{2\pi^5}{15}\right)\frac{(kT)^4}{h^3c^2}\,\cdot \tag{12}$$

To derive Boyle's gas law we look for a relation between P, n_0, k, and T.

$$P = n_0^\alpha k^\beta T^\gamma, \tag{13}$$

$$[\mathrm{ML^{-1}T^{-2}} = \mathrm{M^\beta L^{-3\alpha+2\beta}T^{-2\beta}\Theta^{-\beta+\gamma}}]. \tag{14}$$

Here $\alpha = \beta = \gamma = 1$, and

$$P = n_0 kT. \tag{15}$$

Further illustrations of dimensional analysis appear in the problems at the end of the chapter.

15. Dimensionless combinations. Let us look for a dimensionless combination of the physical constants m, ε, and c.

$$\text{CONST} = m^\alpha \varepsilon^\beta c^\gamma,$$

$$1 = [\mathrm{M^{\alpha+\beta/2}L^{3\beta/2+\gamma}T^{-\beta-\gamma}}].$$

The result gives two equations that are inconsistent with one another:

$$3\beta/2 + \gamma = 0, \qquad \beta + \gamma = 0.$$

Thus no solution is possible. If, in addition, we include Planck's constant, h, we find that

$$1 = \mathrm{M^{\alpha+(\beta/2)+\delta}L^{(3\beta/2)+\gamma+2\delta}T^{-\beta-\gamma-\delta}}. \tag{1}$$

Here we have four variables and only three equations. Regarding δ as fixed and carrying out the solution in terms of δ we find

$$\alpha = 0, \qquad \beta = -2\delta, \qquad \gamma = \delta.$$

Hence the constant is $(hc/\varepsilon^2)^\delta$. For the simplest form we take $\delta = 1$. This constant occurs in numerous physical applications, especially when written in the form $hc/2\pi\varepsilon^2$, where it appears in the theory of atomic spectra, related to the fine structure of lines. Although the numerical value, according to the best determinations, is 137.29 ± 0.11, Eddington has argued that its true value is exactly 137. Since this constant is truly dimensionless, its value is independent of the system of units.

There are many examples where dimensionless combinations of physical constants and parameters may enter. Consider the problem of the distribution of atomic velocities in a gaseous assembly. Let $P(v)\,dv$ be the probability that a given atom will have a velocity between v and $v + dv$. The integral

$$\int_0^\infty P(v)\,dv = 1, \tag{2}$$

since the probability that the atom has *some* velocity is unity. The product

$P(v)\,dv$ must be a pure number, hence $P(v)$ has the physical dimensions of v^{-1}.

$$P(v) = m^{\alpha}v^{\beta}k^{\gamma}T^{\delta}, \tag{3}$$

or

$$[\mathrm{L}^{-1}\mathrm{T} = \mathrm{M}^{\alpha+\gamma}\mathrm{L}^{\beta+2\gamma}\mathrm{T}^{-\beta-2\gamma}\Theta^{-\gamma+\delta}]. \tag{4}$$

The equations are

$$\alpha + \gamma = 0, \quad \beta + 2\gamma = -1, \quad -\beta - 2\gamma = 1, \quad -\gamma + \delta = 0.$$

The second and third equations prove to be identical, so that a unique solution is no longer possible. We have

$$\alpha = -\delta, \qquad \beta = -1 - 2\delta, \qquad \gamma = \delta.$$

Thus

$$P(v) = v^{-1}(mv^2/kT)^{-\delta}. \tag{5}$$

Whenever we have one more unknown than equations, we shall find that a dimensionless factor usually occurs. This factor may be present not only as a multiplying constant but it may also occur as an exponent, as an argument of a logarithm, cosine, etc. We cannot carry the dimensional analysis further in this example. We might conclude, from the condition of equation (2), that δ is negative and that an exponential of the form $e^{(-\mathrm{const}\;mv^2/kT)}$ is involved, to make the definite integral finite at both limits. Later on, by the methods of statistical mechanics, we shall prove that

$$P(v) = 4\pi v^{-1}(mv^2/2\pi kT)^{3/2}e^{-mv^2/2kT}. \tag{6}$$

This problem illustrates of the limitations of dimensional analysis. Nevertheless, even if no unique solution is possible, the general form of the actual equation may often be determined. Let A be the physical quantity under discussion, B any simple combination of parameters of physical dimensions equal to those of A, and X, Y . . . , etc., the simplest dimensionless combinations. Then

$$A = \mathrm{CONST}\, Bf_1(X)f_2(Y)\ldots, \tag{7}$$

where the arbitrary functions, f, may be exponential, logarithmic, etc., or only multiplicative.

16. The choice of constants and parameters. The success of a particular application of dimensional analysis depends on proper choice of constants and variables. The particular problem under investigation generally dictates the choice. We cannot include all conceivable constants for then the equations would be indeterminate. Dimensional analysis is ordinarily employed as a short cut. One must rely upon physical intuition. The following rules are general guides for the choice of the constants to be employed.

Rarely does any question arise as to whether G enters or not. G never comes into atomic or simple radiation problems. The constants n_0 and k enter into problems only where matter is considered in the aggregate. The quantity, k, enters into a problem only when the temperature is involved. Radiation and atomic problems are divisible into two categories: classical and quantum. They differ from one another only in that the Planck constant h appears as an additional factor in the latter. For atomic problems, ε, h, and m are important. For questions involving radiation, c, h, and k are the relevant constants.

17. Numerical coefficients. One rather surprising fact emerges from a comparison of the equations derived by dimensional analysis and the exact equations, derived by other methods. The numerical multiplying factor, indeterminate by dimensional methods is surprisingly close to unity, and usually quite simple in form. 2π occurs very frequently as a factor. The most complicated we have met with so far is $2\pi^5/15$. Occasionally the result is exact, as for the gas law. No general way exists for proving that the numerical coefficients must be nearly unity, but experience indicates that the mathematical processes giving rise to the constant yield factors of no high order of magnitude.

This fortunate circumstance means that the dimensionally derived equation, if unique, is very nearly exact. Very often an exact detailed solution is very difficult. All the labor will be wasted if the theoretical result should turn out to be in disagreement with experiment. We can often save much time by employing dimensional analysis and testing the result with the observational data. If we obtain a rough check, the work of a more detailed analysis is justified.

As an example we may discuss the observed splitting of spectral lines in a magnetic field, as first observed by Zeeman. Suppose that we have quantitative observations showing that the frequency shift, $\Delta\nu$, is proportional to the field. We are considering a detailed study by means of the classical theory of light. The parameters involved are ε, m, c, and H, the last being the magnetic field intensity. We shall express ε and H in Gaussian units:

$$\Delta\nu = \varepsilon^\alpha m^\beta c^\gamma H. \tag{1}$$

The exponent of H is set equal to unity because of the observational result. Dimensional analysis then gives the formula

$$\Delta\nu = (1/4\pi)\varepsilon H/mc. \tag{2}$$

Now, disregarding the numerical factor, which we must consider as undetermined, test the observational data, that the shift is of the order of 10^6 frequency units per gauss, by substituting in the equation:

$$\begin{aligned} \Delta\nu/H \sim 10^6 \sim \varepsilon/mc &= 4.77 \times 10^{-10}/9.03 \times 10^{-28} \times 3 \times 10^{10} \\ &= 1.76 \times 10^7. \end{aligned} \tag{3}$$

The agreement is sufficiently close to be accounted for by a simple numerical constant, and the more detailed investigation is warranted.

The foregoing example illustrates how we may employ experimental data to make the equation determinate. If we had set the exponent of H equal to δ instead of to unity, we should have arrived at the indeterminate result

$$\Delta\nu = \text{CONST}\,(mc^3/\varepsilon^2)(\varepsilon^3 H/m^2c^4)^\delta. \tag{4}$$

SELECTED PROBLEMS FOR PART I

1. The velocity of the sun is 29.8 km sec^{-1}. Find its velocity in parsecs per century.

2. From the known radius of the earth, 6.378×10^8 cm, its mass, 5.983×10^{27} g, and the gravitational constant, find the acceleration of gravity, g, at the surface of the earth. Compute the value of g at the surface of the sun. Find the appropriate conversion factors to reduce to a system expressed in astronomical units per year per year. Prove that g has the physical dimensions of an acceleration. $g = GM/R^2$.

3. Determine the physical dimensions of x in the following equations. (Take ε, the electronic charge, in es units throughout.)

(a) $x = hc/\lambda kT$.
(T = temperature; λ = wavelength; k = Boltzmann's constant)

(b) $x = \pi\varepsilon^2/mc$. (c) $x = (2kT/m)^{1/2}$.

(d) $x = (2h\nu^3/c^2)(e^{h\nu/kT} - 1)^{-1}$. ($\nu$ = frequency)

(e) $x = (2\pi mkT)^{3/2}/h^3$.

4. Which, if any, of the following equations are in error dimensionally?

(a) $md^2x/dt^2 = (2h\nu^3/c^3)(e^{h\nu/kT} - 1)^{-1}$. ($\nu$ = frequency; x = length)

(b) $\alpha_\nu = \pi\varepsilon^2/mc$. ($\alpha_\nu$ = atomic absorption coefficient)

(c) $\lambda = h^3c/2\pi m\varepsilon^4$. ($\lambda$ = wavelength)

(d) $\rho = \rho_0 e^{-2GR/mv^2}$.

(e) $d(\log \rho)/dx = -m^2G/R^2kT$. ($\rho$ and ρ_0 are densities; R = radius; v = velocity)

5. On the supposition that the energy density of ν-radiation in a thermodynamic enclosure depends only on T, k, c, and ν, prove that $\rho_\nu = (8\pi\nu^2/c^3)kT$, the well-known Rayleigh-Jeans law.

6. In electromagnetic theory one may show that a weightless sphere, electrically charged, possesses an effective mass. For a given charge, the mass depends on the radius. On the assumption that the total mass of an electron results from its charge, prove that the relation between the radius and the parameters ε, m, and c is $r = (\frac{2}{3})\varepsilon^2/mc^2$. Compute the numerical value of r in cm.

7. Form the dimensionless products from combinations of the constants h, c, ε, m, and G. (Since the square of any dimensionless product is also dimensionless, give only the lowest powers of the combinations.) Hint: Employ an extension of the discussion of § 15.

8. Assume that the energy of ionization, E, for hydrogen, and radius (a) of an electron's orbit in an atom depend on the constants ε, m, and h. Prove that

$$E = (2\pi) m\varepsilon^4/h^2; \quad a = (\tfrac{1}{4}\pi^2)h^2/m\varepsilon^2.$$

9. If the magnetic moment, μ, of an electron in the first Bohr orbit depends on ε, m, c, and h, show that $\mu = (1/4\pi)\varepsilon h/mc$, where ε is expressed in es units. Note: The dimensions of μ are $[M^{1/2}L^{5/2}T^{-1}]$ in the em system.

10. Select some other physical or astronomical problem and determine the relationship between the parameters and constants by means of dimensional analysis.

PART II

Mechanics and Dynamics

Principles of Mechanics

1. Introduction. Several alternative methods exist for investigating problems of mechanics and dynamics. Each method starts with some basic assumption, from which the mathematical development proceeds toward some result capable of being tested by observation. Which method we use or which assumption we take as fundamental is largely a matter of convenience, since the basic hypothesis of one development appears as a logical corollary in the others. Certain approaches are, of course, more general than others. In all, however, agreement with observation is the final test of the validity of the underlying assumptions.

On the basis of experiments, in particular those performed by Galileo, Newton enunciated his three laws of motion:

I. Every body tends to remain in a state of rest or of uniform rectilinear motion, unless compelled to change its state through the action of an impressed force.

II. The "rate of change of motion," i.e., the rate of change of momentum, is proportional to the impressed force and occurs in the direction of the applied force.

III. To every action there is an equal and opposite reaction, i.e., the mutual actions of two bodies are equal and opposite.

For the present, we shall start from these laws. Other methods of approach, e.g., that of Hamilton, will be discussed presently.

2. The equations of motion. We may write Newton's second law as follows:

$$F = dp/dt. \tag{1}$$

Since the force, F, and momentum, p, are both vector quantities, having direction as well as magnitude, the above expression is incomplete. We may take this fact into account by resolving the force into components parallel to the axes of the coordinate system employed. For a Cartesian system, we may write

$$p_x = mv_x = m\, dx/dt; \quad dp_x/dt = m\, d^2x/dt^2, \tag{2}$$

and similarly for the y- and z-components. For ordinary problems, where the velocity, v, is small compared with that of light, we may disregard variation of mass with velocity. For velocities approaching that of light a relativistic treatment is necessary.

Consider an assembly made up of n particles, appropriately numbered. The equations of motion of the ith particle become

$$m_i\, d^2x_i/dt^2 = X_i, \quad m_i\, d^2y_i/dt^2 = Y_i, \quad m_i\, d^2z_i/dt^2 = Z_i, \tag{3}$$

where X_i, Y_i, and Z_i are the components of force at the position of the particle, resolved along the three Cartesian axes. These equations are of the second order and each will have two arbitrary constants in its solution. The set of $3n$ equations, consequently, involves $6n$ constants, which we identify with the $3n$ initial coordinates and the $3n$ initial velocity components at time $t = 0$.

3. Conservation of energy. To integrate the equations (2.3), multiply the first equation through by dx_i/dt. The left-hand side is immediately integrable.

$$m_i \frac{d^2x_i}{dt^2}\frac{dx_i}{dt} = \frac{1}{2} m_i \frac{d}{dt}\left(\frac{dx_i}{dt}\right)^2 = X_i \frac{dx_i}{dt}. \tag{1}$$

$$\frac{1}{2} m_i \left(\frac{dx_i}{dt}\right)^2 = \int X_i\, dx_i + \text{CONST}. \tag{2}$$

The right-hand side can be integrated further only if we know the functional dependence of X_i on x_i. Summing over all the particles of the assembly and over the coordinates x, y, and z, we have

$$\sum_i \frac{1}{2} m_i \left[\left(\frac{dx_i}{dt}\right)^2 + \left(\frac{dy_i}{dt}\right)^2 + \left(\frac{dz_i}{dt}\right)^2\right] - \sum_i \left[\int X_i\, dx_i + \int Y_i\, dy_i + \int Z_i\, dz_i\right] = E, \tag{3}$$

where E is the sum of all the constants of integration.

Summations like the first, which may be written $\sum_i \frac{1}{2}m_i v_i^2$, where v_i is the space velocity of the ith particle, occur so frequently in physics that we give a special name to them. We call the first summation the kinetic energy of the assembly, and designate it by the letter T. The second summation is the work, W, done by the forces in moving each particle from its initial to its final coordinates. The underlying algebra contains one implicit assumption. We must carry the integration along the actual path followed by the particle, because, if frictional forces are involved, W will depend on the path of integration.

For certain types of force fields, however, W is a constant, independent of the paths followed by the particles and dependent only on the initial and final coordinates. When this condition obtains, energy is said to be conserved for the assembly. In other words, if a body moves from A to B and then back to A, the initial energy conditions repeat themselves. We may then define a function, V, of the coordinates of the particles, so that

$$W = -V + \text{CONST}. \tag{4}$$

We call V the potential energy. Note that when a particle moves in the direction of the force, W is positive but the potential energy decreases. The additive constant is arbitrary and depends on the choice of zero point for V. In place of (3), we may write, symbolically,

$$T - W = T + V = E. \tag{5}$$

The total energy, E, of the system is constant for such an assembly. Further discussion will appear in § (8).

4. Conservation of momentum. Let X_{ij} be the x component of force on the particle i, caused by action of the particle j. Newton's third law of motion requires that the force resulting from the interaction of any two systems be equal and opposite. Hence

$$X_{ij} = -X_{ji}. \tag{1}$$

If the force on a particle i results from the action of all the other particles in the assembly,

$$X_i = X_{i1} + X_{i2} + \ldots X_{ii} + \ldots = \sum_j X_{ij}. \tag{2}$$

Formally we may take care of the meaningless term X_{ii} by defining it to be zero. If, now, we sum both sides of equation (2) with respect to i, we have

$$\sum_i X_i = \sum_i \sum_j X_{ij} = (X_{12} + X_{21}) + (X_{13} + X_{31}) + \ldots$$
$$+ (X_{23} + X_{32}) + \ldots = 0, \tag{3}$$

by (1). (We have omitted the zero terms X_{11}, X_{22}, etc.) Since force is the rate of change of momentum, we express and integrate equation (3) as follows (cf. equation 2.3):

$$\sum_i X_i = \sum_i m_i \frac{d^2x_i}{dt^2} = \frac{d}{dt} \sum_i m_i \frac{dx_i}{dt} = 0. \tag{4}$$

$$\sum_i m_i \frac{dx_i}{dt} = \sum p_i = \frac{d}{dt} \sum m_i x_i = \text{CONST} = a_x. \tag{5}$$

The sum of the x components of momenta is thus conserved in an assembly where no external field is present. We may construct similar proofs for the y and z components.

5. Motion of the center of mass. Equation (4.5) gives, on further integration,

$$\sum_i m_i x_i = a_x t + b_x, \tag{1}$$

where b_x is a second constant of integration. The quantity $m_i x_i$ is the moment of mass of particle i with respect to the yz-plane. The x-coordinate of the center of mass, $\bar{x}$, of the assembly is, by definition,

$$\bar{x} = \sum_i m_i x_i / \sum_i m_i. \tag{2}$$

Equation (1), together with its analogues in y and z, prescribes that the center of mass of the assembly move with uniform velocity in a straight line.

6. The law of areas and conservation of angular momentum. Let P_1P_2 in Fig. 1 be the projection of the trajectory of particle i on the xy-plane. Let the coordinates of the points A and B be respectively (x, y)

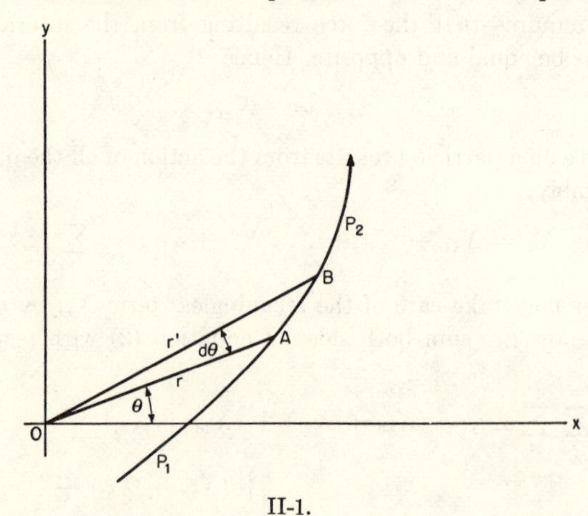

II-1.

and $(x + dx, y + dy)$, or, in polar coordinates, (r, θ) and $(r', \theta + d\theta)$. A and B are supposed to be an infinitesimal distance apart. Under this condition $r' \sim r$; the element of trajectory approaches a straight line, and we may regard the area OAB as a triangle of area $\frac{1}{2}r^2\, d\theta$. Employ the relationships

$$r^2 = x^2 + y^2, \quad \theta = \arc\tan y/x, \quad d\theta = \frac{x\,dy - y\,dx}{x^2 + y^2}. \tag{1}$$

Therefore the triangle has an area

$$\frac{1}{2} r^2\, d\theta = \frac{1}{2}(x\,dy - y\,dx). \tag{2}$$

The moment of inertia of particle i, with respect to the origin is, by definition, $m_i r_i^2$. The particle possesses an angular velocity, $d\theta_i/dt$. Hence the angular momentum, L_i, which by (I–6.14) is the product of these two quantities, becomes

$$L_i = m_i r_i^2 \frac{d\theta_i}{dt} = m_i\left(x_i \frac{dy_i}{dt} - y_i \frac{dx_i}{dt}\right). \tag{3}$$

The total angular momentum, L, is

$$L = \sum_i m_i\left(x_i \frac{dy_i}{dt} - y_i \frac{dx_i}{dt}\right). \tag{4}$$

We are interested in the rate of change of angular momentum with the time,

$$\begin{aligned} T = \frac{dL}{dt} &= \sum_i \left(x_i m_i \frac{d^2 y_i}{dt^2} - y_i m_i \frac{d^2 x_i}{dt^2}\right) \\ &= \sum_i (x_i Y_i - y_i X_i) = \sum_i (x_i \sum_j Y_{ij} - y_i \sum_j X_{ij}), \end{aligned} \tag{5}$$

by equation (4.2). The quantity xY is a "moment" of force; T is the force couple or torque. Suppose that the force is directed along the line joining each particle. The force between each pair of particles will have a component F_{ij} in the xy-plane along the line that joins the points. Let

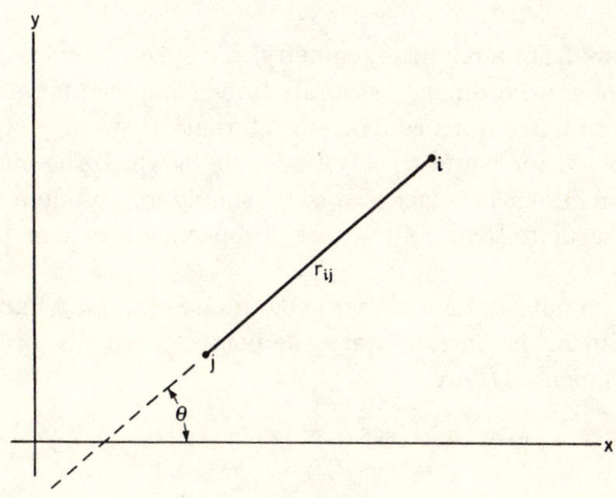

II-2.

r_{ij} be the projected distance, in the xy-plane, between the two points and θ the angle of slope with respect to the x-axis (Fig. 2). Then

$$\cos\theta = (x_i - x_j)/r_{ij}, \quad \sin\theta = (y_i - y_j)/r_{ij}. \tag{6}$$

Also

$$X_{ij} = F_{ij}\cos\theta, \quad Y_{ij} = F_{ij}\sin\theta. \tag{7}$$

If we substitute these expressions into (5) we find for each pair of particles, a symmetrical combination of the sort:

$$x_iY_{ij} - y_iX_{ij} + x_jY_{ji} - y_jX_{ji}$$
$$= F_{ij}[(x_i - x_j) \sin \theta - (y_i - y_j) \cos \theta]$$
$$= F_{ij}r_{ij}(\cos \theta \sin \theta - \sin \theta \cos \theta) = 0,$$

since, as before, $F_{ji} = -F_{ij}$. Each pair of particles thus gives a zero resultant. Hence

$$\frac{dL}{dt} = 0 \tag{8}$$

and

$$L = \text{CONST.} \tag{9}$$

The angular momentum of the assembly about a given axis is thus conserved *if the forces act along the lines joining the particles.* There is no need to specify the nature of the law of force. The choice of axes was arbitrary, so that the result is general. The quantity L_i, by equations (2) and (3), is the product of the mass of a particle by twice its areal velocity. The areal velocity is the rate at which the projected radius vector from the given axis sweeps out an area on a plane normal to the axis.

7. Theorems from analytical geometry. To specify any vector quantity such as the force we ordinarily state its three components relative to some system of coordinates. As a convenient alternative, we may state the value of the unresolved force and specify the angle in which the maximum force is acting. Directions in space are very simply represented by direction cosines, whose definitions and several properties are here presented for reference.

Consider two points A and B with coordinates (x_1, y_1, z_1) and (x_2, y_2, z_2). Let l be the linear distance between the points. Then the direction cosines of the line segment AB are

$$\begin{aligned} \lambda &= \cos \alpha = (x_2 - x_1)/l, \\ \mu &= \cos \beta = (y_2 - y_1)/l, \\ \nu &= \cos \gamma = (z_2 - z_1)/l, \end{aligned} \tag{1}$$

where α, β, and γ are the angles made by the line AB and intersecting lines parallel respectively to the x-, y-, and z-axes (see Fig. 3). Direction cosines obey the Pythagorean relation

$$\lambda^2 + \mu^2 + \nu^2 = 1. \tag{2}$$

Let A and B represent two neighboring points on a curve and the line AB the secant of length l. The coordinates of A and B are, respectively,

(x, y, z) and $(x + \Delta x, y + \Delta y, z + \Delta z)$. By (1) the direction cosines of the secant are

$$\lambda = \Delta x/l, \quad \mu = \Delta y/l, \quad \nu = \Delta z/l. \tag{3}$$

II-3.

Now, as A and B approach each other indefinitely, the secant will approach the tangent to the curve and l will approach the value Δs, where Δs is the distance between A and B measured on the arc of the curve. In the limit we shall have

$$\lambda = dx/ds, \quad \mu = dy/ds, \quad \nu = dz/ds, \tag{4}$$

as the direction cosines of the tangent to the curve.

When two lines with respective direction cosines $(\lambda_1, \mu_1, \nu_1)$ and $(\lambda_2, \mu_2, \nu_2)$ intersect, we shall wish to determine the angle, θ, between them. Translate the origin of coordinates, keeping the axes parallel to their

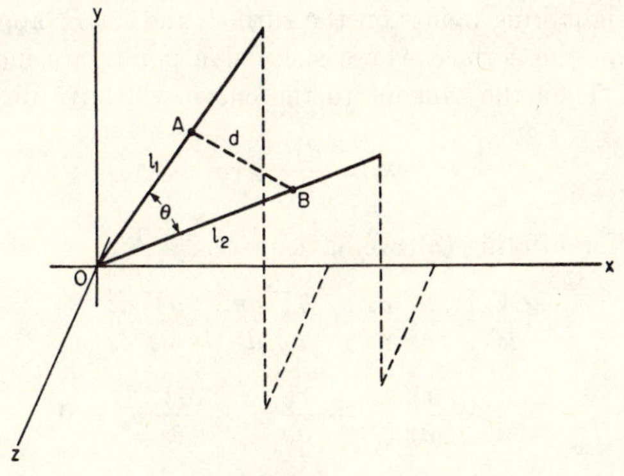

II-4.

original directions, until the point of intersection coincides with the origin. The direction cosines will not be changed by such translation. (Fig. 4.)

Consider the triangle OAB formed by the origin and any two other points on the respective lines. Let

$$\mathrm{OA} = l_1, \quad \mathrm{OB} = l_2, \quad \mathrm{AB} = d.$$

Let the respective coordinates of A and B be (x_1, y_1, z_1) and (x_2, y_2, z_2). Then

$$\begin{aligned} l_1^2 &= x_1^2 + y_1^2 + z_1^2, \\ l_2^2 &= x_2^2 + y_2^2 + z_2^2, \\ d^2 &= (x_2 - x_1)^2 + (y_2 - y_1)^2 + (z_2 - z_1)^2 \\ &= l_1^2 + l_2^2 - 2(x_1x_2 + y_1y_2 + z_1z_2). \end{aligned} \tag{5}$$

But

$$\begin{aligned} x_1 &= l_1\lambda_1. & y_1 &= l_1\mu_1. & z_1 &= l_1\nu_1. \\ x_2 &= l_2\lambda_2. & y_2 &= l_2\mu_2. & z_2 &= l_2\nu_2. \end{aligned} \tag{6}$$

Now

$$d^2 = l_1^2 + l_2^2 - 2l_1l_2(\lambda_1\lambda_2 + \mu_1\mu_2 + \nu_1\nu_2) = l_1^2 + l_2^2 - 2l_1l_2 \cos \theta,$$

by the well-known trigonometric formula, expressing the length of one side of a triangle in terms of the lengths of the other two sides and the cosine of the angle included between them. Hence

$$\cos \theta = \lambda_1\lambda_2 + \mu_1\mu_2 + \nu_1\nu_2. \tag{7}$$

Let $V(xyz)$ be some function of the coordinates. Then

$$V(x, y, z) = \text{CONST.} \tag{8}$$

represents a surface, which we call an "equipotential surface." Let P and P' be two neighboring points on the surface and let P' approach P along some curve on the surface whose successive points are indicated by the coordinate s. Then the tangent to the curve will have direction cosines:

$$\lambda_2 = \frac{dx}{ds}, \text{ ETC.}, \tag{9}$$

as in (4). Differentiating (8), we have

$$\begin{aligned} \frac{dV}{ds} &= \frac{\partial V}{\partial x}\frac{dx}{ds} + \frac{\partial V}{\partial y}\frac{dy}{ds} + \frac{\partial V}{\partial z}\frac{dz}{ds} \\ &= \frac{\partial V}{\partial x}\lambda_2 + \frac{\partial V}{\partial y}\mu_2 + \frac{\partial V}{\partial z}\nu_2 = 0, \end{aligned} \tag{10}$$

by (8) and (9). From (7), we see that two mutually perpendicular lines ($\cos \theta = 0$) obey the relation

$$\lambda_1\lambda_2 + \mu_1\mu_2 + \nu_1\nu_2 = 0. \tag{11}$$

Comparison of (10) and (11) shows that the line defined by

$$\lambda_1 \propto \frac{\partial V}{\partial x}, \quad \mu_1 \propto \frac{\partial V}{\partial y}, \quad \nu_1 \propto \frac{\partial V}{\partial z}, \tag{12}$$

is perpendicular to the line s tangent to the surface. Hence, since s is any curve in the surface, we see that the direction cosines (12) define the normal to the surface V. Equation (2) fixes the constant of proportionality. The signs of the constants are arbitrary, and depend to some extent on the physical significance of V. If the equipotential surface of equation (8) is closed, we shall regard the outward normal as fixing the positive direction.

8. Work as a line integral. From the definition of force, we require knowledge of the mass of the moving body in order to calculate the magnitude of the forces acting. In a large number of problems, however, the mass of the body acted upon is of no great importance. The orbits of particles moving around the sun are, for example, independent of the particle mass. For the investigation of various types of motions, we often introduce a test particle, of unit mass, in order to define the field of force. The force, so defined, we shall usually refer to as the "force intensity," the "force vector," or the "force at a point." Further, the potential energy of the unit particle, expressed as a function of the coordinates, we shall refer to as the "potential."*

With respect to this definition we introduce a minor inconsistency which should, however, cause no confusion. Although we have defined our force field in terms of a test particle of unit mass, such a force and its associated potential possess respective physical dimensions of $[MLT^{-2}]$ and $[ML^2T^{-2}]$. We usually omit this test mass from our formulas, so that force intensity and its associated potential may appear to have the dimensions of $[LT^{-2}]$ and $[L^2T^{-2}]$. We shall, in fact, use the term force intensity as synonymous with acceleration, even though we do imply the existence of the fictitious test mass.

We shall now proceed to calculate, in a more general fashion, the work done by a force, as it moves our unit particle along a curved path. Let λ_1, μ_1, and ν_1, be the direction cosines that define the orientation of the force vector F. If θ is the angle between the force vector and the path at some given position of the particle, the component of force along the trajectory is $F \cos \theta$. The total work done, as the particle moves from A to B is

$$W = \int_A^B F \cos \theta \, ds, \tag{1}$$

*Cf. definitions in I, (7).

where ds is an element of the path. We term an integral of this form a *line integral.* Substitute for $\cos \theta$ from (7.7) and for ds from (7.4), and note that

$$F\lambda_1 = X, \quad F\mu_1 = Y, \quad F\nu_1 = Z, \tag{2}$$

are the force components parallel to the three coordinate axes. We thus find that

$$W = \int_A^B (X\,dx + Y\,dy + Z\,dz). \tag{3}$$

This expression is consistent with the definition of W given in (3.3), except that the present expression is referred to unit mass.

The value of the integral will depend on the path of the particle unless certain conditions are fulfilled. Let us suppose that there exists a function V, of the coordinates, known as the *potential function,* such that

$$X = -\frac{\partial V}{\partial x}, \quad Y = -\frac{\partial V}{\partial y}, \quad Z = -\frac{\partial V}{\partial z}. \tag{4}$$

Then, since

$$dV = \frac{\partial V}{\partial x}\,dx + \frac{\partial V}{\partial y}\,dy + \frac{\partial V}{\partial z}\,dz = -(X\,dx + Y\,dy + Z\,dz), \tag{5}$$

the factor in parentheses in equation (3) will be a complete differential, for we may write

$$W = -\int_A^B dV = -\Big[V\Big]_A^B = V_A - V_B. \tag{6}$$

The value of the integral depends only on the limits; it is independent of the path followed. Since

$$W_{AB} = -W_{BA}, \tag{7}$$

the process is reversible. *Conservation of energy in dynamical systems does not enter into physics as a special postulate; conservation follows naturally for any assembly where the forces are "derivable" from a potential.*

9. The force vector. We may divide physical quantities into various classes: scalar, vector, or tensor. A scalar quantity expresses merely magnitude. Mass, length, potential energy, for example, are scalar in nature. Vector quantities imply direction as well. Examples are velocity, momentum, force, etc.

In § (7) we discussed two alternative methods of representing vectors. We may also consider a third type of representation, which is not so cumbersome mathematically. The concept of adding vectors to give a vector resultant should be familiar to everyone. If $\mathbf{F}$ is the force vector and $\mathbf{X}$,

Y, and **Z**, the component vectors along the respective coordinate axes, we may write, symbolically,

$$\mathbf{F} = \mathbf{X} + \mathbf{Y} + \mathbf{Z}. \tag{1}$$

We shall ordinarily indicate vector quantities by bold-face type. The above representation signifies that one measures from the vector origin along the x-axis an amount equal to X, then up parallel to the y-axis an

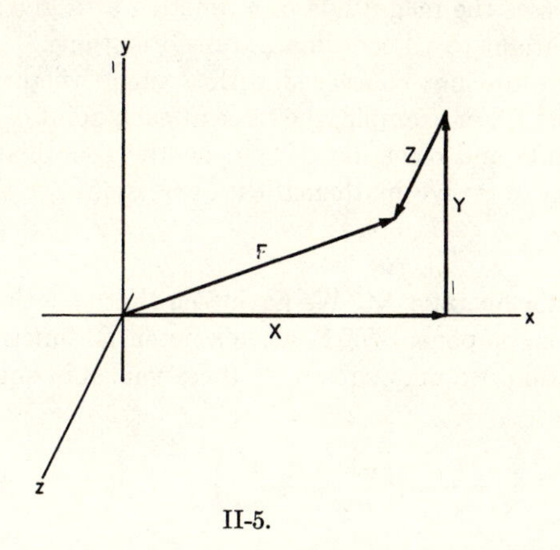

II-5.

amount Y and finally out parallel to the z-axis an amount Z, where X, Y, and Z are the scalar absolute magnitudes of the corresponding vectors. The resultant determines the vector **F**, measured from the initial to the final point. We shall use a so-called right-handed system of coordinates in this book, in the following sense. We extend the thumb and first two fingers of the right hand to indicate roughly three mutually perpendicular directions. We take the thumb as pointing along the x-axis, the forefinger as indicating the y-axis and the middle finger the z-axis.

Vectors obey the ordinary commutative laws for addition and subtraction. Thus

$$\mathbf{F} = \mathbf{Y} + \mathbf{X} + \mathbf{Z} = \mathbf{Z} + \mathbf{X} + \mathbf{Y}, \text{ ETC.} \tag{2}$$

Multiplication by any positive scalar quantity merely alters the magnitude of the vector, without changing its direction. If the scalar is negative the direction of the vector is reversed.

The concept of a unit vector is important. Let **i**, **j**, and **k** be vectors of unit length, measured along the three respective coordinate axes. Then we may express any vector component, **X**, **Y**, or **Z**, in terms of its scalar magnitude X, Y, or Z, as follows:

$$\mathbf{X} = \mathbf{i}X; \quad \mathbf{Y} = \mathbf{j}Y; \quad \mathbf{Z} = \mathbf{k}Z. \tag{3}$$

Cartesian coordinates are not necessary for the representation. We may use any convenient system of axes, such as polar, cylindrical, or parabolic coordinates. Non-orthogonal axes occasionally prove useful.

One of the important properties of a vector is its invariance to coordinate transformations. Rotation of the coordinate system, for example, will alter our concept of the x, y, and z components. But the original vector remains unchanged in magnitude and direction. Since a scalar quantity expresses the magnitude of a function at a given point of space, it must be invariant to all coordinate transformations.

Ordinarily we are not concerned with a single vector, but with a so-called vector field. For example, the force at each point of space is a vector whose magnitude and direction depend on the coordinates of the point. Our problem is to derive mathematical expressions for the vectors at all points of space.

10. The vector operator ∇. We have seen that equation (8.4) expresses the three vector components of **F**, when a potential function exists. Hence, in vector notation, we may substitute these values in equations (9.1) and (9.3) and write

$$\begin{aligned}\mathbf{F} &= -\left(\mathbf{i}\frac{\partial V}{\partial x} + \mathbf{j}\frac{\partial V}{\partial y} + \mathbf{k}\frac{\partial V}{\partial z}\right)\\ &= -\left(\mathbf{i}\frac{\partial}{\partial x} + \mathbf{j}\frac{\partial}{\partial y} + \mathbf{k}\frac{\partial}{\partial z}\right)V = -\nabla V = -\,\mathbf{grad}\ V,\end{aligned} \tag{1}$$

where we have factored out the quantity in parentheses and abbreviated it as ∇ or **grad.** The vector operator ∇, called either "del" or "the gradient of," is frequently written as "**grad**":

$$\mathbf{grad} = \nabla = \mathbf{i}\frac{\partial}{\partial x} + \mathbf{j}\frac{\partial}{\partial y} + \mathbf{k}\frac{\partial}{\partial z}\,. \tag{2}$$

Let us consider the physical significance of this operator, when applied to V. Each term of it directs us to find a certain component of the force along one of the axes. The sum directs us to combine the components vectorially to give the force. Or, if we prefer, we may interpret the symbol ∇ as an order to find in what direction the force is greatest and measure its value in that direction.

The scalar or absolute value of the force vector in the direction of its maximum, is ordinarily written either as F or $|\,\mathbf{F}\,|$. The magnitude of the force is

$$F = |\,\mathbf{F}\,| = \sqrt{(\partial V/\partial x)^2 + (\partial V/\partial y)^2 + (\partial V/\partial z)^2}, \tag{3}$$

with the three vector components compounded according to the Pythagorean theorem. We may regard this expression as a means of determining the gradient of the scalar quantity, V, in the direction where the force is a maximum. We shall show in § (12) that this direction is normal to the surface of constant potential.

We need to know how to express the force in polar coordinates, because the potential frequently appears as a function of r, θ, and ϕ. According to Fig. 6, we may express the rectangular coordinates of the point P in terms of the polar coordinates as follows:

$$x = r \sin \theta \cos \phi\,, \quad y = r \sin \theta \sin \phi, \quad z = r \cos \theta. \tag{4}$$

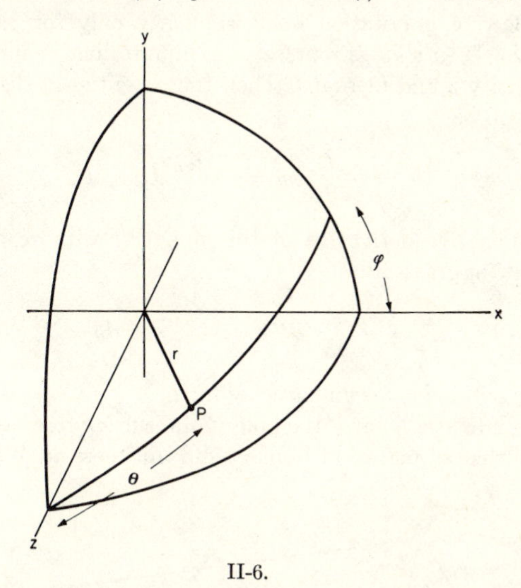

II-6.

Now, ∇V is a vector whose magnitude and direction are independent of the coordinate system employed. The three mutually perpendicular components are arbitrary as long as they add up to give the proper resultant. Each component has a magnitude equal to the rate of change of V with distance in its own direction. The three mutually perpendicular elements in the spherical coordinate system have magnitudes dr, $r\,d\theta$, and $r \sin \theta\, d\phi$ in the respective directions r, θ, and ϕ. Hence, in such a system, the operator ∇ becomes

$$\mathbf{grad} = \nabla = \left(\mathbf{i}_r \frac{\partial}{\partial r} + \mathbf{i}_\theta \frac{1}{r} \frac{\partial}{\partial \theta} + \mathbf{i}_\phi \frac{1}{r \sin \theta} \frac{\partial}{\partial \phi}\right), \tag{5}$$

where $\mathbf{i}_r$, $\mathbf{i}_\theta$, and $\mathbf{i}_\phi$, represent the unit vectors of this system. The student should note that the directions of these unit vectors, unlike those of the rectangular system, vary from point to point. A more general and rigorous derivation of this equation will appear in § (33).

Theory of the Potential

11. The potential of a sphere. The potential V assumes a prominent role in physics because of its scalar nature. When a potential function exists, we can form V by simple addition of the various partial potentials. The derivative will then give the force, whereas calculating the force initially would have required the compounding of numerous vector components.

Newton's law of gravitation holds explicitly only for particles of infinitesimal size. Let $dx\,dy\,dz$ represent an infinitesimal volume containing matter of density ρ and mass dm. Then the potential at distance R from the element will be

$$dV = -\frac{G}{R}\,dm = -\frac{G\rho}{R}\,dx\,dy\,dz, \tag{1}$$

because the negative derivative of this quantity with respect to R expresses the law of gravitation:

$$dF = -\frac{\partial}{\partial R}(dV) = -\frac{G\rho}{R^2}\,dx\,dy\,dz. \tag{2}$$

The negative sign indicates an *attractive* force.

Let us calculate at a point P the potential resulting from a homogeneous thin spherical shell of matter of radius r and thickness dr. We may express

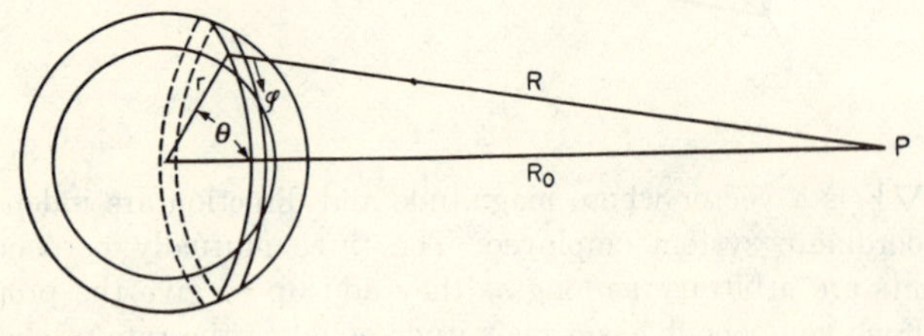

II-7.

the total potential as an integral of the volume element over the spherical shell:

$$V = -G\rho \iiint \frac{dx\,dy\,dz}{R} = -G\rho \int_r^{r+dr} \int_{\theta=0}^{\pi} \int_{\phi=0}^{2\pi} \frac{r^2 \sin^2\theta\,dr\,d\theta\,d\phi}{R}, \tag{3}$$

where the volume element has been expressed in polar coordinates. The elementary lengths are dr, $r\,d\theta$, and $r \sin d\phi$, as noted in § (10). From the "law of cosines," we have the relation

$$R^2 = R_0^2 + r^2 - 2R_0 r \cos\theta. \tag{4}$$

For the thin shell we may regard r as constant and employ (4) to relate R and θ. Also,

$$R\,dR = R_0 r \sin\theta\, d\theta. \tag{5}$$

Let V_i and V_e represent the potentials when P is internal or external, respectively, to the sphere. Then the integrals become

$$V_e = -G\rho \int_r^{r+dr} \int_{R_0-r}^{R_0+r} \int_0^{2\pi} \frac{r}{R_0}\, dr\, dR\, d\phi = -\frac{4\pi G\rho r^2}{R_0}\, dr, \tag{6}$$

$$V_i = -G\rho \int_r^{r+dr} \int_{r-R_0}^{r+R_0} \int_0^{2\pi} \frac{r}{R_0}\, dr\, dR\, d\phi = -4\pi G\rho r\, dr. \tag{7}$$

Since V_i is independent of R_0, its derivative with respect to R_0 is zero. Thus the force at any point inside a hollow spherical shell is zero. The mass M_s of the shell is $4\pi\rho r^2\, dr$. Hence the potential for an external point becomes

$$V_e = -GM_0/R_0. \tag{8}$$

The potential of the shell is thus equal to that produced by a point of mass M_s, located at the center of the sphere.

We may extend the theorem to include the case of a solid sphere, whose density depends only on r.

$$V_e = -\frac{4\pi G}{R_0} \int_0^r \rho r^2\, dr = -\frac{GM}{R_0}, \tag{9}$$

where M is the total mass of the sphere.

The above proof applies to the potential between a sphere and a particle. The equations also hold for the potential between two spheres.

12. Equipotential surfaces. The equation (7.8)

$$V = C,$$

where C is a constant, denotes a surface over every point of which the potential is constant. Such surfaces are important because the force component along such a surface is zero and the work required to move a particle from one point to another on the same surface is zero. Consequently, there is no component of force along the surface and the force vector is perpendicular to a surface of constant potential. This result is not surprising since the direction cosines of the normal to the surface, by (7.12),

$$\lambda \propto \frac{\partial V}{\partial x}, \qquad \mu \propto \frac{\partial V}{\partial y}, \qquad \nu \propto \frac{\partial V}{\partial z}, \tag{1}$$

are proportional to the components of the force vector.

For purposes of evaluating the force vector, we shall often find it convenient to adopt a coordinate system such that one coordinate shall always be perpendicular to the equipotential surfaces. If we denote this coordinate by the letter n, the force is given by the equation

$$F = -\frac{\partial V}{\partial n}. \tag{2}$$

This equation is essentially synonymous with the much more cumbersome expression (10.3). We may further use the quantity $\mathbf{n}$ to represent a unit vector along the normal, and write, in vector notation,

$$\mathbf{F} = -\mathbf{n}\frac{\partial V}{\partial n}, \tag{3}$$

since the partial derivatives of V with respect to either of the two coordinates normal to n, i.e., coordinates lying in the surface, V = constant, vanish.

Thus, for (11.9), since $\mathbf{n} = \mathbf{i}_r$, the unit vector along the radius, the gravitational force external to a sphere becomes

$$\mathbf{F} = -\mathbf{i}_r\frac{GM}{r^2}. \tag{4}$$

Or, since $\mathbf{r}$ is itself a vector, equal to

$$\mathbf{r} = \mathbf{i}_r r = \mathbf{i}x + \mathbf{j}y + \mathbf{k}z, \tag{5}$$

we may write, in place of (4),

$$\mathbf{F} = -\mathbf{r}\frac{GM}{r^3}. \tag{6}$$

13. Surface integral. We may represent an element of some arbitrary surface by two magnitudes, one expressing its area and the other indicating the orientation of the normal to its surface. We may represent both magnitudes simultaneously by a vector along the normal, with its length proportional to the area of the element.

In numerous physical applications, we shall be called on to evaluate what is commonly called the "flux" of some vector field through the surface. As a simple illustration of the procedure we shall calculate the amount of sunlight falling on (or "flowing through") a receiver of arbitrary shape. Let dS be an element of surface whose normal makes an angle θ with respect to the solar beam. Let F be the incident solar energy per cm^2 per sec. We shall suppose that the sun lies far out on the negative portion of the axis. Then the total energy falling per second on the receiver, i.e., the "flux" becomes

$$\phi = \iint F \cos\theta \, dS. \tag{1}$$

But $\cos\theta\, dS = \lambda\, dS = dS' = dy\, dz$, the projection of the area element on the yz-plane normal to the beam; λ is the direction cosine of the surface vector. Since F is a constant, by supposition, we can integrate immediately and write

$$\phi = \iint F\lambda\, dS = \iint F\, dy\, dz = FS', \tag{2}$$

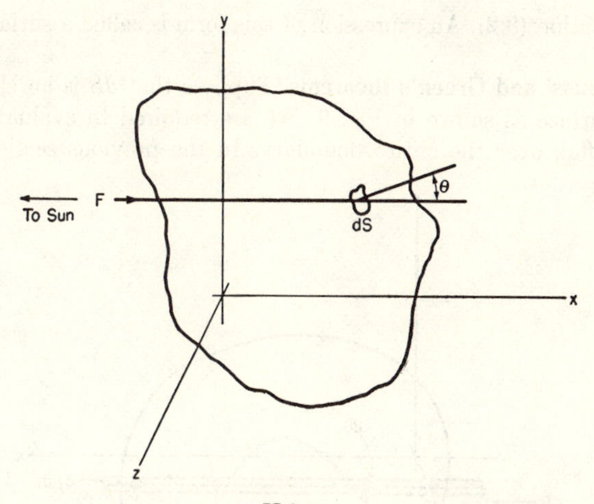

II-8.

where S' is the area of the receiver as projected on the yz-plane. We may, if we wish, interpret S' as the area of the shadow cast by the receiver.

As suggested in the opening paragraph, we might have considered dS as a vector normal to the element. Then $dS \cos\theta$ represents the projection of the vector in the direction of the solar beam. The simplicity of the above problem arose from the constancy of F both in magnitude over the surface and in direction. In many problems we shall have to evaluate integrals of the above variety where F is variable, both in magnitude and direction.

The flux in geometric optics is often allied with the concept of light rays. For example, if F is 6×10^9 ergs, per cm^2 per sec, we may imagine the solar beam to consist of 6×10^9 rays per cm^2. Then a unit surface inclined at angle θ will intercept $6 \times 10^9 \cos\theta$ rays. Rays diverging from a source exhibit the inverse-square-law property and the number intercepting any surface is proportional to the flux. This formal device is useful in gravitational and electromagnetic studies, where rays are called "lines of force," whose density is proportional to the field of force at a given point. We can therefore evaluate the flux in a manner analogous to that used above for light.

Let F be the absolute value of the force vector. Let dS be an element of the surface over which we are to integrate and let θ be the angle between the surface element and the force vector. Equation (1) will hold as before for the flux. Let $(\lambda_1, \mu_1, \nu_1)$ and $(\lambda_2, \mu_2, \nu_2)$ be the respective direction cosines of F and of the normal to dS. Then, by (7.7),

$$\phi = \iint F(\lambda_1\lambda_2 + \mu_1\mu_2 + \nu_1\nu_2)\,dS = \iint (X\lambda_2 + Y\mu_2 + Z\nu_2)\,dS, \quad (3)$$

from equation (8.2). An expression of this form is called a surface integral.

14. Gauss' and Green's theorems. Suppose that dS is an element of a closed surface as shown in Fig. 9. We are required to evaluate the total *outward* flux over the entire boundary. In the previous section we have

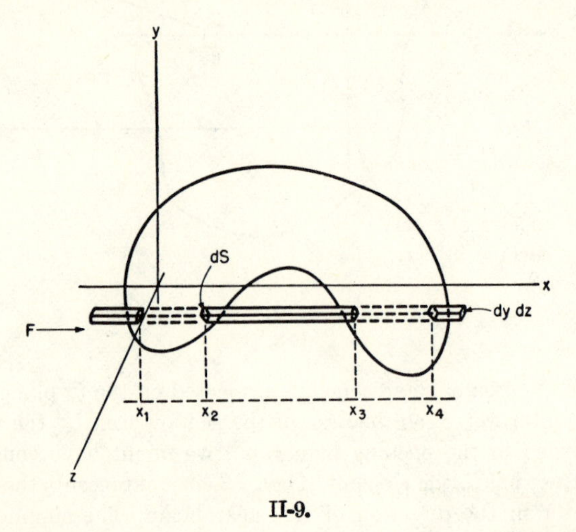

II-9.

reduced the problem to the sum of three integrals. Note the close analogy of each integral in (13.3) to that of (13.2), where F was considered to be parallel to the x-axis. Analogously,

$$\lambda_2\,dS = dy\,dz, \quad (1)$$

the shadow of the area element on the yz-plane. Draw, parallel to the x-axis, a long prism of cross section $dy\,dz$, intersecting the surface at x_1, x_2, x_3, x_4, . . . , etc. The y- and z-coordinates are of no immediate concern. The geometry requires an even number of intersections, because the surface is closed. As in § (13), we may regard this prism as the boundary of a beam of light rays, traveling from left to right.

There are, however, two differences in the present problem as compared with that of the solar beam. In the previous problem we took F as constant. Here X, which replaces F, is a function of x, y, and z. Along the prism, where y and z are constant, we may suppose that X is a function of x alone. If X_1, X_2, etc., are the values of X at the points of intersection, we may write

$$X_2 - X_1 = \int_{x_1}^{x_2} \frac{\partial X}{\partial x}\, dx, \text{ ETC.} \tag{2}$$

If we regard the flux as negative when the beam enters and positive when it leaves, we have

$$\phi_x = \iint X\lambda_2\, dS = -\iint [(X_1 - X_2) + (X_3 - X_4) + \ldots]\, dy\, dz. \tag{3}$$

Substituting from (2), we obtain

$$\phi_x = \iiint \frac{\partial X}{\partial x}\, dx\, dy\, dz. \tag{4}$$

We find similar expressions for y and z. Therefore the complete surface integral (13.3) becomes

$$\phi = \iint F \cos\theta\, dS = \iiint \left(\frac{\partial X}{\partial x} + \frac{\partial Y}{\partial y} + \frac{\partial Z}{\partial z}\right) d\tau, \tag{5}$$

where $d\tau$ is an element of volume. This expression, which enables us to transform a surface integral into a volume integral, we shall refer to as Gauss' theorem, although many writers call it Green's theorem.

We easily derive an extension of this formula, properly known as Green's analytical theorem. Assume that we can represent the components X, Y, and Z, as follows:

$$X = \Phi\frac{\partial\psi}{\partial x}, \quad Y = \Phi\frac{\partial\psi}{\partial y}, \quad Z = \Phi\frac{\partial\psi}{\partial z}, \tag{6}$$

where Φ and ψ are scalar functions of the coordinates. Then, by (13.3),

$$\iint F\cos\theta\, dS = \iint (X\lambda + Y\mu + Z\nu)\, dS$$

$$= \iint \Phi\left(\frac{\partial\psi}{\partial x}\lambda + \frac{\partial\psi}{\partial y}\mu + \frac{\partial\psi}{\partial z}\nu\right) dS, \tag{7}$$

where λ, μ, and ν are the direction cosines of the surface element dS. Let n denote a coordinate along any normal to dS. Then, by (7.4),

$$\lambda = \frac{dx}{dn}, \quad \mu = \frac{dy}{dn}, \quad \nu = \frac{dz}{dn}. \tag{6}$$

But when these quantities are introduced into (7), we find that the function in parentheses is equivalent to

$$\frac{\partial\psi}{\partial x}\frac{dx}{dn} + \frac{\partial\psi}{\partial y}\frac{dy}{dn} + \frac{\partial\psi}{\partial z}\frac{dz}{dn} = \frac{\partial\psi}{\partial n}, \tag{9}$$

by the rules of partial differentiation. Hence

$$\iint F \cos \theta \, dS = \iint \Phi \frac{\partial \psi}{\partial n} \, dS. \tag{10}$$

This expression, in turn, is equal to the volume integral defined in (5). Carrying out the indicated differentiations, we have

$$\iint \Phi \frac{\partial \psi}{\partial n} \, dS = \iiint \left\{ \Phi\left(\frac{\partial^2 \psi}{\partial x^2} + \frac{\partial^2 \psi}{\partial y^2} + \frac{\partial^2 \psi}{\partial z^2} \right) + \left(\frac{\partial \Phi}{\partial x} \frac{\partial \psi}{\partial x} + \frac{\partial \phi}{\partial y} \frac{\partial \psi}{\partial y} + \frac{\partial \phi}{\partial z} \frac{\partial \psi}{\partial z} \right) \right\} d\tau, \tag{11}$$

which is Green's theorem. We may write

$$\frac{\partial^2 \psi}{\partial x^2} + \frac{\partial^2 \psi}{\partial y^2} + \frac{\partial^2 \psi}{\partial z^2} = \left(\frac{\partial^2}{\partial x^2} + \frac{\partial^2}{\partial y^2} + \frac{\partial^2}{\partial z^2} \right) \psi = \nabla^2 \psi, \tag{12}$$

where ∇^2 is an abbreviation for the indicated differentiations. ∇^2 (read "*del* square") is often called the Laplacian operator. It is scalar, whereas ∇ (or **grad**) was vector in character. Since the functions Φ and ψ are arbitrary, we may interchange them in (11) and write the equivalent equation:

$$\iint \psi \frac{\partial \Phi}{\partial n} \, dS = \iiint \left\{ \psi \nabla^2 \Phi + \left(\frac{\partial \Phi}{\partial x} \frac{\partial \psi}{\partial x} + \frac{\partial \Phi}{\partial y} \frac{\partial \psi}{\partial y} + \frac{\partial \Phi}{\partial z} \frac{\partial \psi}{\partial z} \right) \right\} d\tau. \tag{13}$$

Subtracting (13) from (11), we obtain a second form of Green's theorem:

$$\iint \left(\phi \frac{\partial \psi}{\partial n} - \psi \frac{\partial \phi}{\partial n} \right) dS = \iiint (\Phi \nabla^2 \psi - \psi \nabla^2 \Phi) \, d\tau. \tag{14}$$

If we set $\Phi = 1$ and $\psi = V$, then

$$\phi = \iint F \cos \theta \, dS = -\iint \frac{\partial V}{\partial n} \, dS = -\iiint (\nabla^2 V) \, d\tau. \tag{15}$$

15. Gauss' law. Consider any closed surface containing a number of particles of matter attracting each other gravitationally. We are required to calculate the total flux (the surface integral of the force acting on a unit mass) over the boundary of the surface. Let us first determine the value of ϕ when a single particle, of mass M, is located at some point P within the volume. From P draw a small cone to an element of surface dS. Let θ be the angle between the axis of the cone and the outward normal to dS. The total flux is $\iint F \cos \theta \, dS$, as in equation (13.1). About P describe a sphere of radius r, of which the cone intercepts the area, da. Let R be the distance from P to dS. Then, *if the law of force follows the inverse square*, the force intensity, i.e., the force per unit mass, at dS is

$$F = -GM/R^2. \tag{1}$$

Hence
$$\iint F \cos \theta \, dS = -GM \iint \frac{\cos \theta}{R^2} \, dS.$$

But $\cos \theta \, dS$ is merely the projection of dS on the plane normal to R. Therefore

$$\frac{\cos \theta}{R^2} \, dS = \frac{da}{r^2}, \tag{2}$$

by the simple geometry of a sphere.

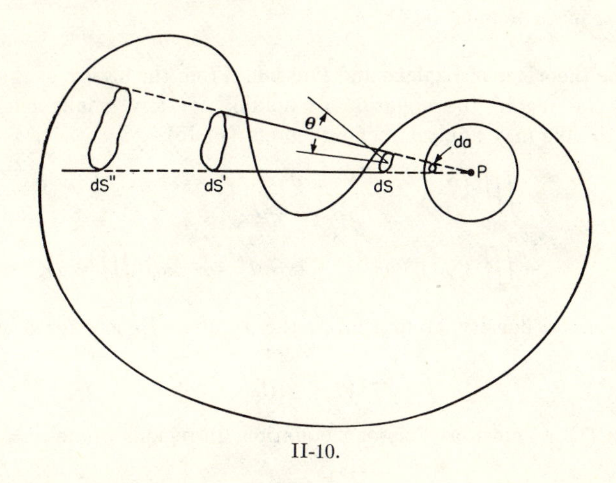

II-10.

The total integral thus becomes

$$\phi = \iint F \cos \theta \, dS = -\frac{GM}{r^2} \iint da = -\frac{GM}{r^2} 4\pi r^2 = -4\pi GM. \tag{3}$$

This equation is known as Gauss' law. If the surface is concave so that the cone enters more than once, as at dS' and dS'', no correction is needed since the contribution at dS', being *inward*, just balances the effect of the *exit* contribution at dS''. If more than one particle exists in the volume, the effects are additive, because (3) is independent of the position of the particle in the volume. Hence equation (3) holds if we interpret M as the total mass contained in the volume. For electric charges we can prove an analogous relation:

$$\phi = 4\pi\varepsilon, \tag{4}$$

where ε represents the total electric charge contained in the volume. The sign is positive because two electric charges of the same sign repel one another, i.e., the electrical analogue of (1) is

$$F = \varepsilon/r^2. \tag{5}$$

Furthermore, if no matter or charges are found in the volume,

$$\phi = 0. \tag{6}$$

Gauss' law has led to an interesting fictitious representation of the force field. During the nineteenth century the concept of "lines (or tubes) of force" played an important role in the development of physical theories. The interpretation was similar to that of light rays and flux, discussed in § (13). If we imagine that $4\pi G$ lines of force emanate from each unit of mass, the number of lines penetrating the closed surface of a volume containing mass M must be $4\pi GM$.

16. The theorems of Laplace and Poisson. From the nature of the proof of (15.3) we regard the equation as holding for any small element of volume, $d\tau$. We may apply Green's theorem (14.15),

$$\begin{aligned}\phi &= -\iiint \left(\frac{\partial^2 V}{\partial x^2} + \frac{\partial^2 V}{\partial y^2} + \frac{\partial^2 V}{\partial z^2}\right) d\tau \\ &= -\iiint (\nabla^2 V)\, d\tau = -4\pi GM = -4\pi G \iiint \rho\, d\tau,\end{aligned}$$

where ρ is the density of matter in the volume. Hence, for a volume element,

$$\nabla^2 V = 4\pi G\rho. \tag{1}$$

Equation (1) is known as Poisson's equation. In regions where $\rho = 0$,

$$\nabla^2 V = 0, \tag{2}$$

by (15.5), or (1), with $\rho = 0$. Equation (2) is due to Laplace. For electric charges, Poisson's equation becomes

$$\nabla^2 V = -4\pi\rho_e, \tag{3}$$

where ρ_e is the density of electric charge per unit volume. Equations (1), (2), and (3) have had many important physical applications. The quantity V is, of course, the potential from which we may calculate the force intensity from the formula

$$\mathbf{F} = -\nabla V. \tag{4}$$

The equations of Laplace and Poisson hold only for inverse-square fields, because their derivation depends on the assumption (15.1). We ordinarily calculate the potential from equation (11.1); thus

$$V = -\iiint \frac{G\rho}{r}\, d\tau, \tag{5}$$

by (1).

Consider the problem of determining the potential at P produced by the irregularly shaped mass M. For convenience we shall adopt a system of coordinates with P at the origin. About P describe a small sphere of radius r_0. Denote the position of any other point, P′ by the spherical polar

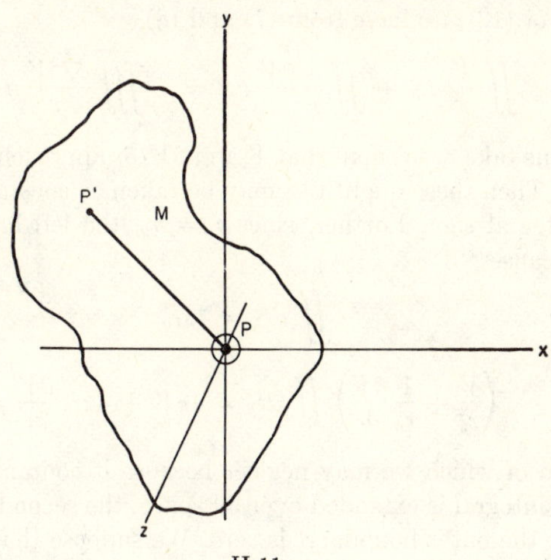

II-11.

coordinates r, θ, and ϕ. In equation (14.14), Green's theorem, set

$$\psi = \frac{1}{r} \quad \text{and} \quad \Phi = V. \tag{6}$$

Then $$\iint \left[V \frac{\partial}{\partial n}\left(\frac{1}{r}\right) - \frac{1}{r}\frac{\partial V}{\partial n} \right] dS = \iiint \left[V\nabla^2\left(\frac{1}{r}\right) - \frac{1}{r}\nabla^2 V \right] d\tau. \tag{7}$$

We shall integrate over the volume and surfaces bounded by two concentric spheres of radius r_0 and R_0. Since n is by definition a coordinate normal to the surface of integration, we have $n = r$, or $-r$ for the respective outer and inner boundaries. The left-hand side of (7) becomes, if for the moment we neglect integration over the outer boundary,

$$-\iint \left[V \frac{\partial}{\partial r}\left(\frac{1}{r}\right) - \frac{1}{r}\frac{\partial V}{\partial r} \right] dS = \iint \frac{V}{r^2}\, dS + \iint \frac{1}{r}\frac{\partial V}{\partial r}\, dS. \tag{8}$$

We have $$r^2 = x^2 + y^2 + z^2. \tag{9}$$

Whence $$\frac{\partial}{\partial x}\left(\frac{1}{r}\right) = -\frac{1}{r^2}\frac{\partial r}{\partial x} = -\frac{x}{r^3} \tag{10}$$

and $$\frac{\partial^2}{\partial x^2}\left(\frac{1}{r}\right) = -\frac{r^2 - 3x^2}{r^5}\,. \tag{11}$$

Similar expressions hold for the y and z derivatives. Therefore

$$\nabla^2\left(\frac{1}{r}\right) = \left(\frac{\partial^2}{\partial x^2} + \frac{\partial^2}{\partial y^2} + \frac{\partial^2}{\partial z^2}\right)\left(\frac{1}{r}\right) = 0. \tag{12}$$

Making use of (12), we have from (7) and (8),

$$\iint \frac{V}{r^2}\, dS + \iint \frac{1}{r}\frac{\partial V}{\partial r}\, dS = -\iiint \frac{\nabla^2 V}{r}\, d\tau. \tag{13}$$

Further, let us take r_0 so small that V and $\partial V/\partial r$ approach their respective values at P. Then these quantities may be taken as constant and removed from the integral sign. Further, since $r = r_0$, the left-hand side of (13) becomes, because

$$\iint dS = 4\pi r_0^2,$$

$$\left(\frac{V}{r_0^2} + \frac{1}{r_0}\frac{\partial V}{\partial r}\right)\iint dS = 4\pi V + 4\pi r_0 \frac{\partial V}{\partial r}, \tag{14}$$

the last term of which we may neglect because it contains r_0 as a factor.

When the integral is extended over all space, the second surface integral of (13), over the outer boundary, is zero. We suppose that R is extremely large compared with the extension of the mass M, so that in the neighborhood of the boundary M acts effectively as a point mass. Then

$$V \sim -\frac{GM}{R} \quad \text{and} \quad \frac{\partial V}{\partial r} \sim \frac{GM}{R^2}\cdot \tag{15}$$

The surface integral over the outer boundary, which is of the form of (14), becomes, to the first approximation,

$$-\frac{4\pi GM}{R} + \frac{4\pi GM}{R} = 0, \tag{16}$$

not merely because of the identity of the two principal terms, but because we may now allow R to become infinite. Combining the results of (13) and (14), we have

$$V = -\frac{1}{4\pi}\iiint \frac{\nabla^2 V}{r}\, d\tau, \tag{17}$$

in agreement with (5).

If we do not wish to carry the integration over all space, we must include the surface integral over the outer boundary, and

$$V = -\frac{1}{4\pi}\iiint \frac{\nabla^2 V}{r}\, d\tau - \iint \left[V\frac{\partial}{\partial n}\left(\frac{1}{r}\right) - \frac{1}{r}\frac{\partial V}{\partial n}\right] dS. \tag{18}$$

We have replaced a portion of the volume integral by an integral over the surface. The function V will, in general, vary from point to point. We located P at the origin only for convenience of notation. As a function of the coordinates, then, V must satisfy the boundary conditions of the surface integral. We may regard (18) as a solution of the differential equation (2).

The function V is determined if we know $\nabla^2 V$ at every point of τ, and the value of V, at every point of the surface enclosing τ. Evaluation of the potential function, with the aid of Poisson's and Laplace's equations becomes a boundary-value problem. The function V, so determined, is unique.

17. Harmonic functions. Any solution of Laplace's equation is known as an harmonic function. When the equation is written in spherical coordinates, the solution comes out in terms of functions known as solid spherical harmonics, of which zonal and tesseral harmonics are examples.

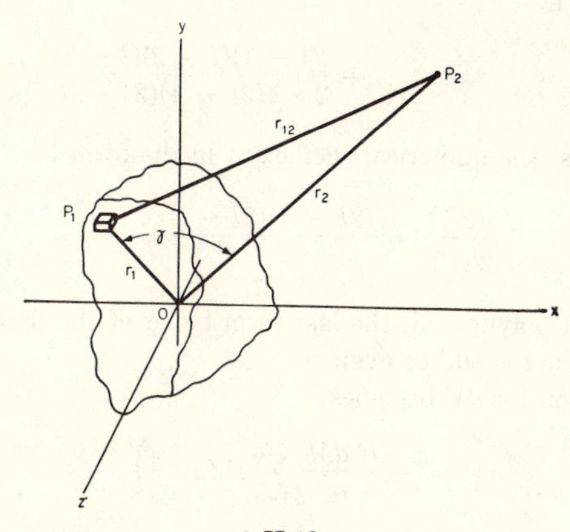

II-12.

We shall briefly indicate here the nature of harmonic functions. Figure 12 depicts the body whose potential we wish to determine. The mass element, dM, located at P_1 (x_1, y_1, z_1), contributes to the potential V, at the point P_2 (x_2, y_2, z_2), an amount

$$dV = -\frac{G\,dM}{r_{12}} = \frac{G\,dM}{(r_2^2 - 2r_1r_2\cos\gamma + r_1^2)^{1/2}}\,. \tag{1}$$

In deriving the foregoing we have made use of the well-known "cosine law" of triangles, that

$$r_{12}^2 = r_1^2 - 2r_1r_2\cos\gamma + r_2^2, \tag{2}$$

where γ is the angle between $OP_1(r_1)$ and $OP_2(r_2)$ and r_{12} the distance P_1P_2.

We first suppose that $r_2 \geq r_1$. Now set

$$\cos\gamma = \mu \quad \text{and} \quad r_1/r_2 = \beta, \tag{3}$$

and expand the brackets in the following by means of the binomial theorem.

$$\begin{aligned} dV &= -\frac{G\,dM}{r_2}[1 - \beta(2\mu - \beta)]^{-1/2} \\ &= -\frac{G\,dM}{r_2}\left[1 + \frac{1}{2}\beta(2\mu - \beta) + \frac{1\cdot 3}{2\cdot 4}\beta^2(2\mu - \beta)^2 + \ldots\right]. \end{aligned} \tag{4}$$

We next expand each binomial factor and collect the resultant series as a power series in β. The coefficient of β^l, which we shall denote by $P_l(\mu)$, proves to be

$$\begin{aligned} P_l(\mu) = \frac{(2l)!}{2^l(l!)^2}\Bigg[\mu^l &- \frac{l(l-1)}{2(2l-1)}\mu^{l-2} \\ &+ \frac{l(l-1)(l-2)(l-3)}{2\cdot 4(2l-1)(2l-3)}\mu^{l-4} - \ldots\Bigg]. \end{aligned} \tag{5}$$

We may express the numerical coefficient in the form

$$\frac{(2l)!}{2^l(l!)^2} = \frac{(2l-1)(2l-3)\ldots 1}{l!}. \tag{6}$$

Here $P_l(\mu)$ is a polynomial, the last term being of the first or zero degree in μ according as l is odd or even.

The expression for dV becomes

$$dV = -\frac{G\,dM}{r_2}\sum_{l=0}^{\infty} P_l(\mu)\left(\frac{r_1}{r_2}\right)^l. \tag{7}$$

These coefficients $P_l(\mu)$, which multiply successive terms of the infinite series, are Legendre's polynomials or zonal harmonics. A few examples follow:

$$\begin{aligned} &P_0 = 1, \quad P_1 = \mu, \quad P_2 = \frac{3}{2}\mu^2 - \frac{1}{2}, \quad P_3 = \frac{5}{2}\mu^3 - \frac{3}{2}\mu, \\ &P_4 = \frac{5\cdot 7}{2\cdot 4}\mu^4 - \frac{3\cdot 5}{2\cdot 4}2\mu^2 + \frac{1\cdot 3}{2\cdot 4}, \text{ ETC.} \end{aligned} \tag{8}$$

Rodrigues gave a simple alternative expression for the coefficients. We establish the identity of (4) and the following, by first expanding the binomial and then performing the indicated differentiation, term by term.

$$P_l(\mu) = \frac{1}{2^l l!}\frac{d^l}{d\mu^l}(\mu^2 - 1)^l. \tag{9}$$

Let the spherical coordinates of the respective points P_1 and P_2 be (r_1, θ_1, ϕ_1) and (r_2, θ_2, ϕ_2). We have the relation

$$r_{12}^2 = (x_2 - x_1)^2 + (y_2 - y_1)^2 + (z_2 - z_1)^2. \tag{10}$$

Substituting for x_1, x_2, etc., from (10.4), and comparing the resulting expression with (2), we identify

$$\cos \gamma = \mu = \cos \theta_1 \cos \theta_2 + \sin \theta_1 \sin \theta_2 \cdot (\cos \phi_1 \cos \phi_2 + \sin \phi_1 \sin \phi_2). \tag{11}$$

A spherical harmonic with μ defined as in (11) is sometimes called a *Laplace* coefficient, as opposed to a *Legendre* coefficient, for which $\theta_2 = 0$.

If we substitute this expression for μ successively into the various equations (8), we find that

$$P_0 = (1)(1). \tag{12}$$

$$\begin{aligned} P_1 = &(\cos \theta_1)(\cos \theta_2) \\ &+ (\sin \theta_1 \cos \phi_1)(\sin \theta_2 \cos \phi_2) \\ &+ (\sin \theta_1 \sin \phi_1)(\sin \theta_2 \sin \phi_2). \end{aligned} \tag{13}$$

$$\begin{aligned} P_2 = &\left(\frac{3}{2} \cos^2 \theta_1 - \frac{1}{2}\right)\left(\frac{3}{2} \cos^2 \theta_2 - \frac{1}{2}\right) \\ &+ \frac{2}{3!}(3 \cos \theta_1 \sin \theta_1 \cos \phi_1)(3 \cos \theta_2 \sin \theta_2 \cos \phi_2) \\ &+ \frac{2}{3!}(3 \cos \theta_1 \sin \theta_1 \sin \phi_1)(3 \cos \theta_2 \sin \theta_2 \sin \phi_2) \\ &+ \frac{2}{4!}(3 \sin^2 \theta_1 \cos 2\phi_1)(3 \sin^2 \theta_2 \cos 2\phi_2) \\ &+ \frac{2}{4!}(3 \sin^2 \theta_1 \sin 2\phi_1)(3 \sin^2 \theta_2 \sin 2\phi_2), \end{aligned} \tag{14}$$

etc. We have employed the following trigonometric expansion:

$$\cos^m \phi = \frac{1}{2^{m-1}}\left[\cos m\phi + m \cos (m-2)\phi + \frac{m(m-1)}{2!} \cos (m-4)\phi + \ldots + L\right]. \tag{15}$$

The coefficients in (15) are those of the binomial theorem. The last term of the series, L, is

$$L = \frac{m!}{2[(m/2)!]^2} \quad \text{or} \quad \frac{m!}{[(m+1)/2]![(m-1)/2]!}, \tag{16}$$

according as m is even or odd.

The developments of further values of P_l become fairly laborious, but the algebra is straightforward. We note especially that the factors involving θ_1 have been completely separated from those involving θ_2. We see further that any given term of the expansion is symmetrical with respect to the angular subscripts 1 and 2. The problem is to find a general expression for the factors in parentheses, in (12), (13), and (14). We note, in advance, that the first term in each equation is a simple Legendre function, like those shown in (8). The others are somewhat more complicated. We group them according to the value of m in $\cos m\phi$ or $\sin m\phi$. The coefficient of $\sin m\phi$ or $\cos m\phi$, in each parentheses, itself a function of θ only, we denote as $P_l^m(\cos\theta)$, where

$$P_l^m(\cos\theta) = P_l^m(\mu) = (1-\mu^2)^{m/2}\frac{d^m}{d\mu^m}P_l(\mu)$$

$$= \frac{1}{2^l l!}(1-\mu^2)^{m/2}\frac{d^{l+m}}{d\mu^{l+m}}(\mu^2-1)^l$$

$$= (1-\mu^2)^{m/2}\frac{(2l)!}{2^l l!(l-m)!}\left[\mu^{l-m} - \frac{(l-m)(l-m-1)}{2(2l-1)}\mu^{l-m-2}\right.$$

$$\left. + \frac{(l-m)(l-m-1)(l-m-2)(l-m-3)}{2\cdot 4\,(2l-1)(2l-3)}\mu^{l-m-4}\cdots\right]. \quad (17)$$

This relationship involving the derivatives is an extension of Rodrigues' formula (9). The quantities $P_l^m(\mu)$ are known as *associated* spherical harmonics. We establish the identities of the functions in (12), (13), (14), etc., by direct differentiation of (17).

Each term of the harmonic functions in (12)–(14) is known as a tesseral harmonic. The order of the harmonic is l, and there are $2l+1$ tesserals for each value of l. The first four orders of tesseral harmonics are:

Order 0: 1.
Order 1: $\cos\theta$, $\sin\theta\cos\phi$, $\sin\theta\sin\phi$.
Order 2: $\frac{3}{2}\cos^2\theta - \frac{1}{2}$, $3\sin\theta\cos\theta\cos\phi$, $3\sin\theta\cos\theta\sin\phi$,
$3\sin^2\theta\cos 2\phi$, $3\sin^2\theta\sin 2\phi$.
Order 3: $\frac{5}{2}\cos^3\theta - \frac{3}{2}\cos\theta$, $\frac{3}{2}\sin\theta(5\cos^2\theta - 1)\cos\phi$, (18)
$\frac{3}{2}\sin\theta(5\cos^2\theta-1)\sin\phi$, $15\sin^2\theta\cos\theta\cos 2\phi$,
$15\sin^2\theta\cos\theta\sin 2\phi$, $15\sin^3\theta\cos 3\phi$, $15\sin^3\theta\sin 3\phi$.

The function, P_l, then becomes

$$P_l(\cos\theta) = \sum_{m=0}^{l}\frac{(l-m)!}{(l+m)!}\frac{2}{a_m}P_l^m(\cos\theta_1)P_l^m(\cos\theta_2)(\cos m\phi_1\cos m\phi_2 + \sin m\phi_1\sin m\phi_2). \quad (19)$$

The quantity

$$a_m = 2, \qquad m = 0,$$
$$a_m = 1, \qquad m \neq 0. \tag{20}$$

The substitution

$$\cos m\phi_1 \cos m\phi_2 + \sin m\phi_1 \sin m\phi_2 = \cos m(\phi_1 - \phi_2) \tag{21}$$

is sometimes useful. Equation (19), which is usually called the addition theorem in spherical harmonics, will be established more rigorously in the next section, equation (18.32).

In deriving equation (7) we assumed that r_1, the distance from the origin to the mass element dM, is less than r_2, the distance from the origin to the point P_2, where the potential is to be calculated. When the reverse is true, i.e., when

$$\frac{r_2}{r_1} \leq 1, \tag{22}$$

we may expand equation (1) in powers of $1/\beta$, equation (3).

Thus
$$dV = -\frac{G\,dM}{r_2}\frac{1}{\beta}\left[1 - \frac{1}{\beta}\left(2\mu - \frac{1}{\beta}\right)\right]^{-1/2}. \tag{23}$$

Comparing this expression with (4), we see that the expansions are identical, except for the extra factor $1/\beta$. Whereas previously $P_l(\mu)$ was the coefficient of β^l, it now appears as the coefficient of $(1/\beta)^{l+1}$. Hence, we may now write the expansion

$$dV = -\frac{G\,dM}{r_2}\sum_{l=0}^{\infty} P_l(\mu)\left(\frac{r_2}{r_1}\right)^{l+1}. \tag{24}$$

18. Orthogonality and further miscellaneous properties of spherical harmonics. Consider the integrals

$$N_a^2 = \int_0^{2\pi} \cos m\phi \cos m'\phi \, d\phi$$
$$= \frac{1}{2}\int_0^{2\pi} [\cos (m + m')\phi + \cos (m - m')\phi] \, d\phi, \tag{1}$$

$$N_b^2 = \int_0^{2\pi} \sin m\phi \cos m'\phi \, d\phi$$
$$= \frac{1}{2}\int_0^{2\pi} [\sin (m + m')\phi + \sin (m - m')\phi] \, d\phi, \tag{2}$$

wherein we have made a simple trigonometrical substitution. Integrals of the cross-product type always vanish whatever value we may assign to m or m'. Hence

$$N_b^2 = 0. \tag{3}$$

Integrals of the first type vanish except when $m' = m$,

$$N_a^2 = \int_0^{2\pi} \cos^2 m\phi \, d\phi = \begin{cases} 2\pi, & m' = m = 0 \\ \pi, & (m' = m),\ m = 1, 2, 3, \ldots \end{cases} \tag{4}$$

$$N_a^2 = 0, \qquad (m' \neq m).$$

Similarly,

$$N_c^2 = \int_0^{2\pi} \sin^2 m\phi \, d\phi = \pi, \quad (m' = m),\ m = 1, 2, 3, \ldots \tag{5}$$

$$N_c^2 = 0, \qquad (m' \neq m).$$

The latter integral is zero when $m = 0$, because the integrand vanishes. Systems of functions possessing properties similar to those demonstrated above for the sine and cosine, are said to be *orthogonal.* Orthogonality requires that the integral of the product of two functions distinguished by some index (like m in the preceding example) vanish except when the two indices are identical. The integration is usually taken over the extreme range of the coordinates involved.

Tesseral harmonics, of the form $P_l^m (\cos \theta) \begin{Bmatrix} \sin m\phi \\ \cos m\phi \end{Bmatrix}$ are well-known examples of orthogonal functions. The notation indicates that either the sine or cosine of $m\phi$ is to be adopted. The harmonics are of the surface variety and the integration is to be over the surface of a unit sphere, the element of whose area is

$$dS = \sin \theta \, d\theta \, d\phi. \tag{6}$$

The expression becomes

$$\int_{\theta=0}^{\pi} \int_{\phi=0}^{2\pi} P_l^m(\cos \theta) P_{l'}^{m'}(\cos \theta) \begin{Bmatrix} \sin \\ \cos \end{Bmatrix} (m\phi) \begin{Bmatrix} \sin \\ \cos \end{Bmatrix} (m'\phi) \sin \theta \, d\theta \, d\phi. \tag{7}$$

The variables θ and ϕ are independent. Integrating first with respect to ϕ, we may apply equations (4) and (5), which require that $m' = m$, if the definite integral is to have a non-zero value. Therefore, removing the factor N_a^2 or N_c^2, we have the analogous quantity for the spherical harmonics:

$$N_d^2 = \int_0^{\pi} P_l^m(\cos \theta) P_{l'}^m(\cos \theta) \sin \theta \, d\theta$$

$$= \int_{-1}^{1} P_l^m(\mu) P_{l'}^m(\mu) \, d\mu. \tag{8}$$

$$N_d^2 = \frac{1}{2^l l!} \frac{1}{2^{l'} l'!} \int_{-1}^{1} \left[\frac{d^{l+m}}{d\mu^{l+m}} (\mu^2 - 1)^l \right] \left[(1 - \mu^2)^m \frac{d^{l'+m}}{d\mu^{l'+m}} (\mu^2 - 1)^{l'} \right] d\mu, \tag{9}$$

wherein we have set $\cos\theta = \mu$ and have substituted from Rodrigues' formula, (17.17). We shall first suppose that $l' < l$. Denote the factor in the second brackets by Y. Y is a polynomial in μ with the highest power equal to $\mu^{l'+m}$. Expanding the factor $(\mu^2 - 1)^{l'}$ by the binomial theorem and performing the indicated differentiations on the term with the highest value of the exponent, we find that

$$Y = (-1)^m \frac{(2l')!}{(l' - m)!} \mu^{l'+m} + \text{terms in lower powers of } \mu. \tag{10}$$

Integrating by parts, we obtain the result

$$N_d^2 = \frac{1}{2^l l!} \frac{1}{2^{l'} l'!} \left\{ \left[Y \frac{d^{l+m-1}}{d\mu^{l+m-1}} (\mu^2 - 1)^l \right]_{-1}^{1} - \int_{-1}^{1} \frac{dY}{d\mu} \frac{d^{l+m-1}}{d\mu^{l+m-1}} (\mu^2 - 1)^l \, d\mu \right\}. \tag{11}$$

The bracketed term vanishes at the limits owing to the presence of the factor $(1 - \mu^2)$ in Y. Successive integration by parts finally gives the result

$$N_d^2 = \frac{1}{2^l l!} \frac{1}{2^{l'} l'!} \int_{-1}^{1} (\mu^2 - 1)^l \frac{d^{l+m}}{d\mu^{l+m}} Y \, d\mu. \tag{12}$$

But if $l' < l$, then $l' + m < l + m$ and the $(l + m)$th derivative of Y vanishes, as one sees from differentiation of (10). Hence, under these circumstances,

$$N_d^2 = 0. \tag{13}$$

If $l' > l$ we have only to interchange the derivatives in equation (9) to arrive at the same result. Thus we prove the orthogonality of the functions.

If, however, $l = l'$, then

$$\frac{d^{l+m}}{du^{l+m}} Y = \frac{(2l)!(l + m)!}{(l - m)!}. \tag{14}$$

The terms of Y with powers of μ less than $l + m$, indicated schematically in (10), vanish. Returning to the variable θ in equation (9), we find

$$N_d^2 = \frac{(2l)!(l + m)!}{2^{2l}(l!)^2(l - m)!} \int_0^{\pi} \sin^{2l+1}\theta \, d\theta = \frac{2(l + m)!}{(l - m)!(2l + 1)}. \tag{15}$$

Equation (8) therefore becomes

$$\int_{-1}^{1} \int_0^{2\pi} \left[P_l^m(\mu) \begin{Bmatrix} \sin \\ \cos \end{Bmatrix} (m\phi) \right]^2 d\mu \, d\phi = \frac{(l + m)!}{(l - m)!} \frac{2\pi a_m}{(2l + 1)}. \tag{16}$$

We shall use the orthogonal properties of the trigonometric and spherical harmonics in Part III as the basis of expansions of functions in Fourier and similar types of series.

Expand the function $(\mu^2 - 1)^l$ by the binomial theorem, differentiate, and then compare the two resultant series within the square brackets, to verify the following expression:

$$\int_\mu^1 P_l(\mu)\,d\mu = \frac{1}{2^l l!}\int_\mu^1 \frac{d^l}{d\mu^l}(\mu^2 - 1)^l\,d\mu = \frac{1}{2^l l!}\left[\frac{d^{l-1}}{d\mu^{l-1}}(\mu^2 - 1)^l\right]_\mu^1$$
$$= \frac{1}{2^l l!}\left[\frac{\mu^2 - 1}{l(l+1)}\frac{d^{l+1}}{d\mu^{l+1}}(\mu^2 - 1)^l\right]_\mu^1 = \frac{(1 - \mu^2)}{l(l+1)}\frac{dP_l(\mu)}{d\mu}, \tag{17}$$

by (17.5) and (17.9).

When l is an even integer, we find that the derivative in (17) becomes:

$$\left[\frac{dP_l(\mu)}{d\mu}\right] = \text{higher powers of } \mu + (-1)^{(l/2)+1}\cdot\frac{1\cdot 3\cdot 5\ldots(l+1)}{2\cdot 4\cdot 6\ldots(l-2)}\,\mu. \tag{18}$$

Similarly, when l is an odd integer,

$$\left[\frac{dP_l(\mu)}{d\mu}\right] = \text{higher powers of } \mu + (-1)^{(l-1)/2}\cdot\frac{1\cdot 3\cdot 5\ldots(l)}{2\cdot 4\cdot 6\ldots(l-1)}\cdot \tag{19}$$

Hence, in (17), setting the lower limit of the integral equal to zero, we find

$$\begin{aligned}\int_0^1 P_l(\mu)\,d\mu &= 0, \quad \text{if } l \text{ is even but } \neq 0,\\ &= 1, \quad \text{if } l = 0,\\ &= (-1)^{(l-1)/2}\frac{1\cdot 3\cdot 5\ldots(l-2)}{2\cdot 4\cdot 6\ldots(l+1)}, \quad \text{if } l \text{ is odd.}\end{aligned} \tag{20}$$

All terms in (18) and all but the last in (19) vanish when $\mu = 0$, as a lower limit.

The function $P_l(\mu)$ is a polynomial of degree (l) in μ. Hence the equation

$$P_l(\mu) = 0 \tag{21}$$

has l roots, all of which are real. In other words there exist l values of θ for which the function $P_l(\mu)$ vanishes. These "zeros" occur on various parallels of latitude on the sphere. The odd-numbered harmonics must vanish at the equator, because they contain the factor μ, which becomes equal to zero for $\theta = 90°$.

Similarly, $P_l^m(\mu)$, vanishes along $(l - m)$ parallels of latitude. Because of the factor $(1 - \mu^2)^{m/2}$ in (17.17), the associated functions vanish at the poles ($\mu = \pm 1$), when $m > 0$. The "zeros" at the poles are of order $m/2$ because of the factor $(1 - \mu)^{m/2}$. Tesseral harmonics, $P_l^m(\mu)\begin{Bmatrix}\sin\\ \cos\end{Bmatrix}(m\phi)$, also vanish along $2m$ meridians. The parallels and meridians on which the sample harmonics $P_4^0(\mu)$, $P_8^3(\mu)\cos 3\phi$, and $P_3^3(\mu)\cos 3\phi$ vanish appear in

Fig. 13. The figures illustrate why we call the Legendre polynomials $P_l^0(\mu)$ *zonal harmonics*, and the functions $P_m^m(\mu)$ cos $m\phi$ *sectorial harmonics*. The zero lines of a function $P_l^m(\mu)$ cos $m\phi$ divide the surface of the sphere into a sort of checkerboard of "squares" (Latin *tesserae*); hence the name *tesseral harmonics*.

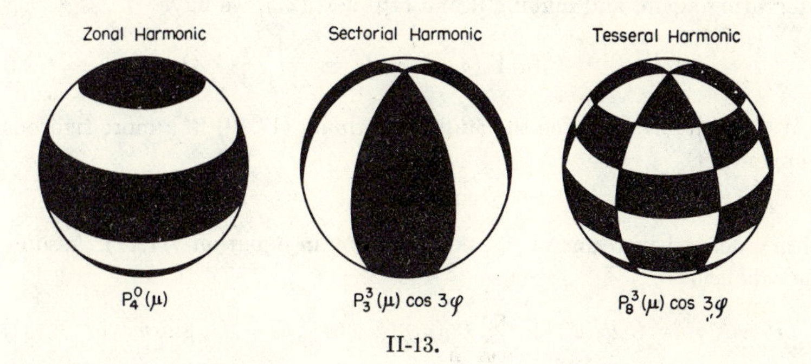

II-13.

We shall now prove that

$$P_l(1) = 1. \tag{22}$$

From (17.1) and (17.7) we have the identity

$$\frac{1}{(r_2^2 - 2r_1r_2\mu + r_1^2)^{1/2}} = \frac{1}{r_2}\sum_{l=0}^{\infty} P_l(\mu)\left(\frac{r_1}{r_2}\right)^l. \tag{23}$$

Setting $\mu = 1$, we have an exact square in the denominator, which expands as follows:

$$\frac{1}{r_2 - r_1} = \frac{1}{r_2}\left[1 + \frac{r_1}{r_2} + \left(\frac{r_1}{r_2}\right)^2 + \ldots\right] = \frac{1}{r_2}\sum_{l=0}^{\infty}\left(\frac{r_1}{r_2}\right)^l. \tag{24}$$

The right-hand sides of (23) and (24) must be identical for $\mu = 1$. This condition requires the relation (22), which is therefore proved.

We shall denote the general surface harmonic of order l by the symbol Y_l.

$$Y_l(\mu, \phi) = A_0P_l^0(\mu) + \sum_{m=1}^{l} P_l^m(\mu)[A_m \cos m\phi + B_m \sin m\phi], \tag{25}$$

where the A_m's and B_m's represent arbitrary numerical constants. Y_l is thus a linear combination of tesseral harmonics of index l. When

$$\theta = 0, \text{ or } \pi; \quad \text{i.e.,} \quad \mu = \cos\theta = \pm 1, \tag{26}$$

the summation in (25) vanishes because of the occurrence of the factor $(1 - \mu^2)^{m/2}$ in the associated harmonic. Hence the value of

$$Y_l(\pm 1, \phi) = A_0P_l^0(\pm 1) = A_0 = Y_l(1), \tag{27}$$

by (22). $Y_l(1)$ is the value assumed by Y_l at the pole. The integral

$$\int_0^{2\pi} Y_l(\mu, \phi)\, d\phi = 2\pi A_0 P_l^0(\mu). \tag{28}$$

The other terms vanish. Further, multiplying both sides of (28) by $P_l^0(\mu)\, d\mu$, integrating again, and making use of (16) and (27), we have

$$\int_{-1}^{1} \int_0^{2\pi} P_l^0(\mu) Y_l(\mu, \phi)\, d\phi\, d\mu = \frac{4\pi}{2l + 1} Y_l(1). \tag{29}$$

We may now determine the addition formula (17.19) in a more rigorous manner. Set

$$\cos \theta_1 = \mu_1, \tag{30}$$

with γ defined in terms of θ_1 and θ_2 and as in equation (17.11). Assume the expansion

$$P_l^0(\cos \gamma) = C_0 P_l^0(\mu_1) + \sum_{m=1}^{l} P_l^m(\mu_1)[C_m \cos m\phi_1 + D_m \sin m\phi_1], \tag{31}$$

where C_m and D_m are constants to be determined. Multiply both sides of (31) by $P_l^m(\mu_1) \cos m\phi_1\, d\mu_1\, d\phi_1$ and integrate as follows:

$$\begin{aligned} \int_0^{2\pi} \int_{-1}^{1} & P_l^0(\cos \gamma) P_l^m(\mu_1) \cos m\phi_1\, d\mu_1\, d\phi_1 \\ &= C_m \int_0^{2\pi} \int_{-1}^{1} [P_l^m(\mu_1) \cos m\phi_1]^2\, d\mu_1\, d\phi_1 \\ &= C_m \frac{(l + m)!}{(l - m)!} \frac{2\pi}{(2l + 1)} a_m, \end{aligned} \tag{32}$$

by (16). a_m is given by (17.20). We still have to evaluate the left-hand side of this equation. We have already shown, in equation (29), that when any zonal harmonic $P_l^0(\mu)$ is multiplied by Y_l, the general surface harmonic of similar degree, and the result integrated over the unit sphere, the result is $4\pi/(2l + 1)$ times the value assumed by Y_l at the pole of the zonal harmonic. The pole of the harmonic P_l^0 $(\cos \gamma)$ is at $\theta_1 = 0$, i.e.,

$$\cos \gamma = \cos \theta_2 = \mu_2 ,$$

by (17.11). The pole lies at θ_2, ϕ_2. Therefore, analogous to (29), we have

$$\int_{-1}^{1} \int_0^{2\pi} P_l(\cos \gamma) Y_l(\mu_1 , \phi_1)\, d\mu_1\, d\phi_1 = \frac{4\pi}{2l+1} Y_l (\mu_2 , \phi_2), \tag{33}$$

and

$$\begin{aligned} \int_{-1}^{1} \int_0^{2\pi} & P_l(\cos \gamma) P_l^m(\mu_1) \cos m\phi_1\, d\mu_1\, d\phi_1 \\ &= \frac{4\pi}{2l + 1} P_l^m(\mu_2) \cos m\phi_2 = C_m \frac{(l + m)!}{(l - m)!} \frac{2\pi}{2l + 1} a_m. \end{aligned} \tag{34}$$

Hence

$$C_m = \frac{(l-m)!}{(l+m)!}\frac{2}{a_m} P_l^m(\mu_2) \cos m\phi_2, \tag{35}$$

and similarly for the terms involving $\sin m\phi_2$. When we introduce these coefficients into (31), we find that the resulting summation agrees with (17.19), which we may now regard as proved.

19. Development of the potential in spherical harmonics. The addition theorem is important for the following reason. The ordinary zonal harmonics are functions of a single angle, γ. These functions will often suffice to describe the potential resulting from an axially symmetrical distribution of matter. We must, however, always measure γ with respect to this axis. When the distribution of matter is unsymmetrical or whenever we wish to use a spherical coordinate inclined to any existing axis of symmetry, we must apply the addition theorem, which is in effect a transformation of coordinates. We may then express the density as a function of variables r_1, θ_1, ϕ_1, and integrate (17.7) to give V as a function of the coordinates r_2, θ_2, ϕ_2. Thus, writing

$$dM = \rho\, d\tau = \rho r_1^2 \sin \theta_1\, dr_1\, d\theta_1\, d\phi_1, \tag{1}$$

and making use of (17.12), and (17.19), and (17.21), we find

$$\begin{aligned} V(r_2\,,\,\theta_2\,,\,\phi_2) = {} & -\frac{G}{r_2} \iiint\limits_{r_1 \le r_2} \rho r_1^2 \sin \theta_1 \sum_{l=0}^{\infty} \sum_{m=0}^{l} \left(\frac{r_1}{r_2}\right)^l \frac{(l-m)!}{(l+m)!}\frac{2}{a_m} \\ & \cdot P_l^m(\cos \theta_1) P_l^m(\cos \theta_2) \cos m(\phi_1 - \phi_2)\, dr_1\, d\theta_1\, d\phi_1 \\ & -\frac{G}{r_2} \iiint\limits_{r_1 \ge r_2} \rho r_1^2 \sin \theta_1 \sum_{l=0}^{\infty} \sum_{m=0}^{l} \left(\frac{r_2}{r_1}\right)^{l+1} \frac{(l-m)!}{(l+m)!}\frac{2}{a_m} \\ & \cdot P_l^m(\cos \theta_1) P_l^m(\cos \theta_2) \cos m(\phi_1 - \phi_2)\, dr_1\, d\theta_1\, d\phi_1. \end{aligned} \tag{2}$$

The respective integrals are to be taken over the indicated ranges of r_1.

When ρ is independent of ϕ, i.e., when the solid is a figure of revolution about the z-axis, the axis from which θ is measured, all the terms except those for $m = 0$ vanish, because

$$\int_0^{2\pi} \cos m(\phi_1 - \phi_2)\, d\phi_1 = \begin{cases} 2\pi, & m = 0 \\ 0, & m \neq 0. \end{cases} \tag{3}$$

Therefore

$$V = -\frac{G}{r_2} 2\pi \iint \rho r_1^2 \sin \theta_1 \sum_{l=0}^{\infty} \left(\frac{r_1}{r_2}\right)^l P_l(\cos \theta_1) P_l(\cos \theta_2)\, dr_1\, d\theta_1 \tag{4}$$

when $r_1 \le r_2$.

Let us calculate the potential of a homogeneous hemisphere, Fig. 14. We may remove the density, ρ, from the integral sign and carry out immediately the integration with respect to r_1. The integration limits for r_1 are 0 to a, where a is the radius of the sphere. Since we have assumed

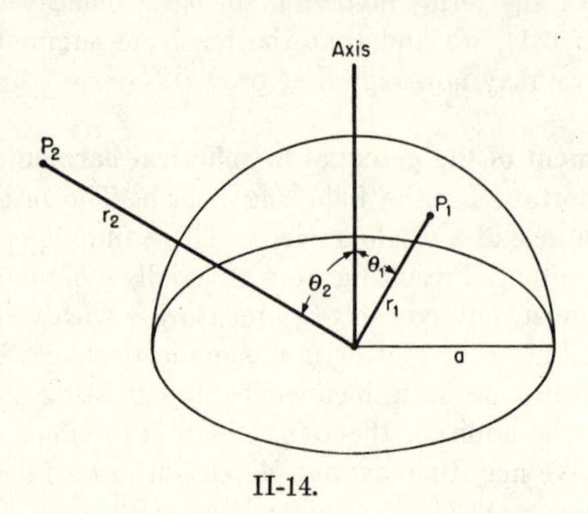

II-14.

that $a < r_2$, we need employ only the first summation in (2). The limits of integration for θ are 0 and $\pi/2$. Setting

$$\cos\theta_1 = \mu_1 \quad \text{and} \quad \cos\theta_2 = \mu_2,$$

we find that (4) assumes the form

$$V = -\frac{2}{3}\pi\rho a^3 \frac{G}{r_2} \sum_{l=0}^{\infty} \frac{3}{l+3}\left(\frac{a}{r_2}\right)^l P_l(\mu_2) \int_0^1 P_l(\mu_1)\, d\mu_1. \tag{5}$$

The latter integral was evaluated in (18.20), whence

$$V = -\frac{GM}{r_2}\left[1 + 3\sum_{k=0}^{\infty}(-1)^k\left(\frac{1\cdot 3\cdot 5\ldots(2k-1)}{2\cdot 4\cdot 6\ldots(2k+4)}\right)\left(\frac{a}{r_2}\right)^{2k+1} P_{2k+1}(\mu_2)\right], \tag{6}$$

where M, the mass of the hemisphere, is

$$M = \frac{2}{3}\pi\rho a^3. \tag{7}$$

We have introduced the index

$$l = 2k + 1 \tag{8}$$

to eliminate all the even-numbered harmonics except that for $l = 0$. From (6) we may derive the force vector ∇V. We note that $\mathbf{F}$ is independent of ϕ, a general result for all figures of revolution.

There is an alternative method of deriving equation (6), which is often extremely useful. Assume that we may express the potential of a solid of revolution in the form

$$V = -\frac{GM}{r_2} \sum_{l=0}^{\infty} A_l \left(\frac{r_1}{r_2}\right)^l P_l(\mu). \tag{9}$$

At any point along the axis of symmetry,

$$P_l(\mu_2) = P_l(1) = 1, \tag{10}$$

from (18.22). We shall calculate the potential along this axis, where, by virtue of the symmetry, the problem is relatively simple. In the previous notation, we have

$$\begin{aligned} V &= -G\rho \int_0^a \int_0^{\pi/2} \int_0^{2\pi} \frac{r_1^2 \sin \theta_1 \, dr_1 \, d\theta_1 \, d\phi_1}{(r_2^2 + r_1^2 - 2r_1r_2 \cos \theta_1)^{1/2}} \\ &= -2\pi G\rho \int_0^a \int_0^1 \frac{r_1^2 \, dr_1 \, d\mu_1}{(r_2^2 + r_1^2 - 2r_1r_2\mu_1)^{1/2}} \\ &= -\frac{2\pi G\rho}{r_2} \int_0^a [r_1r_2 - r_1^2 - r_1(r_2^2 + r_1^2)^{1/2}] \, dr_1 \\ &= -\frac{2\pi G\rho}{r_2} \frac{a^3}{3} \left[1 + \frac{3}{2}\frac{r_2}{a} + \frac{1}{a^3}(r_2^2 + a^2)^{3/2} - \left(\frac{r_2}{a}\right)^3\right]. \end{aligned} \tag{11}$$

Expanding the parentheses and collecting terms, we have

$$\begin{aligned} V = -\frac{GM}{r_2}\bigg\{1 + 3\bigg[\frac{1}{2 \cdot 4}\left(\frac{a}{r_2}\right) \\ - \frac{1}{2 \cdot 4 \cdot 6}\left(\frac{a}{r_2}\right)^2 + \frac{1 \cdot 3}{2 \cdot 4 \cdot 6 \cdot 8}\left(\frac{a}{r_2}\right)^3 - \cdots\bigg]\bigg\}. \end{aligned} \tag{12}$$

Comparing (12) with (9), we find

$$A_0 = 1, \quad A_1 = \frac{3}{2 \cdot 4}, \quad A_3 = -\frac{3}{2 \cdot 4 \cdot 6}, \quad \text{ETC.}, \tag{13}$$

in agreement with equation (6) when $\mu = 1$. By this method we determine the coefficients of the various terms of the potential directly. To calculate the potential at a point where $r_2 < r_1$, we must employ the second integral in (2).

The potential of a homogeneous ellipsoid of revolution follows by an analogous procedure. Let the outer boundary of the ellipsoid be represented by the equation

$$\frac{x^2}{a^2} + \frac{y^2}{a^2} + \frac{z^2}{b^2} = 1. \tag{14}$$

We compute, first of all, the potential at some external point $z = z_0$, along the axis of symmetry. Consider the ellipsoid to be built of superposed circular disks of radius R and thickness dz. If ρ is the density of the ellipsoid, the potential dV produced at z_0 by the disk of radius R and thickness dz is

$$dV = -\int_0^{R_0}\int_0^{2\pi} G\rho\, dz \frac{R\, dR\, d\varphi}{\sqrt{h^2 + R^2}} = -2\pi G\rho[(h^2 - R_0^2)^{1/2} - h]\, dz, \quad (15)$$

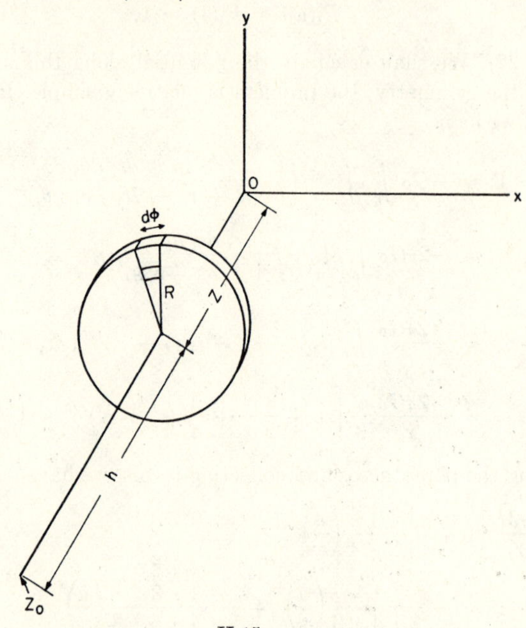

II-15.

where h is the distance from the center of the disk to z_0, i.e.,

$$h = z_0 - z. \quad (16)$$

Now, from (14), the radius R_0 of the element is given by

$$R_0^2 = x^2 + y^2 = a^2 - a^2z^2/c^2. \quad (17)$$

Therefore the complete potential of the ellipsoid is

$$V = -2\pi G\rho \int_{-c}^{c} \{[(z_0 - z)^2 + a^2 - a^2z^2/c^2]^{1/2} - (z_0 - z)\}\, dz. \quad (18)$$

Consider the case, first, of an oblate spheroid, $a > c$. Then, "completing the square" inside the brackets, we find that

$$V = -2\pi G\rho \int_{-c}^{c} \left\{a^2 \frac{a^2 - c^2 + z_0^2}{a^2 - c^2} - \left(\frac{z}{c}\sqrt{a^2 - c^2} + \frac{cz_0}{\sqrt{a^2 - c^2}}\right)^2 - (z_0 - z)\right\} dz. \quad (19)$$

Since

$$\int (b^2 - w^2)^{1/2}\, dw = \frac{w(b^2 - w^2)^{1/2}}{2} + \frac{b^2}{2} \arcsin \frac{w}{b}, \tag{20}$$

we find that

$$V = 2\pi G\rho a^2 c\left[\frac{z_0}{a^2 - c^2} - \frac{a^2 - c^2 + z_0^2}{(a^2 - c^2)^{3/2}} \frac{1}{2}(\theta_1 - \theta_2)\right], \tag{21}$$

where

$$\theta_1 = \arcsin \frac{cz_0 + a^2 - c^2}{a\sqrt{a^2 - c^2 + z_0^2}} = \arccos \frac{(a^2 - c^2)^{1/2}(z_0 - c)}{a\sqrt{a^2 - c^2 + z_0^2}}, \tag{22}$$

$$\theta_2 = \arcsin \frac{cz_0 - a^2 + c^2}{a\sqrt{a^2 - c^2 + z_0^2}} = \arccos \frac{(a^2 - c^2)^{1/2}(z_0 + c)}{a\sqrt{a^2 - c^2 + z_0^2}}. \tag{23}$$

The following trigonometric expansion gives

$$\begin{aligned}\tan \frac{1}{2}(\theta_1 - \theta_2) &= \frac{1 - \cos(\theta_1 - \theta_2)}{\sin(\theta_1 - \theta_2)} \\ &= \frac{1 - \cos\theta_1 \cos\theta_2 - \sin\theta_1 \sin\theta_2}{\sin\theta_1 \cos\theta_2 - \cos\theta_1 \sin\theta_2} \\ &= \frac{(a^2 - c^2)^{1/2}}{z_0},\end{aligned} \tag{24}$$

by (22) and (23). Therefore

$$V = 2\pi G\rho a^2 c\left[\frac{z_0}{a^2 - c^2} - \frac{a^2 - c^2 + z_0^2}{(a^2 - c^2)^{3/2}} \arctan \frac{(a^2 - c^2)^{1/2}}{z_0}\right]. \tag{25}$$

But

$$\arctan \theta = \theta - \frac{\theta^3}{3} + \frac{\theta^5}{5} - \ldots = \sum_{k=0}^{\infty} (-1)^k \frac{\theta^{2k+1}}{(2k+1)}. \tag{26}$$

Furthermore, the mass, M, of the spheroid is

$$M = \frac{4}{3}\pi a^2 c. \tag{27}$$

Hence

$$V = -\frac{GM}{z_0} - 3GM \sum_{k=1}^{\infty} \frac{(-1)^k (a^2 - c^2)^k}{(2k+1)(2k+3)} \frac{1}{z_0^{2k+1}}. \tag{28}$$

For points on the axis $z_0 = r$. Hence the general expansion of V, in terms of zonal harmonics, is

$$V = -\frac{GM}{r} - 3GM \sum_{k=1}^{\infty} \frac{(-1)^k (a^2 - c^2)^k}{(2k+1)(2k+3)} \frac{P_{2k}}{r^{2k+1}}. \tag{29}$$

This series is convergent for values of r such that $(a^2 - c^2)/r^2 < 1$. But $r \geq c$ for all points external to the boundary surface. Hence, if $a^2 < 2c^2$, (29) represents V right down to the surface of the spheroid. When the ellipticity is small, the convergence of (29) is very rapid. The first two terms of (29),

$$V = -\frac{GM}{r} + GM\frac{(a^2 - c^2)}{10}\frac{3\cos^2\theta - 1}{r^3}, \tag{30}$$

are ample for many problems.

Vector Analysis

20. The vector of angular velocity. The potential functions just discussed are scalar in nature. We now return to problems involving vectors. The force fields obtained from an application of the operator ∇ to the scalar potentials represent but one type of physical quantities of vector character. We shall now proceed to develop the vector notation and derive the rules of vector algebra. The reader will require no prior knowledge of vector analysis.

Consider, now, the motion of a particle constrained to move with constant angular velocity, ω in a circular path located in the xy-plane. We may suppose this point to be one of many in a solid disk, rotating around the z-axis. As in §(13), we may represent the area, S, enclosed by the trajectory as a vector of length S, placed normal to the surface, or

$$\mathbf{S} = \mathbf{k}S. \tag{1}$$

The angular velocity, like S itself, is associated with the xy-plane. We may therefore take $\boldsymbol{\omega}$ as a vector normal to the plane; thus

$$\boldsymbol{\omega} = \mathbf{k}\omega. \tag{2}$$

The convention ordinarily adopted for the positive direction of the vector in a right-handed coordinate system is that, when the observer sees the motion as counterclockwise, the vector points toward him, as in Figure 16. Another simple rule is that if the extended thumb of the right hand points along the vector, the rotation is in the direction of the fingers of the clenched fist.

Let x and y be the coordinates of a point on the rotating disk. Then we may represent the motion by the formula:

$$x = r\cos(\omega t + \alpha), \quad y = r\sin(\omega t + \alpha), \tag{3}$$

where α is the angle made by the radius vector with the x-axis at time $t = 0$. The respective velocity components, u and v, are

$$\begin{aligned} u &= -r\omega\sin(\omega t + \alpha) = -\omega y. \\ v &= r\omega\cos(\omega t + \alpha) = \omega x. \end{aligned} \tag{4}$$

We represent the velocity $\mathbf{v}_0$ as the following vector:

$$\mathbf{v}_0 = \mathbf{i}u + \mathbf{j}v. \tag{5}$$

Also, we see that

$$\frac{\partial v}{\partial x} - \frac{\partial u}{\partial y} = 2\omega. \tag{6}$$

This differentiation is essentially scalar, but ω represents the magnitude of a vector.

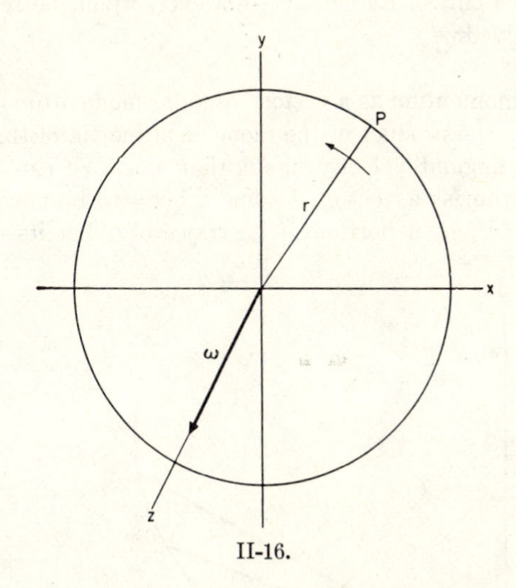

II-16.

If the plane of rotation is inclined to the coordinate axis the normal to the surface of rotation will also be inclined, and the ω-vector will have components along the coordinate axes. The resultant will be the same, of course, and will be given by the equation

$$\sqrt{\left(\frac{\partial v}{\partial x} - \frac{\partial u}{\partial y}\right)^2 + \left(\frac{\partial w}{\partial y} - \frac{\partial v}{\partial z}\right)^2 + \left(\frac{\partial u}{\partial z} - \frac{\partial w}{\partial x}\right)^2} = 2\omega. \tag{7}$$

where w is the z velocity component. Each term denotes a vector component, and the spatial orientation of the resultant vector can be determined by the ratios of the components. The vector itself must be the quantity

$$\mathbf{i}\left(\frac{\partial w}{\partial y} - \frac{\partial v}{\partial z}\right) + \mathbf{j}\left(\frac{\partial u}{\partial z} - \frac{\partial w}{\partial x}\right) + \mathbf{k}\left(\frac{\partial v}{\partial x} - \frac{\partial u}{\partial y}\right) \tag{8}$$

$$= \text{curl } \mathbf{v}_0 = \text{rot } \mathbf{v}_0 = \nabla \times \mathbf{v}_0 = 2\boldsymbol{\omega},$$

where $\mathbf{v}_0$ is the velocity vector. "**rot**" is symbolic for "rotor of v_0"; the significance of the notation $\nabla \times \mathbf{v}_0$ (read "del cross" or "curl") will appear presently. A "curl" of any vector, then, is the vector that results from application of the operator (8), which directs us to find the three components of a new vector associated with the vector operated upon.

The foregoing simple demonstration indicates that the "curl" of a vector has something to do with rotation. The vector $\nabla \times \mathbf{v}_0$, however, will be equal to $2\boldsymbol{\omega}$ *only for the rotation of a rigid body.* A body moving in a straight line, for example, has angular velocity with respect to any given point. The curl of its velocity, however, would be zero because no rotation is involved.

21. Angular momentum as a vector. Angular momentum of an extended body is equal to the product of the moment of inertia of that body about an axis, by its angular velocity about that axis. We can also represent angular momentum as a vector. Assume a body to be moving in the xy-plane and let P_1P_2 be a portion of its trajectory. Let its momentum at

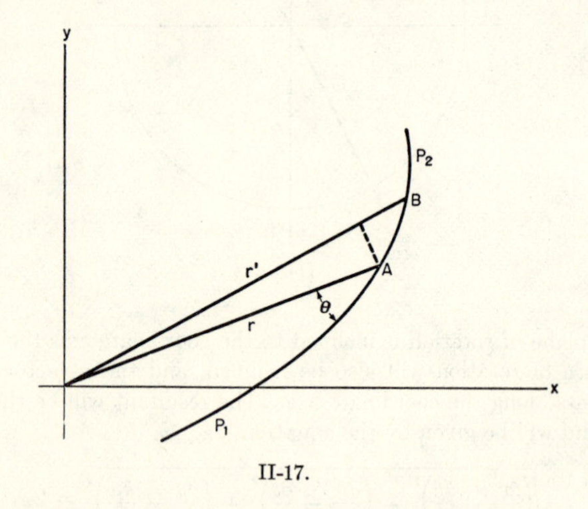

II-17.

the point A be $\mathbf{p}$. Let the momentum vector make an angle θ with respect to the radius vector. Now if B, a neighboring point on the trajectory is allowed to approach indefinitely closely to A, so that $\mathbf{r}' \rightarrow \mathbf{r}$, the component of momentum perpendicular to $\mathbf{r}$ will be $p \sin \theta$ and the moment of momentum L will be

$$L = rp \sin \theta. \tag{1}$$

Since L is associated with the xy-plane, we may represent it as a vector perpendicular to that plane.

We gave a somewhat less restricted definition of angular momentum in § (6). For a single particle, we may write the components of angular momenta along the x-, y-, and z-axes, respectively, as

$$L_x = yp_z - zp_y, \quad L_y = zp_x - xp_z, \quad L_z = xp_y - yp_x, \tag{2}$$

where p_x, p_y, and p_z are the instantaneous components of the momentum. In vector notation we have

$$\mathbf{L} = \mathbf{i}L_x + \mathbf{j}L_y + \mathbf{k}L_z. \tag{3}$$

22. Scalar and vector products. In the foregoing brief introduction to vectors, we have encountered two different ways of combining vectors. We represented the work done by a force, (8.1), as the product of the scalar magnitudes of two vectors times the cosine of the angle between them, thus

$$dW = |\mathbf{F}|\,|d\mathbf{s}| \cos\theta = F \cos\theta\, ds. \tag{1}$$

where $d\mathbf{s}$ is an element of the path. The value of dW is equal to the product of the length of one vector by the projected length of the other vector upon the first. The result is scalar. Hence we call this type of combination the *scalar product*, and indicate the product symbolically by the notation

$$dW = \mathbf{F} \cdot d\mathbf{s}, \tag{2}$$

where the "dot" represents the operation implied in (1), viz., the product of the magnitude of one vector by the magnitude of the projection of the other upon itself. "Scalar product" means "invariant product," invariant, that is, to all coordinate transformations.

The second type of combination is that discussed in the previous section, where the magnitude of the angular-momentum vector proved to be

$$L = rp \sin\theta. \tag{3}$$

Where, in the scalar product, the one vector is projected *along* the other, in the present example the one vector is projected in a direction *perpendicular* to the other. The result of such a product is a vector perpendicular to the two original vectors. The operation, known as taking the *vector product*, we represent symbolically by the formula

$$\mathbf{L} = \mathbf{r} \times \mathbf{p}. \tag{4}$$

The reader who is here meeting the subject of vectors for the first time, may find it helpful to regard the "dot" and "cross" as symbolic, respectively, of $\cos\theta$ and $\sin\theta$.

For the scalar product, we need not distinguish which vector we project on the other, since $\cos\theta = \cos(-\theta)$. Hence the commutative law holds and we have

$$\mathbf{a}\cdot\mathbf{b} = \mathbf{b}\cdot\mathbf{a}. \tag{5}$$

But for the vector product, the projection of $\mathbf{r}$ on a perpendicular to $\mathbf{p}$ is not identical with the projection of $\mathbf{p}$ on a perpendicular to $\mathbf{r}$. Since $\sin\theta = -\sin(-\theta)$, reversing the factors changes the sign,

$$\mathbf{a}\times\mathbf{b} = -\mathbf{b}\times\mathbf{a}. \tag{6}$$

If the right hand with thumb and two fingers are extended to indicate three mutually perpendicular directions, the thumb points along $\mathbf{r}$, the forefinger along $\mathbf{p}$, and the middle finger indicates the direction of $\mathbf{L}$. For both types of products the distributive law holds; thus

$$(\mathbf{a}+\mathbf{b})\cdot(\mathbf{c}+\mathbf{d}) = \mathbf{a}\cdot\mathbf{c}+\mathbf{a}\cdot\mathbf{d}+\mathbf{b}\cdot\mathbf{c}+\mathbf{b}\cdot\mathbf{d}, \tag{7}$$

and similarly for the vector product. Hence we may write, in place of (2),

$$dW = (\mathbf{i}X+\mathbf{j}Y+\mathbf{k}Z)\cdot(\mathbf{i}\,dx+\mathbf{j}\,dy+\mathbf{k}\,dz), \tag{8}$$

where X, Y, Z, dx, dy, and dz measure the magnitudes of the rectangular components of $\mathbf{F}$ and $d\mathbf{s}$, respectively.

The unit vectors, equation (9.3), obey the relations

$$\mathbf{i}\cdot\mathbf{i} = \mathbf{j}\cdot\mathbf{j} = \mathbf{k}\cdot\mathbf{k} = 1, \tag{9}$$

whereas

$$\mathbf{i}\cdot\mathbf{j} = \mathbf{j}\cdot\mathbf{k} = \mathbf{k}\cdot\mathbf{i} = 0,$$

since the projection of a unit vector on itself is unity, and on a perpendicular vector is zero. Hence

$$dW = X\,dx + Y\,dy + Z\,dz, \tag{10}$$

a scalar quantity. The result agrees with that of (8.3).

We have the vector identities

$$d\mathbf{r} = \mathbf{i}\,dx+\mathbf{j}\,dy+\mathbf{k}\,dz, \quad \text{and} \quad \mathbf{r} = \mathbf{i}x+\mathbf{j}y+\mathbf{k}z.$$

Consider the vector product

$$\mathbf{L} = \mathbf{r}\times\mathbf{p} = (\mathbf{i}x+\mathbf{j}y+\mathbf{k}z)\times(\mathbf{i}p_x+\mathbf{j}p_y+\mathbf{k}p_z), \tag{11}$$

$$\mathbf{i}\times\mathbf{i} = \mathbf{j}\times\mathbf{j} = \mathbf{k}\times\mathbf{k} = 0, \tag{12}$$

since any unit vector projected on a line normal to its direction is zero.

But

$$\mathbf{i} \times \mathbf{j} = \mathbf{k}, \quad \mathbf{j} \times \mathbf{k} = \mathbf{i}, \quad \mathbf{k} \times \mathbf{i} = \mathbf{j},$$

and (13)

$$\mathbf{j} \times \mathbf{i} = -\mathbf{k}, \quad \mathbf{k} \times \mathbf{j} = -\mathbf{i}, \quad \mathbf{i} \times \mathbf{k} = -\mathbf{j}.$$

The signs of (13) follow from the right-hand rule, above. The positive directions are consistent with a right-handed system of axes, such as that drawn in Fig. 5. The rotation is counterclockwise.

$$\mathbf{L} = \mathbf{r} \times \mathbf{p} = (yp_z - zp_y)\mathbf{i} + (zp_x - xp_z)\mathbf{j} + (xp_y - yp_x)\mathbf{k}, \tag{14}$$

a vector quantity, whose scalar components are in accord with (21.2).

The operation of multiplying a vector by a scalar quantity, a, is extremely simple. The length of the vector is merely altered by the factor a. We may write

$$a\mathbf{r} = \mathbf{i}(ax) + \mathbf{j}(ay) + \mathbf{k}(az). \tag{15}$$

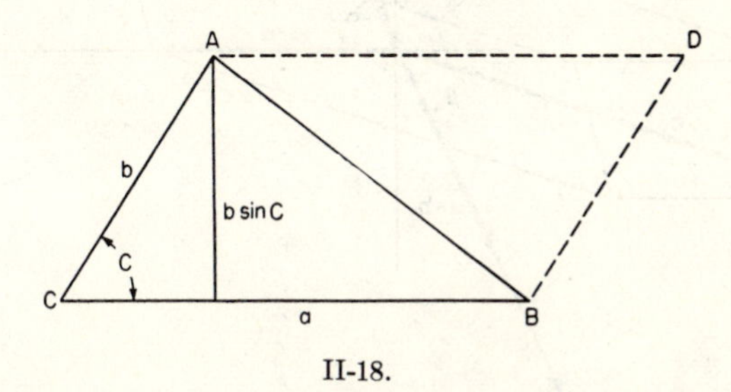

II-18.

Let two sides, a and b, of a plane triangle include the angle C between them. If a is the base, the altitude will be $b \sin C$ and the area (Fig. 18),

$$\text{area} \quad (ABC) = \frac{1}{2} ab \sin C. \tag{16}$$

The sides of the triangle are vector quantities. In accord with equations (3) and (4), we may represent the area as follows:

$$\textbf{area} \quad (ABC) = \frac{1}{2} \mathbf{a} \times \mathbf{b}. \tag{17}$$

The vector representing the area is perpendicular to the surface. Similarly, the area of the parallelogram, $ACBD$ (Fig. 18),

$$\textbf{area} \quad (ACBD) = \mathbf{a} \times \mathbf{b}. \tag{18}$$

The volume of a parallelopiped is equal to the area of the base times the altitude. The value of the latter is equal to $c \cos \beta$. As shown in Fig. 19, β is the angle between the vector **c** and the normal to the parallelogram formed by the vectors **a** and **b**.

$$\text{VOLUME} = abc \sin \alpha \cos \beta = (\mathbf{a} \times \mathbf{b}) \cdot \mathbf{c} \qquad (19)$$

$$= (\mathbf{b} \times \mathbf{c}) \cdot \mathbf{a} = (\mathbf{c} \times \mathbf{a}) \cdot \mathbf{b},$$

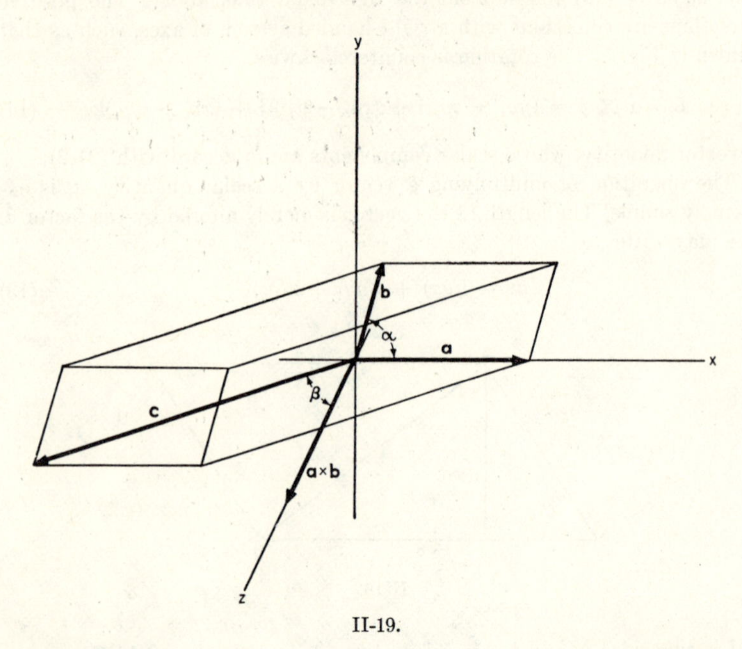

II-19.

since we may use any pair of the three vectors **a**, **b**, and **c**, to form the parallelogram base of the prism. Quantities of the form of (19) we call the "triple scalar product." In the scalar product, as we have seen, we may reverse the order of the factors. From (19) we get

$$(\mathbf{b} \times \mathbf{c}) \cdot \mathbf{a} = \mathbf{a} \cdot (\mathbf{b} \times \mathbf{c}) = (\mathbf{a} \times \mathbf{b}) \cdot \mathbf{c} = \begin{vmatrix} a_x & a_y & a_z \\ b_x & b_y & b_z \\ c_x & c_y & c_z \end{vmatrix}. \qquad (20)$$

The dot and cross may thus be exchanged in the triple scalar product. Since an expression of the form $\mathbf{a} \times (\mathbf{b} \cdot \mathbf{c})$ has no physical vector interpretation, since $\mathbf{b} \cdot \mathbf{c}$ is a scalar, we may drop the parentheses from the triple scalar product. Transposition of two adjacent vectors changes the sign of the product in (19). The determinant equivalent is extremely useful in many cases.

We may reduce the triple vector product, $(\mathbf{a} \times \mathbf{b}) \times \mathbf{c}$, to simpler form. For sake of simplicity and with no loss of generality, we adopt coordinate axes such that

$$\mathbf{a} = a\mathbf{i}, \quad \mathbf{b} = b_1\mathbf{i} + b_2\mathbf{j}, \qquad \mathbf{c} = c_1\mathbf{i} + c_2\mathbf{j} + c_3\mathbf{k}. \tag{21}$$

which is to say that we take the x-axis along $\mathbf{a}$ and set $\mathbf{b}$ in the xy-plane. Then $\mathbf{a} \times \mathbf{b}$ will lie along the z-axis. Finally, the vector $(\mathbf{a} \times \mathbf{b}) \times \mathbf{c}$, which must be perpendicular to $(\mathbf{a} \times \mathbf{b})$, as well as to $\mathbf{c}$, must lie in the xy-plane, Fig. 20. From (21),

$$(\mathbf{a} \times \mathbf{b}) = ab_2\mathbf{k} \tag{22}$$

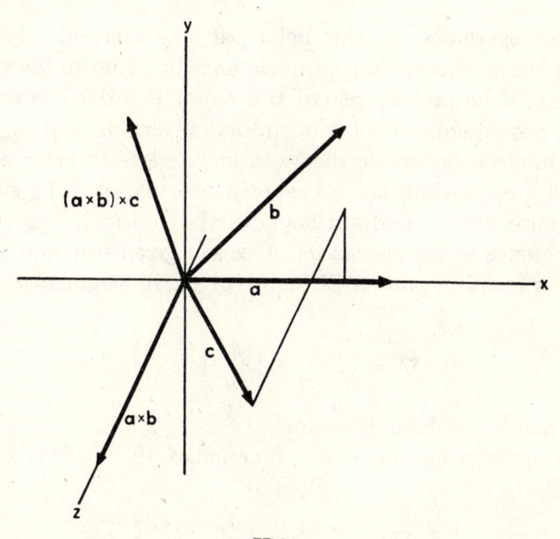

II-20.

and

$$(\mathbf{a} \times \mathbf{b}) \times \mathbf{c} = ab_2c_1\mathbf{j} - ab_2c_2\mathbf{i}. \tag{23}$$

To the right-hand side of (23) add and subtract the vector $ac_1b_1\mathbf{i}$, and factorize as follows:

$$\begin{aligned}(\mathbf{a} \times \mathbf{b}) \times \mathbf{c} &= ac_1(b_1\mathbf{i} + b_2\mathbf{j}) - (b_1c_1 + b_2c_2)a\mathbf{i} \\ &= ac_1\mathbf{b} - (b_1c_1 + b_2c_2)\mathbf{a} \\ &= \mathbf{a} \cdot \mathbf{c}\mathbf{b} - \mathbf{b} \cdot \mathbf{c}\mathbf{a} = \mathbf{c} \cdot \mathbf{a}\mathbf{b} - \mathbf{c} \cdot \mathbf{b}\mathbf{a},\end{aligned} \tag{24}$$

since

$$\mathbf{a} \cdot \mathbf{c} = ac_1 \quad \text{and} \quad \mathbf{b} \cdot \mathbf{c} = b_1c_1 + b_2c_2.$$

We cannot drop the parentheses from the triple vector product. Equation (24) shows that the final vector can be expressed in terms of the vectors in parentheses, with their magnitudes altered by the scalars $(\mathbf{a} \cdot \mathbf{c})$ and $(-\mathbf{b} \cdot \mathbf{c})$. The triple vector product may also be written in the convenient symbolic form:

$$(\mathbf{a} \times \mathbf{b}) \times \mathbf{c} = \mathbf{c} \cdot (\mathbf{ab} - \mathbf{ba}). \tag{25}$$

The quantity $(\mathbf{ab} - \mathbf{ba})$, each member of which is formed by the juxtaposition of two vectors, is called a "dyadic." We shall return to consider the properties of these functions in § (27).

We call a vector like $\mathbf{L}$, which results from the cross-product of two other vectors, an "axial vector," to distinguish it from the ordinary, or "polar vector."

23. Vector operators. At this point we may conveniently summarize and extend the various vector formulae introduced up to the present. The student may, if he prefers, regard the vector notation merely as an abbreviation, not dissimilar to the operators of calculus. But in many problems, e.g., in electromagnetic theory, a knowledge of vector operations is helpful and greatly shortens the labor of calculation. The simplification results because we can perform many of the indicated vector operations without reference to any special set of axes or given system of coordinates.

The vector operator **grad** or ∇ (del) is given, in rectangular coordinates, by (9.5):

$$\nabla = \left(\mathbf{i}\frac{\partial}{\partial x} + \mathbf{j}\frac{\partial}{\partial y} + \mathbf{k}\frac{\partial}{\partial z}\right), \tag{1}$$

where $\mathbf{i}$, $\mathbf{j}$, and $\mathbf{k}$ are the unit vectors.

When ∇ operates on some scalar function, ϕ, the result is a vector, and expressed thus:

$$\nabla\phi = \mathbf{i}\frac{\partial\phi}{\partial x} + \mathbf{j}\frac{\partial\phi}{\partial y} + \mathbf{k}\frac{\partial\phi}{\partial z}. \tag{2}$$

We call this quantity "the *gradient* of ϕ." The gradient of a scalar is thus a vector quantity.

The differential sign in calculus is subject to symbolic manipulation; ∇ may be similarly employed. Let $\mathbf{Q}$ be a vector, with components X, Y, and Z, such that

$$\mathbf{Q} = \mathbf{i}X + \mathbf{j}Y + \mathbf{k}Z. \tag{3}$$

We may then consider $\nabla\mathbf{Q}$ as a "product" of the operator by the vector. But in vector analysis there are two kinds of product, the "dot" and "cross." If we adopt the former, we may write

$$\nabla \cdot \mathbf{Q} = \left(\mathbf{i}\frac{\partial}{\partial x} + \mathbf{j}\frac{\partial}{\partial y} + \mathbf{k}\frac{\partial}{\partial z}\right) \cdot (\mathbf{i}X + \mathbf{j}Y + \mathbf{k}Z). \tag{4}$$

Carry out the symbolic multiplication and note the identities (22.9). Then

$$\nabla \cdot \mathbf{Q} = \operatorname{div} Q = \frac{\partial X}{\partial x} + \frac{\partial Y}{\partial y} + \frac{\partial Z}{\partial z}, \tag{5}$$

a vector operation previously encountered in the derivation of Gauss' theorem, equation (14.5). The operator $\nabla \cdot$ is called the "divergence of", for reasons that will appear in § (34). The result is scalar.

If we adopt the vector product, we obtain the vector $\mathbf{Q}'$; by (22.12) and (22.13),

$$\mathbf{Q}' = \nabla \times \mathbf{Q} = \operatorname{curl} \mathbf{Q} = \mathbf{i}\left(\frac{\partial Z}{\partial y} - \frac{\partial Y}{\partial z}\right) + \mathbf{j}\left(\frac{\partial X}{\partial z} - \frac{\partial Z}{\partial x}\right) + \mathbf{k}\left(\frac{\partial Y}{\partial x} - \frac{\partial X}{\partial y}\right), \tag{6}$$

a vector already discussed in § (20). If we write symbolically,

$$\nabla \cdot \nabla = \nabla^2 = \frac{\partial^2}{\partial x^2} + \frac{\partial^2}{\partial y^2} + \frac{\partial^2}{\partial z^2}, \tag{7}$$

we obtain the Laplacian operator, discussed in § (14),

$$\nabla^2 \phi = \frac{\partial^2 \phi}{\partial x^2} + \frac{\partial^2 \phi}{\partial y^2} + \frac{\partial^2 \phi}{\partial z^2}. \tag{8}$$

When ∇^2 operates on a vector we have the result

$$\nabla^2 \mathbf{Q} = \mathbf{i}\nabla^2 X + \mathbf{j}\nabla^2 Y + \mathbf{k}\nabla^2 Z. \tag{9}$$

The above equations, when extended, form the basis of vector analysis. When $\mathbf{Q}$ is the resultant of two vectors $\mathbf{Q}_1$ and $\mathbf{Q}_2$, we may write

$$\begin{aligned} \nabla \cdot (\mathbf{Q}_1 \pm \mathbf{Q}_2) &= \left(\frac{\partial X_1}{\partial x} + \frac{\partial Y_1}{\partial y} + \frac{\partial Z_1}{\partial z}\right) \pm \left(\frac{\partial X_2}{\partial x} + \frac{\partial Y_2}{\partial y} + \frac{\partial Z_2}{\partial z}\right) \\ &= \nabla \cdot \mathbf{Q}_1 \pm \nabla \cdot \mathbf{Q}_2. \end{aligned} \tag{10}$$

Similarly,

$$\nabla(\phi_1 \pm \phi_2) = \nabla \phi_1 \pm \nabla \phi_2 \tag{11}$$

and

$$\nabla \times (\mathbf{Q}_1 \pm \mathbf{Q}_2) = \nabla \times \mathbf{Q}_1 \pm \nabla \times \mathbf{Q}_2. \tag{12}$$

By carrying out the operations indicated we may show that

$$\nabla \cdot \nabla \times \mathbf{Q} = 0. \tag{13}$$

$$\nabla \times \nabla \phi = 0. \tag{14}$$

$$\begin{aligned} \nabla(\nabla \cdot \mathbf{Q}) = \mathbf{i}\left(\frac{\partial^2 X}{\partial x^2} + \frac{\partial^2 Y}{\partial y\, \partial x} + \frac{\partial^2 Z}{\partial z\, \partial x}\right) + \mathbf{j}\left(\frac{\partial^2 X}{\partial x\, \partial y} + \frac{\partial^2 Y}{\partial y^2} + \frac{\partial^2 Z}{\partial z\, \partial y}\right) \\ + \mathbf{k}\left(\frac{\partial^2 X}{\partial x\, \partial z} + \frac{\partial^2 Y}{\partial y\, \partial z} + \frac{\partial^2 Z}{\partial z^2}\right). \end{aligned} \tag{15}$$

$$\mathbf{Q}'' = \nabla \times \nabla \times \mathbf{Q} = \nabla \times \mathbf{Q}' = \mathbf{i}\left(\frac{\partial Z'}{\partial y} - \frac{\partial Y'}{\partial z}\right)$$
$$+ \mathbf{j}\left(\frac{\partial X'}{\partial z} - \frac{\partial Z'}{\partial x}\right) + \mathbf{k}\left(\frac{\partial Y'}{\partial x} - \frac{\partial X'}{\partial y}\right).$$

But

$$X' = \left(\frac{\partial Z}{\partial y} - \frac{\partial Y}{\partial z}\right), \quad Y' = \left(\frac{\partial X}{\partial z} - \frac{\partial Z}{\partial x}\right), \quad Z' = \left(\frac{\partial Y}{\partial x} - \frac{\partial X}{\partial y}\right).$$

Thus

$$\left(\frac{\partial Z'}{\partial y} - \frac{\partial Y'}{\partial z}\right) = \frac{\partial^2 Y}{\partial y\, \partial x} - \frac{\partial^2 X}{\partial y^2} - \frac{\partial^2 X}{\partial z^2} + \frac{\partial^2 Z}{\partial z\, \partial x}$$
$$= \left(\frac{\partial^2 X}{\partial x^2} + \frac{\partial^2 Y}{\partial y\, \partial x} + \frac{\partial^2 Z}{\partial z\, \partial x}\right) - \left(\frac{\partial^2 X}{\partial x^2} + \frac{\partial^2 X}{\partial y^2} + \frac{\partial^2 X}{\partial z^2}\right),$$

etc., for the **j** and **k** components. The sum of the positive elements is equal to those of (15). The negative elements comprise the sum of the Laplacians of the vector components, (9). Hence

$$\nabla \times \nabla \times \mathbf{Q} = \nabla(\nabla \cdot \mathbf{Q}) - \nabla^2 \mathbf{Q}. \tag{16}$$

We may also summarize various other formulae. The force, **F**, is

$$\mathbf{F} = -\nabla V, \tag{17}$$

where V is the potential. The work, as a line integral, assumes the form:

$$W = \int \mathbf{F} \cdot d\mathbf{s}. \tag{18}$$

Gauss' theorem becomes

$$\iint \mathbf{F} \cdot d\mathbf{S} = \iiint \nabla \cdot \mathbf{F}\, d\tau \tag{19}$$

where $d\tau$ is an element of volume.

24. Differentiation and integration of vectors. We shall also have to consider vectors that are functions of the time, t, as well as of the coordinates. Let **r** be a vector,

$$\mathbf{r} = \mathbf{i}x + \mathbf{j}y + \mathbf{k}z, \tag{1}$$

where

$$x = x(t), \quad y = y(t), \quad z = z(t). \tag{2}$$

The tip of the vector will describe a trajectory, as P_1P_2, Fig. 21. At the time t_1, let $\mathbf{r} = \mathbf{r}_1$, and at time $t_2 = t_1 + \Delta t$, let $\mathbf{r} = \mathbf{r}_2 = \mathbf{r}_1 + \Delta\mathbf{r}$.

We shall define the time derivative of $\mathbf{r}$ as follows:

$$\frac{d\mathbf{r}}{dt} = \lim_{\Delta t \to 0} \frac{\mathbf{r}_2 - \mathbf{r}_1}{\Delta t} = \lim_{\Delta t \to 0} \frac{\Delta\mathbf{r}}{\Delta t}. \tag{3}$$

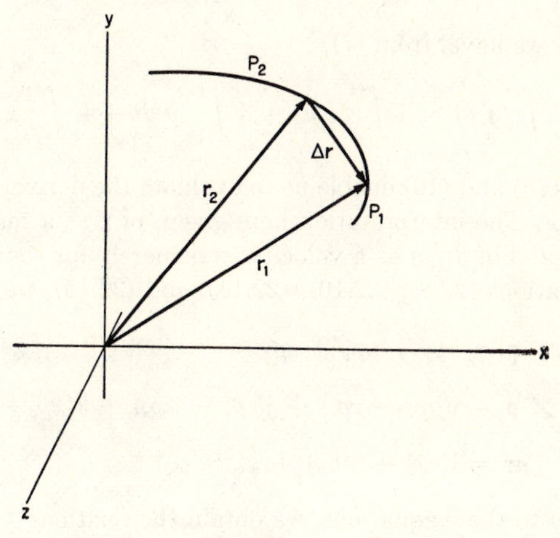

II-21.

The argument is analogous to that used for derivatives of ordinary scalar functions. But

$$\mathbf{r}_1 + \Delta\mathbf{r} = \mathbf{i}(x_1 + \Delta x) + \mathbf{j}(y_1 + \Delta y) + \mathbf{k}(z_1 + \Delta z). \tag{4}$$

Hence

$$\frac{d\mathbf{r}}{dt} = \mathbf{i}\frac{dx}{dt} + \mathbf{j}\frac{dy}{dt} + \mathbf{k}\frac{dz}{dt} = \mathbf{v}_0, \tag{5}$$

where $\mathbf{v}_0$ is the tangential velocity of the point on the trajectory indicated by $\mathbf{r}$. *Note that* $d\mathbf{r}$ *is not necessarily parallel to* $\mathbf{r}$.

In analogous fashion we define the integral of a vector, which process is the inverse of differentiation. Divide the interval from $t = t_1$ to $t = t_2$ into n parts such that

$$t_2 - t_1 = \sum_{k=1}^{n} \Delta t_k. \tag{6}$$

For each element of time, Δt_k, we let the value of $\mathbf{r}$ somewhere in the interval be $\mathbf{r}_k$. Then we define the definite integral of $\mathbf{r}$ from t_1 to t_2 by the equation

$$\int_{t_1}^{t_2} \mathbf{r}\, dt = \lim_{n \to \infty} \sum_{k=1}^{n} \mathbf{r}_k\, \Delta t_k. \tag{7}$$

But

$$\mathbf{r}_k \, \Delta t_k = \mathbf{i} x_k \, \Delta t_k + \mathbf{j} y_k \, \Delta t_k + \mathbf{k} z_k \, \Delta t_k, \tag{8}$$

and

$$\sum_{k=1}^{n} \mathbf{r}_k \, \Delta t_k = \mathbf{i} \sum_{k=1}^{n} x_k \, \Delta t_k + \mathbf{j} \sum_{k=1}^{n} y_k \, \Delta t_k + \mathbf{k} \sum_{k=1}^{n} z_k \, \Delta t_k. \tag{9}$$

In the limit, we have, from (7),

$$\int_{t_1}^{t_2} \mathbf{r} \, dt = \mathbf{i} \int_{t_1}^{t_2} x \, dt + \mathbf{j} \int_{t_1}^{t_2} y \, dt + \mathbf{k} \int_{t_1}^{t_2} z \, dt. \tag{10}$$

Equations (5) and (10) enable us to evaluate the derivative and integral of any vector. The interpretation here given, of $\mathbf{r}$ as a radius vector, of t as the time, and of $d\mathbf{r}/dt$ as a velocity, was merely for convenience.

From equations (22.8), (22.10), (22.14), and (22.15), we have

$$\mathbf{r} \cdot \mathbf{p} = xp_x + yp_y + zp_z \,. \tag{11}$$

$$\mathbf{r} \times \mathbf{p} = \mathbf{i}(yp_z - zp_y) + \mathbf{j}(zp_x - xp_z) + \mathbf{k}(xp_y - yp_x). \tag{12}$$

$$a\mathbf{r} = \mathbf{i}(ax) + \mathbf{j}(ay) + k(az). \tag{13}$$

Applying (5) to these equations, we obtain the relations

$$\frac{d}{dt}(\mathbf{r} \cdot \mathbf{p}) = \frac{d\mathbf{r}}{dt} \cdot \mathbf{p} + \mathbf{r} \cdot \frac{d\mathbf{p}}{dt}. \tag{14}$$

$$\frac{d}{dt}(\mathbf{r} \times \mathbf{p}) = \frac{d\mathbf{r}}{dt} \times \mathbf{p} + \mathbf{r} \times \frac{d\mathbf{p}}{dt}. \tag{15}$$

$$\frac{d}{dt}(a\mathbf{r}) = \frac{da}{dt} \mathbf{p} + a \frac{d\mathbf{p}}{dt}. \tag{16}$$

The reader will note the analogy between these expressions and the ordinary derivative of the product of two scalar variables:

$$\frac{d}{dt}(uv) = \frac{du}{dt} v + u \frac{dv}{dt}. \tag{17}$$

In equation (15), one must not change the order of the various factors.

25. Vector transformations to moving coordinate systems. Consider two rectangular coordinate systems with a common origin. Let one of these, wherein we denote the coordinates by x_0, y_0, and z_0, be a fixed (inertial) system. We shall use the subscript, zero, to indicate the fixed system in all that follows. The second system, x, y, z, we shall suppose to be rotating with a constant angular velocity, $\boldsymbol{\omega}$, with respect to an axis that passes through the origin of the fixed system (Fig. 22). The radius vector, velocity, and acceleration, with respect to the fixed axes, of a point P, are given, respectively, by the equations

$$\mathbf{r}_0 = \mathbf{i}_0 x_0 + \mathbf{j}_0 y_0 + \mathbf{k}_0 z_0. \tag{1}$$

$$\mathbf{v}_0 = \frac{d\mathbf{r}_0}{dt} = \mathbf{i}_0 \frac{dx_0}{dt} + \mathbf{j}_0 \frac{dy_0}{dt} + \mathbf{k}_0 \frac{dz_0}{dt} = \mathbf{i}_0 v_{0x} + \mathbf{j}_0 v_{0y} + \mathbf{k}_0 v_{0z}. \tag{2}$$

$$\mathbf{a}_0 = \frac{d\mathbf{v}_0}{dt} = \mathbf{i}_0 \frac{d^2 x_0}{dt^2} + \mathbf{j}_0 \frac{d^2 y_0}{dt^2} + \mathbf{k}_0 \frac{d^2 z_0}{dt^2} = \mathbf{i}_0 a_{0x} + \mathbf{j}_0 a_{0y} + \mathbf{k}_0 a_{0z}. \tag{3}$$

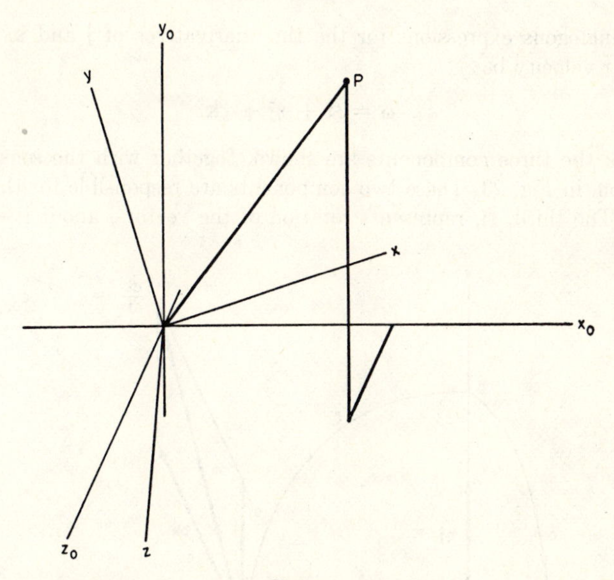

II-22.

The analogous equations, with respect to the moving axes, are

$$\mathbf{r} = \mathbf{i}x + \mathbf{j}y + \mathbf{k}z. \tag{4}$$

$$\mathbf{v} = \mathbf{i}v_x + \mathbf{j}v_y + \mathbf{k}v_z. \tag{5}$$

$$\mathbf{a} = \mathbf{i}a_x + \mathbf{j}a_y + \mathbf{k}a_z. \tag{6}$$

We seek expressions for $\mathbf{v}_0$ and $\mathbf{a}_0$ in terms of $\mathbf{r}$, $\mathbf{v}$, $\mathbf{a}$, and $\boldsymbol{\omega}$. Since

$$\mathbf{r} = \mathbf{r}_0, \tag{7}$$

we have

$$\begin{aligned} \mathbf{v}_0 = \frac{d\mathbf{r}}{dt} &= \mathbf{i}\frac{dx}{dt} + x\frac{d\mathbf{i}}{dt} + \mathbf{j}\frac{dy}{dt} + y\frac{d\mathbf{j}}{dt} + \mathbf{k}\frac{dz}{dt} + z\frac{d\mathbf{k}}{dt} \\ &= \mathbf{v} + x\frac{d\mathbf{i}}{dt} + y\frac{d\mathbf{j}}{dt} + z\frac{d\mathbf{k}}{dt}, \end{aligned} \tag{8}$$

by (24.16) and (5). The unit vectors, **i**, **j**, and **k**, are variable in direction, so that their time derivatives do not vanish in (8). For this reason we did not set $\mathbf{v} = d\mathbf{r}/dt$ in (5).

The unit vectors **i**, **j**, and **k** are, by definition, constant in length. Hence the vector $d\mathbf{i}/dt$, representing the rate of change of the unit vector, must be perpendicular to that vector. It will, therefore, have only y and z components, or

$$\frac{d\mathbf{i}}{dt} = a\mathbf{j} + b\mathbf{k}, \tag{9}$$

with analogous expressions for the time derivatives of **j** and **k**. Let the angular velocity be

$$\boldsymbol{\omega} = \xi\mathbf{i} + \eta\mathbf{j} + \zeta\mathbf{k}. \tag{10}$$

Two of the three components are shown, together with the sense of the rotation, in Fig. 23. These two components are responsible for the vector $d\mathbf{i}/dt$. The third, $\xi\mathbf{i}$, represents rotation of the vector **i** about itself as an

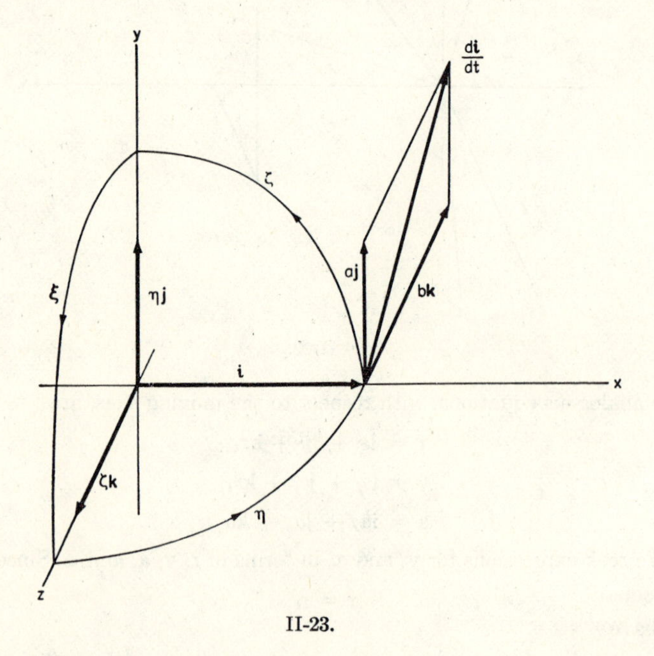

II-23.

axis and does not contribute to $d\mathbf{i}$. From Fig. 23 we see that the vector $a\mathbf{j}$ exists because of the rotation about the z-axis. Hence, because $|\,\mathbf{i}\,| = 1$,

$$a = \zeta \quad \text{and} \quad b = -\eta. \tag{11}$$

We adopt the minus sign because $b\mathbf{k}$, as in the figure, points in the negative direction. Therefore

$$\frac{d\mathbf{i}}{dt} = \zeta\mathbf{j} - \eta\mathbf{k}.$$

Similarly

$$\frac{d\mathbf{j}}{dt} = \xi\mathbf{k} - \zeta\mathbf{i}. \tag{12}$$

$$\frac{d\mathbf{k}}{dt} = \eta\mathbf{i} - \xi\mathbf{j}.$$

From (4) and (10) we have

$$\begin{aligned} \boldsymbol{\omega} \times \mathbf{r} &= (\eta z - \zeta y)\mathbf{i} + (\zeta x - \xi z)\mathbf{j} + (\xi y - \eta x)\mathbf{k} \\ &= x\frac{d\mathbf{i}}{dt} + y\frac{d\mathbf{j}}{dt} + z\frac{d\mathbf{k}}{dt}, \end{aligned} \tag{13}$$

by (12). Then, by (8), we have

$$\mathbf{v}_0 = \mathbf{v} + \boldsymbol{\omega} \times \mathbf{r}. \tag{14}$$

Differentiate (14), and employ (3):

$$\begin{aligned} \mathbf{a}_0 &= \frac{d\mathbf{v}_0}{dt} = \frac{d}{dt}\mathbf{v} + \frac{d}{dt}\boldsymbol{\omega} \times \mathbf{r} = \frac{d}{dt}(\mathbf{i}v_x + \mathbf{j}v_y + \mathbf{k}v_z) + \frac{d}{dt}\boldsymbol{\omega} \times \mathbf{r} \\ &= \left(\mathbf{i}\frac{dv_x}{dt} + \mathbf{j}\frac{dv_y}{dt} + \mathbf{k}\frac{dv_z}{dt}\right) + \left(v_x\frac{d\mathbf{i}}{dt} + v_y\frac{d\mathbf{j}}{dt} + v_z\frac{d\mathbf{k}}{dt}\right) \\ &\qquad + \frac{d\boldsymbol{\omega}}{dt} \times \mathbf{r} + \boldsymbol{\omega} \times \frac{d\mathbf{r}}{dt}. \end{aligned} \tag{15}$$

We identify the first parenthesis as the acceleration, $\mathbf{a}$, from (6). The second, by a legitimate analogy with (13), proves to be $\boldsymbol{\omega} \times \mathbf{v}$. The quantity, $d\boldsymbol{\omega}/dt = 0$, since $\boldsymbol{\omega}$ is constant, by hypothesis. The last term on the right becomes [cf. equations (8), (14)]:

$$\boldsymbol{\omega} \times \frac{d\mathbf{r}}{dt} = \boldsymbol{\omega} \times \mathbf{v}_0 = \boldsymbol{\omega} \times (\mathbf{v} + \boldsymbol{\omega} \times \mathbf{r}) = \boldsymbol{\omega} \times \mathbf{v} + \boldsymbol{\omega} \times (\boldsymbol{\omega} \times \mathbf{r}). \tag{16}$$

Therefore

$$\mathbf{a}_0 = \mathbf{a} + \boldsymbol{\omega} \times (\boldsymbol{\omega} \times \mathbf{r}) + 2\boldsymbol{\omega} \times \mathbf{v}. \tag{17}$$

In equation (17) the second term on the right is usually called the centrifugal acceleration; the third is the acceleration of Coriolis.

26. The problem of two bodies. We shall now consider the relative motion of two bodies, under the action of their mutual gravitational forces. Although we shall presently discuss the problem from the standpoint of Hamilton's principle, we first give the conventional approach, via Newton's laws. The derivation, stated in vector form, constitutes a useful demonstration of vector procedures.

We suppose the bodies to be rigid spheres of masses m_1 and m_2, so that in accord with § (11) we may regard the action as proceeding from their centers. Let the instantaneous positions, P_1 and P_2 of the centers of the two bodies be the respective vectors $\mathbf{r}_1$ and $\mathbf{r}_2$. The vector $\mathbf{r}$, joining the centers is given by the equation

$$\mathbf{r}_1 - \mathbf{r}_2 = \mathbf{r}. \tag{1}$$

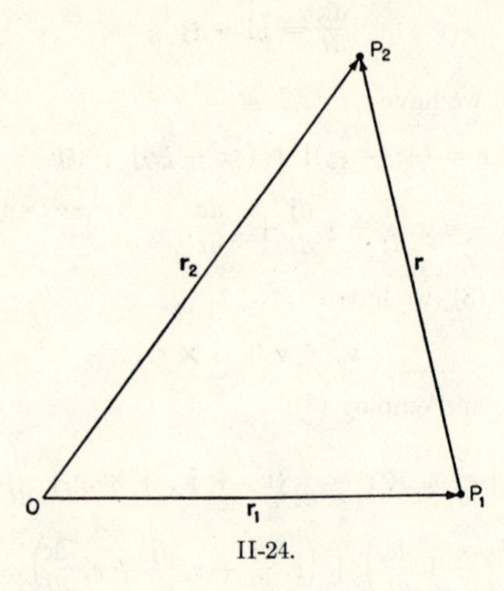

II-24.

The forces are directed along $\mathbf{r}$, by Newton's laws. Further, if $\mathbf{i}_r$ be a unit vector along $\mathbf{r}$, so that

$$\mathbf{r} = \mathbf{i}_r r, \tag{2}$$

where $r = |\mathbf{r}|$, the inverse-square law of force becomes, as in (13.6),

$$m_1 \frac{d^2\mathbf{r}_1}{dt^2} = -G \frac{m_1 m_2}{r^2} \mathbf{i}_r = -G \frac{m_1 m_2}{r^3} \mathbf{r}, \tag{3}$$

and

$$m_2 \frac{d^2\mathbf{r}_2}{dt^2} = -G \frac{m_1 m_2}{r^3} (-\mathbf{r}). \tag{4}$$

Equation (3) represents the force acting on m_1 because of the presence of m_2. The negative sign in (4) arises from the fact that $\mathbf{r}$ is reversed in direction, i.e., the forces in (3) and (4) must be equal and opposite. Adding them, we have

$$m_1 \frac{d^2\mathbf{r}_1}{dt^2} + m_2 \frac{d^2\mathbf{r}_2}{dt^2} = 0. \tag{5}$$

Integrating this equation directly, we have

$$m_1 \frac{d\mathbf{r}_1}{dt} + m_2 \frac{d\mathbf{r}_2}{dt} = \mathbf{p}_0, \tag{6}$$

where $\mathbf{p}_0$ is a constant of integration. Integrating again, we find

$$m_1\mathbf{r}_1 + m_2\mathbf{r}_2 = \mathbf{p}_0 t + \mathbf{a}_0, \tag{7}$$

where $\mathbf{a}_0$ is another constant. We recognize equations (6) and (7) as special cases of the general theorems derived in §§ (4) and (5), conservation of momentum, and uniform rectilinear motion of the center of mass.

Since each of the two vector differential equations (3) and (4) has three components, and since each component equation is of the second order, we shall require twelve constants of integration in the general solution. Equation (7) implies six such constants, because $\mathbf{p}_0$ and $\mathbf{a}_0$ are vectors. Consequently we require only six more constants to determine the orbit.

Having fixed the motion of the center of mass we need now determine only the relative orbit of one body with respect to the other. Dividing equations (3) and (4) by m_1 and m_2, respectively, and taking their difference, we find, on making use of (1), that

$$\frac{d^2\mathbf{r}}{dt^2} = -G(m_1 + m_2)\frac{\mathbf{r}}{r^3} = -GM\frac{\mathbf{r}}{r^3}. \quad M = m_1 + m_2. \tag{8}$$

Now take the vector product of $\mathbf{r}$ into (8), or

$$\mathbf{r} \times \frac{d^2\mathbf{r}}{dt^2} = -\frac{GM}{r}\mathbf{r} \times \mathbf{r} = 0, \tag{9}$$

since $\mathbf{r} \times \mathbf{r} = 0$. We shall find it instructive to write (9) in its expanded form. Setting

$$\mathbf{r} = \mathbf{i}x + \mathbf{j}y + \mathbf{k}z \tag{10}$$

and performing the indicated vector multiplication according to the rules of § (22), we find

$$\mathbf{i}\left(y\frac{d^2z}{dt^2} - z\frac{d^2y}{dt^2}\right) + \mathbf{j}\left(z\frac{d^2x}{dt^2} - x\frac{d^2z}{dt^2}\right) + \mathbf{k}\left(x\frac{d^2y}{dt^2} - y\frac{d^2x}{dt^2}\right) = 0. \tag{11}$$

Integrating directly with respect to t, we obtain

$$\mathbf{i}\left(y\frac{dz}{dt} - z\frac{dy}{dt}\right) + \mathbf{j}\left(z\frac{dx}{dt} - x\frac{dz}{dt}\right) + \mathbf{k}\left(x\frac{dy}{dt} - y\frac{dx}{dt}\right) = \mathbf{c}. \tag{12}$$

One readily checks this result by differentiating it. The integration constant

$$\mathbf{c} = \mathbf{i}c_x + \mathbf{j}c_y + \mathbf{k}c_z \tag{13}$$

contains three arbitrary constants as indicated. Perform, now, the scalar product of (12) with $\mathbf{r}$, making use of (10). The left-hand side vanishes identically, leaving

$$\mathbf{c} \cdot \mathbf{r} = c_x x + c_y y + c_z z = 0. \tag{14}$$

This is the equation of a plane. In fact it represents the plane of the orbital motion. Two of these constants fix the position of the plane so that the original twelfth-order problem reduces to one of the fourth order. Reference to § (6) identifies the components of (12) as comprising the vector of twice the areal velocity. Two of the three components of **c** simultaneously fix the orientation of the orbit plane. The third determines the magnitude of the areal velocity, which becomes

$$\frac{1}{2}|\mathbf{c}| = \frac{1}{2}c = \frac{1}{2}\sqrt{c_x^2 + c_y^2 + c_z^2}. \tag{15}$$

Note that we may obtain the result from (9), by direct vector methods. We may write, in place of (9), cf. equation (24.15),

$$\frac{d}{dt}\left(\mathbf{r} \times \frac{d\mathbf{r}}{dt}\right) = \frac{d\mathbf{r}}{dt} \times \frac{d\mathbf{r}}{dt} + \mathbf{r} \times \frac{d^2\mathbf{r}}{dt^2} = \mathbf{r} \times \frac{d^2\mathbf{r}}{dt^2} = 0, \tag{16}$$

because the vector product of a vector by itself vanishes. Hence

$$\mathbf{r} \times \frac{d\mathbf{r}}{dt} = \mathbf{c}. \tag{17}$$

Multiplying (17) by m_1, we have

$$\mathbf{r} \times \left(m_1 \frac{d\mathbf{r}}{dt}\right) = \mathbf{r} \times \mathbf{p}_1 = m_1\mathbf{c} = \mathbf{L}, \tag{18}$$

where $\mathbf{p}_1$ is the momentum and **L**, by (22.4), the angular momentum, a constant as required by (6.9). We also see that the areal velocity becomes, cf. equation (22.17),

$$\frac{1}{2}\mathbf{r} \times \mathbf{v} = \frac{1}{2}\mathbf{c}, \tag{19}$$

a constant, in accord with Kepler's second law. Here **c** is a vector perpendicular to **r** and **v**, i.e., it is perpendicular to the orbit plane.

Returning now to equation (8), and taking twice the scalar product of (8) by $d\mathbf{r}/dt$, we have

$$2\frac{d\mathbf{r}}{dt} \cdot \frac{d^2\mathbf{r}}{dt^2} = -G\frac{2M}{r^3}\mathbf{r} \cdot \frac{d\mathbf{r}}{dt} = \frac{d}{dt}\left(\frac{d\mathbf{r}}{dt} \cdot \frac{d\mathbf{r}}{dt}\right) = -G\frac{M}{r^3}\frac{d}{dt}(\mathbf{r} \cdot \mathbf{r}), \tag{20}$$

by (24.14). But

$$\frac{d\mathbf{r}}{dt} = \mathbf{i}_r\frac{dr}{dt} + \mathbf{i}_\theta r\frac{d\theta}{dt} + \mathbf{i}_\varphi r \sin\theta\frac{d\varphi}{dt}. \tag{21}$$

Hence (20) becomes

$$\frac{d}{dt}\left[\left(\frac{dr}{dt}\right)^2 + r^2\left(\frac{d\theta}{dt}\right)^2 + r^2\sin^2\theta\left(\frac{d\varphi}{dt}\right)^2\right]$$
$$= -G\frac{M}{r^3}\frac{d}{dt}r^2 = -G\frac{2M}{r^2}\frac{dr}{dt} = 2GM\frac{d}{dt}\left(\frac{1}{r}\right). \tag{22}$$

Integrating (22) we find that

$$\left(\frac{dr}{dt}\right)^2 + r^2\left(\frac{d\theta}{dt}\right)^2 + r^2\sin^2\theta\left(\frac{d\varphi}{dt}\right)^2 = \frac{2GM}{r} + c', \tag{23}$$

where c', another constant of integration, equals the total energy per unit mass. (22) is the well-known "vis viva" integral, expressing the kinetic energy (per unit mass) in terms of the potential.

We find it convenient at this time to orientate the system of coordinates so that the plane, $\theta = \pi/2$, coincides with the plane of the orbit. Then (23) becomes, since θ is ɪ ow a constant,

$$\left(\frac{dr}{dt}\right)^2 + r^2\left(\frac{d\varphi}{dt}\right)^2 = \frac{2GM}{r} + c'; \tag{24}$$

From (17), we obtain

$$\left|\mathbf{r} \times \frac{d\mathbf{r}}{dt}\right| = r^2\frac{d\varphi}{dt} = |\mathbf{c}| = c. \tag{25}$$

Also

$$\frac{dr}{dt} = \frac{dr}{d\varphi}\frac{d\varphi}{dt}. \tag{26}$$

Therefore, using (25) and (26) to eliminate dt from (24), we find

$$d\varphi = \frac{c\,dr}{r\sqrt{-c^2 + 2GMr + c'r^2}} \tag{27}$$

We may perform this integration with the change of variable $r = 1/u$, or take the result directly from integral tables. We find that

$$\varphi = \text{arc cos}\,\frac{GMr - c^2}{r\sqrt{c^2c' + G^2M^2}} + c'', \tag{28}$$

or

$$r = \frac{c^2}{GM - \sqrt{c^2c' + G^2M^2}\cos(\varphi - c'')}, \tag{29}$$

which is the polar equation of a conic with the origin at a focus. The constant c'' defines the position of the apse, i.e., the major axis. If we measure ϕ from the point of nearest approach, $c'' = 0$. To simplify (28) we set

$$p = c^2/GM \tag{30}$$

and

$$e^2 = 1 + c^2c'/G^2M^2. \tag{31}$$

Then (29) becomes

$$r = p/(1 + e \cos \varphi). \tag{32}$$

Equations (30) and (31) enable us to replace the integration constants c and c' with two others. Here p, the semi-parameter or semi-latus rectum determines the size of the ellipse, whereas e, the eccentricity, determines its shape. The constant c'', which really defines the position of the line of apsides in the orbit plane, is the third constant. These constants, together with the pair defining the orientation of the orbit planes, are all that we require for a complete specification of the orbit. A final constant specifies the position of the body in the orbit. For this purpose we ordinarily employ T, the time of some perihelion passage, i.e., when the planet is closest to the sun.

Dyadics, Matrices, and Tensors

27. Introduction to dyadics or tensors. We have just discussed the relative motion of two spherical masses. We might expect that our next logical step would involve a discussion of the motion when the mass distribution in one of the bodies departs from spherical symmetry. There prove, however, to be several intermediate steps. When the body considered is not spherical, we must consider the question of its own rotation. The resultant torques that arise from lack of symmetry may cause a shifting of the body with respect to the axis of rotation and may also produce a shifting of the position of the axis in space (precession). The mathematics of the discussion are greatly shortened by the use of a special notation, an extension of the vector treatment already given.

Consider a point-mass P attached to the centers of three mutually perpendicular stretched rubber bands and held in static equilibrium by the force of elasticity. Take a set of Cartesian axes parallel to the bands. The mass, if displaced a small distance x parallel to the x-axis, will experience a restoring force proportional to and opposite in sign to the displacement, i.e.,

$$\mathbf{F} = -\mathbf{i}a_x x, \tag{1}$$

where a_x measures the tension of the band on the x-axis. We suppose the bands to be so long compared with the displacement that the restoring force in the x-direction produced by the crossed yz-bands is negligibly small.

In general, the tensions a_x, a_y, and a_z will not be equal so that when the displacement is $\mathbf{r}$, where

$$\mathbf{r} = \mathbf{i}x + \mathbf{j}y + \mathbf{k}z, \tag{2}$$

the force becomes

$$\mathbf{F} = -(\mathbf{i}a_x x + \mathbf{j}a_y y + \mathbf{k}a_z z). \tag{3}$$

We shall now express the vector **F** in a new notation that will be particularly useful in certain types of vector problems. The scalars x, y, and z may be written

$$x = \mathbf{i} \cdot \mathbf{r}, \quad y = \mathbf{j} \cdot \mathbf{r}, \quad z = \mathbf{k} \cdot \mathbf{r}. \tag{4}$$

Therefore (3) becomes

$$\mathbf{F} = -[a_x\mathbf{i}(\mathbf{i} \cdot \mathbf{r}) + a_y\mathbf{j}(\mathbf{j} \cdot \mathbf{r}) + a_z\mathbf{k}(\mathbf{k} \cdot \mathbf{r})]. \tag{5}$$

Since quantities of the form **ii**, **jj**, etc., have no interpretation except when one or the other of the two juxtaposed vectors enters as a scalar product, the parentheses in (5) are not really necessary. Hence we may factor out **r**, and write symbolically,

$$\mathbf{F} = -(a_x\mathbf{ii} + a_y\mathbf{jj} + a_z\mathbf{kk}) \cdot \mathbf{r} = -\mathbf{\Phi} \cdot \mathbf{r}, \tag{6}$$

where the quantity,

$$\mathbf{\Phi} = a_x\mathbf{ii} + a_j\mathbf{jj} + a_z\mathbf{kk}, \tag{7}$$

plays a role analogous to the operators of calculus and vector analysis. The quantity $\mathbf{\Phi}$ is called a dyadic or tensor of the second rank and the individual components are known as dyads. It is a quantity that, when "dotted" into a vector, yields another vector.

When the coordinate axes are not specially chosen to lie parallel to the elastic bands, $\mathbf{\Phi}$ assumes a somewhat different form. Let λ_1, μ_1, and ν_1 be the direction cosines of the vector **i** in the new coordinate system, etc., for the vectors **j** and **k**. Then

$$\begin{aligned} \mathbf{i} &= \lambda_1\mathbf{i}' + \mu_1\mathbf{j}' + \nu_1\mathbf{k}', \\ \mathbf{j} &= \lambda_2\mathbf{i}' + \mu_2\mathbf{j}' + \nu_2\mathbf{k}', \\ \mathbf{k} &= \lambda_3\mathbf{i}' + \mu_3\mathbf{j}' + \nu_3\mathbf{k}', \end{aligned} \tag{8}$$

where **i**′, **j**′, and **k**′ are unit vectors in the new system. Substituting these values into (7) and carrying out the expansion we find that

$$\begin{aligned} \mathbf{\Phi} = {} & a_{11}\mathbf{ii} + a_{12}\mathbf{ij} + a_{13}\mathbf{ik} \\ & + a_{21}\mathbf{ji} + a_{22}\mathbf{jj} + a_{23}\mathbf{jk} \\ & + a_{31}\mathbf{ki} + a_{32}\mathbf{kj} + a_{33}\mathbf{kk}, \end{aligned} \tag{9}$$

in which equation we have, for simplicity, dropped the primes because they are no longer needed. The coefficients are

$$\begin{aligned} a_{11} &= \lambda_1^2 a_x + \lambda_2^2 a_y + \lambda_3^2 a_z \text{ , ETC.} \\ a_{12} &= \lambda_1\mu_1 a_x + \lambda_2\mu_2 a_y + \lambda_3\mu_3 a_z = a_{21} \text{ , ETC.} \end{aligned} \tag{10}$$

Note that the general dyadic has nine components, whereas a vector possesses only three.

We now define the so-called conjugate dyadic, $\mathbf{\Phi}_c$, obtained by permuting the unit vectors in each element of $\mathbf{\Phi}$.

$$\begin{aligned}\mathbf{\Phi}_c &= a_{11}\mathbf{ii} + a_{12}\mathbf{ji} + a_{13}\mathbf{ki}\\ &+ a_{21}\mathbf{ij} + a_{22}\mathbf{jj} + a_{23}\mathbf{kj}\\ &+ a_{31}\mathbf{ik} + a_{32}\mathbf{jk} + a_{33}\mathbf{kk} = \mathbf{\Phi},\end{aligned} \tag{11}$$

because, in the present example,

$$a_{ij} = a_{ji}, \tag{12}$$

by (10). Dyadics that are equal to their conjugates are said to be symmetric, whereas those that are equal to the negatives of their conjugates, with

$$a_{ij} = -a_{ji}, \tag{13}$$

are called anti-symmetric (or skew-symmetric).

To sum two or more dyadics we merely take the sum of each element. Consequently the sum of two or more dyadics is itself a dyadic. The compound dyadic $\mathbf{\Phi} + \mathbf{\Phi}_c$ has for its conjugate $\mathbf{\Phi}_c + \mathbf{\Phi} = \mathbf{\Phi} + \mathbf{\Phi}_c$. In other words, this dyadic is symmetric. Analogously, the dyadic $\mathbf{\Phi} - \mathbf{\Phi}_c$ has $\mathbf{\Phi}_c - \mathbf{\Phi} = -(\mathbf{\Phi} - \mathbf{\Phi}_c)$ for a conjugate. It therefore is anti-symmetric. Consequently we may express any dyadic in the form

$$\mathbf{\Phi} = \frac{1}{2}(\mathbf{\Phi} + \mathbf{\Phi}_c) + \frac{1}{2}(\mathbf{\Phi} - \mathbf{\Phi}_c), \tag{14}$$

i.e., as the sum of symmetric and anti-symmetric dyadics.

We may show, by reversing the argument leading to equation (10), that if a dyadic is symmetric, we may by simple rotation of axes cause the non-diagonal elements in (9) to vanish, so that the dyadic takes the form of (7).

The operator

$$\mathbf{ii} + \mathbf{jj} + \mathbf{kk} = \mathfrak{I} \tag{15}$$

is often termed the "idemfactor," or identical dyadic. When $\mathfrak{I}$ is multiplied by any vector, either as a pre- or post-factor, the vector is unchanged, i.e.,

$$\mathbf{F} \cdot \mathfrak{I} = (\mathbf{i}X + \mathbf{j}Y + \mathbf{k}Z) \cdot \mathfrak{I} = \mathbf{i} \cdot \mathbf{ii}X + \mathbf{j} \cdot \mathbf{jj}Y + \mathbf{k} \cdot \mathbf{kk}Z = \mathbf{F}. \tag{16}$$

The expression $\boldsymbol{\alpha} \cdot \mathbf{\Phi} \cdot \boldsymbol{\beta}$ is scalar in nature, as also is $\mathbf{\Phi} : \mathbf{\Psi}$, where $\mathbf{\Psi}$ is another dyadic. An expression of this type is called the "double-dot product," which signifies that the two dyadics are to be expanded by the rules of distributive multiplication and then each dyad pair combined according to the rule

$$(\mathbf{ab}) : (\mathbf{cd}) = (\mathbf{a}\cdot\mathbf{c})(\mathbf{b}\cdot\mathbf{d}). \tag{17}$$

Note that, from equations (9) and (11),

$$\begin{aligned}\mathbf{\Phi} : \mathbf{\Phi}_c = a_{11}^2 &+ a_{12}a_{21} + a_{13}a_{31} \\ &+ a_{21}a_{12} + a_{22}^2 + a_{23}a_{32} \\ &+ a_{31}a_{13} + a_{32}a_{23} + a_{33}^2.\end{aligned} \tag{18}$$

28. Dyadic transformations. The process of multiplying a vector by a dyadic transforms the vector into a new vector. The simplest transformation is the multiplication by the unit dyadic $\mathfrak{I}$ which carries the vector into itself, according to (27.16). The unit dyadic, transformed by simple rotation of the coordinate system, becomes:

$$\begin{aligned}\mathfrak{I} = \mathbf{ii} + \mathbf{jj} + \mathbf{kk} &= \begin{bmatrix} a_{11}\mathbf{i}'\mathbf{i}' + a_{12}\mathbf{i}'\mathbf{j}' + a_{13}\mathbf{i}'\mathbf{k}' \\ +a_{21}\mathbf{j}'\mathbf{i}' + a_{22}\mathbf{j}'\mathbf{j}' + a_{23}\mathbf{j}'\mathbf{k}' \\ +a_{31}\mathbf{k}'\mathbf{i}' + a_{32}\mathbf{k}'\mathbf{j}' + a_{33}\mathbf{k}'\mathbf{k}' \end{bmatrix} \\ &= \mathbf{i}'\mathbf{i}' + \mathbf{j}'\mathbf{j}' + \mathbf{k}'\mathbf{k}' = \mathfrak{I}',\end{aligned} \tag{1}$$

wherein we have substituted for **i**, **j** and **k** from (27.8). The quantities:

$$\begin{aligned}a_{11} &= \lambda_1^2 + \lambda_2^2 + \lambda_3^2 = 1, \\ a_{12} &= \lambda_1\mu_1 + \lambda_2\mu_2 + \lambda_3\mu_3 = 0, \quad \text{ETC.},\end{aligned} \tag{2}$$

for a Cartesian coordinate system. Thus, if we have a vector, **F**, which we wish to transform to a new coordinate system, we can carry out the multiplication:

$$\begin{aligned}\mathbf{F} = \mathfrak{I}\cdot\mathbf{F} = \mathfrak{I}'\cdot\mathbf{F} &= \mathfrak{I}'\cdot(\mathbf{i}X + \mathbf{j}Y + \mathbf{k}Z) \\ &= \mathbf{i}'(X\mathbf{i}'\cdot\mathbf{i} + Y\mathbf{i}'\cdot\mathbf{j} + Z\mathbf{i}'\cdot\mathbf{k}) \\ &+ \mathbf{j}'(X\mathbf{j}'\cdot\mathbf{i} + Y\mathbf{j}'\cdot\mathbf{j} + Z\mathbf{j}'\cdot\mathbf{k}) \\ &+ \mathbf{k}'(X\mathbf{k}'\cdot\mathbf{i} + Y\mathbf{k}'\cdot\mathbf{j} + Z\mathbf{k}'\cdot\mathbf{k}).\end{aligned} \tag{3}$$

But
$$\mathbf{i}'\cdot\mathbf{i} = \lambda_1, \quad \mathbf{i}'\cdot\mathbf{j} = \mu_1, \quad \text{ETC.} \tag{4}$$

Thus
$$\begin{aligned}\mathbf{F} &= \mathbf{i}'(X\lambda_1 + Y\mu_1 + Z\nu_1) + \mathbf{j}'(X\lambda_2 + Y\mu_2 + Z\nu_2) \\ &+ \mathbf{k}'(X\lambda_3 + Y\mu_3 + Z\nu_3) = \mathbf{i}'X' + \mathbf{j}Y' + \mathbf{k}Z'.\end{aligned} \tag{5}$$

The foregoing transformation leaves the original vector unchanged.

Consider, now, the more general transformation:

$$\mathbf{G} = \mathbf{\Phi}\cdot\mathbf{F} = \mathbf{\Phi}\cdot(\mathbf{i}X + \mathbf{j}Y + \mathbf{k}Z), \tag{6}$$

where
$$\mathbf{\Phi} = \begin{bmatrix} a_{11}\mathbf{ii} + a_{12}\mathbf{ij} + a_{13}\mathbf{ik} \\ + a_{21}\mathbf{ji} + a_{22}\mathbf{jj} + a_{23}\mathbf{jk} \\ + a_{31}\mathbf{ki} + a_{32}\mathbf{kj} + a_{33}\mathbf{kk} \end{bmatrix}. \tag{7}$$

Thus
$$\mathbf{G} = \begin{bmatrix} (a_{11}X + a_{12}Y + a_{13}Z)\mathbf{i} \\ + (a_{21}X + a_{22}Y + a_{23}Z)\mathbf{j} \\ + (a_{31}X + a_{32}Y + a_{33}Z)\mathbf{k} \end{bmatrix}. \tag{8}$$

We can regard the new vector, $\mathbf{G}$, as obtained from the old by a process of distortion and rotation. For this reason a dyadic is equivalent to a tensor of the second rank. We shall find later on that, for three definite directions of the vector $\mathbf{r}$, the operation $\mathbf{\Phi} \cdot \mathbf{r}_1$ represents only a stretching of $\mathbf{r}_1$. In other words,

$$\mathbf{\Phi} \cdot \mathbf{r}_1 = \lambda \mathbf{r}_1, \tag{9}$$

where λ is a numerical constant.

Let $\mathbf{r}_1$ be a vector that is to be transformed by pure rotation into a vector $\mathbf{r}_2$ by the equation

$$\mathbf{r}_2 = \mathbf{\Phi} \cdot \mathbf{r}_1. \tag{10}$$

Specify the axis of rotation by a unit vector, $\boldsymbol{\varrho}$, not necessarily perpendicular to $\mathbf{r}_1$ or $\mathbf{r}_2$. Define

$$\boldsymbol{\varrho} = \mathbf{i}\lambda + \mathbf{j}\mu + \mathbf{k}\nu, \tag{11}$$

in terms of its direction cosines. The component of $\mathbf{r}_1$, along $\boldsymbol{\varrho}$, viz., $\mathbf{r}_1 \cdot \boldsymbol{\varrho\varrho}$, is unaltered by the rotation. The component perpendicular to $\boldsymbol{\varrho}$ has the magnitude $|\, \boldsymbol{\varrho} \times \mathbf{r}_1 \,|$. However, $\boldsymbol{\varrho} \times \mathbf{r}_1$ is perpendicular to both $\boldsymbol{\varrho}$ and $\mathbf{r}_1$. Hence the component perpendicular to $\boldsymbol{\varrho}$, in the plane defined by $\boldsymbol{\varrho}$, $\mathbf{r}_1$, is

$$(\boldsymbol{\varrho} \times \mathbf{r}_1) \times \boldsymbol{\varrho} = (\mathbf{r}_1\boldsymbol{\varrho} - \boldsymbol{\varrho}\mathbf{r}_1) \cdot \boldsymbol{\varrho}, \tag{12}$$

by (22.25). The vector of the foregoing equation is the component prior to rotation. If θ is the angle turned through, we get

$$\mathbf{r}_2 = \mathbf{r}_1 \cdot \boldsymbol{\varrho\varrho} + (\mathbf{r}_1\boldsymbol{\varrho} - \boldsymbol{\varrho}\mathbf{r}_1) \cdot \boldsymbol{\varrho} \cos\theta + \boldsymbol{\varrho} \times \mathbf{r}_1 \sin\theta. \tag{13}$$

We can write

$$\boldsymbol{\varrho} \times \mathbf{r}_1 = \mathfrak{J} \cdot \boldsymbol{\varrho} \times \mathbf{r}_1 = \mathfrak{J} \times \boldsymbol{\varrho} \cdot \mathbf{r}_1. \tag{14}$$

$$(\mathbf{r}_1\boldsymbol{\varrho} - \boldsymbol{\varrho}\mathbf{r}_1) \cdot \boldsymbol{\varrho} = \mathbf{r}_1 - \boldsymbol{\varrho\varrho} \cdot \mathbf{r}_1 = (\mathfrak{J} - \boldsymbol{\varrho\varrho}) \cdot \mathbf{r}_1. \tag{15}$$

Therefore

$$\mathbf{r}_2 = [\boldsymbol{\varrho\varrho}(1 - \cos\theta) + \mathfrak{J}\cos\theta + \mathfrak{J} \times \boldsymbol{\varrho}\sin\theta] \cdot \mathbf{r}_1. \tag{16}$$

The dyadic, in the brackets, is independent of the original vector $\mathbf{r}_1$. We identify it with the $\mathbf{\Phi}$ of equation (10). Thus (11) and (16) give

$$\mathbf{\Phi} = \begin{cases} \mathbf{ii}[\lambda^2(1 - \cos\theta) + \cos\theta] \\ + \mathbf{ij}[\lambda\mu(1 - \cos\theta) - \nu\sin\theta] \\ + \mathbf{ik}[\lambda\nu(1 - \cos\theta) + \mu\sin\theta] \\ + \mathbf{ji}[\lambda\mu(1 - \cos\theta) + \nu\sin\theta] \\ + \mathbf{jj}[\mu^2(1 - \cos\theta) + \cos\theta] \\ + \mathbf{jk}[\mu\nu(1 - \cos\theta) - \lambda\sin\theta] \\ + \mathbf{ki}[\lambda\nu(1 - \cos\theta) - \mu\sin\theta] \\ + \mathbf{kj}[\mu\nu(1 - \cos\theta) + \lambda\sin\theta] \\ + \mathbf{kk}[\nu^2(1 - \cos\theta) + \cos\theta] \end{cases} \tag{17}$$

We can break this dyadic into a pair, one of which is symmetric and the other skew-symmetric. Thus

$$\mathbf{\Phi} = \mathbf{S} + \mathbf{\Omega}. \tag{18}$$

$$\mathbf{S} = \begin{bmatrix} \mathbf{ii}[\lambda^2(1 - \cos\theta) + \cos\theta] + \mathbf{ij}\lambda\mu(1 - \cos\theta) + \mathbf{ik}\lambda\nu(1 - \cos\theta) \\ + \mathbf{ji}\lambda\mu(1 - \cos\theta) + \mathbf{jj}[\mu^2(1 - \cos\theta) + \cos\theta] + \mathbf{jk}\mu\nu(1 - \cos\theta) \\ + \mathbf{ki}\lambda\nu(1 - \cos\theta) + \mathbf{kj}\mu\nu(1 - \cos\theta) + \mathbf{kk}[\nu^2(1 - \cos\theta) + \cos\theta] \end{bmatrix}. \tag{19}$$

$$\mathbf{\Omega} = \begin{bmatrix} 0 & - \mathbf{ij}\nu & + \mathbf{ik}\mu \\ + \mathbf{ji}\nu & + 0 & - \mathbf{jk}\lambda \\ - \mathbf{ki}\mu & + \mathbf{kj}\lambda & + 0 \end{bmatrix}. \tag{20}$$

Note that for small rotations,

$$\mathbf{S} \sim \mathfrak{I}. \tag{21}$$

The more general dyadic transformation of equations (6) and (7) represents compression or strain, in addition to rotation of the original vector.

Consider the application of two successive dyadic transformations of a vector, as follows:

$$\mathbf{G} = \mathbf{\Psi} \cdot \mathbf{\Phi} \cdot \mathbf{F}, \tag{22}$$

where $\mathbf{\Phi}$ is given by (7) and $\mathbf{\Psi}$ is a similar dyadic:

$$\mathbf{\Psi} = \begin{bmatrix} b_{11}\mathbf{ii} + b_{12}\mathbf{ij} + b_{13}\mathbf{ik} \\ + b_{21}\mathbf{ji} + b_{22}\mathbf{jj} + b_{23}\mathbf{jk} \\ + b_{31}\mathbf{ki} + b_{32}\mathbf{kj} + b_{33}\mathbf{kk} \end{bmatrix}. \tag{23}$$

For convenience in a later interpretation we shall write the components of **F** in a vertical column,

$$\mathbf{F} = \begin{bmatrix} \mathrm{i}X \\ \mathrm{j}Y \\ \mathrm{k}Z \end{bmatrix}. \tag{24}$$

To simplify our notation, we may omit the factors that represent the unit vectors, in view of the fact that the positions of the components uniquely specify the quantities. Also, we may omit the addition signs. In this fashion we shall write:

$$\begin{aligned}\mathbf{G} = \mathbf{\Psi}\cdot\mathbf{\Phi}\cdot\mathbf{F} &= \begin{bmatrix} b_{11} & b_{12} & b_{13} \\ b_{21} & b_{22} & b_{23} \\ b_{31} & b_{32} & b_{33} \end{bmatrix} \cdot \begin{bmatrix} a_{11} & a_{12} & a_{13} \\ a_{21} & a_{22} & a_{23} \\ a_{31} & a_{32} & a_{33} \end{bmatrix} \cdot \begin{bmatrix} X \\ Y \\ Z \end{bmatrix} \\ &= \begin{bmatrix} b_{11} & b_{12} & b_{13} \\ b_{21} & b_{22} & b_{23} \\ b_{31} & b_{32} & b_{33} \end{bmatrix} \cdot \begin{bmatrix} a_{11}X + a_{12}Y + a_{13}Z \\ a_{21}X + a_{22}Y + a_{23}Z \\ a_{31}X + a_{32}Y + a_{33}Z \end{bmatrix}. \end{aligned} \tag{25}$$

We have carried through the multiplication, exactly as we should have done if the dyad elements were present. Continuing the process, we get

$$\mathbf{G} = \begin{bmatrix} (b_{11}a_{11} + b_{12}a_{21} + b_{13}a_{31})X + (b_{11}a_{12} + b_{12}a_{22} + b_{13}a_{32})Y \\ \qquad + (b_{11}a_{13} + b_{12}a_{23} + b_{13}a_{33})Z \\ (b_{21}a_{11} + b_{22}a_{21} + b_{23}a_{31})X + (b_{21}a_{12} + b_{22}a_{22} + b_{23}a_{32})Y \\ \qquad + (b_{21}a_{13} + b_{22}a_{23} + b_{23}a_{33})Z \\ (b_{31}a_{11} + b_{32}a_{21} + b_{33}a_{31})X + (b_{31}a_{12} + b_{32}a_{22} + b_{33}a_{32})Y \\ \qquad + (b_{31}a_{13} + b_{32}a_{23} + b_{33}a_{33})Z \end{bmatrix} \tag{26}$$

The respective rows are the x, y, and z components of the vector **G**.
We find that

$$\mathbf{G} = \mathbf{\Psi}\cdot(\mathbf{\Phi}\cdot\mathbf{F}) = (\mathbf{\Psi}\cdot\mathbf{\Phi})\cdot\mathbf{F}. \tag{27}$$

The order of multiplication, therefore, is not significant. The dyadic

$$\chi = \psi \cdot \mathbf{\Phi} = \begin{bmatrix} (b_{11}a_{11}+b_{12}a_{21}+b_{13}a_{31}) & (b_{11}a_{12}+b_{12}a_{22}+b_{13}a_{32}) & (b_{11}a_{13}+b_{12}a_{23}+b_{13}a_{33}) \\ (b_{21}a_{11}+b_{22}a_{21}+b_{23}a_{31}) & (b_{21}a_{12}+b_{22}a_{22}+b_{23}a_{32}) & (b_{21}a_{13}+b_{22}a_{23}+b_{23}a_{33}) \\ (b_{31}a_{11}+b_{32}a_{21}+b_{33}a_{31}) & (b_{31}a_{12}+b_{32}a_{22}+b_{33}a_{32}) & (b_{31}a_{13}+b_{32}a_{23}+b_{33}a_{33}) \end{bmatrix}. \tag{28}$$

Note carefully the construction of each of these nine terms. Denote, by ψ_{ij}, the element in row i and column j of the dyadic $\mathbf{\Psi}$ and, by ϕ_{jk}, the term in row j and column k of the dyadic $\mathbf{\Phi}$. Then the term χ_{ik} in row i, column k of the dyadic $\mathbf{X}$, is

$$\chi_{ik} = \sum_j \psi_{ij}\phi_{jk}. \tag{29}$$

In other words, we multiply the terms of row i of the first dyadic by the corresponding terms in column k of the second dyadic, add the individual terms and enter the result in row i and column k of the result.

The above rule for multiplication of two dyadics is identical with that for multiplication of two determinants or two matrices. Indeed, the notation introduced in (9) is the same as that used for matrices. The great majority of matrix theorems hold also for dyadics and vectors.

29. Introduction to the theory of matrices. It is interesting, as well as useful, to consider at this point the general algebra of matrices. The applications of matrix theory to physics are many. Among the newer uses of the methods is the so-called matrix mechanics, employed in conjunction with wave mechanics for the analysis of atomic properties.

The procedures of matrix algebra seem unduly formal and artificial unless one understands the origin and development of the subject. The simplest approach is from the standpoint of simultaneous linear equations. Consider the set of equations

$$\begin{aligned} a_{11}x_1 + a_{12}x_2 + a_{13}x_3 &= d_1, \\ a_{21}x_1 + a_{22}x_2 + a_{23}x_3 &= d_2, \\ a_{31}x_1 + a_{32}x_2 + a_{33}x_3 &= d_3. \end{aligned} \tag{1}$$

Using the multiplication rule previously defined, we can represent these three equations in matrix form:

$$\begin{bmatrix} a_{11} & a_{12} & a_{13} \\ a_{21} & a_{22} & a_{23} \\ a_{31} & a_{32} & a_{33} \end{bmatrix} \cdot \begin{bmatrix} x_1 \\ x_2 \\ x_3 \end{bmatrix} = \begin{bmatrix} d_1 \\ d_2 \\ d_3 \end{bmatrix}. \tag{2}$$

Expansion of the left-hand side gives

$$\begin{bmatrix} a_{11}x_1 + a_{12}x_2 + a_{13}x_3 \\ a_{21}x_1 + a_{22}x_2 + a_{23}x_3 \\ a_{31}x_1 + a_{32}x_2 + a_{33}x_3 \end{bmatrix} = \begin{bmatrix} d_1 \\ d_2 \\ d_3 \end{bmatrix}, \tag{3}$$

which is equivalent to the original equations, (1).

If, now, we designate the matrix of the coefficients by a, and the column matrices, $\{x_1, x_2, x_3\}$ and $\{d_1, d_2, d_3\}$, by x and d, respectively, we can abbreviate our equations in the form

$$ax = d, \tag{4}$$

which is not limited to any specific number of equations. Our matrix can be of any order.

In general, our matrices need not be square. For example, the equations,

$$2x + 3y = 8,$$
$$3x - 4y = -5,$$
$$x + 5y = 11,$$

are equivalent to

$$\begin{bmatrix} 2 & 3 \\ 3 & -4 \\ 1 & 5 \end{bmatrix} \cdot \begin{bmatrix} x \\ y \end{bmatrix} = \begin{bmatrix} 8 \\ -5 \\ 11 \end{bmatrix}. \tag{5}$$

In multiplication, however, the matrices must be "conformable," i.e., the number of columns of the first matrix must be equal to the number of rows of the second. In the foregoing example we have three equations and two unknowns. The three equations, therefore, are not linearly independent. Multiply the first by 19, the last by 17, subtract and divide by -7. The result is identical with the middle equation.

Return now to equations (1)–(4). Effect a change of variable, such that

$$\begin{aligned} x_1 &= b_{11}\xi_1 + b_{12}\xi_2 + b_{13}\xi_3, \\ x_2 &= b_{21}\xi_1 + b_{22}\xi_2 + b_{23}\xi_3, \\ x_3 &= b_{31}\xi_1 + b_{32}\xi_2 + b_{33}\xi_3, \end{aligned} \tag{6}$$

or, in matrix notation,

$$x = b\xi. \tag{7}$$

We can represent the resulting linear equations in ξ by the matrix equations

$$ax = ab\xi = c\xi = d, \tag{8}$$

where c is the matrix product, ab, calculated by the rules of matrix multiplication.

Suppose, now, we solve the equations (6) simultaneously to give

$$\begin{aligned} \xi_1 &= e_{11}x_1 + e_{12}x_2 + e_{13}x_3, \\ \xi_2 &= e_{21}x_1 + e_{22}x_2 + e_{23}x_3, \\ \xi_3 &= e_{31}x_1 + e_{32}x_2 + e_{33}x_3, \end{aligned} \tag{9}$$

so that

$$\xi = ex. \tag{10}$$

Substitute this result in (7), and obtain

$$x = bex = Ix, \tag{11}$$

where

$$I = \begin{bmatrix} 1 & 0 & 0 \\ 0 & 1 & 0 \\ 0 & 0 & 1 \end{bmatrix}, \tag{12}$$

equivalent to the idemfactor or unit dyadic $\mathfrak{J}$.

When two matrices bear such a relationship to one another that their product is equal to I, we say that one is the reciprocal of the other. These matrices cannot be singular, because we should then find the determinant of the coefficients vanishing.

Thus, since

$$be = I, \quad b^{-1}be = e = b^{-1}I = b^{-1}, \tag{13}$$

since multiplication by I gives the original matrix. Therefore, to divide by a matrix, we multiply by its reciprocal, which we denote by the above notation, e.g., the reciprocal of matrix b is the matrix b^{-1}. Only square matrices possess reciprocals.

To calculate the reciprocal, we note that the solution of equations (6) in the form of determinants, is

$$\xi_1 = \frac{\begin{vmatrix} x_1 & b_{12} & b_{13} \\ x_2 & b_{22} & b_{23} \\ x_3 & b_{32} & b_{33} \end{vmatrix}}{\begin{vmatrix} b_{11} & b_{12} & b_{13} \\ b_{21} & b_{22} & b_{23} \\ b_{31} & b_{32} & b_{33} \end{vmatrix}} = \frac{\begin{vmatrix} x_1 & x_2 & x_3 \\ b_{12} & b_{22} & b_{32} \\ b_{13} & b_{23} & b_{33} \end{vmatrix}}{\begin{vmatrix} b_{11} & b_{21} & b_{31} \\ b_{12} & b_{22} & b_{32} \\ b_{13} & b_{23} & b_{33} \end{vmatrix}}, \tag{14}$$

where we have interchanged the rows and columns of the determinants. Similar expressions hold for ξ_2 and ξ_3. Call the determinant in the denominator Δ_b. Then, expanding the numerator in terms of the minors of x_1, x_2, and x_3, we obtain

$$\Delta_b\xi_1 = \begin{vmatrix} x_1 & x_2 & x_3 \\ b_{12} & b_{22} & b_{32} \\ b_{13} & b_{23} & b_{33} \end{vmatrix} = \begin{vmatrix} b_{22} & b_{32} \\ b_{23} & b_{33} \end{vmatrix} x_1 - \begin{vmatrix} b_{12} & b_{32} \\ b_{13} & b_{33} \end{vmatrix} x_2 + \begin{vmatrix} b_{12} & b_{22} \\ b_{13} & b_{23} \end{vmatrix} x_3$$

$$= B_{11}x_1 + B_{21}x_2 + B_{31}x_3. \tag{15}$$

Similarly,

$$\begin{aligned} \Delta_b\xi_2 &= B_{12}x_1 + B_{22}x_2 + B_{32}x_3, \\ \Delta_b\xi_3 &= B_{13}x_1 + B_{23}x_2 + B_{33}x_3, \end{aligned} \tag{16}$$

where B_{ij} is the determinant that we obtain by suppressing the ith row and jth column of the original matrix b. To this determinant we attach the positive sign when $i + j$ is even and the negative sign when $i + j$ is odd. This procedure goes under the name of Cramer's rule.

Expressing our three resultant equations in matrix form, we have

$$\begin{bmatrix} \xi_1 \\ \xi_2 \\ \xi_3 \end{bmatrix} = \frac{1}{\Delta_b} \begin{bmatrix} B_{11} & B_{21} & B_{31} \\ B_{12} & B_{22} & B_{32} \\ B_{13} & B_{23} & B_{33} \end{bmatrix} \cdot \begin{bmatrix} x_1 \\ x_2 \\ x_3 \end{bmatrix}. \tag{17}$$

This equation shows the indeterminacy that arises when the matrix b is singular, because Δ_b will then be zero. Comparing the terms of (17) with those of equation (9), we note that

$$e_{11} = \frac{B_{11}}{\Delta_b}, \quad e_{12} = \frac{B_{21}}{\Delta_b}, \quad e_{13} = \frac{B_{31}}{\Delta_b}. \tag{18}$$

Note especially the reversal of indices in the B's, as compared with the e's.

Although we have carried through the foregoing development for a 3 by 3 matrix, the results must be applicable to a matrix of any order. If a is any square matrix,

$$a = \begin{bmatrix} a_{11} & a_{12} & \cdots & a_{1n} \\ a_{21} & a_{22} & \cdots & a_{2n} \\ \cdot & \cdot & & \cdot \\ \cdot & \cdot & & \cdot \\ \cdot & \cdot & & \cdot \\ a_{n1} & a_{n2} & \cdots & a_{nn} \end{bmatrix}, \tag{19}$$

its reciprocal is

$$a^{-1} = \frac{1}{\Delta_a} \begin{bmatrix} A_{11} & A_{21} & \cdots & A_{n1} \\ A_{12} & A_{22} & \cdots & A_{n2} \\ \cdot & \cdot & & \cdot \\ \cdot & \cdot & & \cdot \\ \cdot & \cdot & & \cdot \\ A_{1n} & A_{2n} & \cdots & A_{nn} \end{bmatrix}. \tag{20}$$

We call the matrix of the A's, whose components are defined as were the B's of equation (15), the adjoint (or adjunct) of matrix a. Thus

$$a_{\text{adj}} = \begin{bmatrix} A_{11} & A_{21} & \cdots & A_{n1} \\ A_{12} & A_{22} & \cdots & A_{n2} \\ \cdot & \cdot & & \cdot \\ \cdot & \cdot & & \cdot \\ \cdot & \cdot & & \cdot \\ A_{1n} & A_{2n} & \cdots & A_{nn} \end{bmatrix}, \tag{21}$$

so that
$$a \cdot a_{\mathrm{adj}} = \Delta_a I. \tag{22}$$

We can calculate the reciprocal of a matrix only if $\Delta_a \neq 0$. When $\Delta_a = 0$, the matrix is said to be singular. Two examples of singular matrices follow:

$$A = \begin{bmatrix} 2 & 3 & 1 \\ -1 & -2 & 2 \\ 1 & 1 & 3 \end{bmatrix}, \quad B = \begin{bmatrix} 2 & 2 & 1 \\ 3 & 3 & 3 \\ 5 & 5 & 7 \end{bmatrix}$$

When multiplied into the column matrix, $\{x, y, z\}$, these matrices produce the equations

$$\begin{array}{cc} A & B \\ 2x + 3y + z = d_1. & 2x + 2y + z = d_1. \\ -x - 2y + 2z = d_2. & 3x + 3y + 3z = d_2. \\ x + y + z = d_3. & 5x + 5y + 7z = d_3. \end{array}$$

The numbers d_1, d_2, and d_3 do not concern us here, except that their values must be consistent with the equations represented. In example A, the third equation is equal to the sum of the first two. In example B, the variables x and y are not independent. They occur in such a form that we may replace the quantity $x + y$ by a single variable w. The determinants of both these matrices vanish. Such a matrix is said to be degenerate. If several such relationships exist the matrix may be multiply degenerate.

If we multiply a matrix through by a constant, we multiply each of its components by that constant. Equality of two matrices implies that each pair of terms in corresponding positions shall be equal. We construct the conjugate (or transpose) of a matrix in the same way that we defined the conjugate of a dyadic, by interchanging the rows and columns. Thus

$$a_c = \tilde{a} = \begin{bmatrix} a_{11} & a_{21} & a_{31} \\ a_{12} & a_{22} & a_{32} \\ a_{13} & a_{23} & a_{33} \end{bmatrix}. \tag{23}$$

If a matrix is equal to its transpose,

$$a = \tilde{a}, \tag{24}$$

we call the matrix symmetrical. This condition implies that

$$a_{12} = a_{21}, \quad a_{23} = a_{32}, \quad a_{13} = a_{31}. \tag{25}$$

If a matrix is equal to the negative of its transpose, so that

$$a = -\tilde{a}, \tag{26}$$

we say that the matrix is skew-symmetric. In such a matrix the terms on the main diagonal are equal to zero. For the others,

$$a_{12} = -a_{21}, \quad a_{23} = -a_{32}, \quad a_{13} = -a_{31}. \tag{27}$$

The properties and definitions are identical with those employed previously for dyadics.

Matrices and dyadics may have complex as well as real elements. Vectors whose components are complex are called "bivectors," with one element in real and one in imaginary space. Later on we shall see that such notation is useful for certain types of physical problems. If, for example,

$$a = \begin{bmatrix} 1 & 1+i & 0 \\ 4i & 3+2i & 1 \\ 1-3i & 7 & 4-3i \end{bmatrix}, \tag{28}$$

$$\tilde{a} = \begin{bmatrix} 1 & 4i & 1-3i \\ 1+i & 3+2i & 7 \\ 0 & 1 & 4-3i \end{bmatrix}. \tag{29}$$

We define the Hermitian conjugate of a as $a^\dagger$, where

$$a^\dagger = \begin{bmatrix} a_{11}^* & a_{21}^* & a_{31}^* \\ a_{12}^* & a_{22}^* & a_{32}^* \\ a_{13}^* & a_{23}^* & a_{33}^* \end{bmatrix}, \tag{30}$$

where a_{ij}^* is the complex conjugate, in the usual sense, of a_{ij}. To find the complex conjugate of a matrix, interchange rows and columns and replace i by minus i, everywhere that i occurs.

Thus, for the a of equation (28), we get

$$a^\dagger = \begin{bmatrix} 1 & -4i & 1+3i \\ 1-i & 3-2i & 7 \\ 0 & 1 & 4+3i \end{bmatrix}. \tag{31}$$

Note that the product,

$$aa^\dagger = a^\dagger a, \tag{32}$$

is real.

If

$$a = a^\dagger, \tag{33}$$

the matrix is Hermitian. An example of an Hermitian matrix is

$$a = \begin{bmatrix} 4 & 3i & 1-2i \\ -3i & -2 & -6 \\ 1+2i & -6 & 0 \end{bmatrix}. \tag{34}$$

The main diagonal is real and elements symmetrically located above or below the diagonal are complex conjugates of one another.

Let us return to the properties of various matrices. First of all, in matrix multiplication, the factors do not ordinarily commute. In fact, some of the commutations are meaningless. For example,

$$\begin{bmatrix} x_1 \\ x_2 \\ x_3 \end{bmatrix} \begin{bmatrix} a_{11} & a_{12} & a_{13} \\ a_{21} & a_{22} & a_{23} \\ a_{31} & a_{32} & a_{33} \end{bmatrix} \tag{35}$$

has no interpretation in terms of our notation. The two matrices are not conformable. The number of columns of the first is not equal to the number of rows of the second. However, we do note that

$$[x_1 \quad x_2 \quad x_3] \begin{bmatrix} a_{11} & a_{21} & a_{31} \\ a_{12} & a_{22} & a_{32} \\ a_{13} & a_{23} & a_{33} \end{bmatrix} = \tilde{x}\tilde{a} = \widetilde{ax}, \tag{36}$$

$$ax = \widetilde{\tilde{x}\tilde{a}}. \tag{37}$$

More generally, we have for complex matrices,

$$ax = (a^{\dagger} x^{\dagger})^{\dagger}. \tag{38}$$

To give some of our matrix procedures definite geometrical meaning, let us now consider the dyadic:

$$\begin{aligned} \mathbf{\Phi} &= \begin{bmatrix} a_{11}\mathbf{ii} + a_{12}\mathbf{ij} + a_{13}\mathbf{ik} \\ + a_{21}\mathbf{ji} + a_{22}\mathbf{jj} + a_{23}\mathbf{jk} \\ + a_{31}\mathbf{ki} + a_{32}\mathbf{kj} + a_{33}\mathbf{kk} \end{bmatrix} \\ &= \mathbf{B}_1\mathbf{i} + \mathbf{B}_2\mathbf{j} + \mathbf{B}_3\mathbf{k}. \end{aligned} \tag{39}$$

We have substituted the vector quantities $\mathbf{B}_1$, $\mathbf{B}_2$, $\mathbf{B}_3$, for their equivalents,

$$\begin{aligned} \mathbf{B}_1 &= \mathbf{i}a_{11} + \mathbf{j}a_{21} + \mathbf{k}a_{31} = \mathbf{\Phi} \cdot \mathbf{i}. \\ \mathbf{B}_2 &= \mathbf{i}a_{12} + \mathbf{j}a_{22} + \mathbf{k}a_{32} = \mathbf{\Phi} \cdot \mathbf{j}. \\ \mathbf{B}_3 &= \mathbf{i}a_{13} + \mathbf{j}a_{23} + \mathbf{k}a_{33} = \mathbf{\Phi} \cdot \mathbf{k}. \end{aligned} \tag{40}$$

Let us regard (40) as three equations, to be solved simultaneously for the vectors $\mathbf{i}$, $\mathbf{j}$, and $\mathbf{k}$.

$$\mathbf{i} = \frac{\begin{vmatrix} \mathbf{B}_1 & \mathbf{B}_2 & \mathbf{B}_3 \\ a_{21} & a_{22} & a_{23} \\ a_{31} & a_{32} & a_{33} \end{vmatrix}}{\begin{vmatrix} a_{11} & a_{12} & a_{13} \\ a_{21} & a_{22} & a_{23} \\ a_{31} & a_{32} & a_{33} \end{vmatrix}}, \quad \mathbf{j} = \frac{\begin{vmatrix} \mathbf{B}_1 & \mathbf{B}_2 & \mathbf{B}_3 \\ a_{11} & a_{12} & a_{13} \\ a_{31} & a_{32} & a_{33} \end{vmatrix}}{\begin{vmatrix} a_{11} & a_{12} & a_{13} \\ a_{21} & a_{22} & a_{23} \\ a_{31} & a_{32} & a_{33} \end{vmatrix}}, \quad \text{ETC.} \tag{41}$$

In the notation of equation (15), we expand the determinant of the numerator, getting

$$\mathbf{i} = \frac{1}{\Delta_a}(A_{11}\mathbf{B}_1 + A_{12}\mathbf{B}_2 + A_{13}\mathbf{B}_3),$$

$$\mathbf{j} = \frac{1}{\Delta_a}(A_{21}\mathbf{B}_1 + A_{22}\mathbf{B}_2 + A_{23}\mathbf{B}_3), \tag{42}$$

$$\mathbf{k} = \frac{1}{\Delta_a}(A_{31}\mathbf{B}_1 + A_{32}\mathbf{B}_2 + A_{33}\mathbf{B}_3),$$

where Δ_a is the determinant of the denominator of (41). Also, let us define the reciprocal dyadic

$$\mathbf{\Phi}^{-1} = \mathbf{ib}^1 + \mathbf{jb}^2 + \mathbf{kb}^3, \tag{43}$$

so that

$$\mathbf{\Phi} \cdot \mathbf{\Phi}^{-1} = \mathfrak{I}. \tag{44}$$

The superscripts of b are *indices*, not exponents. By equations (39) and (43),

$$\begin{aligned} \mathbf{\Phi} \cdot \mathbf{\Phi}^{-1} &= \mathbf{B}_1\mathbf{b}^1 + \mathbf{B}_2\mathbf{b}^2 + \mathbf{B}_3\mathbf{b}^3 \\ &= \begin{bmatrix} \mathbf{b}^1 \cdot \mathbf{B}_1\mathbf{ii} + \mathbf{b}^1 \cdot \mathbf{B}_2\mathbf{ij} + \mathbf{b}^1 \cdot \mathbf{B}_3\mathbf{ik} \\ + \mathbf{b}^2 \cdot \mathbf{B}_1\mathbf{ji} + \mathbf{b}^2 \cdot \mathbf{B}_2\mathbf{jj} + \mathbf{b}^2 \cdot \mathbf{B}_3\mathbf{jk} \\ + \mathbf{b}^3 \cdot \mathbf{B}_1\mathbf{ki} + \mathbf{b}^3 \cdot \mathbf{B}_2\mathbf{kj} + \mathbf{b}^3 \cdot \mathbf{B}_3\mathbf{kk} \end{bmatrix} \\ &= \mathbf{\Phi}^{-1} \cdot \mathbf{\Phi} = \mathbf{ii} + \mathbf{jj} + \mathbf{kk} = \mathfrak{I}. \end{aligned} \tag{45}$$

Equating coefficients of the two dyadics, we get

$$\mathbf{b}^1 \cdot \mathbf{B}_1 = \mathbf{b}^2 \cdot \mathbf{B}_2 = \mathbf{b}^3 \cdot \mathbf{B}_3 = 1.$$

$$\mathbf{b}^1 \cdot \mathbf{B}_2 = \mathbf{b}^2 \cdot \mathbf{B}_1 = \mathbf{b}^2 \cdot \mathbf{B}_3 = \mathbf{b}^3 \cdot \mathbf{B}_2 = \mathbf{b}^3 \cdot \mathbf{B}_1 = \mathbf{b}^1 \cdot \mathbf{B}_3 = 0. \tag{46}$$

The vector $\mathbf{b}^1$ is perpendicular to vectors $\mathbf{B}_2$ and $\mathbf{B}_3$. It is not necessarily parallel to $\mathbf{B}_1$, however.

$$\mathbf{b}^1 \cdot \mathbf{i} = \frac{A_{11}}{\Delta_a}\mathbf{b}^1 \cdot \mathbf{B}_1 = \frac{A_{11}}{\Delta_a},$$

$$\mathbf{b}^2 \cdot \mathbf{i} = \frac{A_{12}}{\Delta_a}, \quad \mathbf{b}^1 \cdot \mathbf{j} = \frac{A_{21}}{\Delta_a}, \quad \text{ETC.} \tag{47}$$

Therefore

$$\mathbf{b}^1 = \frac{1}{\Delta_a}(A_{11}\mathbf{i} + A_{21}\mathbf{j} + A_{31}\mathbf{k}),$$

$$\mathbf{b}^2 = \frac{1}{\Delta_a}(A_{12}\mathbf{i} + A_{22}\mathbf{j} + A_{32}\mathbf{k}), \tag{48}$$

$$\mathbf{b}^3 = \frac{1}{\Delta_a}(A_{13}\mathbf{i} + A_{23}\mathbf{j} + A_{33}\mathbf{k}),$$

and
$$\Phi^{-1} = \frac{1}{\Delta_a}\begin{bmatrix} A_{11} & A_{21} & A_{31} \\ A_{12} & A_{22} & A_{32} \\ A_{13} & A_{23} & A_{33} \end{bmatrix} = \frac{1}{\Delta_a}\Phi_{\text{adj}} \,. \tag{49}$$

We could have inferred this result directly from equation (20), since the elements of the dyadic transform like those of a matrix. The alternative demonstration should be useful in guiding the student to an understanding of the relationship between matrices and dyadics.

Let the determinant

$$\Delta_A = \begin{vmatrix} A_{11} & A_{21} & A_{31} \\ A_{12} & A_{22} & A_{32} \\ A_{13} & A_{13} & A_{33} \end{vmatrix}. \tag{50}$$

We shall prove the relation that

$$\Delta_a^2 = \Delta_A \,. \tag{51}$$

Write Δ_a^2 as $\Delta_a\Delta_a$, and expand in minors around the first and second rows respectively. Assume that (51) is correct. Then

$$\begin{aligned}\Delta_a^2 &= (a_{11}A_{11} + a_{12}A_{12} + a_{13}A_{13})(a_{21}A_{21} + a_{22}A_{22} + a_{23}A_{23}) \\ &= \Delta_A = A_{11}(A_{22}A_{32} - A_{32}A_{23}) + A_{21}(A_{32}A_{13} - A_{12}A_{33}) \\ &\qquad + A_{31}(A_{12}A_{23} - A_{22}A_{13}).\end{aligned} \tag{52}$$

On the right-hand side substitute the equivalents

$$\begin{aligned} A_{31} &= a_{12}a_{23} - a_{13}a_{22} \,. \\ A_{32} &= a_{13}a_{21} - a_{11}a_{23} \,. \\ A_{33} &= a_{11}a_{22} - a_{12}a_{21} \,. \end{aligned} \tag{53}$$

Multiply out and cancel. The remaining nine terms factor as follows:

$$(a_{11}A_{21} + a_{12}A_{22} + a_{13}A_{23})(a_{21}A_{11} + a_{22}A_{12} + a_{23}A_{13}) = 0. \tag{54}$$

Both of these factors are zero. The first, for example, becomes

$$-a_{11}\begin{vmatrix} a_{12} & a_{13} \\ a_{32} & a_{33} \end{vmatrix} + a_{12}\begin{vmatrix} a_{11} & a_{13} \\ a_{31} & a_{33} \end{vmatrix} - a_{13}\begin{vmatrix} a_{11} & a_{12} \\ a_{31} & a_{32} \end{vmatrix} = -\begin{vmatrix} a_{11} & a_{12} & a_{13} \\ a_{11} & a_{12} & a_{13} \\ a_{31} & a_{32} & a_{33} \end{vmatrix} = 0, \tag{55}$$

because the first two rows are identical. The identity (51) is therefore established,

The three vectors $\mathbf{b}^1$, $\mathbf{b}^2$, and $\mathbf{b}^3$ define a parallelopiped whose volume is

$$v = \mathbf{b}^1 \cdot \mathbf{b}^2 \times \mathbf{b}^3 = \Delta_A/\Delta_a^3 , \tag{56}$$

by the rule (21.19) for the triple scalar product. Similarly, the vectors $\mathbf{B}_1$, $\mathbf{B}_2$, and $\mathbf{B}_3$ define the volume

$$V = \mathbf{B}_1 \cdot \mathbf{B}_2 \times \mathbf{B}_3 = \Delta_a . \tag{57}$$

The relation (51) shows that

$$vV = 1 \tag{58}$$

The vectors $\mathbf{b}^1$, $\mathbf{b}^2$, $\mathbf{b}^3$ and $\mathbf{B}_1$, $\mathbf{B}_2$, $\mathbf{B}_3$ define so-called reciprocal systems.

30. Matrix and dyadic transformations. In general, as we have seen, the operation of a dyadic $\mathbf{\Phi}$ upon a vector $\mathbf{r}$ results in a distortion of $\mathbf{r}$ both in direction and magnitude. There are, however, certain unique orientations of the vector $\mathbf{r}$, for which the distortion may be one of stretching alone. We call such values of $\mathbf{r}$ "eigenvectors" or "eigenrays" of the associated dyadic or matrix. The condition, thus expressed, is

$$\mathbf{\Phi} \cdot \mathbf{r} = \lambda \mathbf{r}, \tag{1}$$

where λ is a numerical constant. We can write the equation in the form

$$\mathbf{\Phi} \cdot \mathbf{r} - \lambda \mathbf{r} = \mathbf{\Phi} \cdot \mathbf{r} - \lambda \mathfrak{I} \cdot \mathbf{r} = (\mathbf{\Phi} - \lambda \mathfrak{I}) \cdot \mathbf{r} = 0. \tag{2}$$

Where $\mathbf{\Phi}$ has the form of equation (28.7), we can write (2), in matrix notation, as

$$\begin{bmatrix} a_{11} - \lambda & a_{12} & a_{13} \\ a_{21} & a_{22} - \lambda & a_{23} \\ a_{31} & a_{32} & a_{33} - \lambda \end{bmatrix} \cdot \begin{bmatrix} x \\ y \\ z \end{bmatrix} = 0. \tag{3}$$

Comparing these equations with (29.2), which are similar in form except for the fact that the latter have a non-zero value on the right, we see that the determinantal solution of (3) must be

$$x = \frac{\begin{vmatrix} 0 & 0 & 0 \\ a_{12} & a_{22} - \lambda & a_{32} \\ a_{13} & a_{23} & a_{33} - \lambda \end{vmatrix}}{\begin{vmatrix} a_{11} - \lambda & a_{12} & a_{13} \\ a_{21} & a_{22} - \lambda & a_{23} \\ a_{31} & a_{32} & a_{33} - \lambda \end{vmatrix}}, \quad \text{ETC., FOR } y \text{ AND } z. \tag{4}$$

The numerator is equal to zero. Hence, if we are obtaining a solution other than the trivial expression, $x = y = z = 0$, the determinant in the denominator must also vanish. This relationship, known as the "secular equation,"

$$\begin{vmatrix} a_{11}-\lambda & a_{12} & a_{13} \\ a_{21} & a_{22}-\lambda & a_{23} \\ a_{31} & a_{32} & a_{33}-\lambda \end{vmatrix} = 0, \tag{5}$$

is a cubic in λ. Solving it, we obtain three roots, λ_1, λ_2, and λ_3.

Substitute one of these roots, say λ_1, into equation (3). The three resultant equations, which we write out for purposes of demonstration, are

$$\begin{aligned} (a_{11}-\lambda_1)x + a_{12}y + a_{13}z &= 0. \\ a_{21}x + (a_{22}-\lambda_1)y + a_{23}z &= 0. \\ a_{31}x + a_{32}y + (a_{33}-\lambda_1)z &= 0. \end{aligned} \tag{6}$$

Because we have employed (5) to determine λ_1, these equations are not linearly independent. We may assign any arbitrary value we choose to one of the three variables x, y, or z, and determine the two remaining parameters from any pair of the equations. The third merely provides a numerical check on the process.

Let us consider a concrete example, by determining the eigenrays of the dyadic or matrix:

$$\Phi = \begin{bmatrix} 11 & -6 & 2 \\ -6 & 10 & -4 \\ 2 & -4 & 6 \end{bmatrix}. \tag{7}$$

The determinant of the matrix

$$\begin{vmatrix} 11-\lambda & -6 & 2 \\ -6 & 10-\lambda & -4 \\ 2 & -4 & 6-\lambda \end{vmatrix} = 0, \tag{8}$$

leads to the secular equation

$$324 - 180\lambda + 27\lambda^2 - \lambda^3 = 0, \tag{9}$$

which has the roots $\lambda_1 = 3$; $\lambda_2 = 6$; $\lambda_3 = 18$.

Thus, for λ_1, we get the equivalent of (6):

$$\begin{aligned} 8x_1 - 6y_1 + 2z_1 &= 0. \\ -6x_1 + 7y_1 - 4z_1 &= 0. \\ 2x_1 - 4y_1 + 3z_1 &= 0. \end{aligned} \tag{10}$$

Set $x_1 = c$; then we find that $y_1 = z_1 = 2c$. Our first eigenray is the vector with components

$$\begin{bmatrix} x \\ y \\ z \end{bmatrix} = \begin{bmatrix} 1 \\ 2 \\ 2 \end{bmatrix} c,$$

where c is some arbitrary constant, which we temporarily set equal to unity, for convenience. Later on we shall set $c = 1/\sqrt{1 + 2^2 + 2^2} = 1/3$ so that the eigenray is of unit length.

As a check on our calculations, we readily find that

$$\begin{bmatrix} 11 & -6 & 2 \\ -6 & 10 & -4 \\ 2 & -4 & 6 \end{bmatrix} \begin{bmatrix} 1 \\ 2 \\ 2 \end{bmatrix} = \begin{bmatrix} 3 \\ 6 \\ 6 \end{bmatrix} = 3 \begin{bmatrix} 1 \\ 2 \\ 2 \end{bmatrix}, \tag{11}$$

in accordance with our original condition (1). The vector $\{1, 2, 2\}$ from the origin or any other vector in this direction, irrespective of its length, when operated on by the given matrix, increases its length by a factor of 3.

Now, proceeding in analogous fashion for λ_2 and λ_3 we determine two other eigenrays, respectively $\{2, 1, -2\}$ and $\{2, -2, 1\}$. These three vectors,

$$\begin{aligned} \mathbf{r}_1 &= \mathbf{i} + 2\mathbf{j} + 2\mathbf{k}, \\ \mathbf{r}_2 &= 2\mathbf{i} + \mathbf{j} - 2\mathbf{k}, \\ \mathbf{r}_3 &= 2\mathbf{i} - 2\mathbf{j} + \mathbf{k}, \end{aligned} \tag{12}$$

are mutually perpendicular, because the scalar product of one with either of the others is zero. The vectors will be orthogonal, however, only when the matrices are Hermitian.

Now let us form a matrix (or dyadic) from each of our three sets of eigenrays:

$$\mathbf{K} = \begin{bmatrix} x_1 & x_2 & x_3 \\ y_1 & y_2 & y_3 \\ z_1 & z_2 & z_3 \end{bmatrix}. \tag{13}$$

Because of the special character of these vectors, the product

$$\mathbf{\Phi} \cdot \mathbf{K} = \begin{bmatrix} \lambda_1 x_1 & \lambda_2 x_2 & \lambda_3 x_3 \\ \lambda_1 y_1 & \lambda_2 y_2 & \lambda_3 y_3 \\ \lambda_1 z_1 & \lambda_2 z_2 & \lambda_3 z_3 \end{bmatrix}. \tag{14}$$

Each column of this matrix fulfills the condition (1). Let $\mathbf{\Lambda}$ be the diagonal matrix of the eigenvalues.

$$\mathbf{\Lambda} = \begin{bmatrix} \lambda_2 & 0 & 0 \\ 0 & \lambda_2 & 0 \\ 0 & 0 & \lambda_3 \end{bmatrix}. \tag{15}$$

Then one readily proves the identity

$$\mathbf{\Phi} \cdot \mathbf{K} = \mathbf{K} \cdot \mathbf{\Lambda}, \tag{16}$$

by direct substitution. Multiplying both sides of this equation from the right by the reciprocal matrix, $\mathbf{K}^{-1}$, we have

$$\mathbf{\Phi K K}^{-1} = \mathbf{\Phi} = \mathbf{K} \cdot \mathbf{\Lambda} \cdot \mathbf{K}^{-1}. \tag{17}$$

The square of a matrix,

$$\begin{aligned} \mathbf{\Phi}^2 &= \mathbf{K\Lambda K}^{-1}\mathbf{K\Lambda K}^{-1} = \mathbf{K\Lambda}^2\mathbf{K}^{-1} \\ &= \mathbf{K} \cdot \begin{bmatrix} \lambda_1^2 & 0 & 0 \\ 0 & \lambda_2^2 & 0 \\ 0 & 0 & \lambda_3^2 \end{bmatrix} \cdot \mathbf{K}^{-1}, \end{aligned} \tag{18}$$

and similarly for any higher power of a matrix. In fact, if we have some function, a polynomial in $\mathbf{\Phi}$, such as

$$P(\mathbf{\Phi}) = a\mathbf{I} + b\mathbf{\Phi} + c\mathbf{\Phi}^2 + d\mathbf{\Phi}^3 + \ldots, \tag{19}$$

we can express it in terms of the related polynomial,

$$P(\mathbf{\Lambda}) = \begin{bmatrix} P(\lambda_1) & 0 & 0 \\ 0 & P(\lambda_2) & 0 \\ 0 & 0 & P(\lambda_3) \end{bmatrix}, \tag{20}$$

by the equation

$$P(\mathbf{\Phi}) = \mathbf{K} \cdot [P(\mathbf{\Lambda})] \cdot \mathbf{K}^{-}. \tag{21}$$

For $P(\lambda)$, the secular equation, we adopt the form

$$P(\lambda) = \begin{vmatrix} a_{11} - \lambda & a_{12} & a_{13} \\ a_{21} & a_{22} - \lambda & a_{23} \\ a_{31} & a_{32} & a_{33} - \lambda \end{vmatrix} = a + b\lambda + b\lambda^2 + \ldots, \tag{22}$$

so that

$$\begin{aligned} P(\mathbf{\Phi}) &= \begin{vmatrix} a_{11}\mathbf{I} - \mathbf{\Phi} & a_{12}\mathbf{I} & a_{13}\mathbf{I} \\ a_{21}\mathbf{I} & a_{22}\mathbf{I} - \mathbf{\Phi} & a_{23}\mathbf{I} \\ a_{31}\mathbf{I} & a_{32}\mathbf{I} & a_{33}\mathbf{I} - \mathbf{\Phi} \end{vmatrix} \\ &= a\mathbf{I} + b\mathbf{\Phi} + c\mathbf{\Phi}^2 + \ldots \end{aligned} \tag{23}$$

Then, by (20), we obtain the result that

$$P(\mathbf{\Phi}) = \mathbf{K}[P(\mathbf{\Lambda})]\mathbf{K}^{-1} = \mathbf{K} \cdot \begin{bmatrix} P(\lambda_1) & 0 & 0 \\ 0 & P(\lambda_2) & 0 \\ 0 & 0 & P(\lambda_3) \end{bmatrix} \cdot \mathbf{K}^{-1}$$

$$= \mathbf{K} \cdot \begin{bmatrix} 0 & 0 & 0 \\ 0 & 0 & 0 \\ 0 & 0 & 0 \end{bmatrix} \cdot \mathbf{K}^{-1}, \tag{24}$$

since $P(\lambda) = 0$ for the values λ_1, λ_2, and λ_3. Thus

$$P(\mathbf{\Phi}) = \mathbf{0}, \tag{25}$$

where $\mathbf{0}$ is the zero matrix. This equation is a matrix equation whose algebraic form is identical with that of the secular equation, with the matrix $\mathbf{\Phi}$ substituted for the λ of the original and $\mathbf{I}$ (unit matrix) introduced as a factor for the term independent of $\mathbf{\Phi}$. We call the equation thus transformed from the variable λ to the matrix variable $\mathbf{\Phi}$, the "Hamilton-Cayley equation." Speaking somewhat loosely, we may say that $\mathbf{\Phi}$ satisfies its own secular equation.

Although we have carried out the proofs of the foregoing matrix-dyadic relationships in terms of square three-by-three matrices, the nature of the proofs in no way depends upon this limitation. We can apply the results directly to a square matrix of any number of columns.

In many physical problems, the diagonalization of matrices is of great importance. The general $\mathbf{K}$ matrix, whose columns consist of the vector components of the eigenrays, plays a significant role in matrix mechanics. We term such an array a "modal or transformation matrix."

We may adopt one convention that will uniquely determine, except for sign, the basic eigenvectors. We choose the constant c, (equation 10 *et. seq.*) in such a way that the vector has unit length. For the numerical example given above, we can set $c = 1/3$. Then

$$\mathbf{K} = \frac{1}{3}\begin{bmatrix} 1 & 2 & 2 \\ 2 & 1 & -2 \\ 2 & -2 & 1 \end{bmatrix}. \tag{26}$$

The factor, 1/3, by definition, multiples each of the matrix components. The matrix, so determined, is a special example of a unitary matrix. The individual components are the direction cosines of the unit vectors. When $\mathbf{K}$ is complex, we say that it is a unitary matrix when $\mathbf{K}\mathbf{K}^{\dagger} = \mathbf{K}\mathbf{K}^{-1} = \mathbf{I}$, cf. equations (30) and (33). In words, we term $\mathbf{K}$ unitary when its Hermitian conjugate is also its reciprocal.

When the vectors comprising the matrix are orthogonal, we can immediately write the reciprocal by mere interchange of rows and columns. Thus

$$\mathbf{K}^{-1} = \frac{1}{3}\begin{bmatrix} 1 & 2 & 2 \\ 2 & 1 & -2 \\ 2 & -2 & 1 \end{bmatrix} = \mathbf{K}. \tag{27}$$

In the foregoing example, the matrix is its own reciprocal. The reader may wish to check the fact that

$$\mathbf{K} \cdot \mathbf{K}^{-1} = \mathbf{I}. \tag{28}$$

He will also wish to test the relationship of equation (24) as well as its inverse, viz., that

$$\mathbf{K}^{-1} \cdot \mathbf{\Phi} \cdot \mathbf{K} = \mathbf{\Lambda}. \tag{29}$$

Under a unitary transformation, the "trace" of the matrix, viz., the sum of the diagonal elements is constant and hence equal to the sum of the roots.

One sometimes finds that the matrix has a repeated root. When this condition obtains, the matrix may not be reducible to diagonal form. However, one may, in some cases, find two independent vectors that will satisfy the transformation matrix. Any Hermitian or unitary matrix can always be diagonalized, but the transformation matrix is not necessarily unique.

Matrices or dyadics often have a mechanical or geometrical interpretation. A rigid body having axes of symmetry has certain mechanical properties that we can express in dyadic form. Usually the dyadic is symmetrical. By a proper choice of axes we can express the dyadic in diagonal form:

$$\mathbf{\Phi} = a_{11}\mathbf{ii} + a_{22}\mathbf{jj} + a_{33}\mathbf{kk}. \tag{30}$$

In other words, we can reduce the general dyadic to principal axes. The procedure for effecting such a transformation is identical with that given above. The vectors of the modal matrix define the orientations of the principal axes. If, in (29), we take $\mathbf{\Phi}$ and $\mathbf{\Lambda}$ as any matrices satisfying the relation, whether $\mathbf{\Lambda}$ is diagonal or not, if $\mathbf{K}$ is a unitary matrix, we say that the operation is a "unitary transformation of the matrix $\mathbf{\Phi}$."

Tensor Analysis

31. General theory of tensors. Certain extensions and generalizations of the vector-tensor operations previously discussed lead to a shortened notation, though the individual operations are not thereby abbreviated or altered.

Consider the equations

$$q^1 = Q^1(x^1, x^2, x^3); \quad q^2 = Q^2(x^1, x^2, x^3); \quad q^3 = Q^3(x^1, x^2, x^3); \tag{1}$$

and $$x^1 = X^1(q^1, q^2, q^3); \quad x^2 = X^2(q^1, q^2, q^3); \quad x^3 = X^3(q^1, q^2, q^3); \tag{2}$$

where the superscripts are indices, not exponents. In this notation we employ x^1, x^2, x^3 as replacements for the more conventional rectangular Cartesian coordinates: x, y, z. The new notation has the advantage of symmetry. A single equation represents three above, *sic:*

$$q^i = Q^i(x^1, x^2, x^3), \tag{3}$$

$$x^i = X^i(q^1, q^2, q^3). \tag{4}$$

where i stands in turn for the numerals 1, 2, or 3.

These equations define a new system of curvilinear coordinates: q^1, q^2, q^3.

We now specify a vector, $\mathbf{r}$, in the conventional sense, such that

$$\mathbf{r} = \mathbf{i}_1 x^1 + \mathbf{i}_2 x^2 + \mathbf{i}_3 x^3, \tag{5}$$

where $\mathbf{i}_1$, $\mathbf{i}_2$, and $\mathbf{i}_3$ are the unit vectors. Then

$$\begin{aligned} d\mathbf{r} &= \left(\mathbf{i}_1 \frac{\partial x^1}{\partial q^1} + \mathbf{i}_2 \frac{\partial x^2}{\partial q^1} + \mathbf{i}_3 \frac{\partial x^3}{\partial q^1}\right) dq^1 \\ &\quad + \left(\mathbf{i}_1 \frac{\partial x^1}{\partial q^2} + \mathbf{i}_2 \frac{\partial x^2}{\partial q^2} + \mathbf{i}_3 \frac{\partial x^3}{\partial q^2}\right) dq^2 \\ &\quad + \left(\mathbf{i}_1 \frac{\partial x^1}{\partial q^3} + \mathbf{i}_2 \frac{\partial x^2}{\partial q^3} + \mathbf{i}_3 \frac{\partial x^3}{\partial q^3}\right) dq^3 \\ &= \mathbf{e}_1\, dq^1 + \mathbf{e}_2\, dq^2 + \mathbf{e}_3\, dq^3, \end{aligned} \tag{6}$$

wherein $\mathbf{e}_1$, $\mathbf{e}_2$, and $\mathbf{e}_3$ assume the role previously played by unit vectors. By virtue of (6),

$$\mathbf{e}_1 = \frac{\partial \mathbf{r}}{\partial q^1}; \quad \mathbf{e}_2 = \frac{\partial \mathbf{r}}{\partial q^2}; \quad \mathbf{e}_3 = \frac{\partial \mathbf{r}}{\partial q^3}. \tag{7}$$

We say that these vectors are "unitary," although they ordinarily do not possess unit length. We shall refer to them as the "basic" vectors of the new coordinate system. They depend, in general, upon the particular point in space.

In prior work we have usually introduced the condition that $\mathbf{e}_1$, $\mathbf{e}_2$, and $\mathbf{e}_3$ be orthogonal. We shall now drop this restriction and merely require that the three vectors be not coplanar. Our earlier studies have indicated the possibility of setting up a reciprocal vector system. The parallelopiped whose sides are the unitary vectors will possess a volume

$$V = \mathbf{e}_1 \cdot \mathbf{e}_2 \times \mathbf{e}_3, \tag{8}$$

equation (22.20). Now define a new set of three vectors, $\mathbf{e}^1$, $\mathbf{e}^2$, $\mathbf{e}^3$, such that

$$\mathbf{e}^1 = \mathbf{e}_2 \times \mathbf{e}_3/V; \quad \mathbf{e}^2 = \mathbf{e}_3 \times \mathbf{e}_1/V; \quad \mathbf{e}^3 = \mathbf{e}_1 \times \mathbf{e}_2/V. \tag{9}$$

Note that $\mathbf{e}^1$ is perpendicular to the plane defined by $\mathbf{e}_2$ and $\mathbf{e}_3$, etc. Let

$$\begin{aligned} v &= \mathbf{e}^1 \cdot \mathbf{e}^2 \times \mathbf{e}^3 = (\mathbf{e}_2 \times \mathbf{e}_3) \times [\mathbf{e}_3 \times \mathbf{e}_1] \cdot (\mathbf{e}_1 \times \mathbf{e}_2)/V^3 \\ &= [\mathbf{e}_3 \times \mathbf{e}_1] \cdot (\mathbf{e}_2\mathbf{e}_3 - \mathbf{e}_3\mathbf{e}_2) \cdot (\mathbf{e}_1 \times \mathbf{e}_2)/V^3 \\ &= \{[\mathbf{e}_3 \times \mathbf{e}_1] \cdot \mathbf{e}_2\}\{\mathbf{e}_3 \cdot (\mathbf{e}_1 \times \mathbf{e}_2)\} = (\mathbf{e}_1 \cdot \mathbf{e}_2 \times \mathbf{e}_3)^2/V^3 = 1/V, \end{aligned} \tag{10}$$

wherein we have expanded by means of the triple vector product. We thus get

$$vV = 1, \tag{11}$$

as in equation (29 58).

Associated with this system of reciprocal vectors is a new set of coordinates q_1, q_2, q_3. As we have seen, the coordinate curves q_1, like $\mathbf{e}^1$, lie perpendicular to the surface defined by the coordinates q^2 and q^3 or their associated unitary vectors $\mathbf{e}_2$ and $\mathbf{e}_3$. Similar conditions hold for q_2 and q_3. Any quantity that we wish to define, be it scalar, vector, or tensor, must be independent of the special coordinate system. However, its representation will depend on the particular system. We shall adopt as fundamental that system whose unitary vectors possess subscripts and whose coordinates q possess superscripts. Therefore

$$d\mathbf{r} = \mathbf{e}_1\, dq^1 + \mathbf{e}_2\, dq^2 + \mathbf{e}_3\, dq^3. \tag{12}$$

However, the reciprocal system, though secondary by definition, is equally satisfactory for representing vectors. In terms of the second system, the position vector becomes

$$d\mathbf{r} = \mathbf{e}^1\, dq_1 + \mathbf{e}^2\, dq_2 + \mathbf{e}^3\, dq_3. \tag{13}$$

We can write the two equations above as

$$d\mathbf{r} = \sum_{i=1}^{3} \mathbf{e}_i\, dq^i = \sum_{i=1}^{3} \mathbf{e}^i\, dq_i. \tag{14}$$

Note that the summation index, i, appears twice in each of the sums. We can further simplify our notation if we drop the summation sign, and agree to perform a summation, over a previously specified range, for all literal indices that appear twice in any one term. Here the range of summation is 1 to 3. With this "summation convention," we abbreviate the foregoing equations to

$$d\mathbf{r} = \mathbf{e}^i\, dq_i = \mathbf{e}_i\, dq^i. \tag{15}$$

Since the letter used for the summation index has no special significance, we can replace it at will by j, k, or any other convenient letter. We term such a letter a "dummy index."

The elementary scalar, $(dr)^2$, takes three alternative forms. The first, by equation (12), becomes

$$\begin{aligned}(dr)^2 = d\mathbf{r} \cdot d\mathbf{r} &= \mathbf{e}_1 \cdot \mathbf{e}_1\, dq^1\, dq^1 + \mathbf{e}_1 \cdot \mathbf{e}_2\, dq^1\, dq^2 + \mathbf{e}_1 \cdot \mathbf{e}_3\, dq^1\, dq^3 \\ &+ \mathbf{e}_2 \cdot \mathbf{e}_1\, dq^2\, dq^1 + \mathbf{e}_2 \cdot \mathbf{e}_2\, dq^2\, dq^2 + \mathbf{e}_2 \cdot \mathbf{e}_3\, dq^2\, dq^3 \\ &+ \mathbf{e}_3 \cdot \mathbf{e}_1\, dq^3\, dq^1 + \mathbf{e}_3 \cdot \mathbf{e}_2\, dq^3\, dq^2 + \mathbf{e}_3 \cdot \mathbf{e}_3\, dq^3\, dq^3 \\ &= \sum_{i=1}^{3} \sum_{j=1}^{3} \mathbf{e}_i \cdot \mathbf{e}_j\, dq^i\, dq^j = \mathbf{e}_i \cdot \mathbf{e}_j\, dq^i\, dq^j. \end{aligned} \tag{16}$$

To simplify this notation, we set

$$\mathbf{e}_i \cdot \mathbf{e}_j = g_{ij} = g_{ji}, \tag{17}$$

the scalar coefficient in the ith row and jth column of the expanded form shown above. We can, therefore, write

$$(dr)^2 = \sum_{i=1}^{3} \sum_{j=1}^{3} g_{ij}\, dq^i\, dq^j = g_{ij}\, dq^i\, dq^j, \tag{18}$$

by the summation convention. Analogously, from equation (13), we have

$$(dr)^2 = \sum_{i=1}^{3} \sum_{j=1}^{3} \mathbf{e}^i \cdot \mathbf{e}^j\, dq_i\, dq_j = g^{ij}\, dq_i\, dq_j. \tag{19}$$

The third form results when we take the dot product of (12) and (13). Then

$$(dr)^2 = \sum_{i=1}^{3} \sum_{j=1}^{3} \mathbf{e}_i \cdot \mathbf{e}^j\, dq^i\, dq_j = g_i^j\, dq^i\, dq_j. \tag{20}$$

The reciprocal system is based on the vectors of the original system. When $j = i$, then $\mathbf{e}^j = \mathbf{e}_r \times \mathbf{e}_s/V$, where r and s differ from i. Thus we find that

$$q_i^j = 1, \quad \text{when} \quad j = i. \tag{21}$$

Note that we cannot write (21) as g_i^i because this notation requires a summation, and is merely an abbreviation for $g_1^1 + g_2^2 + g_3^3 = 3$. When $j \neq i$, the cross product that replaces $\mathbf{e}^j$ must have $\mathbf{e}_i$ as one of its components. Hence, momentarily suspending the summation convention, we get

$$\mathbf{e}_i \cdot \mathbf{e}^j = (\mathbf{e}_i \cdot \mathbf{e}_i \times \mathbf{e}_s)/V = 0, \tag{22}$$

because we can interchange the dot and cross. In general,

$$g_i^j = \delta_i^j, \tag{23}$$

where

$$\delta^j_i = \begin{cases} 1 & j = i, \\ 0 & j \neq i, \end{cases} \tag{24}$$

the well-known "Kronecker delta."

Later on, in § (33), we shall introduce a related quantity h_i, specifically for an orthogonal curvilinear system,

$$\sqrt{g_{11}} = 1/h_1; \quad \sqrt{g_{22}} = 1/h_2; \quad \sqrt{g_{33}} = 1/h_3, \tag{25}$$

where h_i measures the length of the special unitary vector. Also, for orthogonal systems only,

$$g_{ij} = g_{ji} = 0; \qquad i \neq j. \tag{26}$$

Consider any vector **u**. We may express **u** in terms of the basic unitary vectors $\mathbf{e}_i$ as

$$\mathbf{u} = \mathbf{e}_i u^i, \tag{27}$$

or in terms of the reciprocal vectors $\mathbf{e}^j$ as

$$\mathbf{u} = \mathbf{e}^j u_j. \tag{28}$$

We say that (27) expresses **u** in terms of its *contravariant* components u^i, which term signifies that the index, i, in terms of its location as sub- or superscript, runs *counter to* or is unlike the index of the base vector $\mathbf{e}_i$. Analogously we say that (28) gives **u** in terms of its *covariant* components u_j, whose index is a subscript and therefore like that of the base vector $\mathbf{e}_i$. The vector **u**, being invariant by definition, is neither covariant nor contravariant.

Although we agree to use the two reciprocal sets of unit vectors, we shall henceforth employ only one coordinate system, viz., the q^i. Hence we express any vector or tensor field, irrespective of covariant or contravariant character, as a function of q^i.

Take the dot product of $\mathbf{e}^j$ into **u**. Then

$$\mathbf{e}^j \cdot \mathbf{u} = \mathbf{e}^j \cdot \mathbf{e}_i u^i = u^i \delta^j_i \tag{29}$$

or

$$\mathbf{e}^i \cdot \mathbf{u} = u_i. \tag{30}$$

Resubstituting this value of u^i into (27), we have

$$\mathbf{u} = \mathbf{e}_i u^i = \mathbf{e}_i \mathbf{e}^i \cdot \mathbf{u}. \tag{31}$$

Now, let

$$\mathbf{u} = \mathbf{e}^j, \tag{32}$$

and we get

$$\mathbf{e}^j = \mathbf{e}_i \mathbf{e}^i \cdot \mathbf{e}^j = \mathbf{e}_i g^{ij}. \tag{33}$$

Thus, if we have a vector given in terms of its covariant components we can immediately transform it to the contravariant type, a process called "raising the index." We substitute (33) into (28) and get

$$\mathbf{u} = \mathbf{e}^j u_j = \mathbf{e}_i g^{ij} u_j = \mathbf{e}_i u^i, \tag{34}$$

so that we may write

$$u^i = g^{ij} u_j. \tag{35}$$

We can omit the unitary vectors $\mathbf{e}_i$ from our equations and deal directly with the components. Note, however, that we can dispense with the unit vector only because we have indicated which component we are dealing with by its appropriate index, i, j, or equivalent, applied to the component itself. Many treatises on vector and tensor calculus do not even bother to express the basic vectors and confine their attention wholly to components.

By reversing the roles of equations (28) and (27) we obtain the equations for transforming contravariant to covariant components, i.e., 'for lowering the index."

$$\mathbf{u} = \mathbf{e}_i u^i = \mathbf{e}^j g_{ij} u^i = \mathbf{e}^j u_j. \tag{36}$$

Hence

$$u_j = g_{ij} u^i. \tag{37}$$

In matrix notation the transformation equations (37) and (35) become

$$\begin{bmatrix} u_1 \\ u_2 \\ u_3 \end{bmatrix} = \begin{bmatrix} g_{11} & g_{21} & g_{31} \\ g_{12} & g_{22} & g_{32} \\ g_{13} & g_{23} & g_{33} \end{bmatrix} \begin{bmatrix} u^1 \\ u^2 \\ u^3 \end{bmatrix} = G \begin{bmatrix} u^1 \\ u^2 \\ u^3 \end{bmatrix} \tag{38}$$

and

$$\begin{bmatrix} u^1 \\ u^2 \\ u^3 \end{bmatrix} = \begin{bmatrix} g^{11} & g^{21} & g^{31} \\ g^{12} & g^{22} & g^{32} \\ g^{13} & g^{23} & g^{33} \end{bmatrix} \begin{bmatrix} u_1 \\ u_2 \\ u_3 \end{bmatrix} = G' \begin{bmatrix} u_1 \\ u_2 \\ u_3 \end{bmatrix}. \tag{39}$$

These equations imply that

$$|G| = g = \text{DET } g_{ij} \neq 0, \quad \text{and} \quad |G'| = g' = \text{DET } g^{ij} \neq 0. \tag{40}$$

Furthermore, the reciprocal character of these equations requires that

$$\begin{bmatrix} u_1 \\ u_2 \\ u_3 \end{bmatrix} = (GG') \begin{bmatrix} u_1 \\ u_2 \\ u_3 \end{bmatrix}, \tag{41}$$

so that the matrix

$$GG' = I, \quad \text{or} \quad G' = G^{-1}. \tag{42}$$

In tensor notation, we have

$$g_{ir} g^{jr} = g_{ri} g^{rj} = g^{jr} g_{ir} = g^{rj} g_{ri} = \delta_i^j. \tag{43}$$

Let us examine the g_{ij} in greater detail. We have defined the system x^k, in equation (2) as Cartesian, so that the unit vectors $\mathbf{i}_k$ are independent of position. Then, by (5), (7), and (17) we have

$$\mathbf{r} = \mathbf{i}_k x^k. \tag{44}$$

$$\begin{aligned} g_{ij} &= \mathbf{e}_i \cdot \mathbf{e}_j = \frac{\partial \mathbf{r}}{\partial q^i} \cdot \frac{\partial \mathbf{r}}{\partial q^j} = \frac{\partial x^k}{\partial q^i}\frac{\partial x^l}{\partial q^j}\,\mathbf{i}_k \cdot \mathbf{i}_l \\ &= \frac{\partial x^k}{\partial q^i}\frac{\partial x^l}{\partial q^j}\,\delta_{kl} = \frac{\partial x^k}{\partial q^i}\frac{\partial x^k}{\partial q^j} = g_{ji}. \end{aligned} \tag{45}$$

In this equation the appearance of k twice as an index signifies the summation from 1 to 3 as before. The factor δ_{kl} is a Kronecker symbol.

If the system q^i is also orthogonal, g_{ij}, which is proportional to the cosine of the angle between the unitary vectors $\mathbf{e}_i$ and $\mathbf{e}_j$, must vanish when $i \neq j$. Similarly, if the system q^i is orthogonal, the reciprocal system must likewise be orthogonal.

$$g_{ij} = \begin{cases} 1/(h_i)^2, & j = i \\ 0 & j \neq i \end{cases} \tag{46}$$

where h_i is the factor defined in (25). Then

$$g^{ij} = \begin{cases} (h_i)^2 & j = i \\ 0 & j \neq i \end{cases} \tag{47}$$

For orthogonal systems, therefore, part of the basis for distinguishing between contravariant and covariant components disappears, because the unitary vectors of the basic and reciprocal systems are parallel. However, they need not have the same length.

The foregoing equations define the g_{ij}'s and g^{ij}'s, which provide a ready means of transforming a vector from covariant to contravariant form or vice versa, all within the same basic coordinate system, q^i. We now wish to transform a vector from one coordinate system to an entirely new one. For example, consider two different and independent sets of unit vectors $\mathbf{e}_i$ and $\bar{\mathbf{e}}_k$, associated with the respective coordinates q^i and $\bar{q}^k$. We are required to determine the rules for transforming a vector from one system to the other. Let

$$\mathbf{u} = \mathbf{e}_i u^i = \bar{\mathbf{e}}_k \bar{u}^k = \bar{\mathbf{u}} \tag{48}$$

in terms of contravariant indices. Of course $\mathbf{u}$ and $\bar{\mathbf{u}}$ are identical vectors. The use of the bar merely indicates that we are to adopt the components of $\mathbf{u}$ in the coordinate system of $\bar{q}$ rather than in q. Furthermore, suppose that we have the equations

$$q^i = Q^i(\bar{q}^1, \bar{q}^2, \bar{q}^3), \quad \bar{q}^k = \bar{Q}^k(q^1, q^2, q^3). \tag{49}$$

Differentiating these equations, we get

$$d\bar{q}^k = \frac{\partial \bar{q}^k}{\partial q^1} dq^1 + \frac{\partial \bar{q}^k}{\partial q^2} dq^2 + \frac{\partial \bar{q}^k}{\partial q^3} dq^3 = \frac{\partial \bar{q}^k}{\partial q^i} dq^i. \tag{50}$$

Now

$$d\mathbf{r} = \mathbf{e}_i \, dq^i = \bar{\mathbf{e}}_k \, d\bar{q}^k = \bar{\mathbf{e}}_k \frac{\partial \bar{q}^k}{\partial q^i} dq^i. \tag{51}$$

Hence

$$\mathbf{e}_i = \bar{\mathbf{e}}_k \frac{\partial \bar{q}^k}{\partial q^i}. \tag{52}$$

Substituting for $\mathbf{e}_i$ in (48), we get

$$\mathbf{u} = \bar{\mathbf{e}}_k u^i \frac{\partial \bar{q}^k}{\partial q^i} = \bar{\mathbf{e}}_k \bar{u}^k. \tag{53}$$

Thus the components of a contravariant vector obey the transformation law

$$\bar{u}^k = u^i \frac{\partial \bar{q}^k}{\partial q^i} = \gamma_i^k u^i, \tag{54}$$

where

$$\gamma_i^k = \frac{\partial \bar{q}^k}{\partial q^i}. \tag{55}$$

Note that equation (54) represents a set of linear equations in the u's which take the form

$$\begin{bmatrix} \bar{u}^1 \\ \bar{u}^2 \\ \bar{u}^3 \end{bmatrix} = \begin{bmatrix} \gamma_1^1 & \gamma_2^1 & \gamma_3^1 \\ \gamma_1^2 & \gamma_2^2 & \gamma_3^2 \\ \gamma_1^3 & \gamma_2^3 & \gamma_3^3 \end{bmatrix} \begin{bmatrix} u^1 \\ u^2 \\ u^3 \end{bmatrix} \tag{56}$$

in matrix notation. Where, in our earlier development, we limited such transformations to orthogonal systems, we now perceive that the linear matrix form applies also to oblique coordinate systems. For this representation to be significant, the determinant of the coefficients must not vanish, i.e.,

$$\gamma = \det \gamma_i^k = \begin{vmatrix} \gamma_1^1 & \gamma_2^1 & \gamma_3^1 \\ \gamma_1^2 & \gamma_2^2 & \gamma_3^2 \\ \gamma_1^3 & \gamma_2^3 & \gamma_3^3 \end{vmatrix} \neq 0. \tag{57}$$

Let $\bar{U}^\circ$ and U° represent the column matrices of (56), the symbol $^\circ$ indicating the contravariant nature of the component and let Γ be the transformation matrix, so that we can write

$$\bar{U}^\circ = \Gamma U^\circ. \tag{58}$$

The equation implies the existence of the reciprocal relation

$$U^\circ = \Gamma^{-1}\overline{U}^\circ = \tilde{C}\overline{U}^\circ, \tag{59}$$

where C is a new matrix whose transpose, obtained from an interchange of rows and columns, is the reciprocal of Γ. Our reason for this choice of notation will appear in a moment.

Now, we set

$$\begin{bmatrix} \overline{\mathbf{e}}_1 \\ \overline{\mathbf{e}}_2 \\ \overline{\mathbf{e}}_3 \end{bmatrix} = \begin{bmatrix} c_1^1 & c_1^2 & c_1^3 \\ c_2^1 & c_2^2 & c_2^3 \\ c_3^1 & c_3^2 & c_3^3 \end{bmatrix} \begin{bmatrix} \mathbf{e}_1 \\ \mathbf{e}_2 \\ \mathbf{e}_3 \end{bmatrix}. \tag{60}$$

Let $\overline{\mathbf{E}}_0$ and $\mathbf{E}_0$ represent the two column matrices and C the transformation matrix, above, so that

$$\overline{\mathbf{E}}_0 = C\mathbf{E}_0, \tag{61}$$

which equation corresponds to the abbreviated tensor form

$$\overline{\mathbf{e}}_k = \frac{\partial q^i}{\partial \overline{q}^k}\,\mathbf{e}_i = c_k^i \mathbf{e}_i. \tag{62}$$

As in (57), the determinant

$$c = \det c_k^i \neq 0. \tag{63}$$

Now, using (61) and (54) successively, we obtain

$$\mathbf{u} = \overline{\mathbf{e}}_k \overline{u}^k = c_k^i \mathbf{e}_i \overline{u}^k = c_k^i \mathbf{e}_i \gamma_j^k u^j = \mathbf{e}_i u^i. \tag{64}$$

Therefore

$$c_k^i \gamma_j^k u^j = u^i. \tag{65}$$

This condition requires that

$$c_k^i \gamma_j^k = \delta_j^i, \tag{66}$$

which equation is equivalent to the matrix equation previously written:

$$\tilde{C}\Gamma = \Gamma\tilde{C} = I, \tag{67}$$

where I is the unit matrix.

We shall now write

$$\overline{\mathbf{E}}^\circ = \Gamma \mathbf{E}^\circ, \tag{68}$$

as a matrix equation analogous to (61). In its transposed form the equation becomes

$$\{\overline{\mathbf{E}}^\circ\} = \{\mathbf{E}^\circ\}\tilde{\Gamma}, \tag{69}$$

where the symbol $\{\mathbf{E}^\circ\}$ now denotes a row matrix. Taking the scalar product, $\mathbf{e}_i \cdot \mathbf{e}^i$, we get the matrix

$$\overline{\mathbf{E}}_0\{\overline{\mathbf{E}}^\circ\} = I = C\mathbf{E}_0\{\mathbf{E}^\circ\}\tilde{\Gamma} = CI\tilde{\Gamma} = C\tilde{\Gamma} = I, \tag{70}$$

in view of the fact that

$$\mathbf{e}_i \cdot \mathbf{e}^j = \delta_i^j. \tag{71}$$

Analogous to (58), we also have

$$\overline{U}_0 = CU_0. \tag{72}$$

Let us now summarize and emphasize the most significant facts of the foregoing discussion of tensor formulae. Thus far we have limited ourselves to the vector $\mathbf{u}$, which must be invariant under any coordinate transformation. We adopt as basic some coordinate system wherein the base vectors, $\mathbf{e}_1$, $\mathbf{e}_2$, $\mathbf{e}_3$, are arbitrary except for the condition that they be independent, i.e., not coplanar. These basic vectors are, in general, functions of position. We further introduce the reciprocal vectors $\mathbf{e}^1$, $\mathbf{e}^2$, $\mathbf{e}^3$, respectively normal to the planes defined by the vector pairs (2, 3), (3, 1), and (1, 2). Although our formulas apply strictly only to three-dimensional space, we can readily generalize the basic concept of reciprocal vectors to a space of n dimensions.

We express the invariant $\mathbf{u}$ either in terms of its contravariant components u^i or its covariant components u_i:

$$\mathbf{u} = \mathbf{e}_i u^i = \mathbf{e}^j u_j. \tag{73}$$

Defining,

$$g^{ij} = \mathbf{e}^i \cdot \mathbf{e}^j = g^{ji}. \tag{74}$$

$$g_{ij} = \mathbf{e}_i \cdot \mathbf{e}_j = g_{ji}. \tag{75}$$

$$g_j^i = \mathbf{e}^i \cdot \mathbf{e}_j = \delta_j^i = g_{j\cdot}^{i}. \tag{76}$$

$$g^{ij} g_{jk} = g_{ij} g^{jk} = \delta_k^i = \delta_{i\cdot}^{k}. \tag{77}$$

We can always raise or lower the index of the unit vector by the transformation

$$\mathbf{e}^j = g^{ij}\mathbf{e}_i \quad \text{and} \quad \mathbf{e}_i = g_{ij}\mathbf{e}^j. \tag{78}$$

Similarly, for the components:

$$u^i = g^{ij}u_j \quad \text{and} \quad u_j = g_{ij}u^i. \tag{79}$$

If, now, we transform from our initial coordinates q^i to a new system $\bar{q}^k$, the unit vectors transform as follows:

$$\bar{\mathbf{e}}_k = c_k^i \mathbf{e}_i \tag{80}$$

and

$$\bar{\mathbf{e}}^k = \gamma_i^k \mathbf{e}^i, \tag{81}$$

where

$$c_k^i = \frac{\partial q^i}{\partial \bar{q}^k} \quad \text{and} \quad \gamma_i^k = \frac{\partial \bar{q}^k}{\partial q^i}, \tag{82}$$

subject to

$$c_k^i \gamma_j^k = \delta_j^i. \tag{83}$$

The associated vector components transform as:

contravariant, $$\bar{u}^k = \gamma^k_i u^i. \tag{84}$$

covariant, $$\bar{u}_k = c^i_k u_i. \tag{85}$$

We say that such transformations are *contragredient* when the matrices C and Γ obey equation (67). Note especially that, in c^i_k the new variable $\bar{q}^k$ appears in the denominator where, for the γ^k_i, $\bar{q}^k$ appears in the numerator. Some theorists prefer to use these transformation properties themselves as fundamental in defining covariant and contravariant components, without reference to the unitary vectors.

The above formulas, derived for vectors, provide the basic rules for transforming tensors in general. Consider, for example, a dyadic

$$\mathbf{T} = \begin{bmatrix} \mathbf{e}_1\mathbf{e}_1T^{11} + \mathbf{e}_1\mathbf{e}_2T^{12} + \mathbf{e}_1\mathbf{e}_3T^{13} \\ + \mathbf{e}_2\mathbf{e}_1T^{21} + \mathbf{e}_2\mathbf{e}_2T^{22} + \mathbf{e}_2\mathbf{e}_3T^{23} \\ + \mathbf{e}_3\mathbf{e}_1T^{31} + \mathbf{e}_3\mathbf{e}_2T^{32} + \mathbf{e}_3\mathbf{e}_3T^{33} \end{bmatrix}, \tag{86}$$

which takes the abbreviated form with our summation convention,

$$\mathbf{T} = \sum_{i=1}^{3} \sum_{j=1}^{3} \mathbf{e}_i\mathbf{e}_jT^{ij} = \mathbf{e}_i\mathbf{e}_jT^{ij}. \tag{87}$$

T, of course, is an invariant concept, but the components T^{ij} will change as we transform from one coordinate system to another, or when we shift to the reciprocal system of basic vectors. T^{ij}, by virtue of our previous definitions, we term a contravariant tensor. The rank or valence is 2, by which we mean that the components involve two indices. We can also express **T** in terms of the reciprocal basic vectors,

$$\mathbf{T} = \mathbf{e}^i\mathbf{e}^jT_{ij}, \tag{88}$$

where T_{ij} is a covariant tensor of rank 2. Or we can use some permutation of the two varieties, such as

$$\mathbf{T} = \mathbf{e}^i\mathbf{e}_jT_{i\cdot}^{\cdot j} = \mathbf{e}_i\mathbf{e}^jT^{i\cdot}_{\cdot j}. \tag{89}$$

Quantities like $T_{i\cdot}^{\cdot j}$ or $T^{i\cdot}_{\cdot j}$ we call "mixed" tensors. Since the order of sub- and superscripts may be important, we have put in the dots to make the order unambiguous, which it would not be if we merely wrote T^i_j.

At this point we may generalize still further, to consider tensors of even higher rank, e.g., the contravariant tensor T^{ijk}, the covariant one T_{ijk}, or any of the six mixed varieties such as $T^{ij\cdot}_{\cdot\cdot k}$, $T^{i\cdot\cdot}_{\cdot jk}$, . . . , etc.

We can raise or lower the index of any tensor component by merely applying the formulae for vectors, one index (or basic vector) at a time.

$$\begin{aligned} T^{ij\cdot}_{\cdot\cdot k} &= g_{kl}T^{ijl}, \\ T^{ijk} &= g^{km}T^{ij\cdot}_{\cdot\cdot m} = g^{km}g^{jl}T^{i\cdot\cdot}_{\cdot lm}, \quad \text{ETC.} \end{aligned} \tag{90}$$

Thus, given a tensor of any type, we can readily transform it, step by step, to any other desired type, but still of the same rank.

To transform from a coordinate system q to one $\bar{q}$, we make use of the relations (80)—(85). For a contravariant triadic tensor we have

$$\bar{T}^{ijk} = \gamma^i_r \gamma^j_s \gamma^k_t T^{rst}. \tag{91}$$

For the covariant type we have

$$\bar{T}_{ijk} = c^r_i c^s_j c^t_k T_{rst}. \tag{92}$$

The number of factors γ^i_r or c^r_i equals the rank of the tensor. For the mixed type we get

$$\bar{T}^{ij\cdot}_{\cdot\cdot k} = \gamma^i_r \gamma^j_s c^t_k T^{rs\cdot}_{\cdot\cdot t}. \tag{93}$$

For triadics and tensors of higher rank, the matrix notation is increasingly cumbersome, because the triadic itself comprises a cubic or hypercubic rather than the familiar square array of the dyadic, as in (86).

Consider the dyadic, $\mathbf{T}$, formed by multiplication of the two vectors

$$\mathbf{u} = \mathbf{e}_i u^i \quad \text{and} \quad \mathbf{v} = \mathbf{e}^j v_j, \tag{94}$$

so that

$$\mathbf{T} = \mathbf{e}_i \mathbf{e}^j u^i v_j = \mathbf{e}_i \mathbf{e}^j T^{i\cdot}_{\cdot j}. \tag{95}$$

The scalar product,

$$T = \mathbf{u} \cdot \mathbf{v} = \mathbf{e}_i \cdot \mathbf{e}^j T^{i\cdot}_{\cdot j} = \delta^j_i T^{i\cdot}_{\cdot j} = T^{i\cdot}_{\cdot i} = u^i v_i. \tag{96}$$

We term the foregoing process "contraction" of the tensor, and comment that the procedure involves "putting a dot" between some pair of the vectors. Before contracting the tensor, for a given pair of indices, we must transform the tensor, if necessary, so that one of the two indices is contravariant and the other covariant. Note that contraction of a tensor of rank two gives a scalar, i.e., a tensor of rank zero. Contraction of a tensor of rank three yields a tensor of rank one, i.e., a vector. Hence contraction always reduces the rank by two.

A knowledge of the quantities, $c^k_i \ldots, \gamma^r_i \ldots, g_{ij} \ldots, g^{kr}$, provides all the necessary information for transforming tensor components from one coordinate system to another. We still have the problem of differentiating a tensor with respect to one of the coordinates, e.g., q^i. Suppose that we are given the derivative

$$\frac{\partial T^{rs\ldots}_{tu\ldots}}{\partial q^i}, \tag{97}$$

and wish to determine its relation to the corresponding quantities in the new coordinate system $\bar{q}^k$, viz,

$$\frac{\partial \bar{T}^{rs\ldots}_{tu\ldots}}{\partial \bar{q}^k}. \tag{98}$$

As we shall subsequently see, neither of the quantities in (97) nor (98) is the component of a tensor.

To indicate the covariant nature of the differential operator, set

$$\frac{\partial}{\partial q^i} = D_i. \tag{99}$$

When D_i operates on a tensor component we obtain another component, $D_i T^{rs}_{tu}{}^{\cdot\cdot}_{\cdot\cdot}$, whose covariant rank is one unit higher than that of the original tensor.

Consider now the action of the operator D_i upon the vector $\mathbf{e}_j u^j$. Differentiate as a product

$$D_i(\mathbf{e}_j u^j) = \mathbf{e}_j D_i u^j + u^j D_i \mathbf{e}_j. \tag{100}$$

The operation $D_i u^j$ amounts to ordinary differentiation of the function u^j with respect to the coordinate q^i and hence presents no unfamiliar problems. However, to complete our knowledge of the derivatives we must further simplify the quantity $D_i \mathbf{e}_j$ and also determine its transformation characteristics.

In a Cartesian system, the direction of the unit vectors is invariant and the derivatives in question vanish. More generally, such a derivative gives an expression that depends on all the unit vectors of the system. Thus let us set

$$\begin{aligned} D_i \mathbf{e}_j &= \Gamma^1_{ij}\mathbf{e}_1 + \Gamma^2_{ij}\mathbf{e}_2 + \Gamma^3_{ij}\mathbf{e}_3 = \Gamma^k_{ij}\mathbf{e}_k \\ &= \Gamma^k_{ij} g_{kl} \mathbf{e}^l = \Gamma_{ij,k}\mathbf{e}^l, \end{aligned} \tag{101}$$

where, by definition,

$$\Gamma_{ij,l} = g_{kl}\Gamma^k_{ij}. \tag{102}$$

The Γ's, the so-called coefficients of the affine connection, are to be determined. The subscripts i and j indicate that they refer to the directions of the basic vectors $\mathbf{e}_i$ and $\mathbf{e}_j$; the index k or l indicates which unit vector applies in the expanded form. We should emphasize that these Γ's are *not* tensors, despite the notation. The reason that they are not tensors (except in special cases) lies in the fact that they do not obey the mathematical transformation laws for tensors. We shall therefore think of them as coefficients or merely as "symbols." Indeed, Γ^k_{ij} frequently appears in the literature as $\left\{ {k \atop ij} \right\}$, the so-called Christoffel symbol of the second kind and $\Gamma_{ij,l}$ as $[ij,l]$, the Christoffel symbol of the first kind. Multiplying (101) through by $\mathbf{e}^k \cdot$, we get

$$\Gamma^k_{ij} = \mathbf{e}^k \cdot D_i \mathbf{e}_j. \tag{103}$$

Now expand the product differential

$$D_i \mathbf{e}^k \cdot \mathbf{e}_j = D_i\ \delta^k_j = \mathbf{e}^k \cdot D_i \mathbf{e}_j + \mathbf{e}_j \cdot D_i \mathbf{e}^k = 0. \tag{104}$$

Hence $$\Gamma^k_{ij} = -\mathbf{e}_j \cdot D_i\mathbf{e}^k \tag{105}$$

and $$D_i\mathbf{e}^k = -\Gamma^k_{ij}\mathbf{e}^j. \tag{106}$$

$$\mathbf{e}_j = \frac{\partial \mathbf{r}}{\partial q^j} = D_j\mathbf{r}, \tag{107}$$

by equations (7) and (99). Then (105) becomes

$$\begin{aligned}\Gamma^k_{ij} &= \mathbf{e}^k \cdot D_iD_j\mathbf{r} = \mathbf{e}^k \cdot D_jD_i\mathbf{r} \\ &= \mathbf{e}^k \cdot D_j\mathbf{e}_i = \Gamma^k_{ji}.\end{aligned} \tag{108}$$

Hence the gammas are symmetrical in the two lower indices, ij. Consequently, by (102),

$$\Gamma_{ij,l} = \Gamma_{ji,l}. \tag{109}$$

Differentiate the product

$$D_i\mathbf{e}_l \cdot \mathbf{e}_j = D_ig_{jl} = \mathbf{e}_l \cdot D_i\mathbf{e}_j + \mathbf{e}_j \cdot D_i\mathbf{e}_l = \Gamma_{ij,l} + \Gamma_{il,j}. \tag{110}$$

Permuting the indices i, j, k in cyclic fashion, we get the equivalent expressions

$$D_jg_{li} = \Gamma_{jl,i} + \Gamma_{ji,l}. \tag{111}$$

$$D_lg_{ij} = \Gamma_{li,j} + \Gamma_{lj,i}. \tag{112}$$

Add equations (110) and (111), subtract (112), and make use of (109). Then

$$\left.\begin{aligned}\Gamma_{ij,l} &= \frac{1}{2}(D_ig_{jk} + D_jg_{li} - D_lg_{ij}) = g_{kl}\Gamma^k_{ij} \\ \text{and} \qquad \Gamma^k_{ij} &= g^{kl}\Gamma_{ij,l} = g^{kl}(D_ig_{jl} + D_jg_{li} - D_lg_{ij}).\end{aligned}\right\} \tag{113}$$

These equations illustrate, among various facts, the enormous compressive power of the tensor notation. For three-space, ijk (or l) independently assume the values 1, 2, and 3. Hence there are 27 values of $\Gamma_{ij,l}$. Because of the symmetry in ij, however, only 18 are distinct. More generally, for n space, the number of permutations of ij is $n(n + 1)/2$. Hence the total number of independent Γ^k_{ij}'s is $n^2(n + 1)/2$.

Since the g_{ij}'s are defined in terms of the adopted coordinate system, equations (124) and (125) permit calculation of the significant Γ's, however cumbersome actual computation may prove to be. In many applications, however, the symmetrical forms, with $\Gamma^k_{ij} = 0$, for $j \neq k$, assume special importance. From (17), (40), and (8), we have

$$g = \det g_{ij} = \det(\mathbf{e}_i \cdot \mathbf{e}_j) = (\mathbf{e}_1 \cdot \mathbf{e}_2 \times \mathbf{e}_3)^2 = V^2, \tag{114}$$

$$D_iV = D_i(\mathbf{e}_1 \cdot \mathbf{e}_2 \times \mathbf{e}_3) = (\Gamma^1_{i1} + \Gamma^2_{i2} + \Gamma^3_{i3})(\mathbf{e}_1 \cdot \mathbf{e}_2 \times \mathbf{e}_3) = \Gamma^k_{ik}\sqrt{g}. \tag{115}$$

all values of Γ vanishing except those for which $j = k$, because the triple scalar product vanishes when any two indices are identical. For a more detailed derivation of (114), see equation (137).

We are now ready to define the familiar operator ∇, in generalized coordinates. Since the differentiation is a covariant process, we must balance it by a contravariant base vector so that the operator will be an invariant, as required. Thus

$$\nabla = \bar{\mathbf{e}}^k \bar{D}_j = \mathbf{e}^i \gamma^k_j c^i_k D_i = \mathbf{e}^i \delta^i_j D_i = \mathbf{e}^i D_i, \tag{116}$$

by (80) and (98).

The simplest operation is that of ∇ upon an absolute scalar, ϕ.

$$\nabla \phi = \mathbf{e}^i D_i \phi = \bar{\mathbf{e}}^k D_k \phi = \mathbf{e}^i u_i = \mathbf{u}, \tag{117}$$

an invariant as required. This procedure immediately gives a vector $\mathbf{u}$ in terms of its covariant components. To obtain $\mathbf{u}$ in terms of its contravariant indices we must raise the index by methods previously given.

Now consider the operation, $\nabla \mathbf{u}$, with $\mathbf{u}$ written in the contravariant form

$$\nabla \mathbf{u} = \mathbf{e}^i D_i(\mathbf{e}_j u^j). \tag{118}$$

Differentiate as a product. Then, by (100) and (101) we get

$$\begin{aligned} \nabla \mathbf{u} &= \mathbf{e}^i \mathbf{e}_j (D_i u^j) + \mathbf{e}^i u^j \Gamma^k_{ij} \mathbf{e}_k \\ &= \mathbf{e}^i \mathbf{e}_j (D_i u^j + u^k \Gamma^j_{ik}), \end{aligned} \tag{119}$$

through an interchange of dummy indices.

Now, if we wish to calculate the divergence of $\mathbf{u}$, we contract the resulting tensor as follows:

$$\nabla \cdot \mathbf{u} = \mathbf{e}^i \cdot \mathbf{e}_j (D_i u^j + u^k \Gamma^j_{ik}) = (D_i + \Gamma^k_{ik}) u^i. \tag{120}$$

$$\nabla \cdot \mathbf{u} = D_i u^i + \frac{u^i}{\sqrt{g}} D_i \sqrt{g} = \frac{1}{\sqrt{g}} D_i(\sqrt{g}\, u^i), \tag{121}$$

by (115). To test the equivalence of these last two forms, merely differentiate $\sqrt{g}\, u^i$ as a product.

If $\mathbf{u}$ is the gradient of a scalar function, ϕ, as in (117), we raise the index of u_i to give

$$u^i = g^{ij} u_j = g^{ij} D_j \phi, \tag{122}$$

$$\nabla \cdot \mathbf{u} = \nabla \cdot \nabla \phi = \nabla^2 \phi = \frac{1}{\sqrt{g}} D_i(\sqrt{g}\, g^{ij} D_j \phi), \tag{123}$$

the Laplacian in generalized coordinates. Note that the summation is over both i and j, so that the generalized Laplacian possesses nine terms, unless orthogonality reduces the nu[illegible]

Now consider the operation $\nabla\mathbf{T}$, where $\mathbf{T}$ is a dyadic. First adopt the contravariant form of $\mathbf{T}$:

$$\nabla\mathbf{T} = \mathbf{e}^i D_i(\mathbf{e}_j\mathbf{e}_k T^{jk}). \tag{124}$$

Differentiate as a triple product, holding the alternative pairs of factors a constant. Then

$$\nabla\mathbf{T} = \mathbf{e}^i\mathbf{e}_j\mathbf{e}_k D_i T^{jk} + \mathbf{e}^i\mathbf{e}_l\mathbf{e}_k \Gamma^l_{ij} T^{jk} + \mathbf{e}^i\mathbf{e}_j\mathbf{e}_l \Gamma^l_{ik} T^{jk}. \tag{125}$$

Interchange the dummy indices j and l in the second term on the right-hand side of (125) and k and l in the third term. Hence

$$\nabla\mathbf{T} = \mathbf{e}^i\mathbf{e}_j\mathbf{e}_k[D_i T^{jk} + \Gamma^j_{il} T^{lk} + \Gamma^k_{il} T^{jl}]. \tag{126}$$

Note that the original vector prefix, $\mathbf{e}^i\mathbf{e}_j\mathbf{e}_k$, which obtained before the differentiation, has now reappeared as the result of the index shifting. We can further simplify our notation by writing

$$\nabla\mathbf{T} = \mathbf{e}^i\mathbf{e}_j\mathbf{e}_k \nabla_i T^{jk}, \tag{127}$$

where ∇_i operates only upon the components T^{jk}, with

$$\nabla_i T^{jk} = D_i T^{jk} + \Gamma^j_{il} T^{lk} + \Gamma^k_{il} T^{jl} = T^{jk}_{\cdot\cdot,i} \tag{128}$$

since the derivative is itself a tensor component. We call such a process "covariant differentiation."

To obtain the divergence of $\mathbf{T}$, contract the foregoing triadic on the first and last indices, i and k. Thus, since $\mathbf{e}^i \cdot \mathbf{e}_k = \delta^i_k$, we get

$$\begin{aligned}\nabla \cdot \mathbf{T} &= \mathbf{e}_j[D_i T^{ji} + \Gamma^j_{il} T^{li} + \Gamma^i_{il} T^{jl}] \\ &= \mathbf{e}_k\left[\frac{1}{\sqrt{g}} D_i(\sqrt{g}\, T^{ki}) + \Gamma^k_{li} T^{li}\right].\end{aligned} \tag{129}$$

Consider, now, the scalar volume $\bar{V}$ of the parallelopiped formed by the unitary vectors $\bar{\mathbf{e}}_1$, $\bar{\mathbf{e}}_2$, and $\bar{\mathbf{e}}_3$, in a new coordinate system. By (62) we get

$$\bar{V} = \bar{\mathbf{e}}_1 \cdot \bar{\mathbf{e}}_2 \times \bar{\mathbf{e}}_3 = c^i_1 c^j_2 c^k_3\, \mathbf{e}_i \cdot \mathbf{e}_j \times \mathbf{e}_k. \tag{130}$$

In the factor $\mathbf{e}_i \cdot \mathbf{e}_j \times \mathbf{e}_k$, each of the indices, by the summation convention, assumes in turn all values from 1 to 3. However, when any two of the indices are identical, e.g., when $i = 1$, $j = 1$, $k = 1$, 2, or 3 the triple scalar product vanishes, because $\mathbf{e}_1 \cdot \mathbf{e}_1 \times \mathbf{e}_k = \mathbf{e}_1 \times \mathbf{e}_1 \cdot \mathbf{e}_k = 0$, etc., by interchange of dot and cross. When the indices i, j, k are cyclic, the triple scalar product equals V. When the cyclic order reverses, the product equals $-V$. Therefore we can set

$$\mathbf{e}_i \cdot \mathbf{e}_j \times \mathbf{e}_k = \epsilon_{ijk}\mathbf{e}_1 \cdot \mathbf{e}_2 \times \mathbf{e}_3 = \epsilon_{ijk} V, \tag{131}$$

where

$$\begin{aligned}\epsilon_{123} &= \epsilon_{231} = \epsilon_{321} = 1, \\ \epsilon_{111} &= \epsilon_{112} = \epsilon_{113} = \epsilon_{122}, \text{ etc} = 0, \\ \epsilon_{132} &= \epsilon_{321} = \epsilon_{213} = -1.\end{aligned} \tag{132}$$

Then we can write, from (130)—(132),

$$\bar{V} = c_1^i c_2^j c_3^k \epsilon_{ijk} V. \tag{133}$$

Note that, from (60) and (63),

$$c = \begin{vmatrix} c_1^1 & c_1^2 & c_1^3 \\ c_2^1 & c_2^2 & c_2^3 \\ c_3^1 & c_3^2 & c_3^3 \end{vmatrix} = c_1^i c_2^j c_3^k \epsilon_{ijk}. \tag{134}$$

thus

$$\bar{V} = cV. \tag{135}$$

From (8), (33) and (11),

$$\begin{aligned} V = \mathbf{e}_1 \cdot \mathbf{e}_2 \times \mathbf{e}_3 &= g_{1i} g_{2j} g_{3k} \mathbf{e}^i \cdot \mathbf{e}^j \times \mathbf{e}^k \\ &= g_{1i} g_{2j} g_{3k} \epsilon^{ijk} \mathbf{e}^1 \cdot \mathbf{e}^2 \times \mathbf{e}^3 = gv, \end{aligned} \tag{136}$$

by a reasoning similar to that of equations (131)—(134). The symbol ϵ^{ijk} has properties identical with that of ϵ_{ijk}, equation (132). The quantity g is the determinant defined in (40). Using (11) we get

$$V = \sqrt{g}, \tag{137}$$

a relation previously given in equation (114).

Reconsider the elementary vector product $\mathbf{u} \times \mathbf{v}$, with the vectors $\mathbf{u}$ and $\mathbf{v}$ written in contravariant form:

$$\begin{aligned} \mathbf{u} \times \mathbf{v} = \mathbf{e}_i u^i \times \mathbf{e}_j v^j &= \mathbf{e}_i \times \mathbf{e}_j T^{ij} \\ &= \mathbf{e}^k \epsilon_{ijk} V T^{ij} = \mathbf{e}^k U_k, \end{aligned} \tag{138}$$

by (9) or (131). When we compute the vector product of a pair of contravariant vectors, we obtain a vector component wherein the volume element V appears in the numerator. Hence the quantity is not a vector component in the ordinary sense because V, as well as U_k, varies in any coordinate transformation.

Such quantities are not uncommon in physics. We do meet with certain quantities, like scalar potentials, which are invariant to any coordinate transformation. Such quantities are absolute scalars. On the other hand, a quantity like density (mass per unit volume) must depend on the coordinate system, because the volume element V may change with the coordinate system. An absolute scalar obeys the transformation law

$$\bar{\phi} = \phi. \tag{139}$$

A scalar density will contain the volume element in the denominator so that we have the transformation law

$$\bar{\phi}/\bar{V} = \phi/V. \tag{140}$$

We thus consider the volume dependence separately from the absolute scalar function ϕ which measures, for ordinary density, the distribution of mass with respect to the coordinate system. If we adopt a still more general law, and make use of (102), we have

$$\bar{\phi} = \left(\frac{\bar{V}}{V}\right)^N \phi = c^N \phi, \tag{141}$$

where we term N the "weight" of the scalar quantity. $N = 0$ corresponds to an absolute scalar, $N = 1$ to an ordinary density, etc. Negative values of N are not excluded.

Similarly we get vector densities or tensor densities in general. The most general transformation law then assumes the form

$$\bar{\mathfrak{T}}^{rs\cdot\cdot}_{tu\cdot\cdot} = c^N \gamma^r_i \gamma^s_j \ldots c^k_t c^l_u \ldots \mathfrak{T}^{ij\cdot\cdot}_{kl\cdot\cdot}. \tag{142}$$

With this definition we note that the "vector" $\mathbf{u} \times \mathbf{v}$, whose properties we examined in (138), is not an absolute vector but a tensor density of weight 1.

Although the cross product is a convenient device for representing an area of magnitude $uv \sin \theta$, with its orientation indicated by the vector perpendicular to that surface, we note that the cross-product notation fails for spaces of more than three dimensions. The failure is due to the fact that two vectors do not uniquely determine a normal to the surface element except for a space of three dimensions. To represent properly a surface element, we may project it in turn upon each basic surface element specified by a pair of vectors $\mathbf{e}_i$, $\mathbf{e}_j$. Thus a surface element becomes, in general, a tensor, whose components, T_{ij}, represent the orientation of the surface element. This tensor replaces the axial vector, whose orientation is defined specifically only for three coordinates.

When we operate upon a vector with ∇, we cannot employ the curl, $\nabla \times$, for spaces of more than three dimensions. Instead we operate directly upon the vector or tensor

$$\nabla \mathbf{u} \quad \text{or} \quad \nabla \mathbf{T}, \tag{143}$$

and interpret the resulting components by rules previously given.

Let us determine the gradient of a tensor density, $\mathfrak{T}^{ik} = T^{ik} V^N$ of weight N. The invariant concept applies to the tensor

$$\mathbf{T} = \mathfrak{T}/V^N = \mathbf{e}_k \mathbf{e}_i \mathfrak{T}^{ik}/V^N,$$

so that $\mathfrak{T}^{ik}/V^N$ denotes an absolute tensor. We now operate with ∇ on this quantity and then multiply again by V^n to produce a tensor of proper weight. Thus

$$\begin{aligned} \nabla \mathfrak{T} &= V^N \mathbf{e}^i D_i (V^{-N} \mathfrak{T}) = g^{N/2} \mathbf{e}^i \mathfrak{T} (-N V^{-N-1} D_i V) + \mathbf{e}^i D_i \mathfrak{T} \\ &= \mathbf{e}^i (D_i - N \Gamma^k_{ik}) \mathfrak{T}. \end{aligned} \tag{144}$$

Now let **T** be a dyadic, and we get

$$\nabla \mathfrak{T} = \mathbf{e}^i \mathbf{e}_j \mathbf{e}_k \nabla_i \mathfrak{T}^{jk}, \tag{145}$$

where ∇_i is an operator:

$$\nabla_i \mathfrak{T}^{jk} = D_i \mathfrak{T}^{jk} + \Gamma^j_{il} \mathfrak{T}^{lk} + \Gamma^k_{il} \mathfrak{T}^{jl} - N \Gamma^l_{il} \mathfrak{T}^{jk}, \tag{146}$$

which operates only on scalars, as in (127) and (128) for absolute tensors. For a covariant dyadic we have

$$\nabla_i \mathfrak{T}_{jk} = D_i \mathfrak{T}_{jk} - \Gamma^l_{ij} \mathfrak{T}_{lk} - \Gamma^l_{ik} \mathfrak{T}_{jl} - N \Gamma^l_{il} \mathfrak{T}_{jk}, \tag{147}$$

and for a mixed dyadic,

$$\nabla_i \mathfrak{T}^j_{.k} = D_i \mathfrak{T}^j_{.k} + \Gamma^j_{il} \mathfrak{T}^l_{.k} - \Gamma^l_{ik} \mathfrak{T}^j_{.l} - N \Gamma^l_{il} \mathfrak{T}^j_{.k}. \tag{148}$$

The above formulae provide the basic rules for covariant differentiation.

In three-dimensional space, the operation, $\nabla \times \mathbf{u}$, is analogous to the vector product. We write

$$\nabla \times \mathbf{u} = \mathbf{e}^i D_i \times \mathbf{e}^j u_j = \mathbf{e}^i \times \mathbf{e}^j \nabla_i u_j = \mathbf{e}^i \times \mathbf{e}^j u_{j,i}, \tag{149}$$

where the covariant derivative

$$u_{j,i} = D_i u_j - \Gamma^l_{ij} u_l - N \Gamma^l_{il} u_j. \tag{150}$$

Noting that the idemfactor I takes the alternative forms

$$I = \mathbf{e}^k \mathbf{e}_k = \mathbf{e}_k \mathbf{e}^k = g_{jk} \mathbf{e}^j \mathbf{e}^k = g^{jk} \mathbf{e}_k \mathbf{e}_j, \tag{151}$$

we can set

$$\mathbf{e}^i \times \mathbf{e}^j = \mathbf{e}^i \times \mathbf{e}^j \cdot \mathbf{e}^k \mathbf{e}_k = \epsilon^{ijk} v \mathbf{e}_k, \tag{152}$$

in which case we can write **curl u** in the more familiar form

$$\nabla \times \mathbf{u} = \mathbf{e}_k \frac{\epsilon^{ijk}}{V} u_{j,i}, \tag{153}$$

where the component is a tensor density of weight -1.

To ascertain the superiority of the tensor notation for the curl, and, at the same time, to extend the analysis to a space of more than three dimensions, apply the result to Stokes's theorem (32.8), which states that

$$\int \nabla \times \mathbf{u} \cdot d\mathbf{S} = \int \mathbf{u} \cdot d\mathbf{s}. \tag{154}$$

Let us omit the cross, according to the agreement of (143), and write $\nabla \mathbf{u}$ from (149). Also represent dS in terms of its contravariant tensor components,

$$d\mathbf{S} = \mathbf{e}_r \mathbf{e}_s \, dS^{rs}. \tag{155}$$

Then we get, in place of the left-hand side of (154),

$$\int \mathbf{e}^i \mathbf{e}^j \mathbf{e}_r \mathbf{e}_s u_{j,i} \, dS^{rs}. \tag{156}$$

Now contract this tensor over the indices j, r and i, s. Since this procedure is equivalent to taking the scalar product $\mathbf{e}^i \cdot \mathbf{e}_s = \delta^i_s$, we are summing the components of $u_{j,i}$ through the face of dS^{ij}. In other words we are computing the flux of the tensor $u_{j,i}$ through the entire surface. And (156) measures the flux through the entire boundary surface. We perform the summation $u_{j,i}\, dS^{ij}$. The surface element, $dS^{ij} = -dS^{ji}$, is antisymmetric. We can, therefore, write the integral in the form

$$\int u_{i,j}\, dS^{ji} = \frac{1}{2}\int u_{i,j}(dS^{ji} + dS^{ji}) = \frac{1}{2}\int u_{i,j}(dS^{ji} - dS^{ij}). \tag{157}$$

Now interchange the dummy indices on the second term, and we continue,

$$= \frac{1}{2}\int (u_{i,j}\, dS^{ji} - u_{j,i}\, dS^{ji}) = \frac{1}{2}\int (u_{i,j} - u_{j,i})\, dS^{ji}. \tag{158}$$

The differential area dS^{ji} consists of an elementary parallelogram, $dq^i\, dq^j$. Continuing, we now divide the integral into two parts:

$$\frac{1}{2}\iint \left(\frac{\partial u_i}{\partial q^j}\, dq^j\right) dq^i - \frac{1}{2}\iint \left(\frac{\partial u_j}{\partial q^i}\, dq^i\right) dq^j. \tag{159}$$

At this point let us suspend the summation rule, and calculate (156) over some specified pair of coordinates, say q^1 and q^2. Then

$$\frac{1}{2}\iint \frac{\partial u_1}{\partial q^2}\, dq^2\, dq^1 - \frac{1}{2}\iint \frac{\partial u_2}{\partial q^1}\, dq^1\, dq^2.$$

To evaluate this entity we have to assume some limits for our integration. Over the elementary parallelogram, as we hold q^2 constant, we suppose that q^1 ranges from a to b. Similarly, let q^2 range from c to d. The quantities u_1 and u_2 are functions of the coordinates q_1 and q_2, as well as of the other coordinates, in general.

$$u_1 = u_1(q^1, q^2), \qquad u_2 = u_2(q^1, q^2). \tag{160}$$

Now, holding q_1 constant, we get

$$\int_c^d \frac{\partial u_1}{\partial q^2}\, dq^2 = u_1(q^1, d) - u_1(q^1, c) = (u_1)^d - (u_1)^c.$$

Similarly,

$$\int_a^b \frac{\partial u_2}{\partial q^1}\, dq^1 = u_2(b, q^2) - u_2(a, q^2) = (u_2)^b - (u_2)^a. \tag{161}$$

Hence

$$\frac{1}{2}\int_a^b \int_c^d (u_{1,2} - u_{2,1})\, dq^1\, dq^2$$

$$= \int_a^b [(u_1)^d - (u_1)^c]\, dq^1 - \int_c^d [(u_2)^b - (u_2)^a]\, dq^2$$

$$= \int_a^b (u_1)^d\, dq^1 - \int_c^d (u_2)^b\, dq^2 - \int_a^b (u_1)^c\, dq^1 + \int_c^d (u_2)^a\, dq^2. \tag{162}$$

These last four terms represent the line integral of u around the circuit *abcd*. Hence, returning to the summation convention, we get, finally,

$$-\frac{1}{2}\int\left(\frac{\partial u_i}{\partial q^i}-\frac{\partial u_j}{\partial q^i}\right)dS^{ii}=\int u_i\,dq^i. \tag{163}$$

The right-hand side of (163) represents the line integral of the scalar quantity $u(q^iq^i)$ around the closed circuit *adbc*. The left-hand side is an integral over the area bounded by the circuit. This expression enables us to extend Stokes's theorem, and indeed other vector operations, to a space of n dimensions. We have done so by dropping the vector product and redefining equation (9), which specifies the reciprocal unitary vectors. As before, our basic coordinate system, $q^1, q^2, \ldots, q^n$ defines n unitary base vectors $\mathbf{e}_1, \mathbf{e}_2, \ldots, \mathbf{e}_n$. Then the vector $\mathbf{e}^i$, reciprocal to the base vector $\mathbf{e}_i$, is normal to all the unitary base vectors except $\mathbf{e}_i$. Further, the n base vectors define a volume V in n dimensions and the reciprocal vectors a volume, v, so that (11) still holds. Given $\mathbf{e}_i$, we choose the length of $\mathbf{e}^i$ so that $\mathbf{e}_i \cdot \mathbf{e}^j = \delta_i^j$. We introduce the metric tensor g_{ij}, as in (17) and employ (18) except that our summation convention must apply to the range 1 to n instead of 1 to 3, as before. The definitions of contravariance and covariance do not change, but the transformation matrices, $G(38)$, $G'(39)$, $\Gamma(58)$, and $C(61)$, become n-dimensional. All of the essential remaining formulas, except those containing vector products, are unaltered. Whenever a symbol $\times$ appears in any formula, we omit it, forming a tensor from the pair of base-vectors originally separated by the symbols. We have to extend our definition of epsilons to n subscripts, each of which may range from 1 to n.

$$\epsilon_{ijkl\ldots}=\begin{cases}0\\ 1.\\ -1\end{cases} \tag{164}$$

The epsilons are zero whenever any pair of indices is duplicated. In other words, ϵ differs from zero only when the subscripts $i, j, k, l, \ldots$, consist of some permutation of the numbers 1 to n. We start with

$$\epsilon_{123\ldots n}=1. \tag{165}$$

The other values of ϵ have as their indices some permutation of the integers 1 to n. If we form this arbitrary sequence by the successive operation of changing adjacent letters, the number of permutations determines the sign of ϵ. If we require an even number of permutations to derive the given ϵ from (165), we assign to ϵ the value $+1$. For odd permutations, $\epsilon = -1$.

Equation (150) gives the basic rule for covariant differentiation of a covariant vector component. Omit the cross, as irrelevant.

$$\nabla\mathbf{u}=\mathbf{e}^j\mathbf{e}^i\nabla_j u_i=\mathbf{e}^j\mathbf{e}^i u_{i,j}=\mathbf{e}^j\mathbf{e}^i[D_j u_i-\Gamma_{ji}^l u_l], \tag{166}$$

wherein we have set $N = 0$. Now form the second covariant derivative

$$\nabla\nabla\mathbf{u} = \mathbf{e}^k \frac{\partial}{\partial q^k} \{\mathbf{e}^j\mathbf{e}^i[D_ju_i - \Gamma^l_{ji}u_l]\}. \tag{167}$$

The quantity in square brackets is a doubly covariant tensor, hence we employ (147) to get the second covariant derivative

$$\nabla\nabla\mathbf{u} = \mathbf{e}^k\mathbf{e}^j\mathbf{e}^i\{D_k[D_ju_i - \Gamma^l_{ji}u_l] - \Gamma^m_{kj}[D_mu_i - \Gamma^l_{mi}u_l] - \Gamma^m_{ki}[D_ju_m - \Gamma^l_{jm}u_l]\}. \tag{168}$$

Differentiating the product term and changing the dummy index from l to m in the first square bracket, we get

$$u_{i,jk} = D_kD_ju_i - [\Gamma^m_{ji}D_ku_m + \Gamma^m_{kj}D_mu_i + \Gamma^m_{ki}D_ju_m] + \Gamma^m_{kj}\Gamma^l_{mi}u_l + u_l[\Gamma^m_{ki}\Gamma^l_{jm} - D_k\Gamma^l_{ji}]. \tag{169}$$

Now reverse the order of differentiation, changing the indices jk,

$$u_{i,kj} = D_jD_ku_i - [\Gamma^m_{ki}D_ju_m + \Gamma^m_{jk}D_mu_i + \Gamma^m_{ji}D_ku_m] + \Gamma^m_{jk}\Gamma^l_{mi}u_l + u_l[\Gamma^m_{ji}\Gamma^l_{km} - D_j\Gamma^l_{ki}], \tag{170}$$

and subtract (170) from (169). The first five terms cancel identically, with the use of (108). Hence

$$u_{i,jk} - u_{i,kj} = u_l\{\Gamma^m_{ki}\Gamma^l_{jm} - \Gamma^m_{ji}\Gamma^l_{km} - D_k\Gamma^l_{ji} + D_j\Gamma^l_{ki}\}. \tag{171}$$

The quantity on the left-hand side represents a triply covariant tensor. Hence the quantity on the right must also be triply covariant. Non-repeated indices within the braces include the covariant indices ijk and the contravariant index l. The dummy index, m, does not count because it is repeated in both positions. Hence we represent the tensor in braces by the symbol $B^{\cdot\cdot\cdot l}_{ijk\cdot}$,

$$B^{\cdot\cdot\cdot l}_{ijk\cdot} = \{\Gamma^m_{ki}\Gamma^l_{jm} - \Gamma^m_{ji}\Gamma^l_{km} - D_k\Gamma^l_{ji} + D_j\Gamma^l_{ki}\}, \tag{172}$$

the famous Riemann-Christoffel curvature tensor. We can lower the index l as follows:

$$B_{ijkl} = g_{rl}B^{\cdot\cdot\cdot l}_{ijk\cdot}. \tag{173}$$

Through (169) and (114), this procedure gives

$$B_{ijkl} = \{\Gamma^m_{ki}\Gamma_{jm,l} - \Gamma^m_{ji}\Gamma_{km,l} - D_k\Gamma_{ji,l} + D_j\Gamma_{ki,l} - \Gamma^m_{ji}D_jg_{lm} + \Gamma^m_{ki}D_kg_{lm}\}. \tag{174}$$

Using (111)—(113), carry out the indicated derivations, canceling various terms in the process. We get, finally,

$$B_{ijkl} = -\Gamma_{ik}^{m}\Gamma_{jl,m} + \Gamma_{ij}^{m}\Gamma_{kl,m}$$

$$+ \frac{1}{2}\,[D_iD_jg_{lk} + D_kD_lg_{ij} - D_iD_kg_{lj} - D_jD_lg_{ik}]. \tag{175}$$

This tensor is antisymmetric in the indices j, k, and in the indices i, l, so that

$$B_{ikjl} = B_{ljki} = -B_{ijkl}. \tag{176}$$

Also, permuting the last three indices cyclically, we get

$$B_{ijkl} + B_{iklj} + B_{iljk} = 0. \tag{177}$$

For a space of four dimensions, this tensor has 4^4 or 256 components. The enormous economy of the tensor notation is immediately evident, when we realize that (173) or (174) represents 256 separate equations, despite the fact that (176) and (177) severely reduce the number of independent relations to only twenty.

Having derived this tensor, let us now consider its physical significance. We started with a vector, **u**, and formed, in turn, the tensors $\nabla\mathbf{u}$ and $\nabla\nabla\mathbf{u}$. This latter tensor is invariant, but its components possess some unusual features. As we perform covariant differentiation, to get the tensors $\mathbf{e}^k\mathbf{e}^j\mathbf{e}^i u_{i,jk}$ and $\mathbf{e}^k\mathbf{e}^j\mathbf{e}^i u_{i,kj}$, these tensors are not necessarily identical. In other words, the components of the tensor $\nabla\nabla\mathbf{u}$ are not symmetrical in the indices jk. The difference,

$$u_{i,jk} - u_{i,kj} = u_l B_{ijk\cdot}^{\cdot\cdot\cdot l}, \tag{178}$$

depends on $B_{ijk\cdot}^{\cdot\cdot\cdot l}$, component of the Riemann-Christoffel tensor. Only when $B_{ijk\cdot}^{\cdot\cdot\cdot l}$ is zero, can we interchange the order of covariant differentiation. And when this tensor vanishes, the components of $\nabla\nabla\mathbf{u}$ are symmetrical in the indices jk.

Of special significance is the fact that $B_{ijk\cdot}^{\cdot\cdot\cdot l}$ in no way depends on the vector **u** under consideration. Its components and those of the related B_{ijkl} depend only on the Γ_{ij}^{k}'s and thus on the g_{ij}'s. The Riemann-Christoffel tensor expresses a particular property of the space itself. For example if the components vanish in one coordinate system they must vanish in all coordinate systems for that space; We are required to find this property of the space.

Compare (162), Stokes's theorem,

$$\oint u_i\, dq^i = \frac{1}{2}\int (u_{i,j} - u_{j,i})\, dS^{ji}, \tag{179}$$

with the equation (172), also integrated around a similar contour,

$$\delta u_i = \int u_l B_{ijk.}^{\cdot\cdot\cdot l}\, dq^j = \frac{1}{2}\int [(u_i)_{j,k} - (u_i)_{k,j}]\, dS^{jk}. \tag{180}$$

If we identify (u_i) in the latter equation with u of the first, and then replace i, j in the first by j, k, the correspondence becomes exact. Thus, even though (180) is a tensor whose rank is one higher than that of (179), the index i remains constant throughout the integration, and hence this subscript has no special significance other than to identify which component of the basic tensor we are applying the theorem to. Whether we write the pair of subscripts $(, jk)$, (j, k), or omit the comma altogether is immaterial to the argument, since the subscripts j and k both refer to differentiation. From equation (180) we can infer directly that if the right-hand side of the equation equals zero, this vanishing property depends fully as much on the character of the tensor $B_{ijk.}^{\cdot\cdot\cdot l}$ as upon the character of u_l. In other words, since $B_{ijk.}^{\cdot\cdot\cdot l}$ derives all of its properties from the space, the value of the integral will depend on those properties. However, the derivation indicates that the equation (180) expresses some property of the vector u_i, which should be independent of the coordinate system. Here is a seeming inconsistency.

Although $\nabla\nabla\mathbf{u}$ is an invariant, independent of the coordinate system, this invariance holds only at a given point or over an infinitesimal area surrounding that point, unless the vector **u** possesses very special properties. When we displace **u** around some arbitrary finite curve or, what is the same thing, study the character of the vector field over a finite region, by means of the equation (180), we are essentially trying to balance out the component $(u_i)_{j,k}$ by subtracting the quantity $(u_i)_{k,j}$. But these are not necessarily identical, because the order of covariant differentiation is not commutable unless

$$B_{ijk.}^{\cdot\cdot\cdot l} = 0. \tag{181}$$

Then and only then does the integral in (180) become independent of the path.

If we are to take the integral of u_i around some arbitrary curve, we must regard u_i as being some function of a parameter p, which defines this curve. Many varieties of curves exist over which we can perform the integration, but one type deserves special consideration, because it is essentially unique. Of all paths joining two points in space, only one is "shortest," in the Euclidean sense of the word. More accurately, we can define a "straightest" path, i.e., one along which the unit vectors show the least possible change.

The vanishing or non-vanishing of the Riemann-Christoffel tensor is an intrinsic property of space itself rather than of the coordinate system. When $B_{ijk}^{\cdot\cdot\cdot l}$ vanishes, we say that the space it represents is flat or Euclidean.

We then can always fit some Cartesian coordinate system into that space. We are, however, not limited to Cartesian systems, and the $B_{ijk.}^{\cdot\cdot\cdot l}$ will vanish for all possible coordinate systems in that space.

A non-vanishing $B_{ijk.}^{\cdot\cdot\cdot l}$ indicates that space is non-Euclidean and that no Cartesian system can be fitted within it. Such a system is the two-dimensional space on the surface of a sphere, whether we define its metric in the conventional system of latitude and longitude or in terms of any other coordinates. In a space of two dimensions, of the sixteen (2^4) possible components of the $B_{ijk.}^{\cdot\cdot\cdot l}$ or its covariant equivalent B_{ijkl}, with indices ranging from (1111), (1211), . . . , (2222), only one is non-vanishing or independent, viz., B_{1212}.

The foregoing analysis indicates why we term the Riemann-Christoffel tensor the curvature tensor. When it does not vanish, the space in which we make our calculations becomes curved. Curvature implies the existence of a radius of curvature. Thus we find that the g_{ij}'s determine the actual metrical properties of the system.

When we perform an integration like that indicated in (180), carrying a vector around some closed curve, we must define some convention for the procedure. We call the process "parallel displacement," a definition that is unique for Euclidean space. But how are we to define a system of parallel vectors for a Riemannian space, e.g., the two-dimensional surface of a sphere. If we "view" this surface from the outside, i.e., as part of a Euclidean space of three dimensions, we readily see when two tangents to the surface are indeed parallel.

If the surface of the earth were itself a plane we should have no great difficulty in keeping track of absolute directions. Sailing a boat, for example, we could complete a closed triangular path by executing three turns, the sum of whose angles would add to 180 degrees. On the surface of the earth, however, we should be sailing along the sides of a spherical triangle the sum of whose angles would exceed 180 degrees. The excess would depend upon the path traversed. Hence parallel displacement, as defined by the integral (180), will depend upon the path unless $B_{ijk}^{\cdot\cdot\cdot l} = 0$, i.e., unless the space is Euclidean.

However, of all the paths that join two points in space, only one is "shortest" in the quasi-Euclidean sense of the word. More accurately, we can define a "straightest" path as the one along which the unit vectors show the least possible change. In a pure Cartesian system, the unit vectors do not alter direction along a straight line, and the $\Gamma_{ij}^{\cdot\cdot k}$ all vanish; the curvature tensor $B_{ijk.}^{\cdot\cdot\cdot l}$ vanishes if we arbitrarily imbed some curved coordinate system in a space originally Euclidean.

We now suppose that $\mathbf{u}$ is a "velocity," defined as the derivative of the coordinates; thus

$$\mathbf{u} = \mathbf{e}_i \frac{dq^i}{dp}. \tag{182}$$

Then

$$\frac{d\mathbf{u}}{dp} = \mathbf{e}_i\left(\frac{d^2q^i}{dp^2} + \Gamma^i_{jk}\frac{dq^j}{dp}\frac{dq^k}{dp}\right), \tag{183}$$

a sort of generalized "acceleration" in terms of the arbitrary parameter p. If we regard p as measuring the displacement along some arbitrary curve s, such that

$$ds^2 = g_{ij}\,dq^i\,dq^j = g_{ij}\frac{dq^i}{dp}\frac{dq^j}{dp}\,dp^2, \tag{184}$$

then dividing the above equation by ds^2, we must have

$$g_{ij}\frac{dq^i}{ds}\frac{dq^j}{ds} = 1. \tag{185}$$

Thus, to maintain $\mathbf{u}$ constant as we displace it along the curve s, we must have

$$\frac{d^2q^i}{ds^2} + \Gamma^i_{jk}\frac{dq^j}{ds}\frac{dq^k}{ds} = 0, \tag{186}$$

as the differential equation of this curve. For Cartesian space, where $\Gamma^i_{jk} = 0$, the integral becomes

$$q^i = as + b = a'p + b', \tag{187}$$

the equation of a straight line. Thus (186) is the generalized expression for the straightest possible path in the space under consideration. We call such a path a "geodesic."

Generalized tensor analysis, with the added theorems on curvature and geodesics, provides us with a mathematical tool especially useful in relativity. In Part V, we shall return to make practical use of this generalized geometry, whose notation is essentially independent of the exact nature of the metric or type of surface. We have been seeking a system that is invariant, not merely to changes of scale, but to the entire basic mesh underlying the idea of measurement. We call this basic measurement system the "gauge system." We have worked out a basic and very general notation that is independent of the gauge and, hence, gauge invariant. Individual vector or tensor components are covariant, contravariant, or mixed, as the case may be. But the basic quantity that they represent, together with the derived transformation properties of the components, have constituted a gauge-invariant system.

Vector Operations

32. Stokes's theorem. Just as Green's theorem, § (14) evaluates a double integral of some function taken over a closed surface in terms of a volume integral, Stokes's theorem expresses a line integral of a function around a closed curve in terms of an area integral. For example, the work done by the forces that move a particle from one point to another is the line integral of the force function along the path. Suppose the path p to be closed and let F be the value of the force at any point on it. Then we seek to evaluate the integral (8.1)

$$\int_p F \cos \theta \, ds = \int \mathbf{F} \cdot d\mathbf{s} \tag{1}$$

around this contour. For simplicity of argument we shall assume that we are to take the integral over a path that lies entirely in the xy-plane. Let P be a point within the area enclosed by p, Fig. 25. Then we may write the vector function $\mathbf{F}(x, y)$,

$$\mathbf{F}(x, y) = \mathbf{i}X(x, y) + \mathbf{j}Y(x, y), \tag{2}$$

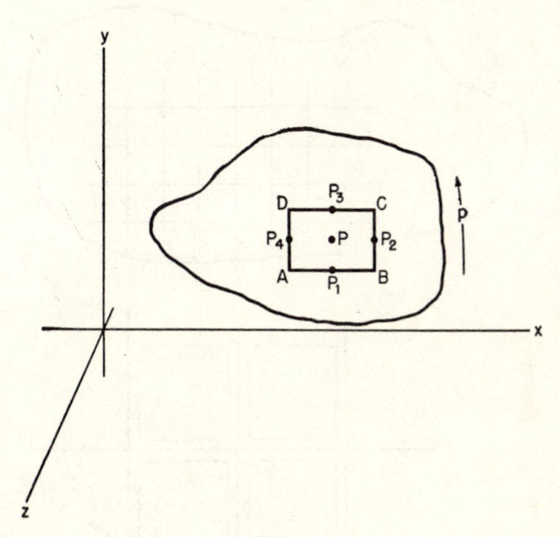

II-25.

where X and Y give the values of the x and y components of $\mathbf{F}$ at any point x, y. We start by evaluating the work done as a unit particle moves about an elementary rectangle surrounding P. Then, by addition of rectangles, we shall find the work accomplished as the particle moves around the contour p. We shall shortly establish the validity of the summation.

The average value of the x-component of the force along AB is $X(P_1)$, etc.

$$\begin{aligned} X(P_1) &= X - \frac{\partial X}{\partial y}\frac{dy}{2}, \quad & Y(P_2) &= Y + \frac{\partial Y}{\partial x}\frac{dx}{2}, \\ X(P_3) &= X + \frac{\partial X}{\partial y}\frac{dy}{2}, \quad & Y(P_4) &= Y - \frac{\partial Y}{\partial x}\frac{dx}{2}. \end{aligned} \tag{3}$$

We make these approximations by assuming that X and Y are linear in x and y over the short distances along the sides of the rectangle, in accord with the customary calculus procedure. In moving the particle from A to B, the y-component of force does no work; similarly from B to C, the x-component makes no contribution to the work. Hence the total work spent as the unit particle moves completely around the rectangular circuit, is

$$X(P_1)\,dx + Y(P_2)\,dy - X(P_3)\,dx - Y(P_4)\,dy = \left(\frac{\partial Y}{\partial x} - \frac{\partial X}{\partial y}\right) dx\,dy. \quad (4)$$

We now divide the entire area enclosed by the curve p into elementary rectangles and sum. Since every line in the diagram except the periphery is traversed twice and in opposite directions, the contributions to the

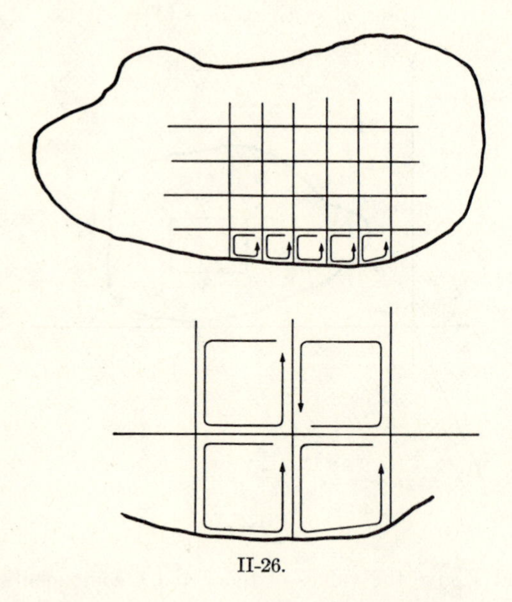

II-26.

integral cancel. Hence the work required to move the particle around all the individual rectangles of the network equals that for the peripheral path, p, or

$$\iint \left(\frac{\partial Y}{\partial x} - \frac{\partial X}{\partial y}\right) dx\,dy = \int_p F \cos\theta\,ds = \int_p (X\,dx + Y\,dy). \quad (5)$$

If the area to be integrated is of arbitrary shape and not wholly in the xy-plane, we project the path on three fundamental planes in turn and sum the result.

$$\int_p F \cos \theta \, ds = \int_p (X \, dx + Y \, dy + Z \, dz)$$

$$= \iint \left[\left(\frac{\partial Y}{\partial x} - \frac{\partial X}{\partial y} \right) dx \, dy + \left(\frac{\partial Z}{\partial y} - \frac{\partial Y}{\partial z} \right) dy \, dz + \left(\frac{\partial X}{\partial z} - \frac{\partial Z}{\partial x} \right) dz \, dx \right]$$

$$= \iint \left[\lambda \left(\frac{\partial Y}{\partial x} - \frac{\partial X}{\partial y} \right) + \mu \left(\frac{\partial Z}{\partial y} - \frac{\partial Y}{\partial z} \right) + \nu \left(\frac{\partial X}{\partial z} - \frac{\partial Z}{\partial x} \right) \right] dS, \tag{6}$$

since

$$\lambda \, dS = dx \, dy, \text{ ETC.}, \tag{7}$$

where λ, μ, ν, are the direction cosines of the normal to the surface element dS. In vector notation, Stokes's theorem takes the beautifully simple form

$$\int_p \mathbf{F} \cdot d\mathbf{s} = \int \nabla \times \mathbf{F} \cdot d\mathbf{S}, \tag{8}$$

as we see from (23.6), the definition of the curl. For comparison we add Gauss' theorem (23.19):

$$\int \mathbf{F} \cdot d\mathbf{S} = \int \nabla \cdot \mathbf{F} \, d\tau. \tag{9}$$

33. Vector operators in general orthogonal curvilinear coordinates. Since the various functions on which we shall require the vector operator ∇ to act may appear in polar, cylindrical, or types of coordinate systems other than Cartesian, we must interpret ∇ in terms of appropriate variables. Here we treat special cases of the general formulas of § (31). Consider the three equations

$$q_1 = Q_1(x, y, z), \quad q_2 = Q_2(x, y, z), \quad q_3 = Q_3(x, y, z), \tag{1}$$

where Q_1, Q_2, and Q_3 are functions of the rectangular coordinates x, y, and z. Further, let us suppose we have solved these equations simultaneously so that we have the unique solutions

$$x = X(q_1, q_2, q_3), \quad y = Y(q_1, q_2, q_3), \quad z = Z(q_1, q_2, q_3). \tag{2}$$

The equations

$$q_1 = a, \quad q_2 = b, \quad \text{and} \quad q_3 = c, \tag{3}$$

where a, b, and c are constants, are the equations of three surfaces. In the first surface q_1 is constant, while q_2 and q_3 vary, etc.

The three curved surfaces defined by (3) intersect at the point P, as shown in Fig. 27. Along the lines that form the intersection of two surfaces only one parameter varies while the other two remain constant. For example along q_1, we hold q_2 and q_3 constant, whereas q_1 varies. Let us consider the possibility of adopting q_1, q_2, and q_3 as a system of coordinates. To any point x, y, z, we may, through equations (1) and (2), assign a

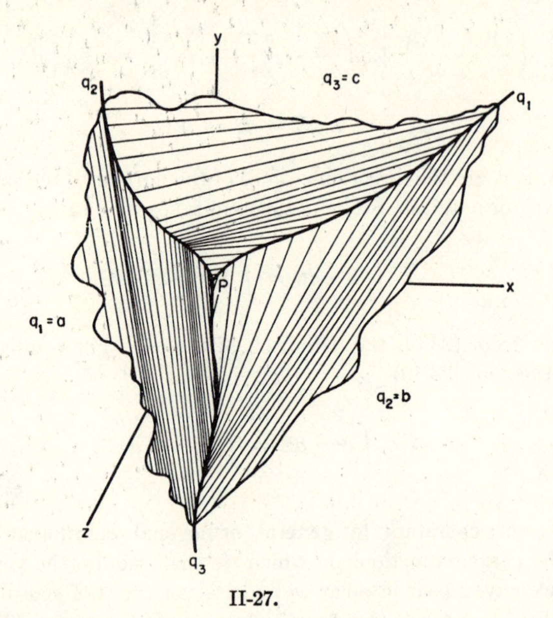

II-27.

corresponding set of coordinates: q_1, q_2, q_3. These alternative coordinate systems are most useful when the three curves, at any point, are orthogonal to one another. This condition places a restriction on the character of the functions in (1) and (2).

We may regard the line q_1 as determined by the equations

$$x = X(q_1, b, c),$$
$$y = Y(q_1, b, c), \tag{4}$$
$$z = Z(q_1, b, c),$$

where b and c are constant values assigned to q_2 and q_3 as in (3).

We assume that the functions Q_1, Q_2, Q_3, X, Y, Z are continuously differentiable. Since dq is not necessarily a linear distance (e.g., it may be an angle) we shall measure lengths in the q coordinate system by the relations

$$ds_1 = h_1\, dq_1, \quad ds_2 = h_2\, dq_2, \quad ds_3 = h_3\, dq_3, \tag{5}$$

so that ds_2, etc., are infinitesimal displacements along the respective coordinated curves. The direction cosines of the element ds_1 at P are, by (7.4),

$$\lambda_1 = \frac{dx}{ds_1} = \frac{dX(q_1, b, c)}{dq_1}\frac{dq_1}{ds_1} = \frac{\partial X(q_1, q_2, q_3)}{\partial q_1}\frac{1}{h} = \frac{1}{h_1}\frac{\partial X}{\partial q_1} = \frac{X_1'}{h_1}. \tag{6}$$

Similarly,

$$\mu_1 = \frac{1}{h_1}\frac{\partial Y}{\partial q_1} = \frac{Y_1'}{h_1}, \quad \nu_1 = \frac{Z_1'}{h_1}, \tag{7}$$

with similar equations for q_2 and q_3. For an orthogonal system we must have

$$\begin{aligned} h_1^2 &= X_1'^2 + Y_1'^2 + Z_1'^2. \\ h_2^2 &= X_2'^2 + Y_2'^2 + Z_2'^2. \\ h_2^2 &= X_3'^2 + Y_3'^2 + Z_3'^2. \\ 0 &= X_1'X_2' + Y_1'Y_2' + Z_1'Z_2', \quad \text{ETC, BY (7.11)}. \end{aligned} \tag{8}$$

Further, expressing dx, dy, and dz as total differentials, we have

$$dx = \frac{\partial X}{\partial q_1}dq_1 + \frac{\partial X}{\partial q_2}dq_2 + \frac{\partial X}{\partial q_3}dq_3,$$

or

$$\begin{aligned} dx &= X_1'\,dq_1 + X_2'\,dq_2 + X_3'\,dq_3. \\ dy &= Y_1'\,dq_1 + Y_2'\,dq_2 + Y_3'\,dq_3. \\ dz &= Z_1'\,dq_1 + Z_2'\,dq_2 + Z_3'\,dq_3. \end{aligned} \tag{9}$$

The element $d\mathbf{s}$ must be independent of the coordinate system. Thus, in vector notation, we obtain

$$d\mathbf{s} = \mathbf{i}\,dx + \mathbf{j}\,dy + \mathbf{k}\,dz = \mathbf{i}_1 h_1\,dq_1 + \mathbf{i}_2 h_2\,dq_2 + \mathbf{i}_3 h_3\,dq_3, \tag{10}$$

where $\mathbf{i}_1$, $\mathbf{i}_2$, and $\mathbf{i}_3$ are the unit vectors at P in the curvilinear system. Substituting from (9) and equating coefficients of the various dq's, we find that

$$\begin{aligned} d\mathbf{s}_1 &= \mathbf{i}_1 h_1\,dq_1 = (\mathbf{i}X_1' + \mathbf{j}Y_1' + \mathbf{k}Z_1')\,dq_1. \\ d\mathbf{s}_2 &= \mathbf{i}_2 h_2\,dq_2 = (\mathbf{i}X_2' + \mathbf{j}Y_2' + \mathbf{k}Z_2')\,dq_2. \\ d\mathbf{s}_3 &= \mathbf{i}_3 h_3\,dq_2 = (\mathbf{i}X_3' + \mathbf{j}Y_3' + \mathbf{k}Z_3')\,dq_3. \end{aligned}$$

The volume of the rectangular parallelopiped formed by the elements $d\mathbf{s}_1$, $d\mathbf{s}_2$ and $d\mathbf{s}_3$ is

$$\begin{aligned} d\tau &= d\mathbf{s}_1 \cdot d\mathbf{s}_2 \times d\mathbf{s}_3 = h_1h_2h_3\,dq_1\,dq_2\,dq_3 \\ &= [X_1'(Y_2'Z_3' - Y_3'Z_2') + Y_1'(Z_2'X_3' - Z_3'X_2') \\ &\qquad + Z_1'(X_2'Y_3' - X_3'Y_2')]\,dq_1\,dq_2\,dq_3 \end{aligned} \tag{11}$$

by (22.19). If we introduce the well-known Jacobian determinant,

$$J = \begin{vmatrix} X_1' & Y_1' & Z_1' \\ X_2' & Y_2' & Z_2' \\ X_3' & Y_3' & Z_3' \end{vmatrix}, \tag{12}$$

then

$$d\tau = J\, dq_1\, dq_2\, dq_3, \tag{13}$$

a general relationship which, unlike (8), holds whether the q coordinate system is orthogonal or not, although the proof here given is restricted to the orthogonal case. Equation (13) is a very important mathematical relationship whenever one wishes to change variables in an integration. It can be applied to any number of coordinates.

For orthogonal systems, the element ds obeys the relation

$$ds^2 = h_1^2\, dq_1^2 + h_2^2\, dq_2^2 + h_3^2\, dq_3^2. \tag{14}$$

For rectangular coordinates, x, y, z, we have

$$h_1 = h_2 = h_3 = 1, \quad d\tau = dx\, dy\, dz, \quad ds^2 = dx^2 + dy^2 + dz^2. \tag{15}$$

For cylindrical coordinates, r, θ, z, we have

$$x = r \cos\theta, \quad y = r \sin\theta, \quad z = z.$$

$$X_1' = \frac{\partial x}{\partial r} = \cos\theta, \quad Y_1' = \frac{\partial y}{\partial r} = \sin\theta, \quad Z_1' = \frac{\partial z}{\partial r} = 0, \quad \text{ETC.} \tag{16}$$

Hence, by (9),

$$h_1^2 = X_1'^2 + Y_1'^2 + Z_1'^2 = \cos^2\theta + \sin^2\theta = 1.$$

Similarly,

$$h_2^2 = r^2, \quad h_3^2 = 1.$$

Therefore

$$\begin{gathered} h_1 = 1, \quad h_2 = r, \quad h_3 = 1. \quad d\tau = r\, dr\, d\theta\, dz, \\ ds^2 = dr^2 + r^2\, d\theta^2 + dz^2. \end{gathered} \tag{17}$$

Analogously for spherical coordinates, r, θ, ϕ, we find

$$x = r \sin\theta \cos\phi, \quad y = r \sin\theta \sin\phi, \quad z = r \cos\theta. \tag{18}$$

Whence

$$\begin{gathered} h_1 = 1, \quad h_2 = r, \quad h_3 = r \sin\theta, \\ d\tau = r^2 \sin\theta\, dr\, d\theta\, d\phi \end{gathered} \tag{19}$$

and

$$ds^2 = dr^2 + r^2\, d\theta^2 + r^2 \sin^2\theta\, d\phi^2.$$

For parabolic coordinates ξ, η, ϕ, we have

$$x = \sqrt{\xi\eta}\cos\phi, \quad y = \sqrt{\xi\eta}\sin\phi, \quad z = \frac{1}{2}(\xi - \eta). \tag{20}$$

Therefore

$$h_1 = \frac{1}{2}\sqrt{\xi + \eta/\xi}, \quad h_2 = \frac{1}{2}\sqrt{\xi + \eta/\eta}, \quad h_3 = \sqrt{\xi\eta},$$

$$d\tau = \frac{1}{4}(\xi + \eta)\, d\xi\, d\eta\, d\phi, \tag{21}$$

$$ds^2 = \frac{1}{4}\left(\frac{\xi + \eta}{\xi}\right) d\xi^2 + \frac{1}{4}\left(\frac{\xi + \eta}{\eta}\right) d\eta^2 + \xi\eta\, d\phi^2.$$

There are numerous other varieties of orthogonal coordinate systems, but the ones here given will suffice for many problems.

We first derive an expression for the gradient in generalized coordinates. The components of the gradient of any scalar function V are simply the derivatives of V with respect to three mutually perpendicular directions. We must, therefore, have

$$\frac{\partial V}{\partial s_1} = \frac{1}{h_1}\frac{\partial V}{\partial q_1}, \quad \frac{\partial V}{\partial s_2} = \frac{1}{h_2}\frac{\partial V}{\partial q_2}, \quad \text{ETC.} \tag{22}$$

Hence

$$\nabla = \left(\frac{\mathbf{i}_1}{h_1}\frac{\partial}{\partial q_1} + \frac{\mathbf{i}_2}{h_2}\frac{\partial}{\partial q_2} + \frac{\mathbf{i}_3}{h_3}\frac{\partial}{\partial q_3}\right), \tag{23}$$

where $\mathbf{i}_1$, $\mathbf{i}_2$, and $\mathbf{i}_3$ are unit vectors as in (10). Thus, for spherical coordinates, the gradient becomes, from (19),

$$\nabla = \left(\mathbf{i}_r\frac{\partial}{\partial r} + \mathbf{i}_\theta\frac{1}{r}\frac{\partial}{\partial\theta} + \mathbf{i}_\phi\frac{1}{r\sin\theta}\frac{\partial}{\partial\phi}\right), \tag{24}$$

in agreement with (10.5).

We evaluate the divergence in general orthogonal coordinates from an application of Gauss' theorem to the volume element

$$d\tau = h_1h_2h_3\, dq_1\, dq_2\, dq_3. \tag{25}$$

Let a vector quantity $\mathbf{F}$ have orthogonal components F_1, F_2, and F_3 along the three coordinate curves q_1, q_2, q_3. We consider the flux of this vector over the boundary surface of the volume element. The flux across the face ABCD, i.e., across the face q_1, of area $h_2h_3\, dq_2\, dq_3$ is $(F_1h_2h_3\, dq_2\, dq_3)$ and the amount leaving by the face $q_1 + dq_1$ is

$$(F_1h_2h_3\, dq_2\, dq_3) + \frac{\partial(F_1h_2h_3\, dq_2\, dq_3)}{\partial q_1}\, dq_1, \tag{26}$$

by Taylor's theorem. Subtracting the incident flux from the emergent flux, we find the excess leaving over the pair of faces to be

$$\frac{\partial(F_1h_2h_3)}{\partial q_1}\,dq_1\,dq_2\,dq_3 = \frac{1}{h_1h_2h_3}\frac{\partial(F_1h_2h_3)}{\partial q_1}\,d\tau, \tag{27}$$

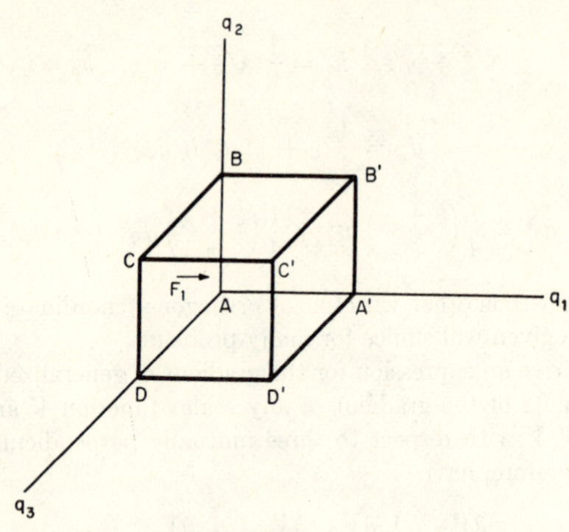

II-28.

by (25). Similar expressions obtain for the other pairs of faces. But the total outward flux, by Gauss' theorem, (23.19), is $\nabla \cdot \mathbf{F}\,d\tau$. Hence, dividing by $d\tau$, we obtain the general expression for the divergence:

$$\nabla \cdot \mathbf{F} = \frac{1}{h_1h_2h_3}\left[\frac{\partial}{\partial q_1}(F_1h_2h_3) + \frac{\partial}{\partial q_2}(F_2h_1h_3) + \frac{\partial}{\partial q_3}(F_3h_1h_2)\right]. \tag{28}$$

We are now in a position to derive the expression for the Laplacian ∇^2. Let ψ be a scalar function such that

$$\mathbf{F} = \mathbf{i}_1F_1 + \mathbf{i}_2F_2 + \mathbf{i}_3F_3 = \nabla\psi. \tag{29}$$

$$\nabla \cdot \mathbf{F} = \nabla \cdot \nabla\psi = \nabla^2\psi. \tag{30}$$

We have only to introduce the expression for F_1, etc., in terms of ψ into (28), to obtain the final answer. But from (22) or (23),

$$F_1 = \frac{1}{h_1}\frac{\partial\psi}{\partial q_1}, \quad \text{ETC.} \tag{31}$$

Therefore

$$\nabla^2\psi = \frac{1}{h_1h_2h_3}\left[\frac{\partial}{\partial q_1}\left(\frac{h_2h_3}{h_1}\frac{\partial\psi}{\partial q_1}\right) + \frac{\partial}{\partial q_2}\left(\frac{h_1h_3}{h_2}\frac{\partial\psi}{\partial q_2}\right) + \frac{\partial}{\partial q_3}\left(\frac{h_1h_2}{h_3}\frac{\partial\psi}{\partial q_3}\right)\right], \tag{32}$$

since $\mathbf{i}_1 \cdot \mathbf{i}_1 = 1, \quad \mathbf{i}_2 \cdot \mathbf{i}_2 = 1, \quad \mathbf{i}_1 \cdot \mathbf{i}_2 = 0,$ ETC.,

as for rectangular coordinates.

In polar coordinates this expression becomes, by (18) and (19),

$$\nabla^2\psi = \frac{1}{r^2 \sin\theta}\left[\frac{\partial}{\partial r}\left(r^2 \sin\theta \frac{\partial\psi}{\partial r}\right) + \frac{\partial}{\partial\theta}\left(\sin\theta\frac{\partial\psi}{\partial\theta}\right) + \frac{\partial}{\partial\phi}\left(\frac{1}{\sin\theta}\frac{\partial\psi}{\partial\phi}\right)\right]$$

$$= \frac{1}{r^2}\frac{\partial}{\partial r}\left(r^2\frac{\partial\psi}{\partial r}\right) + \frac{1}{r^2\sin\theta}\frac{\partial}{\partial\theta}\left(\sin\theta\frac{\partial\psi}{\partial\theta}\right) + \frac{1}{r^2\sin^2\theta}\frac{\partial^2\psi}{\partial\phi^2} \quad (33)$$

We derive the expression for the curl analogously, with the aid of Stokes's theorem. Consider first an element of area in the q_1q_2 plane. We compute the line integral about the circuit, following the general method of § (32). Equations (32.3) and (32.4) apply if we merely replace X by F_1 and Y by F_2, where F_1 and F_2 are force components along ds_1 and ds_2 respectively. Thus in place of (32.4) we have for the work done as a unit particle moves around the rectangle,

$$\frac{\partial F_2}{\partial q_1}dq_1\,ds_2 - \frac{\partial F_1}{\partial q_2}dq_2\,ds_1 = \left[\frac{\partial}{\partial q_1}(h_2F_2) - \frac{\partial}{\partial q_2}(h_1F_1)\right]dq_1\,dq_2, \quad (34)$$

wherein we have employed (13). One must include the parameters h_1 and h_2 under the sign of particle differentiation, because they are not generally constant and may vary with the coordinates.

We may now apply Stokes's theorem directly, which requires that (34) be the q_3th component of the curl, i.e., the component normal to the surface element, times the area $ds_1\,ds_2$ of the element. Hence, in general,

$$\nabla\times\mathbf{F} = \frac{\mathbf{i}_1}{h_2h_3}\left[\frac{\partial}{\partial q_2}(h_3F_3) - \frac{\partial}{\partial q_3}(h_2F_2)\right]$$

$$+ \frac{\mathbf{i}_2}{h_3h_1}\left[\frac{\partial}{\partial q_3}(h_1F_1) - \frac{\partial}{\partial q_1}(h_3F_3)\right]$$

$$+ \frac{\mathbf{i}_3}{h_1h_2}\left[\frac{\partial}{\partial q_1}(h_2F_2) - \frac{\partial}{\partial q_2}(h_1F_1)\right]. \quad (35)$$

Hydrodynamics

34. The equation of continuity. Consider a volume τ fixed in a space through which fluid flows. The mass of fluid flowing past an area element of cross-section dS per unit time will be $\rho v_0 \cos\theta\, dS$ where ρ, the density, is a function of x, y, z, and t. The velocity, $\mathbf{v}_0$ has as its Cartesian components u, in the x-direction, v, in the y-direction, and w, in the z-direction. θ is the angle between the velocity vector $\mathbf{v}_0$ and the normal to dS.

The amount of fluid in the volume τ is $\int \rho\, d\tau$. This quantity may change because of production or destruction of fluid, action of sources or sinks, or flow of material across the boundaries. From the law of conservation of matter, the change in the amount of fluid in the volume τ, must be equal to the sum of the rate of inflow and the rate of production, or, in vector notation,

$$\int \frac{\partial \rho}{\partial t} d\tau = -\int \rho \mathbf{v}_0 \cdot d\mathbf{S} + \int \varepsilon \, d\tau, \tag{1}$$

where ε is the rate of production of fluid per unit volume.

Let us take $d\tau$ in the form of a rectangular parallelepiped bounded by the faces x, $x + dx$, y, $y + dy$, z, $z + dz$. The flow across the face x into $d\tau$ is $(\rho u) \, dy \, dz$; across the face $x + dx$ the flow out of $d\tau$ is $[\rho u + (\partial/\partial x)(\rho u) \, dx] \, dy \, dz$. Hence the net flow into $d\tau$, through the pair of faces x and $x + dx$ is the difference of these quantities, or $-(\partial/\partial x)(\rho u)$ $dx \, dy \, dz$. The reasoning is analogous to that used for the derivation of (33.26). A similar analysis, applied to the y and z faces, yields, as the total flow into τ,

$$-\iiint \left(\frac{\partial}{\partial x} \rho u + \frac{\partial}{\partial y} \rho v + \frac{\partial}{\partial z} \rho w \right) dx \, dy \, dz. \tag{2}$$

The quantity $[(\partial/\partial x)\rho u + (\partial/\partial y)\rho v + (\partial/\partial z)\rho w]$ is merely the divergence of $\rho \mathbf{v}_0$; abbreviated to $\nabla \cdot (\rho \mathbf{v}_0)$; it represents the amount of matter *diverging* from the volume. One readily sees that the derivation above is a proof of Gauss' theorem, which indeed could have been used directly, by virtue of (23.19).

From (1) we now have

$$\int \frac{\partial \rho}{\partial t} d\tau = -\int \nabla \cdot (\rho \mathbf{v}_0) \, d\tau + \int \varepsilon \, d\tau, \tag{3}$$

which we differentiate with respect to volume to give the general equation of continuity:

$$\frac{\partial \rho}{\partial t} = -\nabla \cdot (\rho \mathbf{v}_0) + \varepsilon = -\rho \nabla \cdot \mathbf{v}_0 - \mathbf{v}_0 \cdot \nabla \rho + \varepsilon. \tag{4}$$

There are several special cases of (4). When there are no sources or sinks, the term in ε drops out. Thus

$$\frac{\partial \rho}{\partial t} + \rho \nabla \cdot \mathbf{v}_0 + \mathbf{v}_0 \cdot \nabla \rho = 0. \tag{5}$$

For an incompressible fluid, ρ is constant, and $\partial \rho / \partial t = 0$, so that

$$\rho \nabla \cdot \mathbf{v}_0 = \varepsilon. \tag{6}$$

If, now, there are no sources or sinks, and the fluid is incompressible, the equation of continuity reduces to

$$\nabla \cdot \mathbf{v}_0 = 0. \tag{7}$$

The equation of continuity may assume an alternative form in the absence of sources or sinks. In the equation

$$\frac{\partial \rho}{\partial t} + \rho \nabla \cdot \mathbf{v}_0 + \mathbf{v}_0 \cdot \nabla \rho = 0, \tag{8}$$

we perform the indicated differential operations, obtaining

$$\frac{\partial \rho}{\partial t} + u\frac{\partial \rho}{\partial x} + v\frac{\partial \rho}{\partial y} + w\frac{\partial \rho}{\partial z} + \rho\left(\frac{\partial u}{\partial x} + \frac{\partial v}{\partial y} + \frac{\partial w}{\partial z}\right) = 0. \tag{9}$$

Now the total differential of $\rho = f(t, x, y, z)$ is, by a purely mathematical relation, given by

$$d\rho = \frac{\partial \rho}{\partial t}dt + \frac{\partial \rho}{\partial x}dx + \frac{\partial \rho}{\partial y}dy + \frac{\partial \rho}{\partial z}dz. \tag{10}$$

Dividing by dt, and substituting the values u, v, w, for the velocities expressed by dx/dt, dy/dt, dz/dt, we find the equivalent equation

$$\frac{d\rho}{dt} = \frac{\partial \rho}{\partial t} + u\frac{\partial \rho}{\partial x} + v\frac{\partial \rho}{\partial y} + w\frac{\partial \rho}{\partial z} = \frac{\partial \rho}{\partial t} + \mathbf{v}_0 \cdot \nabla \rho. \tag{11}$$

Substracting (11) from (9), we obtain

$$\frac{1}{\rho}\frac{d\rho}{dt} + \frac{\partial u}{\partial x} + \frac{\partial v}{\partial y} + \frac{\partial w}{\partial z} = 0, \quad \text{or} \quad \frac{1}{\rho}\frac{d\rho}{dt} = -\nabla \cdot \mathbf{v}_0. \tag{12}$$

The partial derivative $(\partial\rho/\partial t)$ denotes the rate of change of density at a given point in the coordinate system. The total derivative $d\rho/dt$ denotes the rate of change of density of some definite material element of the fluid as it flows from point to point.* Let V' denote the volume of such an element. Its mass, m, will be constant as it flows through the fluid, or,

$$\rho V' = m = \text{CONST}. \tag{13}$$

By differentiating logarithmically we obtain

$$\frac{1}{\rho}\frac{d\rho}{dt} + \frac{1}{V'}\frac{dV'}{dt} = 0. \tag{14}$$

From (12), then,

$$\nabla \cdot \mathbf{v}_0 = \frac{1}{V'}\frac{dV'}{dt}, \tag{15}$$

which is the rate of change of volume per unit volume.

35. Velocity potentials and vector potentials. In many problems, we can find a single scalar function, ϕ, such that

$$\mathbf{v}_0 = -\nabla\phi, \tag{1}$$

*Some writers use the symbol $D\rho/Dt$ for this fluid hydrodynamic derivative.

or, in more extended notation,

$$u = -\frac{\partial \phi}{\partial x}, \quad v = -\frac{\partial \phi}{\partial y}, \quad w = -\frac{\partial \phi}{\partial z}. \tag{2}$$

In such examples we may write the equation of continuity (34.4):

$$\frac{\partial \rho}{\partial t} = \varepsilon - \nabla \cdot (\rho \mathbf{v}_0) = \rho \nabla^2 \phi + (\nabla \phi) \cdot (\nabla \rho) + \varepsilon. \tag{3}$$

For an incompressible medium with no sources or sinks, this equation becomes

$$\nabla^2 \phi = 0, \tag{4}$$

the Laplace equation in the theory of potentials.

If such a function ϕ exists,

$$\frac{\partial u}{\partial y} - \frac{\partial v}{\partial x} = \frac{\partial^2 \phi}{\partial x\, \partial y} - \frac{\partial^2 \phi}{\partial y\, \partial x} = 0, \tag{5}$$

etc., which by (23.6) and (23.14) is equivalent to

$$\nabla \times \mathbf{v}_0 = \nabla \times \nabla \phi = 0. \tag{6}$$

By mathematical analogy, the scalar function ϕ, from which the vector velocities may be "derived" in the manner that conservative force fields may be derived from a potential, is usually called the *velocity potential.* The component of the velocity that is derivable from a velocity potential is called the linear or lamellar component, $\mathbf{v}_l$. The possibility of representing the velocities by a single function ϕ is an aid in the analysis.

Sometimes, as for instance in a vortex, we can find no such function ϕ, by whose means we can represent the velocities. Assume, however, that we can define the velocities in vortical flow by a single *vector* $\mathbf{V}$, whose scalar components are X, Y, Z. We must not regard $\mathbf{V}$ as a velocity but merely as a mathematical function from which we may determine the velocities by some mathematical process. By definition,

$$\nabla \times \mathbf{V} = \mathbf{i}\left(\frac{\partial Z}{\partial y} - \frac{\partial Y}{\partial z}\right) + \mathbf{j}\left(\frac{\partial X}{\partial z} - \frac{\partial Z}{\partial x}\right) + \mathbf{k}\left(\frac{\partial Y}{\partial x} - \frac{\partial X}{\partial y}\right). \tag{7}$$

We shall proceed to identify the vortical or solenoidal component of velocity, $\mathbf{v}_s$, with the above vector, i.e.,

$$\mathbf{v}_s = \nabla \times \mathbf{V}, \tag{8}$$

and regard the true velocity as the sum of the lamellar and solenoidal components.

$$\begin{aligned}\mathbf{v}_0 = \mathbf{v}_l + \mathbf{v}_s = \mathbf{i}&\left(-\frac{\partial \phi}{\partial x} + \frac{\partial Z}{\partial y} - \frac{\partial Y}{\partial z}\right) \\ &+ \mathbf{j}\left(-\frac{\partial \phi}{\partial y} + \frac{\partial X}{\partial z} - \frac{\partial Z}{\partial x}\right) + \mathbf{k}\left(-\frac{\partial \phi}{\partial z} + \frac{\partial Y}{\partial x} - \frac{\partial X}{\partial y}\right).\end{aligned} \tag{9}$$

Further, by applications of various theorems from § 23, we easily prove the following relationships:

$$\nabla \times \mathbf{v}_0 = \nabla \times \mathbf{v}_s = \nabla \times \nabla \times \mathbf{V} = \nabla(\nabla \cdot \mathbf{V}) - \nabla^2\mathbf{V}, \tag{10}$$

because

$$\nabla \times \mathbf{v}_l = \nabla \times \nabla\phi = 0. \tag{11}$$

also

$$\nabla \cdot \mathbf{v}_0 = \nabla \cdot \mathbf{v}_l = -\nabla^2\phi, \tag{12}$$

since

$$\nabla \cdot \mathbf{v}_s = \nabla \cdot \nabla \times \mathbf{V} = 0. \tag{13}$$

The descriptive terms "lamellar" and "solenoidal" have been taken over into vector analysis and applied to the vectors themselves. The former adjective suggests flow in sheets, the latter flow in circles. We shall, therefore, speak of lamellar or solenoidal vectors. A lamellar vector is, by definition, one whose curl is zero, as in (11), whereas a solenoidal vector is one whose divergence is zero, as in (13).

The representative vector **V** can be uniquely defined by its curl and its divergence, and, since its curl is not equal to zero, we may conveniently assume its divergence to be zero. It is then a solenoidal vector. The vector function **V** from which we may derive the vortical velocities, as in equation (8), is termed the *vector potential.* The vector potential plays an important role in electromagnetic theory. The vector velocity in general is given by

$$\mathbf{v}_0 = -\nabla\phi + \nabla \times \mathbf{V}, \tag{14}$$

which is equation (9) in full vector notation.

36. Euler's hydrodynamic equations of motion. The relationship between the velocities and densities depends on the equation of continuity. This relationship does not, however, consider the pressures and forces. Two other equations are necessary, one to define the densities in relation with the pressures, and one to define the relationship between the forces and the velocities. The latter is Newton's second law. To find the rate of change of momentum, consider a moving volume element $dx\,dy\,dz$ whose coordinates, x, y, z, and whose velocity components, u, v, w, are functions of the time t. Suppose that the density is ρ, and the pressure p. Let the components of external force, **F**, acting on the volume element be X, Y, and Z. The mass of the fluid element is $\rho\,dx\,dy\,dz$. Then the rate of change of momentum, in the x-direction, is

$$\rho\,dx\,dy\,dz\,\frac{du}{dt}. \tag{1}$$

If we are dealing with "ideal" fluid, i.e., one possessing no viscosity, (1) must be equal to the sum of the x components of the internal and external forces. The external force in the x direction is $X\rho\,dx\,dy\,dz$. The internal forces arise from pressure gradients within the fluid. By reasoning analogous to that used in deriving (34.2), we deduce that the resultant x-component of pressure in the volume element is

$$-\frac{\partial p}{\partial x}\,dx\,dy\,dz. \tag{2}$$

Hence the equation of motion becomes

$$\rho\frac{du}{dt} = \rho X - \frac{\partial p}{\partial x}. \tag{3}$$

Note that ρX has the physical dimensions of force per unit volume. From the calculus theorem giving the total derivative of u with respect to t, and by manipulations similar to those of (34.9) and (34.10), we obtain for the equation of motion in the x-direction,

$$\frac{du}{dt} \equiv \frac{\partial u}{\partial t} + u\frac{\partial u}{\partial x} + v\frac{\partial u}{\partial y} + w\frac{\partial u}{\partial z} = X - \frac{1}{\rho}\frac{\partial p}{\partial x}, \tag{4}$$

or, in symbolic vector notation,

$$\frac{du}{dt} \equiv \frac{\partial u}{\partial t} + \mathbf{v}_0 \cdot \nabla u = X - \frac{1}{\rho}\frac{\partial p}{\partial x}. \tag{5}$$

Similarly, for the y- and z-coordinates,

$$\frac{dv}{dt} \equiv \frac{\partial v}{\partial t} + \mathbf{v}_0 \cdot \nabla v = Y - \frac{1}{\rho}\frac{\partial p}{\partial y}, \quad \text{ETC.}$$

Combining these equations into one vector equation, we get

$$\frac{d\mathbf{v}_0}{dt} \equiv \frac{\partial \mathbf{v}_0}{\partial t} + (\mathbf{v}_0 \cdot \nabla)\mathbf{v}_0 = \mathbf{F} - \frac{1}{\rho}\nabla p. \tag{6}$$

Since $|\mathbf{v}_0|^2 = u^2 + v^2 + w^2 = v_0^2$, we may write the term $(\mathbf{v}_0 \cdot \nabla)\mathbf{v}_0$, using the notation of (20.8) in the expansions and simplifications,

$$\begin{aligned}(\mathbf{v}_0 \cdot \nabla)\mathbf{v}_0 &= \nabla\left(\frac{1}{2}v_0^2\right) + 2\mathbf{i}u(\eta - \zeta) + 2\mathbf{j}v(\zeta - \xi) + 2\mathbf{k}w(\xi - \eta)\\ &= \nabla\left(\frac{1}{2}v_0^2\right) + (\nabla \times \mathbf{v}_0)\,\nabla\,\mathbf{v}_0,\end{aligned} \tag{7}$$

where ξ, η, ζ, represent the components of angular velocity. If the last term is not zero, vortical motion is present.

37. Integrals of the hydrodynamic equations of motion. Under certain conditions we may integrate the equation (36.6). When the motion is lamellar, a velocity potential exists, and the velocity becomes $-\nabla\phi$. The last three terms of (36.7) vanish. If the external forces are conservative, they too may be expressed in terms of a potential, V, as

$$\mathbf{F} = -\nabla V. \tag{1}$$

Then Euler's equation of motion becomes

$$-\frac{\partial}{\partial t}\nabla\phi + \nabla\left(\frac{1}{2}v_0^2\right) = -\nabla V - \frac{1}{\rho}\nabla p. \tag{2}$$

Performing the scalar multiplication of this equation by

$$d\mathbf{r} = \mathbf{i}\,dx + \mathbf{j}\,dy + \mathbf{k}\,dz,$$

we see that the resultant scalar equation consists of the sum of various total differentials, by (8.5). Thus

$$-d\left(\frac{\partial\phi}{\partial t}\right) + d\left(\frac{1}{2}v_0^2\right) = -dV - \frac{1}{\rho}\,dp. \tag{3}$$

This equation integrates to

$$\int\frac{dp}{\rho} = \frac{\partial\phi}{\partial t} - V - \frac{1}{2}v_0^2 + C. \tag{4}$$

If ρ is constant and if the flow is steady so that ϕ is independent of t,

$$p + \rho V + \frac{1}{2}\rho v_0^2 = C\rho. \tag{5}$$

The above equation is known as Bernoulli's theorem. The three terms of the left-hand side are respectively the potential energy of the fluid pressure, the potential energy of the external forces, and the kinetic energy. The sum of all three, as the volume element flows along a stream line, is constant for an incompressible fluid. Equation (5) is thus an expression of the law of conservation of energy.

If the fluid is replaced by a perfect gas at uniform temperature, so that it obeys Boyle's law,

$$p = c'\rho, \tag{6}$$

we have, in place of (5),

$$c'\ln p + V + \frac{1}{2}v_0^2 = C. \tag{7}$$

If, instead of (6), the gas obeys the adiabatic law,

$$p = c''\rho^{\gamma}, \tag{8}$$

where γ is the ratio of the specific heat at constant pressure to that at constant volume, the theorem becomes

$$\frac{\gamma}{\gamma - 1}p + \rho V + \frac{1}{2}\rho v_0^2 = C\rho. \tag{9}$$

The integration constants on the right-hand sides of (4), (5), (7), and (9) are all different; the value of the constant in a given medium will in general change from one stream line to another. Since the integration is carried out with respect to the volume, the quantity C may depend upon the time. Variations in C however, represent additions to or subtractions from the energy content of the medium, as with an external source of heat.

38. Circulation and vortical motion. We now turn our attention to vortical motion, for which we shall define several fundamental concepts. This discussion applies, of course, to motion in which the fluid elements possess angular velocity. A *vortex line* is a curve that has everywhere the direction of the axis of rotation of the fluid elements. If vortex lines are drawn through every point on an infinitesimal closed curve, the fluid contained in the tube so formed is said to constitute a *vortex filament.* The tube is so small in cross section that the angular velocity, ω, is constant over the enclosed area a. The *vorticity* is then ωa. The *circulation*, C, along any path in the fluid is defined as the line integral of the tangential velocity along the path, i.e.,

$$C = \int \mathbf{v}_0 \cdot d\mathbf{s} = \int (u\,dx + v\,dy + w\,dz), \tag{1}$$

where $\mathbf{v}_0$ is a vector function of the form

$$\mathbf{v}_0 = \mathbf{i}u(x, y, z) + \mathbf{j}v(x, y, z) + \mathbf{k}w(x, y, z). \tag{2}$$

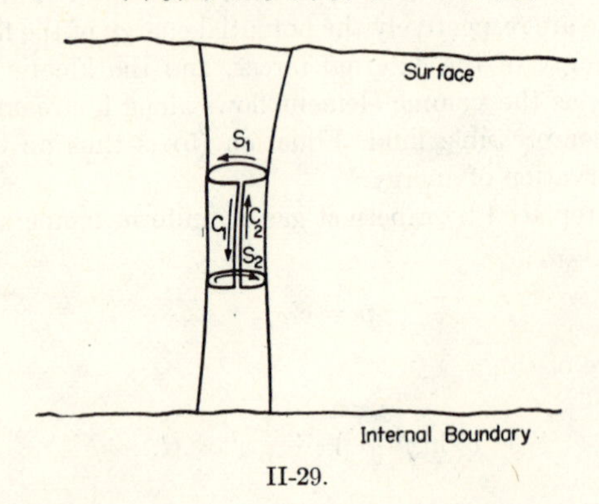

II-29.

Consider the path, Fig. 29, composed of the lines $S_1C_1S_2C_2$, lying wholly upon the surface of the filament and the surface bounded by this path. By Stokes's theorem,

$$C = \int \mathbf{v}_0 \cdot d\mathbf{s} = \iint \nabla \times \mathbf{v}_0 \cdot d\mathbf{S} = \iint 2\boldsymbol{\omega} \cdot d\mathbf{S}, \tag{3}$$

by (20.8). But

$$C = \int_{S_1} \mathbf{v}_0 \cdot d\mathbf{s} + \int_{C_1} \mathbf{v}_0 \cdot d\mathbf{s} + \int_{S_2} \mathbf{v}_0 \cdot d\mathbf{s} + \int_{C_2} \mathbf{v}_0 \cdot d\mathbf{s} = 0, \tag{4}$$

because the vector $\boldsymbol{\omega}$ is everywhere parallel to the surface of the vortex lines, i.e., $\boldsymbol{\omega}$ and $d\mathbf{S}$ are perpendicular, so that $\boldsymbol{\omega} \cdot d\mathbf{S} = 0$. Also

$$\int_{C_1} \mathbf{v}_0 \cdot d\mathbf{s} + \int_{C_2} \mathbf{v}_0 \cdot d\mathbf{s} = 0, \tag{5}$$

because the integrals, taken in opposite directions over the same path, just cancel. Therefore, if the integrals S_1 and S_2 are taken in the same direction, we have

$$\int_{S_1} \mathbf{v}_0 \cdot d\mathbf{s} = \int_{S_2} \mathbf{v}_0 \cdot d\mathbf{s}. \tag{6}$$

The interpretation of this equality is that the circulation about any path enclosing the vortex filament is constant. Applying Stokes's theorem to one such path, S_1, we find

$$C = \int_{S_1} \mathbf{v}_0 \cdot d\mathbf{s} = \iint \nabla \times \mathbf{v}_0 \cdot d\mathbf{S} = \iint 2\boldsymbol{\omega} \cdot d\mathbf{S} = 2\omega a, \tag{7}$$

where a is the area of the tube's cross section, measured perpendicular to the vector $\boldsymbol{\omega}$. Our complete result is that the circulation is constant over the tube and equals $2\omega a$ at any point.

From this law follow several conclusions. First, a vortex filament must end in the surface on some boundary of the liquid or else be re-entrant. Otherwise, if the tube were to end at S_1, below the surface of the fluid as

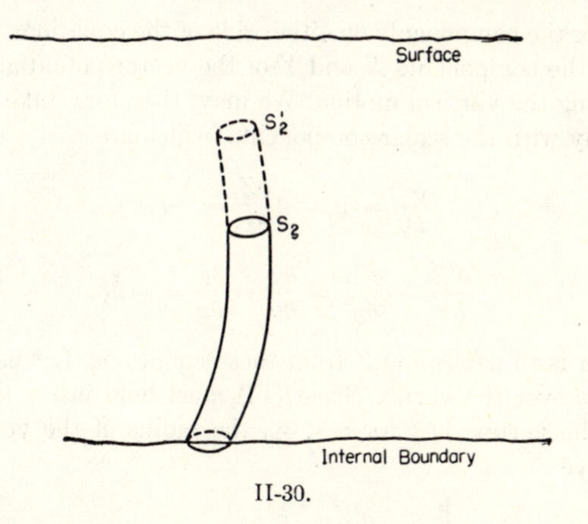

II-30.

in Fig. 30, the line integral $S_1C_1S_1'C_2$, as in (4), would not vanish because the circulation around S' is zero by hypothesis.

And yet equation (4) requires this integral to be zero. The only way to fulfill these conditions is to have S_1 and S_2 lie in a boundary surface of the liquid, be re-entrant to form a vortex ring, or extend to infinity. Second, in the absence of frictional or viscous forces, the vorticity of a tube remains constant with time. Third, since the circulation is zero over the entire surface of the tube, the matter within a given tube will always remain part of the same tube.

We shall now consider the problem of a simple rectilinear vortex, with its axis of rotation along the z-axis, in an incompressible liquid extending to infinity in the xy-plane. Within the vortex, as we have seen, we can represent the velocities in terms of the vector potential $\mathbf{V}$. We assume that $\mathbf{V}$ is solenoidal, so that $\nabla \cdot \mathbf{V} = 0$. Then, by (35.8), (20.8), and (35.10),

$$\mathbf{v}_S = \nabla \times \mathbf{V}, \tag{8}$$

and

$$2\boldsymbol{\omega} = \nabla \times \mathbf{v}_S = -\nabla^2 \mathbf{V}. \tag{9}$$

We set

$$\mathbf{V} = \mathbf{i}X + \mathbf{j}Y + \mathbf{k}Z, \tag{10}$$

and note that the partial derivatives of these components with respect to z must vanish because the vortex is constant along this axis. Hence, by (35.7),

$$\mathbf{v}_S = \mathbf{i}u + \mathbf{j}v + \mathbf{k}w = \mathbf{i}\frac{\partial Z}{\partial y} - \mathbf{j}\frac{\partial Z}{\partial x} + \mathbf{k}\left(\frac{\partial Y}{\partial x} - \frac{\partial X}{\partial y}\right) \tag{11}$$

and

$$2\boldsymbol{\omega} = 2\mathbf{k}\zeta = -\mathbf{i}\left(\frac{\partial^2 X}{\partial x^2} + \frac{\partial^2 X}{\partial y^2}\right) - \mathbf{j}\left(\frac{\partial^2 Y}{\partial x^2} + \frac{\partial^2 Y}{\partial y^2}\right) - \mathbf{k}\left(\frac{\partial^2 Z}{\partial x^2} + \frac{\partial^2 Z}{\partial y^2}\right). \tag{12}$$

Comparing the components on either side of the equations (11) and (12), we see that the components X and Y of the vector potential play no part in determining the vortical motion. We may, therefore, take $X = Y = 0$, and deal only with the scalar components, which are

$$\frac{\partial Z}{\partial y} = u, \qquad \frac{\partial Z}{\partial x} = -v, \tag{13}$$

and

$$\frac{\partial^2 Z}{\partial x^2} + \frac{\partial^2 Z}{\partial y^2} = \frac{\partial u}{\partial y} - \frac{\partial v}{\partial x} = -2\zeta. \tag{14}$$

Our problem is to determine Z from these equations. Let us assume that ζ is constant over the vortex. Then (14) must hold inside the cylindrical column of the vortex, i.e., for $r < r_0$, the radius of the vortex, whereas we must have

$$\frac{\partial^2 Z}{\partial x^2} + \frac{\partial^2 Z}{\partial y^2} = 0 \tag{15}$$

when $r > r_0$ outside the vortex. We are to regard these equations, which we may write in the form

$$\nabla^2 Z = -2\zeta, \quad \text{and} \quad \nabla^2 Z = 0, \tag{16}$$

as the respective analogues of Poisson's and Laplace's equations in two dimensions. The axial symmetry indicates that we should employ cylindrical coordinates. From (33.17) and (33.32), we have

$$\nabla^2 = \frac{1}{r}\left[\frac{\partial}{\partial r}\left(r\frac{\partial}{\partial r}\right)\right], \tag{17}$$

since

$$\frac{\partial Z}{\partial z} = 0, \quad \frac{\partial Z}{\partial \theta} = 0, \tag{18}$$

because of axial symmetry. Since r is the only variable, we may use total derivatives. Equation (16) then becomes

$$\frac{1}{r}\frac{d}{dr}\left(r\frac{dZ}{dr}\right) = -2\zeta \qquad r < r_0. \tag{19}$$

$$\frac{1}{r}\frac{d}{dr}\left(r\frac{dZ}{dr}\right) = 0 \qquad r > r_0. \tag{20}$$

The radial velocity v_r at the point (r, θ) is

$$v_r = u \cos\theta + v \sin\theta = \frac{x}{r}\frac{\partial Z}{\partial y} - \frac{y}{r}\frac{\partial Z}{\partial x} = 0, \tag{21}$$

the radial component vanishing in a cylindrical vortex. The transverse velocity v_θ is

$$v_\theta = -u \sin\theta + v \cos\theta = -\frac{y}{r}\frac{\partial Z}{\partial y} - \frac{x}{r}\frac{\partial Z}{\partial x} = -\frac{dZ}{dr}, \tag{22}$$

since

$$\frac{\partial Z}{\partial x} = \frac{dZ}{dr}\frac{\partial r}{\partial x} = \frac{\partial Z}{dr}\frac{x}{r}, \quad \frac{\partial Z}{\partial y} = \frac{dZ}{dr}\frac{y}{r}.$$

Integrating (19) and (20), we get

$$Z = -\frac{1}{2}\zeta r^2 + A \ln r + B. \qquad r < r_0. \tag{23}$$

$$Z = C \ln r + D. \qquad r > r_0. \tag{24}$$

Within the vortex, which rotates like a rigid cylinder, we must have

$$v_\theta = \zeta r. \tag{25}$$

For (25) to be consistent with (22) and (23) A must be zero, and since the constant B plays no useful role we may conveniently (although not necessarily) take it also to be zero. At the boundary r_0, the two equations (23) and (24) must agree, as must also the tangential components of velocity computed from them by means of (22). Thus

$$-\zeta r_0^2/2 = C \ln r_0 + D, \qquad \zeta r_0 = -C/r_0, \tag{26}$$

which equations define C and D. Hence

$$Z = -\zeta r^2/2, \qquad r \leq r_0, \tag{27}$$

and

$$Z = -\zeta r_0^2 \ln r/r_0 - \zeta r_0^2/2, \qquad r \geq r_0. \tag{28}$$

Outside the vortex, we must not confuse Z with the scalar velocity potential ϕ. The two functions are related as follows:

$$\begin{aligned} u &= -\frac{\partial \varphi}{\partial x} = \frac{\partial Z}{\partial y} = -\zeta y \frac{r_0^2}{r^2}, \\ v &= -\frac{\partial \varphi}{\partial y} = -\frac{\partial Z}{\partial x} = \zeta x \frac{r_0^2}{r^2}, \end{aligned} \tag{29}$$

and

$$\varphi = \zeta r_0^2 \arctan y/x. \tag{30}$$

The tangential velocity

$$v_\theta = \zeta r_0^2/r \tag{31}$$

outside the vortex, is to be compared with the value inside given by (25).

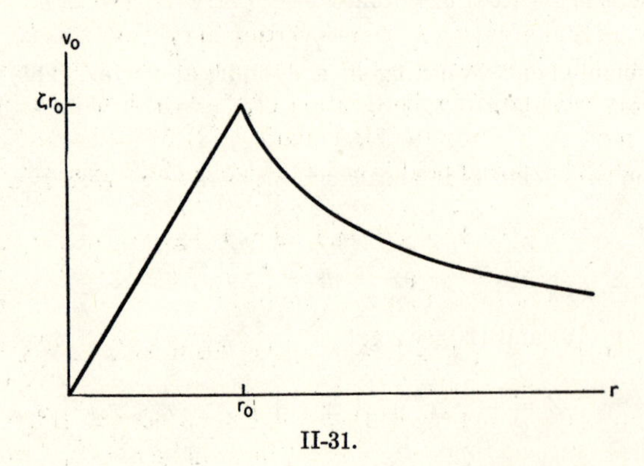

II-31.

A plot of the velocity appears in Fig. 31. Outside the vortex, as within, the motion is circular. The flow outside is nevertheless lamellar, consisting of pure translation. A match floating on the surface will always move parallel to itself, whereas in the vortex the match will rotate and indicate, in turn, all points of the compass.

The function Z has an interesting physical interpretation, apart from its significance as a vector potential. The flux, ψ, across a given surface in the liquid is given by

$$\psi = \iint \mathbf{v} \cdot d\mathbf{S}, \tag{32}$$

where $\mathbf{v}$ is the velocity vector. In the rectilinear vortex we shall consider the surface bounded by a curve AB lying in the surface of the liquid and the unit vector, $\mathbf{k}$, parallel to the z-axis. If $d\mathbf{s}$ is an element of the curve,

$$d\mathbf{S} = d\mathbf{s} \times \mathbf{k} \tag{33}$$

and

$$\psi = \int_{\text{A}}^{\text{B}} \mathbf{v} \cdot d\mathbf{s} \times \mathbf{k} = \int_{\text{A}}^{\text{B}} (-v\,dx + u\,dy), \tag{34}$$

since

$$d\mathbf{s} = \mathbf{i}\,dx + \mathbf{j}\,dy \quad \text{and} \quad \mathbf{v} = \mathbf{i}u + \mathbf{j}v. \tag{35}$$

Making use of (30), we have

$$\psi = \int_{\text{A}}^{\text{B}} \left(\frac{\partial Z}{\partial x}\,dx + \frac{\partial Z}{\partial y}\,dy\right) = \int_{\text{A}}^{\text{B}} dZ = Z_{\text{B}} - Z_{\text{A}}. \tag{36}$$

This equation relates the vector potential to the flux or *current function*, ψ. We see that, in an incompressible liquid, the flux across a surface bounded by $\mathbf{k}$ and any curve whatsoever joining A and B is constant, and equal to the difference between the vector potentials at the points.

We readily calculate the distribution of the liquid. Outside the vortex, since the motion is irrotational, equation (37.5) defines the pressure. Assume that the external field is constant along the z-axis, so that

$$V = gz \quad \text{and} \quad \mathbf{F} = -\mathbf{k}g. \tag{37}$$

Then, introducing $v_0 = v_\theta$ from (31), we have

$$p/\rho = -gz - \zeta^2 r_0^4/2r^2 + c. \tag{38}$$

The pressure vanishes at the surface, and if we assume that the surface at $r = \infty$ lies in the xy-plane, $z = 0$, we must have $c = 0$.

Within the vortex we must return to the equation (36.6), because (37.5) is inapplicable in a vortex. We prove, using (10), that

$$(\mathbf{v}_0 \cdot \nabla)\mathbf{v} = -\zeta^2(\mathbf{i}x + \mathbf{j}y) = -\mathbf{i}_r\zeta^2 r. \tag{39}$$

Also, for steady motion, the partial time derivative vanishes, and (36.6) becomes, in cylindrical coordinates,

$$-\mathbf{i}_r\zeta^2 r + \mathbf{k}g = -\frac{1}{\rho}\left(\mathbf{i}_r\frac{\partial p}{\partial r} + \mathbf{i}_\theta\frac{\partial p}{\partial \theta} + \mathbf{k}\frac{\partial p}{\partial z}\right), \tag{40}$$

which is the equivalent of three partial differential equations. Integrating them successively, we have

$$p/\rho = \zeta^2 r^2/2 + B(\theta, z), \quad p/\rho = B(r, z), \quad p/\rho = -gz + B(r, \theta). \tag{41}$$

The integration constants are, of course, functions of the variables that we treated as constants during the partial differentiation. Combining all three equations, we have

$$p/\rho = -gz + \zeta^2 r^2/2 + B. \tag{42}$$

At the boundary r_0 equations (38) and (42) must give identical values for the pressure. Hence

$$B = -\zeta^2 r_0^2. \tag{43}$$

The pressure vanishes at the surface of the liquid. Therefore the equations of the boundary are

$$z = -\zeta^2(2r_0^2 - r^2)/2g \tag{44}$$

and

$$z = -\zeta^2 r_0^4/2gr^2. \tag{45}$$

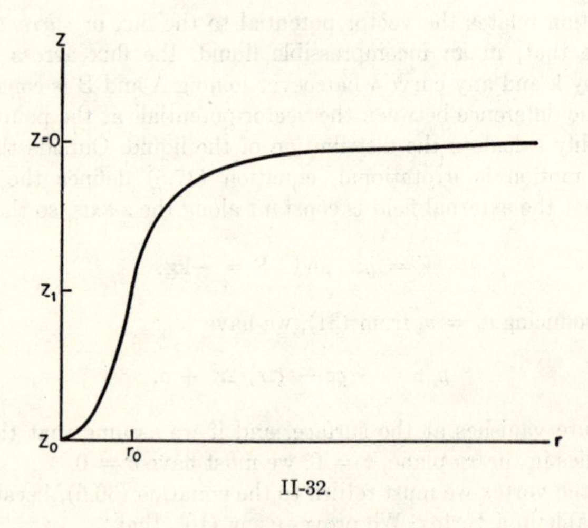

II-32.

The surface profile is shown in Fig. 32. If the bottom of the tank lies above

$$z_0 = -\zeta^2 r_0^2/g, \tag{46}$$

the vortex is partly hollow. If the bottom lies as high as

$$z_1 = -\zeta^2 r_0^2/2g, \tag{47}$$

the entire vortex is hollow, as the diagram indicates.

When the medium contains two or more vortices, the filaments interact to produce relative motion. Consider two equal vortices, rotating in counter directions, and separated by a distance r (Fig. 33). The velocity component contributed by vortex A to the medium at B is $\zeta r_0^2/r$, whereas the joint contribution of both vortices at the point C, halfway between them is $2\zeta r_0^2/(r/2)$, or four times as great. The velocity vectors are perpendicular

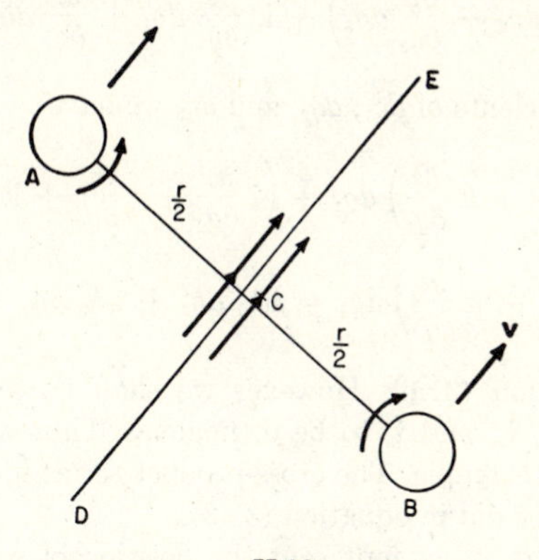

II-33.

to the radius vector. The vortex pair moves parallel to the plane of symmetry DCE, with a velocity one-fourth that of the liquid at C. The flow along the plane, by virtue of symmetry, must be linear, and DCE may be considered a boundary of the fluid. If we remove vortex A entirely and place a wall along DCE, reflection of the liquid from the walls will produce precisely the same effect as the vortex A, and the motion of B will be the same as if A were present. The wall acts like a mirror, the "image" of B taking the place of vortex A.

We repeat that the results of this section hold exactly only for a perfect, i.e., non-viscous liquid.

Principles of Advanced Dynamics

39. Equations of motion in Lagrangian form. In §§ (31) and (33) we introduced the concept of generalized coordinates. From (31.18) we find that an arc ds is given by

$$(ds)^2 = g_{11}\,dq_1^2 + g_{22}\,dq_2^2 + g_{33}\,dq_3^2 + 2g_{12}\,dq_1\,dq_2 + 2g_{23}\,dq_2\,dq_3 + 2g_{31}\,dq_3\,dq_1, \qquad (1)$$

where

$$g_{mn} = \frac{\partial x}{\partial q_m}\frac{\partial x}{\partial q_n} + \frac{\partial y}{\partial q_m}\frac{\partial y}{\partial q_n} + \frac{\partial z}{\partial q_m}\frac{\partial z}{\partial q_n} = g_{nm}. \qquad (2)$$

The vector element, $d\mathbf{s}$, becomes

$$d\mathbf{s} = \mathbf{i}\,dx + \mathbf{j}\,dy + \mathbf{k}\,dz = \mathbf{i}\left(\frac{\partial x}{\partial q_1}dq_1 + \frac{\partial x}{\partial q_2}dq_2 + \frac{\partial x}{\partial q_3}dq_3\right)$$
$$+ \mathbf{j}\left(\frac{\partial y}{\partial q_1}dq_1 + \frac{\partial y}{\partial q_2}dq_2 + \frac{\partial y}{\partial q_3}dq_3\right) + \mathbf{k}\left(\frac{\partial z}{\partial q_1}dq_1 + \frac{\partial z}{\partial q_2}dq_2 + \frac{\partial z}{\partial q_3}dq_3\right). \quad (3)$$

Collecting the coefficients of dq_1, dq_2, and dq_3 we have

$$d\mathbf{s} = \left(\mathbf{i}\frac{\partial x}{\partial q_1} + \mathbf{j}\frac{\partial y}{\partial q_1} + \mathbf{k}\frac{\partial z}{\partial q_1}\right)dq_1 + \left(\mathbf{i}\frac{\partial x}{\partial q_2} + \mathbf{j}\frac{\partial y}{\partial q_2} + \mathbf{k}\frac{\partial z}{\partial q_2}\right)dq_2$$
$$+ \left(\mathbf{i}\frac{\partial x}{\partial q_3} + \mathbf{j}\frac{\partial y}{\partial q_3} + \mathbf{k}\frac{\partial z}{\partial q_3}\right)dq_3 = \mathbf{i}_1 h_1\,dq_1 + \mathbf{i}_2 h_2\,dq_2 + \mathbf{i}_3 h_3\,dq_3, \quad (4)$$

analogous to equation (31.6). However we shall no longer require the new unit vectors $\mathbf{i}_1$, $\mathbf{i}_2$, and $\mathbf{i}_3$ to be orthogonal. Thus we cannot assume that $\mathbf{i}_1 \cdot \mathbf{i}_2 = 0$, for example. The cross-product terms in ds^2 do not necessarily vanish, as they did in equation (33.8).

The coordinates q_1, q_2, q_3 represent the position of a single particle in three-dimensional space. If we are dealing with n particles, instead of with one, we require $3n$ such coordinates.

Suppose, to take a simple example, that we have but two particles, one of which is constrained to move along the x-axis and the other along the y-axis. We require two numbers, x and y, to fix the instantaneous

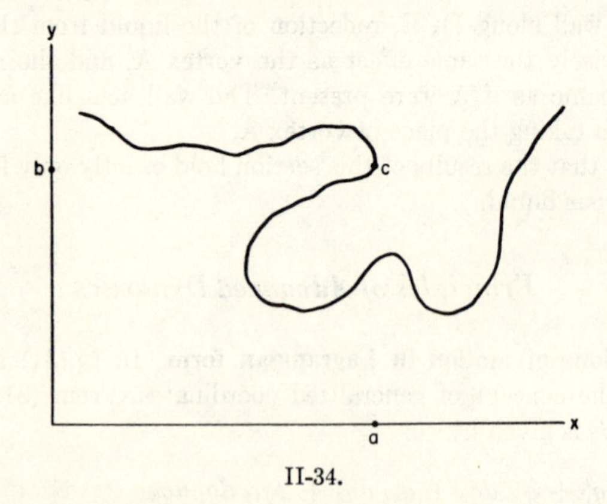

II-34.

position of the particles, a and b, Fig. 34. Instead of representing their positions by *two points* in *one-dimensional* space, we may represent the same information symbolically by *one point* in *two-dimensional* space, as at c. When the particles move, c will describe a path in the plane. The path is clearly not the real path of the particle; nevertheless if the position of c is defined with respect to the time, the motions of a and b are also defined. By means of this artifice, we have to keep track of not two but only one moving point in representing the motions of the particles.

We may extend the argument to the general case of n particles. Instead of having to follow the histories of n points in a space of three dimensions, we represent the instantaneous condition of an assembly of particles by one point in a "configuration" space of $3n$ dimensions. The n-body problem reduces to that of defining a single trajectory. And since our $3n$ spatial dimensions represent the $3n$ coordinates of the particles, the problem will be simplified if we use, not a Euclidean hyperspace, but one wherein the metric conforms most naturally to the motions of the particles. The number of basic points is at most $3n$. Constraints, which reduce the number of degrees of freedom also reduce the number of points required in configuration space.

Fix attention on the ith particle. Instead of representing its position in three-dimensional space, independently of the remaining particles, we suppose that its coordinates conform to the equations ($i = 1$ to n)

$$x_i = X_i(q_1, q_2, \ldots q_{3n}), \quad y_i = Y_i(q_1, q_2, \ldots q_{3n}), \quad z_i = Z_i(q_1, q_2, \ldots q_{3n}), \tag{5}$$

rather than to equations like (33.2). The coordinate x_i depends on the time, as follows:

$$\frac{dx_i}{dt} = \frac{\partial x_i}{\partial q_1}\frac{dq_1}{dt} + \frac{\partial x_i}{\partial q_2}\frac{dq_2}{dt} + \ldots \frac{\partial x_i}{\partial q_{3n}}\frac{dq_{3n}}{dt}, \text{ ETC.,} \tag{6}$$

which we may write in the form

$$\dot{x}_i = \sum_{j=1}^{3n} \frac{\partial x_i}{\partial q_j}\dot{q}_j, \quad \dot{y}_i = \sum_{j=1}^{3n} \frac{\partial y_i}{\partial q_j}\dot{q}_j, \quad \dot{z}_i = \sum_{j=1}^{3n} \frac{\partial z_i}{\partial q_j}\dot{q}_j. \tag{7}$$

The "dot" notation signifies the time derivative. We call the $\dot{q}_j$'s *generalized velocities* even when they do not have the dimensions of a linear velocity. For example, q_j may represent an angle, and $\dot{q}_j$ an angular velocity. The products $(\partial x/\partial q_j)\dot{q}_j$, etc., must have the dimensions $[LT^{-1}]$. We note that

$$\frac{\partial \dot{x}_i}{\partial \dot{q}_j} = \frac{\partial x_i}{\partial q_j}. \tag{8}$$

We have customarily expressed forces as derivatives of the proper potential function, V. By analogy, we introduce the concept of a *generalized force*, so that the component, F_j, along q_j is

$$F_j = -\frac{\partial V}{\partial q_j} = -\frac{\partial V}{\partial x_1}\frac{\partial x_1}{\partial q_j} - \frac{\partial V}{\partial y_1}\frac{\partial y_1}{\partial q_j} - \ldots - \frac{\partial V}{\partial z_n}\frac{\partial z_n}{\partial \dot{q}_j}$$

$$= -\sum_{i=1}^{n}\left(\frac{\partial V}{\partial x_i}\frac{\partial x_i}{\partial q_j} + \frac{\partial V}{\partial y_i}\frac{\partial y_i}{\partial q_j} + \frac{\partial V}{\partial z_i}\frac{\partial z_i}{\partial q_j}\right). \tag{9}$$

The total kinetic energy of the assembly of particles is

$$T = \frac{1}{2} \sum_{i=1}^{n} m_i(\dot{x}_i^2 + \dot{y}_i^2 + \dot{z}_i^2), \tag{10}$$

where m_i is the mass of the ith particle. Form the derivatives

$$\frac{\partial T}{\partial \dot{x}_i} = m_i \dot{x}_i \quad \text{and} \quad \frac{d}{dt}\frac{\partial T}{\partial \dot{x}_i} = \frac{d}{dt}(m_i \dot{x}_i) = X_i, \tag{11}$$

since the time derivative of the momentum is simply the force X_i. But

$$X_i = -\frac{\partial V}{\partial x_i}, \tag{12}$$

so that

$$\frac{d}{dt}\frac{\partial T}{\partial \dot{x}_i} + \frac{\partial V}{\partial x_i} = 0, \tag{13}$$

etc. There are three such equations for every particle. Multiply them by $\partial x_i/\partial q_j$, $\partial y_i/\partial q_j$, $\partial z_i/\partial q_j$, respectively, and add. We get

$$\sum_{i=1}^{n} \left(\frac{\partial x_i}{\partial q_j}\frac{d}{dt}\frac{\partial T}{\partial \dot{x}_i} + \frac{\partial y_i}{\partial q_j}\frac{d}{dt}\frac{\partial T}{\partial \dot{y}_i} + \frac{\partial z_i}{\partial q_j}\frac{d}{dt}\frac{\partial T}{\partial \dot{z}_i} \right) + \frac{\partial V}{\partial q_j} = 0, \tag{14}$$

by (9). We now differentiate (7) partially with respect to a coordinate q_k;

$$\frac{\partial \dot{x}_i}{\partial q_k} = \sum_{j=1}^{3n} \frac{\partial^2 x_i}{\partial q_k\, \partial q_j}\dot{q}_j = \sum_j \frac{\partial}{\partial q_j}\frac{\partial x_i}{\partial q_k}\dot{q}_j = \frac{d}{dt}\frac{\partial x_i}{\partial q_k}. \tag{15}$$

Also the derivative of a product gives

$$\frac{d}{dt}\left(\frac{\partial T}{\partial \dot{x}_i}\frac{\partial x_i}{\partial q_j}\right) = \frac{\partial x_i}{\partial q_j}\frac{d}{dt}\frac{\partial T}{\partial \dot{x}_i} + \frac{\partial T}{\partial \dot{x}_i}\frac{d}{dt}\frac{\partial x_i}{\partial q_j}, \tag{16}$$

whence (14) becomes, by (15), (16) and (8),

$$\sum_i \left[\frac{d}{dt}\left(\frac{\partial T}{\partial \dot{x}_i}\frac{\partial \dot{x}_i}{\partial \dot{q}_j} + \frac{\partial T}{\partial \dot{y}_i}\frac{\partial \dot{y}_i}{\partial \dot{q}_j} + \frac{\partial T}{\partial \dot{z}_i}\frac{\partial \dot{z}_i}{\partial \dot{q}_j} \right) - \left(\frac{\partial T}{\partial \dot{x}_i}\frac{\partial \dot{x}}{\partial q_j} + \frac{\partial T}{\partial \dot{y}_i}\frac{\partial \dot{y}_i}{\partial q_j} + \frac{\partial T}{\partial \dot{z}_i}\frac{\partial \dot{z}_i}{\partial q_j} \right) \right] + \frac{\partial V}{\partial q_j} = 0. \tag{17}$$

By partial differentiation of (10), we find

$$\frac{\partial T}{\partial \dot{q}_j} = \sum_i \left(\frac{\partial T}{\partial \dot{x}_i}\frac{\partial \dot{x}_i}{\partial \dot{q}_j} + \frac{\partial T}{\partial \dot{y}_i}\frac{\partial \dot{y}_i}{\partial \dot{q}_j} + \frac{\partial T}{\partial \dot{z}_i}\frac{\partial \dot{z}_i}{\partial \dot{q}_j} \right). \tag{18}$$

$$\frac{\partial T}{\partial q_j} = \sum_i \left(\frac{\partial T}{\partial \dot{x}_i}\frac{\partial \dot{x}_i}{\partial q_j} + \frac{\partial T}{\partial \dot{y}_i}\frac{\partial \dot{y}_i}{\partial q_j} + \frac{\partial T}{\partial \dot{z}_i}\frac{\partial \dot{z}_i}{\partial q_j} \right). \tag{19}$$

Therefore (17) reduces to

$$\frac{d}{dt}\frac{\partial T}{\partial \dot{q}_j} - \frac{\partial}{\partial q_j}(T - V) = 0, \tag{20}$$

so that if we introduce the *Lagrangian function* or *kinetic potential*,

$$L = T - V, \tag{21}$$

and regard V as a function of the coordinates and not of the velocities, we obtain finally

$$\frac{d}{dt}\frac{\partial L}{\partial \dot{q}_i} - \frac{\partial L}{\partial q_i} = 0. \tag{22}$$

Here L must be expressed as a function of the coordinates and the first time-derivatives. Since the coordinates are general, the Lagrangian equations are independent of the system used. There are three such generalized coordinates for each particle. For many problems, these equations are more convenient than those of Newton, to which they immediately reduce, when we employ a Cartesian system. Making use of (11) and (12), we recover the Newtonian form, cf. (2.3),

$$X_i = \frac{d}{dt}(m\dot{x}_i), \quad \text{ETC.} \tag{23}$$

As an example of the use of generalized coordinates consider the case of orbital motion expressed in plane polar coordinates, r, φ. Then

$$T = \frac{1}{2}m(\dot{r}^2 + r^2\dot{\phi}^2). \tag{24}$$

We may introduce a generalized momentum, p_i:

$$p_i = \frac{\partial T}{\partial \dot{q}_i} = \frac{\partial L}{\partial \dot{q}_i}, \tag{25}$$

for which the respective components are

$$p_r = m\dot{r}, \quad p_\varphi = mr^2\dot{\phi}. \tag{26}$$

The reason for calling L the "kinetic potential" thus becomes apparent, since the momenta are "derived" from it after the manner that the force is derived from the force potential. We obtain the generalized force from (21), (22), and (24):

$$\frac{d}{dt}(m\dot{r}) - mr\dot{\phi}^2 = -\frac{\partial V}{\partial r} = F_r, \tag{27}$$

from (9). If p_r is constant, as for a particle constrained to move in a circle,

$$F_r = -mr\dot{\phi}^2. \tag{28}$$

F_r is termed the *centripetal force*. In this example F_r possesses the appropriate dimensions, but not all forces, so derived, possess the dimensions that we conventionally attribute to this parameter.

40. The Hamiltonian function. We frequently find it advantageous to split the system of second-order Lagrangian differential equations into two systems of first-order equations. We begin by introducing a new function, H, called the Hamiltonian, defined by

$$H = \sum_{j=1}^{3n} p_j\dot{q}_j - L(q_1, q_2 \dots q_{3n}; \dot{q}_1, \dot{q}_2 \dots \dot{q}_{3n}). \tag{1}$$

We shall see later on that H represents the total energy when the assembly is conservative. Since p_j and q_j are related to one another through equation (39.25), we may eliminate the latter variable, if we so desire. The generalized or canonical momenta and the coordinates are then the only parameters required to specify the Hamiltonian. The total differential of H now takes the form

$$dH = \sum_i \left(p_i\, d\dot{q}_i + \dot{q}_i\, dp_i - \frac{\partial L}{\partial q_i}\, dq_i - \frac{\partial L}{\partial \dot{q}_i}\, d\dot{q}_i\right). \tag{2}$$

(For the total differential of a variable, cf. (34.10.) But, from (39.25) and (39.22),

$$p_i = \frac{\partial L}{\partial \dot{q}_i} \quad \text{and} \quad \dot{p}_i = \frac{\partial L}{\partial q_i}. \tag{3}$$

Therefore

$$dH = \sum_i (\dot{q}_i\, dp_i - \dot{p}_i\, dq_i), \tag{4}$$

which we may compare with the identity determined from direct differentiation of H, considered as a function of the p_i's and q_i's.

$$dH = \sum_i \left(\frac{\partial H}{\partial p_i}\, dp_i + \frac{\partial H}{\partial q_i}\, dq_i\right). \tag{5}$$

These two equations will agree only if

$$\partial H/\partial p_i = \dot{q}_i \quad \text{and} \quad \partial H/\partial q_i = -\dot{p}_i, \tag{6}$$

the equations of motion in canonical form. As previously mentioned, these two systems of first-order Hamiltonian equations are equivalent to one system of second-order Lagrangian equations.

Proceeding as in equations (2)–(4), we take the time derivative of H. Thus

$$\frac{dH}{dt} = \sum_i (\dot{q}_i\dot{p}_i - \dot{p}_i\dot{q}_i) = 0. \tag{7}$$

Therefore

$$H = \text{CONST.} \tag{8}$$

in a conservative system.

In rectangular coordinates T is given by (39.10) and $\dot{x}_i$ by (39.7). Substituting the latter into the former, we see that T will be a quadratic function of the generalized velocities, which we write in the form

$$T = \sum_{j=1}^{3n} \sum_{k=1}^{3n} A_{jk}\dot{q}_j\dot{q}_k, \tag{9}$$

where

$$A_{jk} = \sum_i \frac{m_i}{2}\left(\frac{\partial x_i}{\partial q_j}\frac{\partial x_i}{\partial q_k} + \frac{\partial y_i}{\partial q_j}\frac{\partial y_i}{\partial q_k} + \frac{\partial z_i}{\partial q_j}\frac{\partial z_i}{\partial q_k}\right). \tag{10}$$

The terms in the summation for which $j = k$ are to be counted only once. Then

$$\frac{\partial T}{\partial \dot{q}_j} = 2\sum_k A_{jk}\dot{q}_k, \tag{11}$$

the factor 2 appearing because each term of the summation is counted twice, when $j \neq k$. When $j = k$, the resulting single term is $\dot{q}_i^2$, which, when differentiated, again introduces a factor of 2. Then

$$\sum_j p_j\dot{q}_j = \sum_j \frac{\partial L}{\partial \dot{q}_j}\dot{q}_j = \sum_j \frac{\partial T}{\partial \dot{q}_j}\dot{q}_j = 2\sum_k\sum_j A_{jk}\dot{q}_j\dot{q}_k = 2T. \tag{12}$$

Hence, by (39.21), (1) becomes

$$H = 2T - L = T + V = \text{CONST}, \tag{13}$$

the total energy, a fact we wished to prove. If the system is non-conservative, H is no longer the total energy, but merely a function of the coordinates and momenta, as expressed in (1).

41. Poisson brackets. Consider any two variables, u and v, which are continuous and differentiable functions of the coordinates and momenta. We shall introduce the notation

$$[u, v] = \sum_{j=1}^{3n}\left(\frac{\partial u}{\partial p_j}\frac{\partial v}{\partial q_j} - \frac{\partial v}{\partial p_j}\frac{\partial u}{\partial q_j}\right). \tag{1}$$

This expression is known as the *Poisson bracket.*

$$[u, v] = -[v, u]. \tag{2}$$

We have, by direct differentiation and application of (40.6),

$$\dot{u} = \sum_j\left(\frac{\partial u}{\partial q_j}\dot{q}_j + \frac{\partial u}{\partial p_j}\dot{p}_j\right) = \sum_j\left(\frac{\partial u}{\partial q_j}\frac{\partial H}{\partial p_j} - \frac{\partial u}{\partial p_j}\frac{\partial H}{\partial q_j}\right) = [H, u]. \tag{3}$$

In this notation we obtain the canonical equations by setting $u = q_j$ or p_j, in turn, in (3). Thus

$$\dot{q}_j = [H, q_j] \quad\text{and}\quad \dot{p}_j = [H, p_j]. \tag{4}$$

These relations follow directly:

$$\begin{aligned} [q_j, q_k] &= 0 = [q_k, q_j], \\ [p_j, p_k] &= 0 = [p_k, p_j], \\ [p_j, q_k] &= 0 = [q_k, p_j], \qquad j \neq k. \end{aligned} \tag{5}$$

Quantities that obey this rule are said to *commute*, because we may reverse their order in the Poisson bracket, without changing its value. Similarly, we have

$$[p_j, q_j] = 1, \qquad [q_j, p_j] = -1. \tag{6}$$

Variables that do not commute, but whose Poisson bracket is equal to unity, with the proviso that the other combinations, as in (5), are zero, are said to be canonically conjugate. At this point we make use of the notation, originated by Kronecker, which we shall find useful later on. Let δ_{jk} be a number whose value is zero when $k \neq j$, and unity when $k = j$. Then the condition that two systems of variables u_j and v_k be canonically conjugate is

$$[u_j, v_k] = \delta_{jk}, \quad [u_j, u_k] = 0, \quad [v_j, v_k] = 0. \tag{7}$$

The Poisson brackets, which have a limited use in ordinary mechanics, have recently been elevated to a position of considerable importance in wave mechanics. We append for reference several additional relations, the proofs of which are merely extensions of the foregoing. Note the similarity of the formulas to those of ordinary calculus.

$$\begin{aligned} [u, c] &= 0. \quad (c = \text{CONST.}) \\ [u, v + w] &= [u, v] + [u, w]. \\ [u + v, w] &= [u, w] + [v, w]. \\ [uv, w] &= u[v, w] + v[u, w]. \\ [u, vw] &= w[u, v] + v[u, w]. \end{aligned} \tag{8}$$

Consider, now, two sets of functions, P_k and Q_k, related to the dynamical variables so that

$$\begin{aligned} P_k &= P_k(p_1 \ldots p_{3n}, q_1 \ldots q_{3n}), \\ Q_k &= Q_k(p_1 \ldots p_{3n}, q_1 \ldots q_{3n}). \end{aligned} \tag{9}$$

We may suppose these $6n$ equations to be solved simultaneously to give

$$\begin{aligned} p_j &= p_j(P_1 \ldots P_{3n}, Q_1 \ldots Q_{3n}), \\ q_j &= q_j(P_1 \ldots P_{3n}, Q_1 \ldots Q_{3n}). \end{aligned} \tag{10}$$

We form the time derivatives

$$\frac{dP_k}{dt} = \dot{P}_k = \sum_i \left(\frac{\partial P_k}{\partial q_i} \dot{q}_i + \frac{\partial P_k}{\partial p_i} \dot{p}_i \right) = \sum_i \left(\frac{\partial P_k}{\partial q_i} \frac{\partial H}{\partial p_i} - \frac{\partial P_k}{\partial p_i} \frac{\partial H}{\partial q_i} \right), \tag{11}$$

by (40.6). Therefore

$$\dot{P}_k = [H, P_k] \quad \text{and} \quad \dot{Q}_k = [H, Q_k], \tag{12}$$

which are in form similar to (4), the Hamiltonian in bracket notation.

Since H is a function of the p_i's and q_i's, we may also regard it, through (10), as being a function of the P_k's and Q_k's.

Taking the partial derivatives in terms of the new variables, we find

$$\frac{\partial H}{\partial p_j} = \sum_{i=1}^{3n} \left(\frac{\partial H}{\partial P_i} \frac{\partial P_i}{\partial p_j} + \frac{\partial H}{\partial Q_i} \frac{\partial Q_i}{\partial p_j} \right), \tag{13}$$

and similarly for the partial derivative with respect to q_j. (The index i is used as an alternate for k, and does not now distinguish a particle.) We insert these expressions in (11) and rearrange the terms, thus:

$$\dot{P}_k = \sum_i \frac{\partial H}{\partial P_i} \sum_j \left(\frac{\partial P_i}{\partial p_j} \frac{\partial P_k}{\partial q_j} - \frac{\partial P_k}{\partial p_j} \frac{\partial P_i}{\partial q_j} \right) + \sum_i \frac{\partial H}{\partial Q_i} \sum_j \left(\frac{\partial Q_i}{\partial p_j} \frac{\partial P_k}{\partial q_j} - \frac{\partial P_k}{\partial p_j} \frac{\partial Q_i}{\partial q_j} \right). \tag{14}$$

In writing (14) we have reversed the summations over i and j. Also, the derivatives of H, which are independent of j, have been factored from that summation. We note the identity of the parentheses with (1), and accordingly write

$$\dot{P}_k = [H, P_k] = \sum_i \left\{ \frac{\partial H}{\partial P_i} [P_i, P_k] - \frac{\partial H}{\partial Q_i} [P_k, Q_i] \right\}. \tag{15}$$

We shall now restrict our consideration to functions that are canonically conjugate as in (7). Then all the brackets vanish except those for $i = k$, i.e.,

$$[P_k, Q_k] = 1. \tag{16}$$

Therefore

$$\dot{P}_k = -\partial H/\partial Q_k. \tag{17}$$

An analogous derivation for $\dot{Q}_k$ gives

$$\dot{Q}_k = \partial H/\partial P_k. \tag{18}$$

Equations (17) and (18) are thus in canonical form (40.6). They are valid, however, only if the P's and Q's are canonically conjugate, whereas equations (12) are not subject to this restriction.

If, in (12), P_k *commutes* with H,

$$\dot{P}_k = 0, \quad \text{or} \quad P_k = \text{CONST.} \tag{19}$$

Thus a dynamical variable is a constant during the motion of the system if, and only if, it commutes with the Hamiltonian. A transformation of variable such as we have just made, which satisfies the condition (7), is called a *point* or *contact transformation.* The most significant feature of the Hamiltonian equations, which are rarely integrated to give trajectories, is their invariance under a contact transformation.

As a simple example, consider the case of a single particle in a potential field. Try the bracket $[H, L_x]$, where L_x is the x-component of angular momentum for a particle of mass m. By (21.2),

$$L_x = yp_z - zp_y \tag{20}$$

and

$$H = T + V = \frac{1}{2m}(p_x^2 + p_y^2 + p_z^2) + V. \tag{21}$$

In accord with previous notation, we set

$$\partial H/\partial y = \partial V/\partial y = -Y, \quad \text{and} \quad \partial H/\partial z = -Z, \tag{22}$$

the y- and z-components of force. In the bracket the summation is to be carried over x, y, and z, which replace q_1, q_2, and q_3. The number of particles, $n = 1$.

$$[H, L_x] = yZ - zY. \tag{23}$$

Hence H and L_x do not commute unless

$$yZ - zY = 0. \tag{24}$$

But the left-hand side of (24) is merely the x-component of the force moment $\mathbf{r} \times \mathbf{F}$ (cf. equations 22.4 and 6.5), i.e., the torque about the x-axis. Therefore H and L_x commute only when the torque vanishes, and then the angular momentum is constant, as we have previously proved.

The value of the Poisson bracket in quantum mechanics is due in part to symmetry of form and simplicity of notation. But its major use is the investigation of the relations between dynamical variables. Further, in the atomic problem, the major emphasis is placed on observable quantities and these are often constants of motion.

42. Hamilton's principle and least action. We have already noted, in §(1), that there exist numerous alternative approaches to the problems of dynamics. In the foregoing we *assumed* the validity of Newton's equations of motion. Other possible initial assumptions appear as corollaries in the Newtonian method. Analogously, we may deduce the Newtonian equations as corollaries of other approaches.

The broadest of all fundamental dynamical principles is that enunciated by Hamilton, which may be written in the form

$$\delta S = \delta \int_{t_a}^{t_b} L\,dt = \delta \int_{t_a}^{t_b} (T - V)\,dt = 0, \tag{1}$$

where δ symbolizes a variation in S that occurs when we arbitrarily vary the coordinates of the natural path by infinitesimal amounts. This equation requires that S, the time integral of the Lagrangian function for the natural orbit, assume a stationary value. This stationary value is usually a minimum or maximum, compared with that for any neighboring path, traversed in the same time, with the same terminal points.

The physical significance of the problem is best gained from an actual example. Suppose that two identical particles start from a at time $t = t_a$,

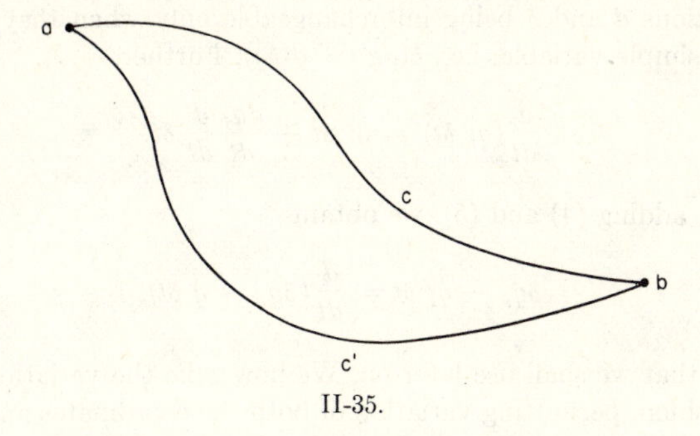

II-35.

the one traversing the natural path acb, Fig. 35, and the other, sliding along a smooth wire, following the "varied" path ac′b, and both arriving simultaneously, at $t = t_b$, at the terminal b. Then Hamilton's principle states that the time integral of L, which is proportional to the average value of the difference of the kinetic and potential energies, must be either *less* or *greater* for the natural trajectory than for any other path. In practice S usually assumes a minimum value.

Hamilton's principle is broader than Newton's because we can apply it more readily to problems where the potential is a function of the time as well as of the coordinates, or to those where no true potential function exists, e.g., when friction is involved.

To show the equivalence of the Hamiltonian and Newtonian points of view we shall derive the former from the latter. Or, more accurately, we shall assume the Lagrangian equations (39.22), which in turn followed from those of Newton, as our basic relationship. The operator δ, in the calculus of variations, is analogous to a differential. It is distinguished from d in that the latter refers to a variation along the trajectory, whereas δ refers to an infinitesimal variation of the position of the trajectory. When the operand is a function of two or more variables, we must apply the rules of calculus in calculating the variation. Thus

$$\delta(uv) = u\,\delta v + v\,\delta u, \tag{2}$$

etc. And if v is expressed as a function of the coordinates q_i and velocities

$\dot{q}$, we determine δv by the calculus rule for determining a total differential:

$$\delta v = \frac{\partial v}{\partial q_1}\,\delta q_1 + \frac{\partial v}{\partial \dot{q}_1}\,\delta \dot{q}_i + \ldots = \sum_i \left(\frac{\partial v}{\partial q_i}\,\delta q_i + \frac{\partial v}{\partial \dot{q}_i}\,\delta \dot{q}_i\right). \tag{3}$$

As an example of the foregoing rule, we may apply (2) to the quantity $\delta \dot{q}_i$:

$$\delta \dot{q}_i = \delta\,\frac{dq_i}{dt} = \frac{1}{dt}\,\delta\,dq_i + dq_i\,\delta\,\frac{1}{dt} = \frac{d(\delta q_i)}{dt} - \frac{dq_i}{dt}\frac{d}{dt}\,\delta t, \tag{4}$$

the operations d and δ being interchangeable only when they operate in turn on a simple variable, i.e., $\delta(dq) = d(\delta q)$. Further,

$$\frac{d}{dt}(\dot{q}_i\,\delta t) = \ddot{q}_i\,\delta t + \frac{dq_i}{dt}\frac{d}{dt}\,\delta t. \tag{5}$$

Therefore, adding (4) and (5), we obtain

$$\delta \dot{q}_i - \ddot{q}_i\,\delta t = \frac{d}{dt}(\delta q_i - \dot{q}\,\delta t), \tag{6}$$

a relation that we shall use later on. We now take the variation of S in a general fashion, permitting variation of both the coordinates and the time.

$$\delta S = \delta \int_{t_a}^{t_b} L\,dt = \int_{t_a}^{t_b} \delta L\,dt + \int_{t_a}^{t_b} L\,d(\delta t). \tag{7}$$

Integrate the last term by parts, to give

$$\begin{aligned}\int_{t_a}^{t_b} L\,d(\delta t) &= [L\,\delta t]_{t_a}^{t_b} - \int_{t_a}^{t_b} dL\,\delta t \\ &= [L\,\delta t]_{t_a}^{t_b} - \int_{t_a}^{t_b} \sum_i \left(\frac{\partial L}{\partial q_i}\,dq_i + \frac{\partial L}{\partial \dot{q}_i}\,d\dot{q}_i\right)\delta t,\end{aligned} \tag{8}$$

wherein we replace the total differential, dL, by the sum of partial derivatives. Similarly,

$$\int_{t_a}^{t_b} \delta L\,dt = \int_{t_a}^{t_b} \sum \left(\frac{\partial L}{\partial q_i}\,\delta q_i + \frac{\partial L}{\partial \dot{q}_i}\,\delta \dot{q}_i\right) dt. \tag{9}$$

In (8) we set

$$dq_i = \frac{dq_i}{dt}\,dt = \dot{q}_i\,dt \quad \text{and} \quad d\dot{q}_i = \ddot{q}_i\,dt, \tag{10}$$

so that the integration is carried out with respect to the time. Then (7) becomes

$$\delta S = \int_{t_a}^{t_b} \sum_i \frac{\partial L}{\partial q_i}(\delta q_i - \dot{q}\,\delta t)\,dt + \int_{t_a}^{t_b} \sum_i \frac{\partial L}{\partial \dot{q}_i}(\delta \dot{q}_i - \ddot{q}_i\,\delta t)\,dt. \tag{11}$$

Evaluate the second integral by parts, with the aid of (6), so that

$$\delta S = \left[L\,\delta t + \sum_i \frac{\partial L}{\partial \dot{q}_i}(\delta q_i - \dot{q}_i\,\delta t) \right]_{t_a}^{t_b}$$

$$+ \int_{t_a}^{t_b} \sum_i \left(\frac{\partial L}{\partial q_i} - \frac{d}{dt}\frac{\partial L}{\partial \dot{q}_i} \right)(\delta q_i - \dot{q}_i\,\delta t)\,dt. \tag{12}$$

Since, in (12), L applies to the actual trajectory, each term of the integrand vanishes, by Lagrange's equation, (39.22). Also, by (39.25) and (40.12),

$$\frac{\partial L}{\partial \dot{q}_i} = p_i, \quad \sum_i \frac{\partial L}{\partial \dot{q}_i}\dot{q} = 2T. \tag{13}$$

Therefore we obtain the final general result that

$$\delta S = [L\,\delta t + \sum_i p_i\,\delta q_i - 2T\,\delta t]_{t_a}^{t_b}. \tag{14}$$

Thus δS is a function only of the variations in time and coordinates of the terminal points. Hence if we set $\delta t = 0$ and $\delta q_i = 0$, as required by Hamilton's principle,

$$\delta S = 0, \tag{15}$$

and S itself is a maximum or minimum along the trajectory. If, therefore, we accept the Lagrangian equations, Hamilton's principle follows as a corollary.

We turn now to an elementary example. Consider a particle moving along the x-axis, in a constant potential field, and proceeding with constant velocity v according to the relation

$$x = vt. \tag{16}$$

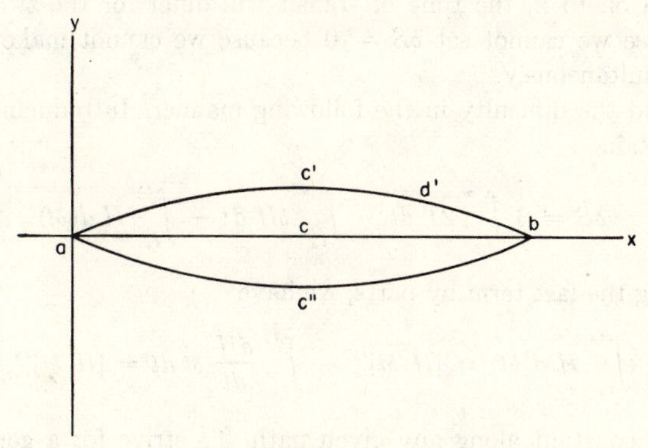

II-36.

A particle moving along one of the varied sinusoidal paths ac′b or ac″b, Fig. 36, described by the equations

$$x = vt, \quad y = \pm B \sin \pi vt/b, \tag{17}$$

will start from the origin at $t_a = 0$ and arrive at the point $x = b$, at time $t_b = b/v$, simultaneously with the particle following path acb. The kinetic potential for the varied paths is

$$L = \frac{m}{2}\left[v^2 + \frac{\pi^2 B^2 v^2}{b^2} \cos^2 \frac{\pi v t}{b}\right] - V. \tag{18}$$

Hence

$$S = \int_0^{b/v} L\, dt = \frac{mvb}{2} + \frac{\pi^2 mv}{4b} B^2 - V \frac{b}{v}, \tag{19}$$

since V is constant. We may vary the path by varying B, or

$$\delta S = (\pi^2 mv/2b) B\, \delta B = 0. \tag{20}$$

Since δB is not zero, this equation requires that $B = 0$, a condition that brings (16) and (17) into agreement, suggesting that (16) represents the natural path of the particle in a field of constant potential.

When the assembly is conservative, we may state Hamilton's principle in somewhat different form. Let us write

$$H = T + V, \quad L = 2T - H. \tag{21}$$

But if we assume that H is the same for the varied as for the natural path, we run into difficulties. The particle moving on the varied path will not be able to reach the terminal point b within the allotted time. In the previous example, the condition (16) requires that the tangential velocity along ac′b be constant. Hence at time $t = t_b$, the particle will be only at d′, where ab = ad′, measured along the curve. Or, if we allow the particle to proceed on to b, the time of transit will differ for the two paths. In consequence we cannot set $\delta S = 0$ because we cannot make δq_i and δt vanish simultaneously.

We avoid the difficulty in the following manner. Introducing (21) into (1), we obtain

$$\delta S = \delta \int_{t_a}^{t_b} 2T\, dt - \int_{t_a}^{t_b} \delta H\, dt - \int_{t_a}^{t_b} H\, d(\delta t). \tag{22}$$

Integrating the last term by parts, we have

$$\int_{t_a}^{t_b} H\, d(\delta t) = [H\, \delta t]_{t_a}^{t_b} - \int_{t_a}^{t_b} \frac{dH}{dt}\, \delta t\, dt = [H\, \delta t]_{t_a}^{t_b}, \tag{23}$$

since H is constant along any given path. To strive for a general result we shall postpone introducing $\delta H = 0$. δH is a constant, however, along the varied trajectory, whose total energy is $H + \delta H$. Then (22) becomes

$$\delta S = \delta \int_{t_a}^{t_b} 2T\, dt - [H\, \delta t]_{t_a}^{t_b} - \delta H(t_b - t_a). \tag{24}$$

But, from (14) and (21), we derive the equivalent expression,

$$\delta S = [-H\,\delta t + \sum_i p_i\,\delta q_i]_{t_a}^{t_b}, \tag{25}$$

which requires that

$$\delta A = \delta \int_{t_a}^{t_b} 2T\,dt = [\sum p_i\,\delta q_i]_{t_a}^{t_b} + \delta H(t_b - t_a). \tag{26}$$

We have now solved our problem, because if we now set $\delta H = 0$,

$$\delta A = 0, \tag{27}$$

independent of the time of transit as long as we make the terminal points coincide, i.e., δq_i must be zero at both limits. The function A is called the *action*, and the principle involved is termed *least action* because it is a minimum (though occasionally a maximum) for the natural path. We usually write A in different form. By (13),

$$A = \int_{t_a}^{t_b} 2T\,dt = \sum_i \int_{q_{ia}}^{q_{ib}} p_i\,dq_i. \tag{28}$$

If n particles are involved and if $\mathbf{p}_i$ is the generalized momentum vector along the tangential orbital element $d\mathbf{s}_i$, of particle i,

$$A = \sum_{i=1}^{n} \int_{s_{ia}}^{s_{ib}} \mathbf{p}_i \cdot d\mathbf{s}_i, \tag{29}$$

the sum of the line integrals of the momenta.

43. Varying action and the Hamilton-Jacobi method. Our application of Hamilton's principle might lead one to conclude that preliminary knowledge of the trajectory is necessary, in order to calculate S and prove its variation zero. We shall now attack the inverse problem and show that we can determine the trajectory itself directly from the principle. We shall rewrite it, however, in somewhat different form. We now assume that the varied trajectory is a natural motion, consistent with the force field. Under these circumstances we can no longer enforce the condition $\delta q_i = 0$, because, even if the initial coordinates agree, the momentum vectors will be different and we cannot force the two natural paths to cross again at any arbitrarily chosen coordinate. Hence, dropping the subscript b in (42.26) as no longer necessary, we have

$$\delta A = \sum_i p_i\,\delta q_i - \sum_i p_{ia}\,\delta q_{ia} + \delta H(t - t_a), \tag{1}$$

instead of $\delta A = 0$. We call (1) the equation of *varied action*.

If we choose the times of transits to be the same for both trajectories, so that A is a function of the initial and final coordinates, the rules of differential calculus require that

$$\delta A = \sum \frac{\partial A}{\partial q_i}\,\delta q_i + \sum \frac{\partial A}{\partial q_{ia}}\,\delta q_a + \frac{\partial A}{\partial H}\,\delta H. \tag{2}$$

Comparison with (1) shows that

$$\partial A/\partial q_i = p_i, \quad \partial A/\partial q_{ia} = -p_{ia}, \quad \partial A/\partial H = t - t_a. \tag{3}$$

Hence we may express the Hamiltonian function, H, which depends on the time, coordinates, and momenta, in the form

$$H = H(q_1 \ldots q_{3n}, p_1 \ldots p_{3n}) = H\left(q_1 \ldots q_{3n}, \frac{\partial A}{\partial q_1} \ldots \frac{\partial A}{\partial q_{3n}}\right), \tag{4}$$

where we have substituted $\partial A/\partial q_i$ for p_i everywhere it appears in the equation of energy. The result is a partial differential equation, which we may solve to give A directly. This method originated with Hamilton and Jacobi. The formulation includes the case where H may be a function of the time. We can best exemplify the procedure by an illustrative case. We again turn to the two-body problem. To shorten the analysis we shall assume in advance that the motion is in a plane, though this result would follow, of course, from a more extended investigation. If M and m are the respective masses of the attracting body and of the particle, we have for the potential energy,

$$V = -GMm/r, \tag{5}$$

and the Hamiltonian becomes, in polar coordinates,

$$\begin{aligned} H = T + V &= \frac{1}{2m}\left(p_r^2 + \frac{1}{r^2}p_\varphi^2\right) - \frac{GMm}{r} \\ &= \frac{1}{2m}\left[\left(\frac{\partial A}{\partial r}\right)^2 + \frac{1}{r^2}\left(\frac{\partial A}{\partial \varphi}\right)^2\right] - \frac{GMm}{r}, \end{aligned} \tag{6}$$

by (3). To effect the solution of this partial differential equation, let us try to guess the functional form of A. For example, test the equation

$$A = R + \Phi, \tag{7}$$

where R and Φ are functions of only r and φ, respectively. Our aim is to achieve complete separation of variables. If (7) does not work we may try more complicated forms, such as $A = R\Phi$, etc. From (7),

$$\partial A/\partial r = dR/dr, \quad \partial A/\partial \varphi = d\Phi/d\varphi, \tag{8}$$

where we use total derivatives because R depends only on r, etc. Hence (6) becomes

$$r^2\left(\frac{dR}{dr}\right)^2 - 2mHr^2 - 2GMm^2r = -\left(\frac{d\Phi}{d\varphi}\right)^2 = -\alpha. \tag{9}$$

The variations in R must be independent of those in Φ. Hence we can satisfy (9) only if both sides are equal to a constant, which we call $-\alpha$. Thus, solving (9) for dR and $d\Phi$, integrating and substituting the results into (7), we obtain

$$A = \int \sqrt{2GMm^2/r - \alpha/r^2 + 2mH}\, dr + \sqrt{\alpha}\, \varphi. \tag{10}$$

We postpone carrying out the integration. Thus far we have employed only the first of the equations (3). We may assume that the quantities q_{ia} and p_{ia} in the second equation represent the initial coordinates and velocities, and therefore play the role of the $6n$ constants of integration demanded in the solution. Thus, to apply the principle of varied action to (10), we should differentiate A partially with respect to the constant representing the initial coordinate and set the result equal to another constant, the negative of the initial momentum. In (10), only one arbitrary constant apart from H appears, i.e., α. Had we not restricted the problem to a plane we should have had still another such constant. The physical dimensions of α, however, are seen to be those of A^2, or $[M^2L^4T^{-2}]$. Hence α can in no way be interpreted as an initial r_a or ϕ_a, though some functional relationship is implied between α and the initial coordinates. For example, $\sqrt{\alpha}$ has the physical dimensions of angular momentum.

Let us suppose that we can express the coordinate q_{ia} as some function, $f_i(c_1 \ldots c_i \ldots)$, of the constants of integration. Then we have, as before,

$$\delta q_{ia} = \sum_k \frac{\partial f_i}{\partial c_k}\, \delta c_k. \tag{11}$$

Substituting this result into (1), we find

$$\delta A = \sum p_i\, \delta q_i - \sum C_i\, \delta c_i + \delta H(t - t_a), \tag{12}$$

where
$$C_i = \sum_k p_{ia} \frac{\partial f_k}{\partial c_i}, \tag{13}$$

another constant. And since we may write

$$\delta A = \sum_i \frac{\partial A}{\partial q_i}\, \delta q_i + \sum \frac{\partial A}{\partial c_i}\, \delta c_i + \frac{\partial A}{\partial H}\, \delta H, \tag{14}$$

analogous to (2), we have as an alternative for the second equation of the principle of varied action, that

$$\frac{\partial A}{\partial c_i} = -C_i. \tag{15}$$

In other words, we are not necessarily required to differentiate with respect to the initial coordinates. We may differentiate A in turn with respect to any constant of integration and set the result equal to another constant. Hence we proceed as follows with (10):

$$\frac{\partial A}{\partial \alpha} = -\frac{1}{2}\int \frac{dr}{\sqrt{-\alpha + 2GMm^2r + 2mHr^2}} + \frac{\varphi}{2\sqrt{\alpha}} = \beta. \tag{16}$$

The integral is of the same form as the one we previously encountered in the two-body problem (26.27). We thus obtain finally for the equation of the orbit,

$$r = \frac{\alpha}{GMm^2 - \sqrt{2\alpha mH + G^2M^2m^4}\cos(\varphi - 2\beta\sqrt{\alpha})}, \tag{17}$$

which agrees with (26.29). To introduce the time into the equations we may employ the third equation of (3) and set

$$\frac{\partial A}{\partial H} = \int \frac{m\,dr}{\sqrt{2GMm^2/r - \alpha/r^2 + 2mH}} = t - t_0, \tag{18}$$

which determines r as a function of the time.

Perhaps one of the most striking facts of the foregoing analysis is the behavior of the action variable A. The action starts out to assume a dominant role. We write down integrals that represent A and go through all the motions of minimizing the action. Then, suddenly, A vanishes from the stage and leaves us with the desired orbit. Had it been necessary for us to determine A explicitly, in terms of the coordinates, we could have done so from equation (10).

The Hamilton-Jacobi method owes its power to the fact that we can represent the dynamical variables as partial derivatives. Also, we work with scalar functions rather than with the vectorial forces of the Newtonian method. For complicated types of forces, we thus can choose coordinate systems natural to the problem.

SELECTED PROBLEMS FOR PART II

1. Three points have coordinates (1, 4, 3), (2, −2, 5), (−3, 1, −3). What are the direction cosines of the lines to these points from the origin? Of the latter two from the first?

2. Two lines have direction cosines $(1/2, \sqrt{6}/4, -\sqrt{6}/4)$ and $(\sqrt{6}/4, 1/4, 3/4)$. Show that they are orthogonal.

3. A potential function $V = x^2 + 2xy + xz =$ const. Find the direction cosines of a normal to the surface at any point (x, y, z).

Ans. $\lambda = \dfrac{2x + 2y + z}{[9x^2 + 4y^2 + z^2 + 4(xy + xz + yz)]^{1/2}}$, etc.

4. Find the force that each of the following potential functions implies:

(a) $V(x) = ax^2$. (d) $V(x) = A \ln \cos ax$.

(b) $V(x, y) = a \cos by + c \sin dx$. (e) $V(x, y, z) = e^{xy}[\tan z - \text{arc cos}(x/y)]$.

(c) $V(x) = -a/x - b/x^2$. (f) $V(r, \theta, \varphi) = (1/r)(e^{\cos\theta} - \ln a\varphi)$.

5. Let the potential of the earth be $V = -GM/r$, where G is the constant of gravitation, M the mass of the earth, and r the distance from the center to some external point. Then, if we set $r = R + h$, where R is the earth's radius and h the height above the surface, show that the force, for small values of h, follows the law:

$$F_h = -\frac{GM}{R^2}\left(1 - 2\frac{h}{R} + \ldots\right).$$

6. Show by direct substitution that the function $P_l^m(\mu)$ satisfies the differential equation

$$(1 - \mu^2)\frac{d^2P}{d\mu^2} - 2\mu\frac{dP}{d\mu} + \left[l(l + 1) - \frac{m^2}{1 - \mu^2}\right]P = 0.$$

7. Calculate the work necessary to move a unit mass along the following paths in the xy-plane from the point (1,0) to (0,1):

(a) counterclockwise in a circle of radius unity.

(b) parallel to the y-axis to the point (1,1) and then parallel to the x-axis.

Assume the following force fields, normal to the xy-plane:

(α) $F = Axy$.

(β) $F = B \ln y$.

(γ) $F = A/\sqrt{x^2 + y^2}$.

(δ) $F = Axyv$, where v is the velocity.

Which of the above are "conservative fields"?

8. Given: $\mathbf{A} = 3\mathbf{i} - 5\mathbf{j} + 2\mathbf{k}$; $\mathbf{B} = 9\mathbf{i} + \mathbf{j} - 4\mathbf{k}$. Find $\mathbf{A} \cdot \mathbf{B}$.

Ans. 14.

9. Given the vector field: $\mathbf{A} = 4x\mathbf{i} - 2xy\mathbf{j} + z^2\mathbf{k}$; $\mathbf{B} = z\mathbf{i} + 2x^2y\mathbf{j} + 3\mathbf{k}$. Find $\nabla(\mathbf{A} \cdot \mathbf{B})$, $\nabla \cdot \mathbf{A}$ and $\nabla \cdot \mathbf{B}$ at the point $(x, y, z) = (1, 1, 3)$.

10. Given: $\mathbf{A} = (x + y)\mathbf{i} + 3z^2\mathbf{j} + \mathbf{k} \cos y$; $\mathbf{B} = 2\mathbf{i} \sin x + 3\mathbf{j}x^2y + \mathbf{k}zx$. Find $\nabla \cdot (\mathbf{A} + \mathbf{B})$ and $\nabla(\mathbf{A} \cdot \mathbf{B})$.

11. Prove $\nabla \cdot u\mathbf{A} = u\nabla \cdot \mathbf{A} + \mathbf{A} \cdot \nabla u$.

12. Find the flux due to the vector $\mathbf{F} = 4xy\mathbf{i} + 3\mathbf{j} + z^3\mathbf{k}$ through the surface of a sphere of radius a whose center is at the origin.

Ans. $\nabla \cdot \mathbf{F} = 4y + 3z^2 = 4r \sin\theta \sin\varphi + 3r^2 \cos\theta$, $\phi = 4\pi a^5/5$.

13. Given: $\phi = y + \sin x$; $\psi = y^2 \cos z$. Verify the second form of Green's analytic theorem for a rectangular box bounded by the planes $x = 0$, $x = a$, $y = 0$, $y = b$, $z = 0$, and $z = c$.

Ans. $$\iint_s \left(\phi\frac{\partial\psi}{\partial n} - \psi\frac{\partial\phi}{\partial n}\right) dS = \iiint (\phi\nabla^2\psi - \psi\nabla^2\phi)\, d\tau;$$

$$\nabla^2\phi = -\sin x; \quad \nabla^2\psi = 2\cos z - y^2 \cos z.$$

14. A particle moves from A $(-a, 0)$ to $B(2a, 0)$ over the rectangular path $(-a, 0)-(-a, a)-(2a, a)-(2a, 0)$ and over the path formed by a straight line from $(-a, 0)$ to $(0, 2a)$ and the arc of a circle from $(0, 2a)$ to $(2a, 0)$. Calculate by integration over the appropriate path (i.e., do not assume a conservative field) the work done in each case if there is present a force $F = k/(x^2 + y^2)$ directed toward the origin.

Ans. $W = +k/2a$.

15. Calculate the work done for the paths of problem (14) if there is no force field present, but the particle moves in a viscous medium which retards the motion of the particle with a force $\mathbf{F} = -k\mathbf{v}$. Assume the particles to move with constant speed.

Ans. (a) $W = 5akv$; (b) $W = 5.38akv$.

16. Find the angle between the vectors: $\mathbf{A}_1 = \mathbf{i} + 2\mathbf{j} + 3\mathbf{k}$; $\mathbf{A}_2 = 2\mathbf{i} + 3\mathbf{j} + 3\mathbf{k}$.

Ans. $\theta = 14°23'$.

17. Given two vectors at the origin: $\mathbf{A} = 2\mathbf{i} + 4\mathbf{j} - \mathbf{k}$; $\mathbf{B} = 3\mathbf{i} - 8\mathbf{j} + 2\mathbf{k}$. Find a unit vector at the origin perpendicular to the given vectors.

Ans. $\mathbf{C} = (1/\sqrt{17})(\mathbf{j} + 4\mathbf{k})$.

18. Find the force field associated with the potential function $V = xy + 2z^2 + 3zyx^3$.

Ans. $\mathbf{F} = -(y + 9zyx^2)\mathbf{i} - (x + 3zx^3)\mathbf{j} - (4z + 3yx^3)\mathbf{k}$.

19. Given the vector function

$$\mathbf{F} = \mathbf{i}2xy + \mathbf{j}(x^2 + 3y^2 \ln z) + \mathbf{k}y^3/z.$$

Find the potential function, V, from which the force is derived.

Ans. $V = -x^2y - y^3 \ln z$.

20. Remove from a uniform thin spherical shell whose center is at the origin, the half of the shell having negative z-coordinates. Calculate the potential at all points P on the z-axis resulting from the remaining hemispherical shell. Let a = radius of spherical shell, M = mass of spherical shell.

Ans.

$$z > a: \quad V = -(MG/z)[\sqrt{1 + (z/a)^2} + 1 - (z/a)];$$

$$z < a: \quad V = -(MG/z)[\sqrt{1 + (z/a)^2} - 1 + (z/a)].$$

21. Find the flux of the vector

$$\mathbf{F} = 2\mathbf{i} \cos y + y^2z\mathbf{j} + z^2\mathbf{k}$$

through a cylinder of radius a and height h whose axis of symmetry extends from $(0, 0, 0)$ to $(0, 0, h)$. Consider the cylinder closed at both ends as well as along the sides.

Ans. $\phi = \pi a^2h^2$.

22. Derive the first three Legendre polynomials from the conditions that $P(x)$ is a polynomial of order l and that

$$\int_{-1}^{1} P_l(x)P_m(x)\, dx = 0 \quad l \neq m,$$

$$\int_{-1}^{1} [P_l(x)]^2\, dx = 2/(2l + 1).$$

23. Find the gravitational potential at a distance R from the center of the base of a solid hemisphere of radius a whose density is directly proportional to the distance from the center of the base. Assume that we can neglect terms beyond those in R^{-2}.

Ans. $V = -(GM/R)(1 + \frac{2}{5} a \cos \theta)$.

24. Prove (a) $\nabla \times \nabla a = 0$. (b) $\nabla \cdot \nabla \times \mathbf{A} = 0$.

25. Prove (a) $\nabla \times (\nabla \times \mathbf{A}) = \nabla\nabla \cdot \mathbf{A} - \nabla^2\mathbf{A}$.
(b) $\nabla \cdot \mathbf{A} \times \mathbf{B} = \mathbf{B} \cdot \nabla \times \mathbf{A} - \mathbf{A} \cdot \nabla \times \mathbf{B}$.

26. Find the area of the triangle defined by the points $(2, -3, 1)$, $(4, -1, 2)$ $(3, 5, -2)$.

Ans. $A = 2\frac{1}{2}$.

27. Given: $\mathbf{A} = 3y\mathbf{i} + 2z^2\mathbf{j} + xy\mathbf{k}$; $\mathbf{B} = x^2\mathbf{i} - 4\mathbf{k}$. Find $\nabla \times (\mathbf{A} \times \mathbf{B})$.

Ans. $(-16z + 4xz^2)\mathbf{j} + 3x^2y\mathbf{k}$.

28. Given the matrices:

$$A = \begin{bmatrix} 0 & 4 & 2 \\ 1 & 3 & 5 \\ 2 & 1 & 2 \end{bmatrix} \quad \text{and} \quad B = \begin{bmatrix} 2 & 3 & 1 \\ 4 & 0 & 2 \\ 1 & 2 & 1 \end{bmatrix}.$$

Find: (a) AB. (b) BA. (c) $A + B$.

Ans.
$$AB = \begin{bmatrix} 18 & 4 & 10 \\ 19 & 13 & 12 \\ 10 & 10 & 6 \end{bmatrix} \text{ etc.}$$

29. (a) Express the following equations in matrix form:

$$\mathbf{v} = (2u_1 + 3u_2 + u_3)\mathbf{i} + 4(u_1 - 3u_2)\mathbf{j} + (u_1 + 2u_2 - u_3)\mathbf{k},$$
$$\mathbf{w} = (3v_1 - 2v_2 + v_3)\mathbf{i} + (v_1 + 3v_2 - v_3)\mathbf{j} + (v_1 - 2v_3)\mathbf{k}.$$

(b) Express in matrix form $\mathbf{w}$ as a function of $\mathbf{u}$.

Ans.
$$\begin{bmatrix} w_1 \\ w_2 \\ w_3 \end{bmatrix} = \begin{bmatrix} -1 & 35 & 2 \\ 13 & -35 & 2 \\ 0 & -1 & 3 \end{bmatrix} \begin{bmatrix} u_1 \\ u_2 \\ u_3 \end{bmatrix}.$$

30. Given two systems of Cartesian coordinates related by the expressions:

$$x' = x \cos \phi + y \sin \phi,$$
$$y' = -x \cos \theta \sin \phi + y \cos \theta \cos \phi + z \sin \theta,$$
$$z' = x \sin \theta \sin \phi - y \sin \theta \cos \phi + z \cos \theta.$$

(a) Prove that the transformation is orthogonal.
(b) Express x, y, and z in terms of x', y', and z'.

31. (a) Prove the distributive law for matrix multiplication:

$$A(B + C) = AB + AC.$$

(b) If $A = I + \alpha$, $B = I + \beta$, where I is the unit matrix and α and β are matrices whose elements α_{ij} and β_{ij} are all infinitesimal, prove $AB = BA$, to first order.

32. Find the eigenvalues of the matrix

$$\begin{bmatrix} 1 & 2 & 0 & 1 \\ 6 & 8 & 2 & 10 \\ 4 & 6 & 1 & 6 \\ 7 & 10 & 2 & 11 \end{bmatrix}.$$

Ans. $0,\ 0,\ (21 + \sqrt{505})/2,\ (21 - \sqrt{505})/2.$

33. Given a triangle with vertices at $(x_1\ y_1)$, $(x_2\ y_2)$, and $(x_3\ y_3)$ prove that its area is

$$A = \pm\tfrac{1}{2}\begin{vmatrix} 1 & 1 & 1 \\ x_1 & x_2 & x_3 \\ y_1 & y_2 & y_3 \end{vmatrix}.$$

34. Given the matrix

$$A = \begin{bmatrix} 1 & 0 & 6 \\ 0 & -2 & 0 \\ 6 & 0 & 6 \end{bmatrix}.$$

(a) Verify the Cayley-Hamilton theorem with respect to this matrix.
(b) Find the adjoint matrix.

Ans.
$$\begin{bmatrix} -12 & 0 & 12 \\ 0 & -30 & 0 \\ 12 & 0 & -2 \end{bmatrix}.$$

(c) Find A^{-1}, (i) using the results of Cayley-Hamilton theorem. (ii) using the adjoint matrix.

Ans.
$$\begin{bmatrix} -1/5 & 0 & 1/5 \\ 0 & -1/2 & 0 \\ 1/5 & 0 & -1/30 \end{bmatrix}.$$

(d) (i) Find the normalized eigenvectors. (ii) Show that these eigenvectors are orthogonal.

Ans.

$$\mathbf{u}_1 = (1/\sqrt{13})(2, 0, 3); \quad \mathbf{u}_2 = (0, 1, 0); \quad \mathbf{u}_3 = (1/\sqrt{13})(-3, 0, 2).$$

(e) Write the diagonal matrix D and the corresponding unitary modal matrix S. Verify by direct multiplication that $S^{-1}AS = D$.

Ans. $$D = \begin{bmatrix} -2 & 0 & 0 \\ 0 & 10 & 0 \\ 0 & 0 & -3 \end{bmatrix}. \quad S = \begin{bmatrix} 2/\sqrt{13} & 0 & -3/\sqrt{13} \\ 0 & 1 & 0 \\ 3/\sqrt{13} & 0 & 2/\sqrt{13} \end{bmatrix}.$$

35. The normalized eigenvectors of problem (34d) define a new set of right-handed Cartesian coordinates with unit vectors $\mathbf{i}'$, $\mathbf{j}'$ and $\mathbf{k}'$. Choose this set in such a manner that the transformation from the $\mathbf{i}\,\mathbf{j}\,\mathbf{k}$ system to the $\mathbf{i}'\,\mathbf{j}'\,\mathbf{k}'$ system represents a rotation about an axis.

(a) Given $\mathbf{r} = 3\mathbf{i} + 4\mathbf{j} + 2\mathbf{k}$, find $\mathbf{r}'$.

Ans. $$\mathbf{r}' = (12/\sqrt{13})\mathbf{i}' + 4\mathbf{j}' - (5/\sqrt{13})\mathbf{k}'.$$

(b) Show that the matrix B of the transformation $\mathbf{r}' = B\mathbf{r}$ is orthogonal.

(c) Show that eigenvectors expressed in the unprimed system transform into unit vectors in the primed system.

(d) Find the angle of rotation.

Ans. $\theta = 33.°7$.

36. Prove that the volume within any closed surface is given by the formula $V = \frac{1}{3} \iint \mathbf{r} \cdot d\mathbf{S}$, where $\mathbf{r}$ is the radius vector and $d\mathbf{S}$ is an element of surface.

37. Describe the trajectory of a particle of charge q and mass m which enters a uniform magnetic field $\mathbf{H}$ with initial velocity $\mathbf{v}$. In vacuum the force on a particle in a magnetic field is $\mathbf{F} = -q\mathbf{H} \times \mathbf{v}$.

Ans. Helix with axis parallel to $\mathbf{H}$; radius $= (vm/qH) \sin \theta$; $\omega = qH/m$; axial velocity $= v \cos \theta$.

38. Prove $\nabla \times (\mathbf{A} \times \mathbf{B}) = (\mathbf{B} \cdot \nabla)\mathbf{A} - \mathbf{B}(\nabla \cdot \mathbf{A}) - (\mathbf{A} \cdot \nabla)\mathbf{B} + \mathbf{A}(\nabla \cdot \mathbf{B})$.

39. Find the gravitational potential of a uniform ring of radius a, at distances R from the center of the ring. Assume $R > a$ and neglect terms beyond those in R^{-3}.

Ans. $V = -(MG/R)[1 + (a^2/2R^2)(\frac{3}{2} \sin^2 \theta - 1)]$.

40. Find the potential at a distance R from the center of the base of a uniformly charged hemisphere of radius a and total charge Q. Neglect terms beyond those in R^{-2}.

Ans. $V = (Q/R)[1 + \frac{3}{8}(a/R) \cos \theta]$.

41. If the hemisphere of problem (40) has a radius of 10 cm and a total charge of $+50$ esu, what force will be exerted on a particle of charge $+20$ esu placed in the equatorial plane of the hemisphere at a distance 80 cm from its center?

Ans. 0.156 dyne at 2°.69 below equatorial plane.

42. Demonstrate that the two following dyadics differ from one another only in respect to the orientations of the coordinate axes. Determine in each case the direction cosines of the three major axes. Reduce to diagonal form.

$$\begin{bmatrix} 28/9 & -16/9 & 2/9 \\ -16/9 & 22/9 & -14/9 \\ 2/9 & -14/9 & 13/9 \end{bmatrix} \quad \text{and} \quad \begin{bmatrix} 2 & 1 & 2 \\ 1 & 3 & 1 \\ 2 & 1 & 2 \end{bmatrix}.$$

Ans. Diagonal matrix $= \begin{bmatrix} 5 & 0 & 0 \\ 0 & 2 & 0 \\ 0 & 0 & 0 \end{bmatrix}.$

Matrix (1) direction cosines: $(2/3, -2/3, 1/3)$;
$(2/3, 1/3, -2/3)$;
$(1/3, 2/3, 2/3)$.

Matrix (2):
$(1/\sqrt{3}, 1/\sqrt{3}, 1/\sqrt{3})$;
$(1/\sqrt{6}, 2/\sqrt{6}, 1/\sqrt{6})$;
$(1/\sqrt{2}, 0, -1/\sqrt{2})$.

43. Transform the dyadic

$$\begin{bmatrix} 1 & 0 & 0 \\ 0 & 2 & 0 \\ 0 & 0 & 3 \end{bmatrix}$$

to axes whose direction cosines are:

$$\left(\frac{1}{\sqrt{3}}, -\frac{1}{\sqrt{3}}, \frac{1}{\sqrt{3}}\right);$$

$$\left(\frac{1+\sqrt{3}}{4}, \frac{1}{2}, \frac{1-\sqrt{5}}{4}\right);$$

and
$$\left(\frac{\sqrt{15}-3\sqrt{3}}{12}, \frac{\sqrt{15}}{6}, \frac{\sqrt{15}+3\sqrt{3}}{12}\right).$$

Ans.
$$\begin{bmatrix} \frac{47-3\sqrt{5}}{24}, & \frac{13-3\sqrt{5}}{24}, & \frac{5}{12} \\ \frac{13-3\sqrt{5}}{24}, & \frac{25}{12}, & \frac{13+3\sqrt{5}}{24} \\ -\frac{5}{12}, & \frac{13+3\sqrt{5}}{24}, & \frac{47+3\sqrt{5}}{24} \end{bmatrix}$$

44. By diagonalization of the inertial dyadic obtain the principal axes of inertia and the momentum of inertia about these axes for a system composed of two similar rods of uniform density extending from $(-3, -4, 0)$ to $(3, 4, 0)$ and $(-4, -3, 0)$ to $(4, 3, 0)$, respectively.

Ans. $I = \frac{50}{3} M$ direction cosines of axis $(0, 0, 1)$;
$I = \frac{24}{3} M$ direction cosines of axis $(1/\sqrt{2}, -1/\sqrt{2}, 0)$;
$I = \frac{1}{3} M$ direction cosines of axis $(1/\sqrt{2}, 1/\sqrt{2}, 0)$;
M is mass of one rod.

45. The earth moves about the sun in an elliptical orbit of eccentricity 0.0167. Calculate the maximum variation of the velocity of the earth in its orbit.

Ans. 1.0 km sec.

46. A plumb line hangs from the ceiling of a plane that is moving due east at a north latitude of 60° with a velocity of 400 mph. What angle does this line make with the direction towards the center of earth?

Ans. 0°.217.

47. Show that a particle moving under the influence of an inverse-square repulsive central force will describe a hyperbola.

48. Find the Laplacian $\nabla^2 V$, given $V = 2r \sin\theta + r^2 \cos\phi$.

Ans. $2 \sin\theta + r \cos\phi(6 \sin^2\theta - 1)$.

49. Given a system of coordinates defined by the relations: $x = 2\sqrt{\xi\eta} \cos\phi$, $y = 2\sqrt{\xi\eta} \sin\phi$, $z = \xi - \eta$ where $0 \leq \phi \leq 2\pi$, $0 \leq \xi$, $0 \leq \eta$.

(a) Show that surfaces of constant ξ and surfaces of constant η represent paraboloids of revolution.

(b) Find the element of distance ds, the element of volume $d\tau$, and the Laplacian operator ∇^2.

Ans.

$$ds = \sqrt{\frac{\xi + \eta}{\xi} d\xi^2 + \frac{\eta + \xi}{\eta} d\eta^2 + 4\xi\eta\, d\varphi^2};$$

$$d\tau = 2(\xi + \eta)\, d\xi\, d\eta\, d\varphi;$$

$$\nabla^2 = \frac{1}{\xi + \eta}\left[\frac{\partial}{\partial\xi}\left(\xi \frac{\partial}{\partial\xi}\right) + \frac{\partial}{\partial\eta}\left(\eta \frac{\partial}{\partial\eta}\right) + \frac{1}{4\xi\eta}\frac{\partial^2}{\partial\varphi^2}\right].$$

50. Given $\mathbf{A} = \mathbf{i}_r r^2 \cos\theta + 2\mathbf{i}_\theta z^2 \sin\theta + \mathbf{i}_z rz$. Verify Gauss' divergence theorem for a cylinder of radius a whose axis of symmetry extends from $z = 0$, $r = 0$, to $z = h$, $r = 0$.

Ans. $\phi = \frac{2}{3}\pi h a^3$.

51. Find the work done by a force field: $\mathbf{F} = 3x^2\mathbf{i} + 2xy\mathbf{j} + x^2z^3\mathbf{k}$ upon a particle moving through one circle of radius a about the origin in the xy-plane.

Ans. Work $= 0$.

52. If the energy of a system is independent of one of the coordinates show that the momentum associated with this coordinate will be constant. Use this result to show that a real velocity is constant in planetary orbits.

53. Obtain the Lagrangian differential equations of motion for a pendulum consisting of a mass suspended by a rigid weightless rod attached to a ball and socket. Neglect any effects of the earth's rotation.

Ans. $a\ddot{\theta} - a\dot{\varphi}^2 \sin\theta \cos\theta + g \sin\theta = 0; \quad \dot{\varphi} \sin^2\theta = \text{const.}$

54. Given: $\mathbf{A} = 2\mathbf{i} + 3\mathbf{j} + \mathbf{k}; \mathbf{B} = -\mathbf{i} + 2\mathbf{j} - 2\mathbf{k}; \mathbf{C} = 2\mathbf{i} - \mathbf{j} + 4\mathbf{k}$. Find $\mathbf{\Phi} = \mathbf{AB}$, in both dyadic and matrix forms. Find $\mathbf{\Phi} \cdot \mathbf{C}$ and $\mathbf{C} \cdot \mathbf{\Phi}$.

55. Given

$$A = \begin{bmatrix} 3 & -2 & 0 \\ 0 & 1 & -1 \\ 3 & -3 & 1 \end{bmatrix}.$$

(a) Find the transposed matrix of A.
Find the adjoint matrix of A.
Find the inverse matrix of A.
(b) Find the diagonal matrix corresponding to A.

56. Given

$$A = \begin{bmatrix} 0 & 1 & 0 & 0 \\ 0 & 0 & 1 & 0 \\ 0 & 0 & 0 & 1 \\ 1 & 0 & 0 & 0 \end{bmatrix}.$$

Show that the diagonal matrix, D, corresponding to A is

$$D = \begin{bmatrix} i & 0 & 0 & 0 \\ 0 & -1 & 0 & 0 \\ 0 & 0 & -i & 0 \\ 0 & 0 & 0 & 1 \end{bmatrix},$$

and find the transformation matrices, T, such that $A = TDT^{-1}$.

57. Given $\mathbf{A} = 2r^2\mathbf{i}_r \cos\theta + r\mathbf{i}_\theta \sin\theta$. Verify Gauss' divergence theorem

$$\int \nabla \cdot \mathbf{A}\, d\tau = \int \mathbf{A} \cdot d\mathbf{S}$$

for a hemispherical shell of radius r_0 bounded by the plane $z = 0$. Hint: Employ spherical coordinates for ∇ and express $\mathbf{A}$ in rectangular coordinates over the plane interface.

58. By diagonalization of the inertial dyadic, obtain the principal axes of inertia and moments of inertia about these axes for a system composed of two similar rods of uniform density extending from $(-12, -5, 0)$ to $(12, 5, 0)$ and from $(-5, -12, 0)$ to $(5, 12, 0)$, respectively. Note: The inertial dyadic is:

$$\mathbf{\Phi} = \iiint \rho(\mathbf{I}r^2 - \mathbf{rr})\, d\tau$$

where $\mathbf{I}$ is the idemfactor.

59. Prove that $\nabla \times a\mathbf{A} = a\nabla \times \mathbf{A} - \mathbf{A} \times \nabla a$, where a is a scalar and $\mathbf{A}$ a vector product.

60. If g is some scalar quantity, prove that

$$\int g\, d\mathbf{s} = -\iint \nabla g \times d\mathbf{S}.$$

Hint: first take the scalar product of the first integral by some constant vector $\mathbf{a}$, thus forming another vector $\mathbf{A} = g\mathbf{a}$. Further reduction depends on Stokes' theorem and simple vector algebra, including the rule for the triple vector product and the theorem for the foregoing problem. Finally, factor out the constant vector.

61. If Φ is some dyadic, prove that Stokes' theorem applies as follows:

$$\int \mathbf{\Phi} \cdot d\mathbf{s} = \iint \nabla \times \mathbf{\Phi} \cdot d\mathbf{S}.$$

62. Prove that the Poisson bracket $[H, t] = 1$.

63. Prove that $[u, (v,w)] + [v, (w,u)] + [w, (u,v)] = 0$.

64. Under what conditions will the angular momentum vector

$$\mathbf{L} = \mathbf{i}L_x + \mathbf{j}L_y + \mathbf{k}L_z$$

be a constant of motion? In other words, show that $\mathbf{L}$ commutes with H, only if the external torque is zero.

65. Show that the Hamiltonian for a particle of mass m, moving in an elliptic orbit in an inverse-square field can be expressed as follows:

$$H = \frac{p_r^2}{2m} + \frac{p_\phi^2}{2r^2 m} - \frac{GMm}{r},$$

where p_r and p_ϕ are the respective radial and angular momenta, and r is the radius vector.

66. Show that, for a particle moving in a plane under the action of a central force, $\dot{\phi}$ is constant only if r is a constant. Take

$$H = 1/2m(p_r^2 + p_\phi^2/r^2) + V(r) \qquad \text{and} \qquad p_\phi = mr^2\dot{\phi}.$$

$V(r)$ is a potential function of r alone.

67. Prove that a pair of uniform, rectilinear vortices, rotating in the same direction and separated by a distance r, will describe circles about their center of gravity with an angular velocity $\omega = 2\zeta r_0^2/r^2$, where r_0 is the vortex radius and ζ the angular velocity of the vortex.

68. Prove, using the method of images, that a linear vortex, located at the point (x,y) in a medium bounded by the planes $x = 0$, $y = 0$, will move in the spiral path $1/x^2 + 1/y^2 = c$. Hint: introduce three "image" vortices symmetrically located in the other quadrants. Evaluate $dx/dy = u/v$.

69. Calculate the Γ^k_{ij} for a system of cylindrical coordinates: $q^1 = r$, $q^2 = \phi$, $q^3 = z$.

Ans. $\Gamma^2_{12} = \Gamma^2_{21} = 1/r$; $\Gamma^1_{22} = -r$; others vanish.

70. Calculate the Γ^k_{ij} for spherical coordinates, $q^1 = r$, $q^2 = \theta$, $q^3 = \phi$.

Ans. $\Gamma^2_{12} = \Gamma^2_{21} = 1/r$; $\Gamma^3_{13} = \Gamma^3_{31} = 1/r$; $\Gamma^1_{22} = -r\sin^2\theta$;

$\Gamma^3_{22} = -\sin\theta\cos\theta$; $\Gamma^2_{23} = \Gamma^2_{32} = \cot\theta$; $\Gamma^1_{33} = -r$; others vanish.

71. Given $\mathbf{v} = \mathbf{e}_i v^i$, and $\mathbf{a} = d\mathbf{v}/dt$, where $\mathbf{v}$ is the velocity and $\mathbf{a}$ the acceleration, prove that

$$\mathbf{a} = \mathbf{e}_i(dv^i/dt + v^j v^k \Gamma^i_{jk}) = \mathbf{e}_i(\ddot{q} + \dot{q}^j \dot{q}^k \Gamma^i_{jk}).$$

72. Combine the results of (69), (70), and (71) to calculate the acceleration terms in (a) cylindrical, and (b) spherical coordinates.

Ans. (a): $a^1 = \ddot{r} - r(\dot{\phi})^2$; $a^2 = \ddot{\phi} + 2\dot{r}\dot{\phi}/r$; $a^3 = \ddot{z}$.

73. Show that $B_{\cdot ijk}^{\cdot\cdot\cdot l}$ vanishes for the coordinate systems of problems (69) and (70). Why?

74. Give the rules for transforming the tensors: T_{ij}; T^{ij}; and T^i_j from a rectangular to a cylindrical coordinate system and thence to a spherical system.

75. Write down the precise formulae for taking (a) the gradient of a scalar function, (b) the divergence of a vector, and (c) the divergence of a tensor in both cylindrical and spherical coordinates.

76. If ϕ is a scalar function in three-space, prove that its gradient can be expressed in the invariant form

$$\nabla\phi = \frac{1}{g}\begin{vmatrix} \mathbf{e}_1 & \mathbf{e}_2 & \mathbf{e}_3 & 0 \\ g_{11} & g_{12} & g_{13} & \dfrac{\partial\phi}{\partial q^1} \\ g_{21} & g_{22} & g_{23} & \dfrac{\partial\phi}{\partial q^2} \\ g_{31} & g_{32} & g_{33} & \dfrac{\partial\varphi}{\partial q^3} \end{vmatrix}.$$

Also show that

$$(\nabla\varphi)\cdot(\nabla\varphi) = \frac{1}{g}\begin{vmatrix} \dfrac{\partial\phi}{\partial q^1} & \dfrac{\partial\phi}{\partial q^2} & \dfrac{\partial\phi}{\partial q^3} & 0 \\ g_{11} & g_{12} & g_{13} & \dfrac{\partial\phi}{\partial q^1} \\ g_{21} & g_{22} & g_{23} & \dfrac{\partial\phi}{\partial q^2} \\ g_{31} & g_{32} & g_{33} & \dfrac{\partial\phi}{\partial q^3} \end{vmatrix}.$$

PART III

Waves and Vibrations

The Wave Equation

1. Derivation of the wave equation. Subject a gaseous medium in equilibrium to a small compressional disturbance. If the original equilibrium was stable, the natural resilience of the gas will return the medium to the initial state. The gas, however, will tend to "overshoot" the equilibrium mark. Oscillations will be set up and the disturbance will spread out from the origin in the form of a wave.

We apply the hydrodynamic equations of motion to calculate the vibrations. Assume that the initial equilibrium density of the medium is ρ_0. As a result of the disturbance the density increases to $\rho_0(1 + \psi)$, where we assume that ψ is small compared with unity. Otherwise the equations would not be linear. Let u, v, and w now refer to the velocity components of the displaced material. Take the velocities so small that we may neglect their squares and products by their spatial derivatives. Then the hydrodynamic equations of motion, (II-36.4), in the absence of an external field of force, become

$$\frac{\partial u}{\partial t} = -\frac{1}{\rho}\frac{\partial p}{\partial x}, \quad \frac{\partial v}{\partial t} = -\frac{1}{\rho}\frac{\partial p}{\partial y}, \quad \frac{\partial w}{\partial t} = -\frac{1}{\rho}\frac{\partial p}{\partial z}. \tag{1}$$

We have also the equation of continuity (II-34.12),

$$\frac{\partial u}{\partial x} + \frac{\partial v}{\partial y} + \frac{\partial w}{\partial z} = -\frac{1}{\rho}\frac{\partial \rho}{\partial t}, \tag{2}$$

and the "equation of state," defining the relation between the pressure and density of the gas. For the present problem we do not require some specific relationship, such as Boyle's law or the adiabatic gas law. We may write, in general,

$$p = f(\rho) = f[\rho_0(1 + \psi)], \tag{3}$$

where f is some function. We expand this formula by Taylor's theorem:

$$p = f(\rho_0) + \rho_0\psi f'(\rho_0) + \frac{\rho_0^2\psi^2}{2!} f''(\rho_0) + \ldots, \tag{4}$$

where f' denotes the first derivative of f with respect to p, etc. Therefore

$$\frac{1}{\rho}\frac{\partial p}{\partial x} \sim \frac{f'(\rho_0)}{1+\psi}\frac{\partial \psi}{\partial x} \sim f'(\rho_0)\frac{\partial \psi}{\partial x} = v^2\frac{\partial \psi}{\partial x}, \tag{5}$$

to the first order of small quantities. We have set

$$f'(\rho_0) = v^2, \tag{6}$$

and similarly for the y and z derivatives. Also

$$\frac{1}{\rho}\frac{\partial \rho}{\partial t} = \frac{\partial \psi}{\partial t}, \quad \text{ETC.} \tag{7}$$

Equations (1) and (2) then become

$$\frac{\partial u}{\partial t} = -v^2\frac{\partial \psi}{\partial x}, \quad \frac{\partial v}{\partial t} = -v^2\frac{\partial \psi}{\partial y}, \quad \frac{\partial w}{\partial t} = -v^2\frac{\partial \psi}{\partial z}, \tag{8}$$

and

$$\frac{\partial u}{\partial x} + \frac{\partial v}{\partial y} + \frac{\partial w}{\partial z} = -\frac{\partial \psi}{\partial t}. \tag{9}$$

Differentiating the three equations (8) with respect to x, y, and z respectively and adding, we obtain the result

$$\frac{\partial}{\partial t}\left(\frac{\partial u}{\partial x} + \frac{\partial v}{\partial y} + \frac{\partial w}{\partial z}\right) = -v^2\left(\frac{\partial^2 \psi}{\partial x^2} + \frac{\partial^2 \psi}{\partial y^2} + \frac{\partial^2 \psi}{\partial z^2}\right). \tag{10}$$

Or, by (9),

$$\frac{\partial^2 \psi}{\partial t^2} = v^2\left(\frac{\partial^2 \psi}{\partial x^2} + \frac{\partial^2 \psi}{\partial y^2} + \frac{\partial^2 \psi}{\partial z^2}\right) = v^2\nabla^2\psi. \tag{11}$$

Assume that a velocity potential ϕ exists for these small displacements. Since

$$u = -\frac{\partial \phi}{\partial x}, \quad \text{ETC.}, \tag{12}$$

equation (8) becomes

$$-\frac{\partial^2 \phi}{\partial x\,\partial t} + v^2\frac{\partial \psi}{\partial x} = \frac{\partial}{\partial x}\left[-\frac{\partial \phi}{\partial t} + v^2\psi\right] = 0.$$

Similarly, we may express the y and z equations of motion as the partial derivatives with respect to the appropriate coordinates of the equation

$$\frac{\partial \phi}{\partial t} - v^2\psi = 0. \tag{13}$$

Differentiate (13) partially with respect to t and make use of (9) and (12). Then

$$\frac{\partial^2 \phi}{\partial t^2} = v^2\nabla^2\phi, \tag{14}$$

an equation of the same form as (11). We call this equation the *wave equation* because its solution indicates that the initial disturbance spreads out as a wave from the origin. Both the velocity potential and the "compression," ψ, obey this wave equation.

Equation (11) is more general than the specific assumptions made above would indicate. The dimensionless quantity ψ may represent a rarefaction as well as a condensation, or it may represent a spatial displacement. The term $(1/\rho)(\partial p/\partial x)$, in equation (5), measures, in general, the acceleration, i.e., the restoring force per unit mass of displaced material. The medium need not be gaseous. The equation of continuity always appears in some form or other. We shall now write u^2 instead of v^2. Then the wave equation becomes

$$\nabla^2\psi = \frac{1}{u^2}\frac{\partial^2\psi}{\partial t^2}. \tag{15}$$

The form taken for ∇^2 depends upon the coordinate system and upon the number of dimensions involved. One may select the appropriate expression for ∇^2 from those given in Part II, § (33).

2. One-dimensional vibrations. For a one-dimensional oscillation, the basic equation is

$$\frac{\partial^2\psi}{\partial x^2} = \frac{1}{u^2}\frac{\partial^2\psi}{\partial t^2}. \tag{1}$$

If the vibrating medium is, for example, the air in an organ pipe, we have, from equation (1.6) that

$$u^2 = f'(\rho). \tag{2}$$

If the vibrations were those of a stretched string we should have had

$$u^2 = T/\rho, \tag{3}$$

where T is the tension and ρ the density of the string. From the form of the wave equation, we perceive that the physical dimensions of u are

$$u = [\mathrm{LT}^{-1}], \tag{4}$$

i.e., u has the dimensions of a velocity.

We have first to show that (1) really represents wave motion. To prove that it does, introduce two new variables, defined by

$$x_1 = x + ut \quad \text{and} \quad x_2 = x - ut. \tag{5}$$

Then

$$\frac{\partial\psi}{\partial x} = \frac{\partial\psi}{\partial x_1}\frac{\partial x_1}{\partial x} + \frac{\partial\psi}{\partial x_2}\frac{\partial x_2}{\partial x} = \frac{\partial\psi}{\partial x_1} + \frac{\partial\psi}{\partial x_2},$$

since
$$\frac{\partial x_1}{\partial x} = \frac{\partial x_2}{\partial x} = 1.$$

$$\frac{\partial^2 \psi}{\partial x^2} = \frac{\partial^2 \psi}{\partial x_1^2} + 2\frac{\partial^2 \psi}{\partial x_1\, \partial x_2} + \frac{\partial^2 \psi}{\partial x_2^2} = \frac{\partial}{\partial x}\left(\frac{\partial \psi}{\partial x_1} + \frac{\partial \psi}{\partial x_2}\right)$$
$$= \frac{\partial}{\partial x_1}\left(\frac{\partial \psi}{\partial x_1} + \frac{\partial \psi}{\partial x_2}\right)\frac{\partial x_1}{\partial x} + \frac{\partial}{\partial x_2}\left(\frac{\partial \psi}{\partial x_1} + \frac{\partial \psi}{\partial x_2}\right)\frac{\partial x_2}{\partial x}.$$

Carrying through a similar reduction for the variable t, we find

$$\frac{1}{u^2}\frac{\partial^2 \psi}{\partial t^2} = \frac{\partial^2 \psi}{\partial x_1^2} - 2\frac{\partial^2 \psi}{\partial x_1\, \partial x_2} + \frac{\partial^2 \psi}{\partial x_2^2}.$$

These two last equations, when substituted in (1), yield the result

$$\frac{\partial^2 \psi}{\partial x_1\, \partial x_2} = 0.$$

The integral of this equation, as one may verify by differentiation, is

$$\psi = f_1(x_1) + f_2(x_2),$$

where f_1 and f_2 are arbitrary functions. Changing to the original variables, we have as the solution of the initial equation,

$$\psi = f_1(x + ut) + f_2(x - ut). \tag{6}$$

Now the function $f_1(x - ut)$ is exactly of the same form as $f_1(x)$, except that all points are shifted a distance ut to the right of the origin. If t is supposed to vary, we may therefore consider that the values of f_1 move to the right with a uniform velocity u; the values of f_2 exhibit a similar progression to the left. These characteristics illustrate why we call (2) the wave equation. Any initial disturbance, periodic or not, will progress from one location to another. This interpretation of the wave equation leads, by equation (2), to a prediction for the velocity of sound in a gas. We have to know merely the form of the equation of state, relating the pressure to the density. If Boyle's law obtains,

$$p = f(\rho) = c'\rho,$$
$$f'(\rho) = c',$$
and
$$u = \sqrt{c'}. \tag{7}$$

We may regard c', a function of the temperature and molecular weight, as known.

If the condensations and rarefactions accompanying the sound waves follow the adiabatic law, in which γ is the ratio of specific heats,

$$p = c''\rho^{\gamma}, \qquad u = \sqrt{\gamma c''\rho^{\gamma-1}}.$$

The latter expression agrees more closely with experiment, whereby we conclude that the pulsations occur so rapidly that the medium has no opportunity to lose energy by radiation or conduction of heat.

3. "Eigenvalues." In our solution thus far we have tacitly assumed the string (or organ pipe) to be infinitely long. Any disturbance whatever will be propagated with velocity u. The medium will sustain a vibration of any frequency. We shall now consider periodic disturbances. When many frequencies are present simultaneously, the resulting ψ will arise from superposition of all of them. We may, however, consider each frequency separately, adding them later, if we wish. This procedure is equivalent to resolving ψ into its individual single periodic components. The displacement for a single periodic vibration may be written

$$\psi = A \sin m(x + ut) + B \cos m(x + ut) + C \sin m(x - ut) + D \cos m(x - ut), \tag{1}$$

where A, B, C, D, are amplitudes. By a simple trigonometric transformation,

$$\psi = (A' \cos mut + B' \sin mut) \sin mx + (C' \sin mut + D' \cos mut) \cos mx, \tag{2}$$

where A', B', C', D' are amplitude constants related to A, B, C, D. Also

$$\frac{\partial\psi}{\partial t} = mu(-A' \sin mut + B' \cos mut) \sin mx + mu(C' \cos mut - D' \sin mut) \cos mx. \tag{3}$$

Thus far we still have the case of the infinite string and there is no restriction on m. But if we consider the string to be rigidly clamped, say at $x = 0$ and $x = X$, at these points both ψ and $\partial\psi/\partial t$ must be zero, no matter what value we assign to t. Since $\cos mx$ is not zero for $x = 0$, we must have

$$C' = D' = 0,$$

The term $\sin mx$ vanishes automatically for $x = 0$. At the point X, the coefficient of $\sin mx$ will not necessarily vanish. Hence

$$\sin mX = 0, \quad mX = \text{arc} \sin 0, \quad mX = k\pi,$$

where

$$k = 0, \pm 1, \pm 2, \quad \text{ETC.} \tag{4}$$

Clamping the wire at two points has forced m and therefore ψ to adopt certain values, which we call "eigenvalues," and eigenfunctions, respectively. If now we substitute for m and sum over all values of k, to allow for simultaneous superposition of displacements from various frequencies, we find

$$\psi = \sum_k \psi_k = \sum_k \sin k\pi \frac{x}{X}\left(A_k \cos \frac{k\pi ut}{X} + B_k \sin \frac{k\pi ut}{X}\right). \tag{5}$$

Since the function ψ_k is a solution of the original equation, subject to the special restrictions imposed on the problem, we may refer to ψ_k as an "eigenfunction," i.e., a function specially selected to fill certain conditions. The quantity ψ is a simple example of a "wave function," formed by superposition of the eigenfunctions. Each term of the summation is periodic with the time, returning to the initial state after an interval t_k defined by

$$\frac{k\pi u t_k}{X} = 2\pi.$$

Hence the frequency, ν_k, of vibration, is

$$\nu_k = \frac{1}{t_k} = \frac{ku}{2X} = \frac{mu}{2\pi}. \tag{6}$$

The condition $k = 0$ corresponds to the state of no vibration; $k = 1$ is the fundamental vibration, with a single loop and nodes at either end of the string. Larger values of k correspond to the overtones. Ordinarily we exclude $k = 0$ from the set of eigenvalues. The negative values of k do not add new vibrational states.

4. Solution in terms of complex variable. There is another more general way of approaching the solution of the wave equation. Our previous result (3.5) showed that each eigenfunction ψ_k was the product of two factors, one, let us say $R(x)$, depending on the coordinates, and the other, $S(t)$, depending only on the time. Thus we may write

$$\psi = R(x)S(t). \tag{1}$$

Taking the second partial derivatives of ψ, first with respect to x and then with respect to t, and introducing the results into (2.1), we find

$$\frac{1}{R}\frac{d^2R}{dx^2} = \frac{1}{u^2S}\frac{d^2S}{dt^2} = -a^2. \tag{2}$$

The first and the second terms now are independent of one another, provided that u does not depend on the coordinates. The derivatives are total, because R is a function of x alone and S a function of t. The variations in S and R are independent. Hence if (2) is to hold, each side must be equal to some constant. Here we shall anticipate the result and set this constant equal to $-a^2$. However, we could have chosen any form and would have found the value $-a^2$ at the end of our solution. The equation for S becomes

$$\frac{d^2S}{dt^2} + a^2u^2S = 0. \tag{3}$$

We recognize (3) as the differential equation of harmonic motion. We may express its solution in the form

$$S = A \cos aut + B \sin aut. \tag{4}$$

Or if we proceed to solve the equation in the normal fashion, by setting $S = e^{mt}$, differentiating, and substituting into (3), we find the auxiliary equation,

$$m^2 + a^2u^2 = 0,$$

or

$$m = \pm aui, \tag{5}$$

where $i = \sqrt{-1}$. We may therefore write

$$S = Ce^{iaut} + De^{-iaut}. \tag{6}$$

The correspondence between (4) and (6) becomes evident when we expand the following functions by Maclaurin's theorem, obtaining the results

$$\left.\begin{aligned} \cos\theta &= 1 - \frac{\theta^2}{2!} + \frac{\theta^4}{4!} - \frac{\theta^6}{6!} + \ldots . \\ \sin\theta &= \theta - \frac{\theta^3}{3!} + \frac{\theta^5}{5!} - \frac{\theta^7}{7!} + \ldots . \\ e^{i\theta} &= 1 + i\theta - \frac{\theta^2}{2!} - i\frac{\theta^3}{3!} + \frac{\theta^4}{4!} + i\frac{\theta^5}{5!} - \ldots . \end{aligned}\right\} \tag{7}$$

Hence

$$\cos\theta + i\sin\theta = e^{i\theta}, \quad \text{and} \quad \cos\theta - i\sin\theta = e^{-i\theta}. \tag{8}$$

Substituting for the exponential in (6), we find that

$$\begin{aligned} S &= C\cos(aut) + D\cos(-aut) + iC\sin(aut) + iD\sin(-aut) \\ &= (C + D)\cos aut + i(C - D)\sin aut. \end{aligned} \tag{9}$$

If we set, in (9),

$$C = C_r + C_i i \quad \text{and} \quad D = D_r + D_i i, \tag{10}$$

where the subscripts r and i denote the real and imaginary parts of the respective coefficients (C_i and D_i are taken as real, however,), we obtain the result

$$\begin{aligned} S = [(C_r + D_r)\cos aut + (D_i - C_i)\sin aut] \\ + i[(C_i + D_i)\cos aut + (C_r - D_r)\sin aut]. \end{aligned} \tag{11}$$

The two terms in brackets, the first representing the real part and the second the imaginary, are individually identical with equation (4), if we take

$$\begin{aligned} (C_r + D_r) = (C_i + D_i) = A. \\ (D_i - C_i) = (C_r - D_r) = B. \end{aligned} \tag{12}$$

Accordingly we may take as our solution for S either the real or the imaginary parts of the resulting equation. Usually we adopt the real part.

We may represent a real number as a point on a line. A single parameter expresses the magnitude of the number. A complex number, on the other hand, by equation (10), is a function of two parameters, one defining the real part and the other the imaginary part. This characteristic suggests that we may possibly find it convenient to represent the number as a point in a plane. Let x denote the real and y the imaginary part of a complex number, z. (The notation z has no relation to the z of three-dimensional Cartesian coordinates.) We shall therefore write

$$z = x + iy, \tag{13}$$

and interpret z as a point in the "complex plane." Or, if we prefer, we may interpret z as a vector. The ordinary vector

$$\mathbf{z} = \mathbf{i}x + \mathbf{j}y \tag{14}$$

has certain features in common with the complex quantity z. Equation (13), however, lends itself to mathematical manipulation. We can treat the algebra of a changing vector in a plane more conveniently by the methods of complex variable than by the methods of vector analysis.

If we denote the distance OZ by r,

$$y = r \sin \theta, \qquad x = r \cos \theta,$$

$$z = x + iy = r(\cos \theta + i \sin \theta) = re^{i\theta}, \tag{15}$$

by (8); hence

$$\theta = \arctan y/x. \tag{16}$$

We call the parameter r the *absolute value* or *modulus* of z. To indicate the operation of finding the *modulus* of a complex quantity we enclose the variable between vertical parallel lines; thus

$$\text{Mod}\, z = |\, z \,| = \sqrt{x^2 + y^2} = r. \tag{17}$$

θ is the so-called *argument*, or *phase*, of z. If $\theta = 0, \pi, 2\pi$, etc., z is real. If $\theta = \pi/2, 3\pi/2, 5\pi/2$, etc., z is a pure imaginary. z goes through a complete cycle when θ changes by 2π.

We shall now show that equation (6), for our purposes, is equivalent to

$$S = Ce^{iaut}, \tag{18}$$

where C itself may be complex. Let

$$C = C_r - iC_i. \tag{19}$$

Substituting from (8), we find that the real part of S,

$$S_r = C_r \cos aut + C_i \sin aut. \tag{20}$$

Thus we obtain no greater generality by including the second term of (6). Only two constants of integration are significant and these are C_r and C_i.

Now, by introducing two new constants, A and α defined by

$$\cos\alpha = C_r/A \qquad \sin\alpha = C_i/A, \tag{21}$$

i.e.,

$$C_r^2 + C_i^2 = A^2, \quad \text{or} \quad A = |S_r|,$$

we may transform equation (20) to

$$S_r = A(\cos aut \cos\alpha + \sin aut \sin\alpha) = A \cos(aut - \alpha), \tag{22}$$

by a simple trigonometric transformation. The symbol α therefore represents the "phase," or the initial angle at $t = 0$. Hence we may write

$$S = Ae^{i(aut-\alpha)}, \tag{23}$$

where A is real. For most problems we may conveniently set $\alpha = 0$. We may also set $A = 1$, since S ordinarily multiplies some other function, whose magnitude changes periodically with the time. We may suppose the factor A to be absorbed in the spatial part of the wave function, because its magnitude is independent of the time.

As we have previously mentioned, S goes through a complete cycle when the value of aut alters by 2π. In wave motion, where the frequency of oscillation is ν times per second, one cycle will occupy ν^{-1} second.

Therefore

$$au\nu^{-1} = 2\pi, \quad \text{or} \quad au = 2\pi\nu. \tag{24}$$

Thus we may also interpret S, in the problem of periodic vibrations, as either the real or the imaginary part of

$$S = e^{2\pi i\nu(t-t_0)}, \tag{25}$$

where

$$\alpha = 2\pi\nu t_0,$$

and t_0 represents the initial time.

We have seen the necessity for evaluating the absolute value and also the real part of a complex quantity. Consider the complex quantities

$$z = x + iy = re^{i\theta}, \qquad z^* = x - iy = re^{-i\theta}.$$

The quantity z^* is known as the *complex conjugate* of z. To find the complex conjugate of any function, replace (i) by $(-i)$ *everywhere* it appears in the function. The square of the absolute value of z is

$$|z|^2 = zz^* = r^2. \tag{26}$$

The real and imaginary parts of z are given, respectively, by the relations

$$\left.\begin{aligned} \mathcal{R}(z) &= \frac{1}{2}(z + z^*). \\ \mathcal{I}(z) &= \frac{1}{2i}(z - z^*). \end{aligned}\right\} \tag{27}$$

5. The form of the wave equation for periodic solutions. When we limit our consideration to only periodic solutions of the wave equation, we may simplify the form of the basic wave equation. In

$$\nabla^2\psi = \frac{1}{u^2}\frac{\partial^2\psi}{\partial t^2}, \tag{1}$$

set

$$\psi = \psi_c S, \tag{2}$$

where ψ_c depends only on the coordinates and S upon the time, and introduce (4.25):

$$S = e^{2\pi i\nu(t-t_0)}. \tag{3}$$

The equation takes the form

$$e^{2\pi i\nu(t-t_0)}\left(\frac{\partial^2\psi_c}{\partial x^2} + \frac{\partial^2\psi_c}{\partial y^2} + \frac{\partial^2\psi_c}{\partial z^2}\right) = -\psi_c 4\pi^2\frac{\nu^2}{u^2}e^{2\pi i\nu(t-t_0)}.$$

Divide this equation through by the common factor. Dropping the subscript c, we shall suppose that ψ refers only to the spatial part of the wave function. The wave equation takes the form

$$\nabla^2\psi + \frac{4\pi^2\nu^2}{u^2}\psi = 0. \tag{4}$$

The time variation will be governed by the real part of (3). The values of ∇^2 in various coordinate systems remain as before.

6. Alternative solution for the vibrating string. The wave equation in one dimension becomes

$$\frac{d^2\psi}{dx^2} + \frac{4\pi^2\nu^2}{u^2}\psi = 0, \tag{1}$$

which is of the form of equation (4.3), the general solution of which was given by the real part of (4.25). Thus we may write

$$\psi = Ae^{2\pi i(\nu/u)(x-x_0)}. \tag{2}$$

The real part of ψ is

$$\psi = A_r\cos 2\pi\frac{\nu}{u}(x - x_0) + A_i\sin 2\pi\frac{\nu}{u}(x - x_0). \tag{3}$$

If the wire is clamped at $x = 0$ and at $x = X$, nodes must occur at these positions, or ψ must be zero. Therefore we must have

$$A_r = 0 \tag{4}$$

and

$$2\pi\frac{\nu}{u}X = k\pi. \tag{5}$$

Then,

$$\psi_k = A_k \sin \frac{k\pi}{X} x, \tag{6}$$

which agrees in giving the spatial part of equation (3.5). Here A_k represents the amplitude of the kth mode of vibration.

7. Fourier series. In part II, § 18, we investigated the orthogonal properties of sine and cosine functions. For example, we showed that

$$N_a^2 = \int_0^{2\pi} \cos m\phi \cos m'\phi \, d\phi = \begin{cases} 0, & m' \neq m, \\ 2\pi, & m' = m = 0, \\ \pi, & m' = m = 1, 2, 3, \ldots, \end{cases} \tag{1}$$

as in (II-18.4). From this property follows an interesting mathematical theorem of far-reaching importance. First, we note that equation (1) possesses similar forms for other limits of integration. Following the method of II, § (18), we prove that

$$N_a^2 = \int_{-\pi}^{\pi} \cos m\phi \cos m'\phi \, d\phi = \begin{cases} 0, & m' \neq m, \\ 2\pi, & m' = m = 0, \\ \pi, & m' = m = 1, 2, 3, \ldots. \end{cases} \tag{2}$$

$$N_b^2 = \int_{-\pi}^{\pi} \sin m\phi \cos m'\phi \, d\phi = 0, \tag{3}$$

$$N_c^2 = \int_{-\pi}^{\pi} \sin m\phi \sin m'\phi \, d\phi = \begin{cases} 0, & m' \neq m, \\ 0, & m' = m = 0, \\ \pi, & m' = m = 1, 2, 3, \ldots. \end{cases} \tag{4}$$

Suppose we are given $f(\phi)$, some arbitrary function of ϕ. For example, we might have $f(\phi) = e^{\alpha\phi}$, or $f(\phi) = a\phi + b\phi^2$. The problem is to represent this function, over an interval 2π, by a series of the form

$$f(\phi) = B_0 + \sum_{m=1}^{\infty} (B_m \cos m\phi + A_m \sin m\phi). \tag{5}$$

This representation is an example of a Fourier series.

We determine the constant coefficients B_m and A_m as follows. Multiply both sides of (5) by $\cos k\phi \, d\phi$ and integrate from $-\pi$ to π. By virtue of the orthogonal properties of the sine and cosine functions, all the integrals of the right-hand side vanish except the cosine term for which $m = k$, and we have, when $k \neq 0$,

$$\int_{-\pi}^{\pi} f(\phi) \cos k\phi \, d\phi = \pi B_k,$$

or

$$B_k = \frac{1}{\pi} \int_{-\pi}^{\pi} f(\phi) \cos k\phi \, d\phi. \tag{6}$$

We must evaluate the constant term B_0 separately. Its value is

$$B_0 = \frac{1}{2\pi}\int_{-\pi}^{\pi} f(\phi)\,d\phi. \tag{7}$$

Similarly, we have

$$A_k = \frac{1}{\pi}\int_{-\pi}^{\pi} f(\phi)\sin k\phi\,d\phi. \tag{8}$$

Substituting these results back into (5) we have

$$f(\phi) = \frac{1}{2\pi}\int_{-\pi}^{\pi} f(\phi)\,d\phi$$

$$+\frac{1}{\pi}\sum_{k=1}^{\infty}\left[\cos k\phi\int_{-\pi}^{\pi} f(\phi)\cos k\phi\,d\phi + \sin k\phi\int_{-\pi}^{\pi} f(\phi)\sin k\phi\,d\phi\right]. \tag{9}$$

Within the definite integrals let us replace ϕ by ψ. This procedure does not change the value of the result, but it enables us to write (9) in an abbreviated form. We obtain, with the aid of the trigonometric expansion for $\cos(\alpha - \beta)$,

$$f(\phi) = \frac{1}{2\pi}\int_{-\pi}^{\pi} f(\psi)\,d\psi + \frac{1}{\pi}\sum_{k=1}^{\infty}\int_{-\pi}^{\pi} f(\psi)\cos k(\psi - \phi)\,d\psi. \tag{10}$$

By changing the limits in equations (2) to (10) from $-\pi$ and π to 0 and 2π, respectively, we easily obtain expansions valid over the latter range. For the theoretical justification and proof of convergence of the Fourier series, we refer the reader to mathematical treatises on the subject. If f is a continuous function in the range $-\pi$ to π, and if $f(-\pi) = f(\pi)$, or $f(0) = f(2\pi)$, the series will be valid over the entire range inclusive of the limits. But if these conditions are not fulfilled, the representation is exclusive of the actual limit. The series will give the mean of the values at the limits. For example, if $f(0) = 0$ and $f(2\pi) = 1$, the representative Fourier series will give the result $f(0) = f(2\pi) = 1/2$. Outside the range of integration the Fourier series cannot represent the function unless $f(\phi)$ is itself periodic with a period of 2π.

We often find it possible to represent a function, $F(x)$, over an arbitrary range of x, say from a to b, by a Fourier series. To derive the series we shall introduce a change of variable. Let

$$F(x) = f(\phi), \quad F(y) = f(\psi), \tag{11}$$

and set

$$x = \frac{(b-a)\phi}{2\pi} + \frac{b+a}{2}, \quad y = \frac{(b-a)\psi}{2\pi} + \frac{b+a}{2}. \tag{12}$$

The variable, y, is to be used in the integration. Note that the limits become

$$\psi = \pi, \quad y = b; \quad \psi = -\pi, \quad y = a;$$
$$dy = \frac{(b-a)}{2\pi}\,d\psi. \tag{13}$$

Substituting these results into (10), we find that

$$F(x) = \frac{1}{b-a}\int_a^b F(y)\,dy + \frac{2}{b-a}\sum_{k=1}^{\infty}\int_a^b F(y)\cos\frac{2\pi k(y-x)}{b-a}\,dy. \tag{14}$$

Expanding the cosine term in the integral, we obtain the respective cosine and sine series for the new range:

$$[F(x)]_{\cos} = \frac{1}{b-a}\int_a^b F(y)\,dy + \frac{2}{b-a}\sum_{k=1}^{\infty}\cos\frac{2k\pi x}{b-a}\int_a^b F(y)\cos\frac{2\pi ky}{b-a}\,dy. \tag{15}$$

$$[F(x)]_{\sin} = \frac{2}{b-a}\sum_{k=1}^{\infty}\sin\frac{2k\pi x}{b-a}\int_a^b F(y)\sin\frac{2k\pi y}{b-a}\,dy. \tag{16}$$

We shall now make a practical application of the foregoing formulae. Equation (6.7) represents the sinusoidal displacement of the string caused by the kth mode of vibration. Each mode gives rise to a pure note of single frequency. Normally, the motion of a vibrating string will consist of the superposition of all possible modes, the so-called "principle of superposition." What modes will be present or what ones will predominate depend on the initial displacement of the string. A musician will recall that when any open string is plucked or bowed with one finger lightly touching the center of the string, the fundamental tone is suppressed and the first harmonic emphasized. The finger at the center induces a node to appear at that point.

We may thus assume, in general, that the actual motion of the string conforms to the equation

$$\psi(x, t) = \sum_{k=0}^{\infty}\psi_k(x)\cos\frac{k\pi ut}{X}, \tag{17}$$

in accord with (3.5). We have taken the amplitude of the sine term equal to zero, to make the velocity $d\psi/dt$ vanish at $t = 0$, so that the initial displacement, by (6.7), is

$$\psi(x) = \sum_{k=1}^{\infty}\psi_k(x) = \sum_{k=1}^{\infty}A_k\sin\frac{k\pi x}{X}. \tag{18}$$

Equation (18) is a Fourier series, for which we are to evaluate the coefficients by the method previously outlined in this section. As we compare the argument of the sine term in (18) with that of (16), we see that

$$2k\pi x/(b-a) = k\pi x/X. \tag{19}$$

Therefore the entire range of integration must be $(b - a) = 2X$, or *twice* the length of the string. We readily resolve this apparent difficulty. We have merely to assume that the string is actually of length $2X$, stretching say from $-X \leq x \leq X$. $\psi(x)$ is given specifically for only the positive segment of the string. However, we assume that $\psi(x)$ also represents the displacements over the negative half, with the appropriate value of $(-x)$, sign included, substituted in the equation.

Consider an initial distortion of the string into the form of an isosceles triangle, with maximum amplitude c. Then we shall have

$$\begin{aligned} \psi(x) &= -2c(X - x)/X, & -X \leq x \leq -X/2. \\ \psi(x) &= 2cx/X, & -X/2 \leq x \leq X/2. \\ \psi(x) &= 2c(X - x)/X, & X/2 \leq x \leq X. \end{aligned} \tag{20}$$

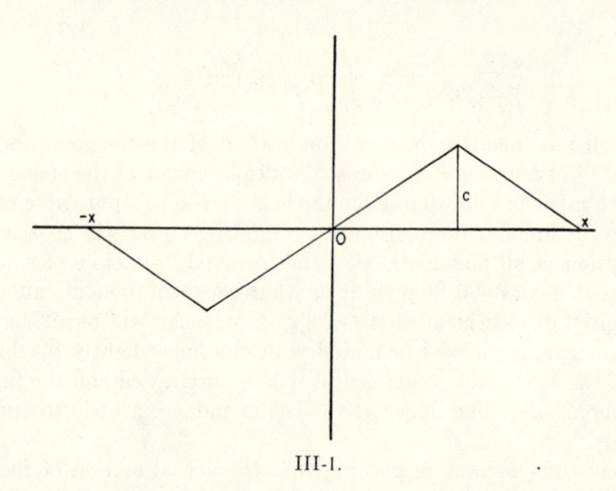

III-1.

The form of the function appears in Fig. 1. Since our final interest is in the positive half, the resulting series is sometimes called a *half-range* Fourier series, to distinguish it from the *full-range* series, where the entire range is necessary to represent the given function.

We have thus broken up the range of integration into three parts. The resulting cosine series (15) vanishes. The sine series (16) becomes

$$\begin{aligned} \psi(x) &= \frac{1}{X} \sum_{k=1}^{\infty} \sin \frac{k\pi x}{X} \left[\int_{-X}^{-X/2} \frac{2c(X - y)}{X} \sin \frac{k\pi y}{X}\, dy \right. \\ &\qquad \left. + \int_{-X/2}^{X/2} \frac{2cy}{X} \sin \frac{k\pi y}{X}\, dy + \int_{X/2}^{X} \frac{2c(X - y)}{X} \sin \frac{k\pi y}{X}\, dy \right] \\ &= \frac{8c}{\pi^2} \sum_{1}^{\infty} \frac{1}{k^2} \sin \frac{k\pi x}{X} \sin \frac{k\pi}{2}. \end{aligned} \tag{21}$$

In evaluating the integrals we have made use of the relation

$$\int \theta \sin \theta \, d\theta = \sin \theta - \theta \cos \theta. \tag{22}$$

The terms of even k vanish, and the series alternates in sign. Thus, finally,

$$\psi(x) = \frac{8c}{\pi^2}\left(\frac{\sin \pi x/X}{1^2} - \frac{\sin 3\pi x/X}{3^2} + \frac{\sin 5\pi x/X}{5^2} - \ldots\right). \quad (23)$$

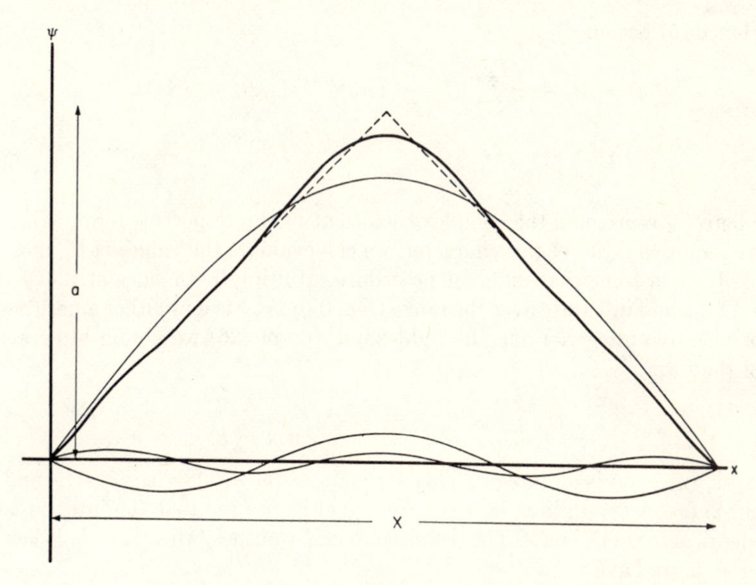

III-2.

The series is rapidly convergent. The first three terms have been plotted in Fig. 2. If the ordinates of even these few terms are added, the resulting displacement approximates very closely the triangle representing the initial ψ.

Fourier series are useful in many types of investigation, especially when one wishes to represent, over a certain range of variable, a function whose derivative is discontinuous, as in the example here considered, where an abrupt change occurs from one line to another. The student should be warned, however, that the representative Fourier series cannot always suffice for all types of mathematical analysis. Differentiation, for example, introduces increasingly large coefficients into the higher terms, which *may* destroy the convergence. Thus if

$$\psi = \sum A_k \cos k\theta, \quad d\psi/d\theta = -\sum kA_k \sin k\theta. \quad (24)$$

One must test the derivatives for convergence. If the original function has finite derivatives, so also has its representative series.

We may obtain a useful, alternative representation of the Fourier series in terms of the complex equivalents of the sine and cosine functions. From (4.8),

$$\cos \phi = \frac{1}{2}(e^{i\phi} + e^{-i\phi}), \quad \sin \phi = \frac{1}{2i}(e^{i\phi} - e^{-i\phi}). \tag{25}$$

Hence (5) becomes

$$f(\phi) = B_0 + \frac{1}{2}\sum_{m=1}^{\infty} [(B_m - iA_m)e^{im\phi} + (B_m + iA_m)e^{-im\phi}]$$

$$= \sum_{-\infty}^{\infty} C_m e^{im\phi}, \tag{26}$$

where C_m represents the complex coefficient of the respective term. When we require a series of this character, we can evaluate the values of C_m most easily by a somewhat different procedure. Multiply both sides of (26) by $e^{-ik\phi}\,d\phi$, and integrate over the range from 0 to 2π; k is now either a positive or negative integer. From the right-hand side of (26) we obtain integrals of the form

$$C_m \int_0^{2\pi} e^{i(m-k)\phi}\,d\phi = C_m \frac{[e^{i(m-k)\phi}]_0^{2\pi}}{i(m-k)}. \tag{27}$$

Now, $(m - k)$ is an integer. Therefore $\cos 2\pi\,(m - k) = \cos 0 = 1$, and $\sin 2\pi(m - k) = \sin 0 = 0$. Hence, by (4.8), we see that the integral is identical for both limits. The definite integral vanishes, when $m \neq k$. When $m = k$, we have

$$C_k \int_0^{2\pi} d\phi = 2\pi C_k\,. \tag{28}$$

Thus
$$C_k = \frac{1}{2\pi}\int_0^{2\pi} f(\phi)e^{-ik\phi}\,d\phi, \tag{29}$$

and
$$f(\phi) = \frac{1}{2\pi}\sum_{k=-\infty}^{\infty}\int_0^{2\pi} f(\psi)e^{ik(\phi-\psi)}\,d\psi. \tag{30}$$

This type of complex Fourier series may be conveniently employed to expand a complex $f(\phi)$. Such a representation is particularly useful in wave mechanics, where we often deal with complex wave functions.

If, now, we attempt to apply the Fourier series to points outside the specific range of calculation, we find that the pattern of the function repeats itself periodically. We may, therefore, legitimately inquire whether there is some way of extending a Fourier series to represent some non-periodic function over the entire range from $-\infty$ to $+\infty$.

To accomplish this result, we first express our function over some large, but finite range, and calculate the Fourier series representative of the function. Then, by proceeding to the limit of an infinite range, we find that the summation essentially goes over to an integration.

Consider some function of x, whose expansion we desire to obtain over the range from $x = -X/2$ to $x = +X/2$, let us say. We shall expect our expansion to take the form

$$f(x) = B_0' + \sum_{k=1}^{\infty} A_k' \sin \frac{2k\pi x}{X} + \sum_{k=1}^{\infty} B_k' \cos \frac{2k\pi x}{X}$$

$$= \sum_{k=-\infty}^{\infty} A_k e^{2\pi ikx/X}, \tag{31}$$

by analogy with equation (18), and in accord with equation (26). We are dealing with a full-range series, rather than with the half-range series of equation (18), which fact accounts for the extra factor of 2. We shall adopt the exponential form for our study. Then, multiplying both sides of equation (31) by $e^{-2\pi ikx/X}$ and integrating from $-X/2$ to $+X/2$, we get

$$A_k = \int_{-X/2}^{X/2} f(\xi)e^{-2\pi ik\xi/X} \frac{d\xi}{X}, \tag{32}$$

wherein we have substituted ξ for x, as the variable of integration. Introducing this equation in (31), we get for our series:

$$f(x) = \sum_{k=-\infty}^{\infty} \int_{-X/2}^{X/2} f(\xi)e^{2\pi ik(x-\xi)/X} \frac{d\xi}{X}. \tag{33}$$

Our arbitrary parameter, X, is much more closely related to k than to either x or ξ. The larger we assume X to be, the larger we must take k in our series, to arrive at significant amplitudes for A_k in our summation. For very large X, the A_k's, for successive values of k, differ insignificantly from one another. Hence we may replace the sum over k by an integral over the same range. Thus

$$f(x) = \int_{-\infty}^{\infty} \int_{-X/2}^{X/2} f(\xi)e^{2\pi ik(x-\xi)/X} \, d\xi \frac{dk}{X}. \tag{34}$$

Moreover, since the significant k's depend upon our choice of X, we now find it convenient to replace our continuous variable k, by a new variable, m, such that

$$m = 2\pi k/X; \qquad dm = 2\pi \, dk/X. \tag{35}$$

Now, letting X approach the limit infinity, we can write our final expression:

$$f(x) = \frac{1}{2\pi} \int_{-\infty}^{\infty} \int_{-\infty}^{\infty} f(\xi)e^{im(x-\xi)} \, d\xi \, dm. \tag{36}$$

This equation, known as a Fourier integral, enables us to express our original function in terms of an integral. The above derivation lacks the finest considerations of rigor and does not indicate the conditions under which we may apply the transformation. Advanced treatises show that the condition depends upon whether the integral

$$\int_{-\infty}^{\infty} | f(x) | \, dx \tag{37}$$

exists or not. We shall give a practical application of this formula in § (14).

As a corollary to the foregoing demonstration, we may write the following relationship between a pair of functions:

$$\begin{aligned} f(x) &= \frac{1}{\sqrt{2\pi}} \int_{-\infty}^{\infty} g(m) e^{imx} \, dm. \\ g(m) &= \frac{1}{\sqrt{2\pi}} \int_{-\infty}^{\infty} f(\xi) e^{-im\xi} \, d\xi. \end{aligned} \tag{38}$$

We say that the two functions $f(x)$ and $g(m)$ are the Fourier transforms of one another. These equations have many physical applications. They are special cases of the more general Fourier-Bessel transforms, which we state for sake of reference.

$$\begin{aligned} f(x) &= \int_0^{\infty} g(m) J_n(mx) m \, dm, \\ g(m) &= \int_0^{\infty} f(x) J_n(mx) x \, dx, \end{aligned} \tag{39}$$

where J_n represents a Bessel function of order n.

8. Normalization of eigenfunctions. We have seen that the eigenfunctions

$$\psi_k = A_k \sin \frac{k\pi x}{X} \tag{1}$$

possess orthogonality properties. If we form the product $\psi_j \psi_k \, dx$ and integrate over the length of the string, we find that

$$\int_0^X \psi_j \psi_k \, dx = \begin{cases} 0, & j \neq k, \\ A_k^2 X/2, & j = k. \end{cases} \tag{2}$$

We have also studied spherical and tesseral harmonics, which behave similarly.

Certain characteristics of the vibrating system, e.g., the potential energy, depend upon the square of the amplitude, integrated over the coordinates. As a matter of convenience, we shall often find it desirable to assign a value to the constant A_k, such that the integral (2) will assume the value of unity, or

$$A_k = \sqrt{2/X}. \tag{3}$$

The process of determining this constant is called normalization. Thus the normalized function is

$$\psi_k = \sqrt{2/X} \sin \frac{k\pi x}{X}. \tag{4}$$

If we employ the complex notation for the wave functions, we have as a representative term of the complex Fourier series, the expression.

$$\psi_k = C_k e^{ik\phi}. \tag{5}$$

Here the procedure for normalization is slightly different. We first form the complex conjugate, obtained as explained in § (4) by replacing i by $-i$ everywhere it appears in the expression. Denote this conjugate by ψ_k^*. The product $\psi_k \psi_k^*$ is real and may be integrated as before over the coordinates

$$\int_0^{2\pi} \psi_k \psi_k^* \, d\phi = C_k C_k^* \int_0^{2\pi} d\phi = 2\pi C_k C_k^* = 1, \tag{6}$$

to give the normalization condition.

We express the complex coefficient C_k in the form

$$C_k = B_k + A_k i = \sqrt{B_k^2 + A_k^2}\, e^{i\alpha}, \tag{7}$$

By (4.15) and (4.16), where

$$\alpha = \text{arc tan} \frac{A_k}{B_k}, \tag{8}$$

$$C^2 = C_k C_k^* = B_k^2 + A_k^2. \tag{9}$$

Then, by (6), (7), and (9), we obtain in place of (5), the normalized complex function

$$\psi_k = \frac{1}{\sqrt{2\pi}} e^{ik\phi + i\alpha}. \tag{10}$$

The phase α is rarely of importance because we can include it as a complex factor in the coefficient.

Vibrations in Mechanical Systems

9. Eigenfunctions for the tensed circular membrane. In two dimensions, we have from (II-33.32) the time-independent wave equation in polar coordinates,

$$\frac{\partial^2 \psi}{dr^2} + \frac{1}{r^2} \frac{\partial^2 \psi}{\partial \theta^2} + \frac{1}{r} \frac{\partial \psi}{\partial r} + \frac{4\pi^2 \nu^2}{u^2} \psi = 0. \tag{1}$$

Guided by our previous experience, we shall now look for a solution in the form

$$\psi = R\Theta, \tag{2}$$

where R and Θ represent functions of r and θ, respectively.

$$\frac{d^2R}{dr^2}\Theta + \frac{R}{r^2}\frac{d^2\Theta}{d\theta^2} + \frac{1}{r}\frac{dR}{dr}\Theta + \frac{4\pi^2\nu^2}{u^2}R\Theta = 0. \tag{3}$$

The equation becomes

$$\frac{r^2}{R}\frac{d^2R}{dr^2} + \frac{r}{R}\frac{dR}{dr} + \frac{4\pi^2\nu^2r^2}{u^2} + \frac{1}{\Theta}\frac{d^2\Theta}{d\theta^2} = 0, \tag{4}$$

where we now employ total instead of partial derivatives. Transposing all but the term in θ to the right-hand side, we see that variations in θ cannot possibly affect those in r. Since the two sides are thus independent, they must be constant, which, for later convenience, we take equal to $-m^2$, as in (4.2). We then have

$$\frac{d^2\Theta}{d\theta^2} + m^2\Theta = 0, \tag{5}$$

whose solution is

$$\Theta = A \sin m\theta + B \cos m\theta = Ce^{im\theta},$$

as in equation (4.22), where C may be complex. As our eigenfunctions, therefore, we adopt

$$\Theta = A\,{}^{\sin}_{\cos}\,(m\theta); \quad \text{or} \quad \Theta = Ae^{im\theta}. \tag{6}$$

The values of Θ must repeat themselves when θ alters by 2π. This condition will obtain only if m is an integer, or

$$m = 0, \quad \pm 1, \quad \pm 2, \quad \ldots . \tag{7}$$

When we adopt the form of sine or cosine for the eigenfunction, we take only positive values of m. In the exponential form of equation (6), we must use both positive and negative values of m to get a complete solution. The essential equivalence of the two procedures follows from the replacement of $e^{im\theta}$ by its equivalent in terms of the angular functions. The above expression for Θ is the angle-dependent factor of an eigenfunction of the original equation.

When we substitute (5) back in (4), we have

$$\frac{d^2R}{dr^2} + \frac{1}{r}\frac{dR}{dr} + \left(\frac{4\pi^2\nu^2}{u^2} - \frac{m^2}{r^2}\right)R = 0. \tag{8}$$

A simple substitution reduces this differential equation for R to the form of Bessel's equation, which mathematicians have widely studied. It has no solution in terms of the simpler functions. The R's are Bessel functions, developed in terms of the variable r. Each value of the constant m yields a distinct equation. Introduce a new variable r' defined by

$$r' = \frac{2\pi\nu}{u} r, \tag{9}$$

$$\frac{dR}{dr} = \frac{dR}{dr'}\frac{dr'}{dr} = \frac{2\pi\nu}{u}\frac{dR}{dr'}, \quad \text{ETC.}, \tag{10}$$

by means of which the transformed equation becomes

$$r'^2 \frac{d^2R}{dr'^2} + r' \frac{dR}{dr'} + (r'^2 - m^2)R = 0. \tag{11}$$

Dropping the primes temporarily, we shall look for a series solution in the form

$$R = \sum_{j=j'}^{\infty} A_j r^j. \tag{12}$$

When we introduce this series in (11), we obtain the following result:

$$\sum_{j=j'}^{\infty} A_j(j^2 - m^2)r^j + \sum_{j=j'} A_j r^{j+2} = 0. \tag{13}$$

Or, collecting coefficients of the same powers of r, we find

$$\sum_{j=j'}^{\infty} [A_j(j^2 - m^2) + A_{j-2}]r^j = 0. \tag{14}$$

The series is an ascending one, with powers of r increasing by jumps of 2. By the theorem of undetermined coefficients, we must set each coefficient individually equal to zero, which gives the following relation between successive values of A_j:

$$A_j = -A_{j-2}/(j^2 - m^2). \tag{15}$$

The ratio between successive terms of the series is

$$-r^2/(j^2 - m^2),$$

which approaches zero as $j \to \infty$; hence the series converges. But the series must not extend to negative values of j; otherwise we should have an infinite value of R at $r = 0$, which condition is physically inadmissible, if we exclude the possibility of a broken drum head. We have supposed that the series begins with $j = j'$; then $A_{j'-2}$ in (14) equals zero by hypothesis, and

$$A_{j'}(j'^2 - m^2) = 0. \tag{16}$$

Now $A_{j'}$, being the coefficient of the lowest power of r, is not zero; hence

$$j'^2 - m^2 = 0, \qquad j' = \pm m. \tag{17}$$

Negative values of m are not physically distinct from the positive values since equations (8) to (16) contain only the square of m. Hence we set $j' = |m|$, and the series becomes

$$R = A_m r'^m \left[1 - \frac{r'^2}{2^2 1!(m+1)} + \frac{r'^4}{2^4 2!(m+1)(m+2)} - \cdots + (-1)^j \frac{r'^{2j}}{2^{2j} j!(m+1)(m+2)\ldots(m+j)} + \ldots \right], \tag{18}$$

where r' has been reinstated as the variable, and m used for $|m|$ when m is negative.

If, now, we set

$$A_m = \frac{1}{2^m m!}, \tag{19}$$

then

$$R = J_m(r'), \tag{20}$$

where $J_m(r')$ is a Bessel's function of order m. From the generating function for the Bessel series, viz.,

$$e^{-\frac{1}{2}z(t-1/t)} = \sum_{m=-\infty}^{\infty} t^m J_m(z),$$

we establish a result that does not follow uniquely from the differential equation,

$$J_{-m}(z) = (-1)^m J_m(z). \tag{21}$$

when m is an integer. To check this result, merely substitute $u = -1/t$ in the generating function.

From the fact that R and d^2R/dr'^2, in (11) must have opposite signs for $r' > m$, we conclude that $J_m(r')$ is an oscillatory function of r', not unlike a sine or cosine function, except that amplitudes of successive waves and the distances between the nodes are not constant. By the same argument we may show that the first node occurs for $r' > m$. Various mathematical tables give graphs or numerical values for selected functions.

If the membrane is infinite we have no restrictions on ν, but if the membrane is clamped, say, at $r = r_1$, then the point r_1 must lie at a node. From the tables we may find that the function $J_m(r') = 0$, say at $r' = a_1, a_2, \ldots a_n, \ldots$ etc.; whence, by (9),

$$\nu_1 = \frac{u}{2\pi}\frac{a_1}{r_1}, \quad \nu_2 = \frac{u}{2\pi}\frac{a_2}{r_2}, \quad \text{ETC.} \tag{22}$$

Since the values a_1, a_2, etc., are not in any simple harmonic ratio to one another, the successive overtones of ν, will not be true harmonics. This analysis shows why a drum gives a noise, not a pleasant musical sound.

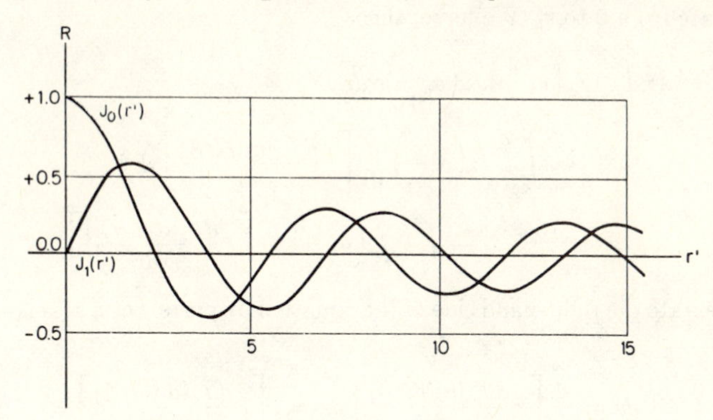

III-3.

By (9) and (22), we restrict r' to the condition

$$r' = a_n \frac{r}{r_1}, \quad \text{or} \quad R = J_m\left(a_n \frac{r}{r_1}\right). \tag{23}$$

We should, perhaps, point out that $J_m(a_n\, r/r_1)$ also depends on m, a fact that the diagram clarifies.

The expression for ψ, therefore, is

$$\psi = \Theta R = Ae^{im\theta} J_m\left(a_n \frac{r}{r_1}\right), \tag{24}$$

by (6) and (20). We shall now proceed to normalize ψ by integrating $\psi\psi^*$ over the area of the membrane. An element of area is $r\, dr\, d\theta$. Hence

$$\int_0^{2\pi}\int_0^{r_1} \psi^2 r\, dr\, d\theta = \int_0^{2\pi} A^2\, d\theta \int_0^{r_1} J_m^2\left(a_n \frac{r}{r_1}\right) r\, dr. \tag{25}$$

The r and θ factors may thus be normalized separately. The normalization, for Θ, derived in the manner of equation (8.6), gives

$$\int_0^{2\pi} \Theta\Theta^*\, d\theta = 2\pi \mid A \mid^2. \tag{26}$$

The functions are orthogonal.

We develop the normalization of R, as well as its orthogonality in the integral above, most easily from the original differential equation, which by (8) and (22) may be written in the form

$$\frac{1}{r}\frac{d}{dr}\left(r\frac{dR_{mn}}{dr}\right) + \left(\frac{a_n^2}{r_1^2} - \frac{m^2}{r^2}\right)R_{mn} = 0. \tag{27}$$

Write a second equation for the same value of m, but with a different value of n, say k. Substitute R from (24), multiply the first equation by $J_m(a_k\, r/r_1)r\, dr$ and the second by $J_m(a_n\, r/r_1)r\, dr$. Subtract the two and integrate from 0 to r_1. We have, then,

$$\begin{aligned}\frac{1}{r_1^2}(a_n^2 - a_k^2)\int_0^{r_1} & J_m\left(a_k\frac{r}{r_1}\right)J_m\left(a_n\frac{r}{r_1}\right)r\, dr \\ &= \int_0^{r_1} J_m\left(a_n\frac{r}{r_1}\right)\frac{d}{dr}\left[r\frac{dJ_m[a_k(r/r_1)]}{dr}\right]dr \\ &\quad - \int_0^{r_1} J_m\left(a_k\frac{r}{r_1}\right)\frac{d}{dr}\left[r\frac{dJ_n[a_n(r/r_1)]}{dr}\right]dr.\end{aligned} \tag{28}$$

Integrate the right-hand side of the equation by parts. Set $u = J_m(a_n\, r/r_1)$ and

$$dv = \frac{d}{dr}\left[r\frac{dJ_m[a_k(r/r_1)]}{dr}\right]dr = d\left[r\frac{dJ_m[a_k(r/r_1)]}{dr}\right].$$

Proceeding in the usual manner, we find that the first of these two integrals becomes

$$\left[rJ_m\left(a_n\frac{r}{r_1}\right)\frac{dJ_m[a_k(r/r_1)]}{dr}\right]_0^{r_1} - \int_0^{r_1} r\frac{dJ_m[a_k(r/r_1)]}{dr}\frac{dJ_m[a_n(r/r_1)]}{dr}dr,$$

while the second takes the same form with the subscripts n and k reversed. The integral, being identical for both, vanishes. The terms are zero at the lower limit because of the factor r, and we have

$$\begin{aligned}\frac{1}{r_1^2}(a_n^2 - a_k^2)\int_0^{r_1} & J_m\left(a_k\frac{r}{r_1}\right)J_m\left(a_n\frac{r}{r_1}\right)r\, dr \\ &= r_1\left[J_m\left(a_n\frac{r}{r_1}\right)\frac{dJ_m[a_k(r/r_1)]}{dr} - J_m\left(a_k\frac{r}{r_1}\right)\frac{dJ_m[a_n(r/r_1)]}{dr}\right]_{r=r_1}\end{aligned} \tag{29}$$

Now, both $J_m(a_n\, r/r_1)$ and $J_m(a_k\, r/r_1)$ are equal to zero when $r = r_1$, by the hypothesis that a node occurs at r_1. Hence the right-hand side of (29) is zero. Either the integral, or the factor $(a_n^2 - a_k^2)$ must vanish. The latter, obviously, does not vanish as long as n and k are different. Therefore

$$\int_0^{r_1} J_m\left(a_k\frac{r}{r_1}\right)J_m\left(a_n\frac{r}{r_1}\right)r\, dr = 0, \qquad n \neq k, \tag{30}$$

which proves the orthogonality of the function.

Dividing through by $(a_n^2 - a_k^2)$, we find that the integral takes the indeterminate form 0/0, as $n = k$. To evaluate the expression we must differentiate numerator and denominator, treating a_n as a variable that will be allowed to approach indefinitely close to a_k.

We find

$$\int_0^{r_1} J_m\left(a_k \frac{r}{r_1}\right) J_m\left(a_n \frac{r}{r_1}\right) r\, dr = \frac{r_1^3}{2a_n}\left[\frac{dJ_m[a_n(r/r_1)]}{da_n}\frac{dJ_m[a_k(r/r_1)]}{dr}\right]_{r=r_1},$$

the second term vanishing because of the factor $J_m(a_k r/r_1)$. Since $J_m(a_n r/r_1)$ is expanded in terms of $a_n r/r_1$, see equation (23), we may write

$$\frac{dJ_m[a_n(r/r_1)]}{da_n} = \frac{r}{a_n}\frac{dJ_m[a_n(r/r_1)]}{dr}.$$

Thus, setting $n = k$, we have

$$\int_0^{r_1}\left[J_m\left(a_n \frac{r}{r_1}\right)\right]^2 r\, dr = \frac{r_1^4}{2a_n^2}\left\{\left[\frac{dJ_m[a_n(r/r_1)]}{dr}\right]_{r=r_1}\right\}^2. \tag{31}$$

We can make one further simplification. Compare the general term in $J_m(r')$, (equation 18),

$$(-1)^j \frac{1}{2^m m!}\frac{1}{2^{2j} j!(m+1)(m+2)\ldots(m+j)} r'^{m+2j},$$

and the analogous term in $J_{m+1}(r')$ with the power r'^{m+2j+1}. We note the identity

$$J_{m+1}(r') = \frac{m}{r'} J_m(r') - \frac{dJ_m(r')}{dr'}.$$

Whence

$$\int_0^{r_1}\left[J_m\left(a_n \frac{r}{r_1}\right)\right]^2 r\, dr = \frac{r_1^2}{2}[J_{m+1}(a_n)]^2, \tag{32}$$

since

$$\frac{dJ_m[a_n(r/r_1)]}{dr} = \frac{a_n}{r_1}\frac{dJ_m(r')}{dr'} \quad \text{and} \quad J_m(a_n) = 0.$$

The normalized eigenfunction for the tensed membrane thus becomes, by (24), (26), and (32),

$$\psi_{m,n} = \frac{1}{\sqrt{2\pi}} e^{im\phi} \frac{\sqrt{2}}{r_1 J_{m+1}(a_n)} J_m\left(a_n \frac{r}{r_1}\right). \tag{33}$$

We may legitimately term the two numbers m and n "quantum numbers" because they have selected, from the double infinity of possible vibration states, certain discrete modes. The locations of the nodes in the vibrating membrane, for several vibrational states, appear in Fig. 4. We build up our complete solution by superposing the eigenfunctions, as before.

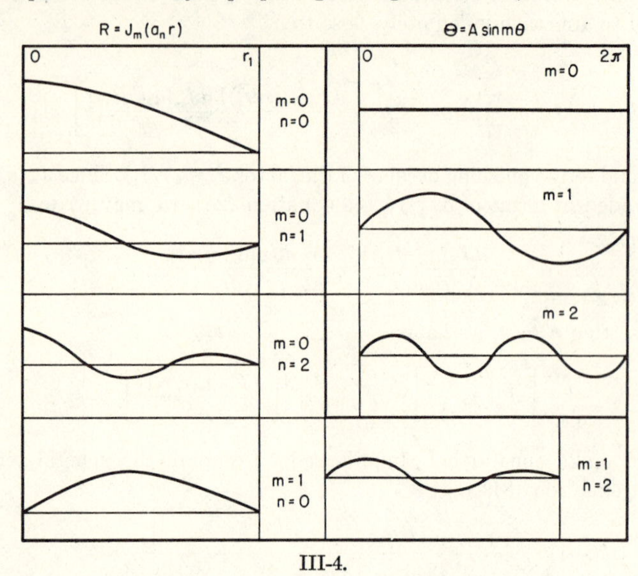

III-4.

10. The wave equation for the vibrating sphere. The wave equation in three dimensions, with the time factor eliminated, and expressed in spherical coordinates, is, from (5.4) and (II-33.33),

$$\frac{1}{r^2}\frac{\partial}{\partial r}\left(r^2\frac{\partial\psi}{\partial r}\right)+\frac{1}{r^2\sin\theta}\frac{\partial}{\partial\theta}\left(\sin\theta\frac{\partial\psi}{\partial\theta}\right)+\frac{1}{r^2\sin^2\theta}\frac{\partial^2\psi}{\partial\phi^2}+\frac{4\pi^2\nu^2\psi}{u^2}=0. \tag{1}$$

For the spherical case we may regard ψ as a measure of the density of the vibrating medium, rather than a spatial displacement. Setting $\psi = R\Theta\Phi$, we have, after multiplying the equation through by $r^2 \sin^2 \theta$, and dividing by ψ,

$$\frac{\sin^2\theta}{R}\frac{d}{dr}\left(r^2\frac{dR}{dr}\right)+\frac{\sin\theta}{\Theta}\frac{d}{d\theta}\left(\sin\theta\frac{d\Theta}{d\theta}\right)+\frac{1}{\Phi}\frac{d^2\Phi}{d\phi^2}+\frac{4\pi^2r^2\sin^2\theta}{u^2}=0. \tag{2}$$

As far as Φ is concerned, we may regard the rest of the equation as a constant, let us say m^2. Then

$$\frac{d^2\Phi}{d\phi^2}+m^2\Phi=0, \tag{3}$$

whose solution we have discussed several times previously.

$$\Phi=\frac{1}{\sqrt{2\pi}}e^{im\phi}, \tag{4}$$

normalized as in equation (9.33).

Since ϕ is cyclic, we have, as in the case of the vibrating membrane, the permitted integral values

$$m = 0, \quad \pm 1, \quad \pm 2, \quad \dots . \tag{5}$$

Dividing equation (2) by $\sin^2 \theta$, and making use of (3), we now find that we can separate the terms in r and θ:

$$\frac{1}{\Theta \sin \theta} \frac{d}{d\theta}\left(\sin \theta \frac{d\Theta}{d\theta}\right) - \frac{m^2}{\sin^2 \theta} = -\frac{1}{R}\frac{d}{dr}\left(r^2 \frac{dR}{dr}\right) - \frac{4\pi^2 \nu^2 r^2}{u^2} = -C, \tag{6}$$

where C is a constant. From the above equation we perceive that positive and negative values of m give essentially equivalent solutions in θ and r.

To solve the equation in θ, first make the substitution

$$\Theta = \sin^{|m|} \theta \, Q(\cos \theta),$$

where Q is some function still to be determined and where $|m|$ denotes the absolute value of m. In all further developments we shall understand that the positive value of m is to be taken. Make the substitution

$$\cos \theta = \mu.$$

Then

$$\frac{dQ}{d\theta} = \frac{dQ}{d\mu}\frac{d\mu}{d\theta} = -\sin \theta \frac{dQ}{d\mu},$$

$$\frac{d^2Q}{d\theta^2} = (1 - \mu^2)\frac{d^2Q}{d\mu^2} - \mu \frac{dQ}{d\mu}.$$

The resulting differential equation is

$$(1 - \mu^2)\frac{d^2Q}{d\mu^2} - (1 + 2m)\mu \frac{dQ}{d\mu} + (C - m - m^2)Q = 0.$$

Let us substitute in the above equation the following assumed solution:

$$Q(\mu) = \mu^q \sum_{j=0}^{\infty} A_j \mu^j.$$

Equating the lowest power of μ to zero, we have as the solutions of the indicial equation,

$$q = 0, \quad \text{or} \quad 1.$$

We thus see that the solution can be expressed as a power series in μ.

Writing now

$$Q(\mu) = \sum_{j=0}^{\infty} A_j \mu^j, \tag{7}$$

where j is integral, and substituting in the above differential equation for Q, we have an identity in μ.

Collecting the coefficients of the same power of μ, we have

$$\sum_{j=0}^{\infty} \{-A_j[j^2 + j(1 + 2m) + m^2 + m - C] + A_{j+2}(j + 1)(j + 2)\}\mu^j = 0. \tag{8}$$

By the theory of undetermined coefficients, each coefficient of μ^j must itself be equal to zero, since μ is not zero, generally. Hence

$$A_{j+2} = A_j \frac{(j + m)(j + m - 1) - C}{(j + 1)(j + 2)}. \tag{9}$$

Note that (9) holds even when consecutive values for j differ by unity, without j being necessarily integral. In that case, the series for finiteness must terminate on the side of lower j on a non-negative value. Now, reversing the relation (9) we see that the series will terminate on the side of lower j only when j is an integer. The series then terminates on the side of lower j at A_0. We are thus again led to the power series solution (7) for $Q(\mu)$.

The usual tests of convergence show that an infinite series with the relation (9) between alternate coefficients converges for $-1 < \mu < 1$, but diverges for $\mu = +1$ or -1. We can therefore get a physically acceptable wave function only when *both* the series with the even and the odd coefficients terminate. Now the condition for termination of either of these series is, from (9),

$$C = (j + m)(j + m + 1). \tag{10}$$

We see from (10) that for no value of C can *both* the even and the odd series terminate. We do have, however, a set of values of C given by (10) which will make either one of these series terminate, and we can make the other series vanish by putting A_0 or A_1 equal to zero.

As j and m are integers, we can put

$$j + m = l \tag{11}$$

and write (10) in the form

$$C = l(l + 1). \tag{12}$$

We perceive that the condition of a physically possible solution, with finite amplitudes at all points, has not only dictated the form of the constant C, that is its eigenvalues, but has limited the solution to certain discrete eigenfunctions.

The relation between successive coefficients is, by (9) and (12),

$$A_j = -A_{j+2} \frac{(j + 2)(j + 1)}{(l - m - j)(l + m + j + 1)}. \tag{13}$$

The series thus becomes

$$Q_l^m(\mu) = A_{l-m}\Bigg[\mu^{l-m} - \frac{(l-m)(l-m-1)}{2^1 1!(2l-1)}\mu^{l-m-2}$$

$$+ \frac{(l-m)(l-m-1)(l-m-2)(l-m-3)}{2^2 2!(2l-1)(2l-2)}\mu^{l-m-4} + \dots \tag{14}$$

$$+ (-1)^k \frac{(l-m)(l-m-1)\dots(l-m-2k+1)}{2^k k!(2l-1)(2l-2)\dots(2l-k)}\mu^{l-m-2k} + \dots\Bigg].$$

The constant A_{l-m} is arbitrary as far as the equation is concerned, but if we take

$$A_{l-m} = \frac{1\cdot 3\cdot 5\dots(2l-1)}{(l-m)!} = \frac{(2l)!}{2^l l!(l-m)!}, \tag{15}$$

Θ becomes

$$\Theta = \sin^{|m|}\theta\, Q_l^{|m|}(\cos\theta) = P_l^m(\cos\theta), \tag{16}$$

where P_l^m is one of the so-called associated Legendre polynomials, as given in equation (II-17.17). We make this identification chiefly for convenience of notation. In practice, we must fix the value of the constant, A_{l-m}, by normalization. By (II-18.15), we have

$$\int_0^\pi [P_l^m(\cos\theta)]^2 \sin\theta\, d\theta = \int_{-1}^1 [P_l^m(\mu)]^2\, d\mu = \frac{2(l+m)!}{(l-m)!(2l+1)!}. \tag{17}$$

Therefore, instead of (16), we adopt the normalized form

$$\Theta(l, m) = \sqrt{\frac{(l-m)!(2l+1)!}{2(l+m)!}}\, P_l^m(\cos\theta). \tag{18}$$

The solution of the wave equation of the sphere is thus seen to be, apart from a normalizing factor, equal to a sine or cosine function, exemplified by the factor $e^{im\phi}$, times an associated spherical harmonic. In other words, for each surface defined by r = const, we can express the displacements or condensations of density in ϕ and θ in terms of tesseral harmonics. The result is not surprising. The potential function V, similarly expressed in Part II, must satisfy Laplace's or Poisson's equation, which are akin to the wave equation.

11. The radial equation for the vibrating sphere and the general normalized eigenfunction. From equation (10.6), we find that the radial function satisfies the equation

$$r^2\frac{d^2R}{dr^2} + 2r\frac{dR}{dr} - \left[l(l+1) - \frac{4\pi^2\nu^2r^2}{u^2}\right]R = 0, \tag{1}$$

wherein we have employed (10.12). If now we substitute

$$R = r^{-1/2}S,$$

S must satisfy the equation

$$\frac{d^2S}{dr^2} + \frac{1}{r}\frac{dS}{dr} + \left[\frac{4\pi^2\nu^2}{u^2} - \frac{(l+\frac{1}{2})^2}{r^2}\right]S = 0. \tag{2}$$

This equation is of the form of (9.8), which had as its solution a Bessel function. Hence

$$R = r^{-1/2}J_{l+1/2}\left(\frac{2\pi\nu}{u}r\right), \tag{3}$$

by (9.23). Since we suppose that the sphere possesses a boundary at $r = r_1$, then R must be zero at that coordinate. If $a_1, a_2, \ldots a_n$, represent successive numerical values where $J_{l+\frac{1}{2}}(a_n\, r/r_1)$ goes to zero, then

$$\nu_n = \frac{u}{2\pi}\frac{a_n}{r_1}, \tag{4}$$

as in (9.22). The frequencies of the vibrating sphere, however, will not agree with those of the membrane because of the occurrence of $(l + \frac{1}{2})$, which gives half-integral rather than integral values of the Bessel functions. These functions possess alternative expressions in terms of sines and cosines, as in problem 12. The function $J_{-l-\frac{1}{2}}$ is infinite at the origin.

We shall require the normalized value of R_{nl} as follows:

$$\int_0^{r_1} R_{nl}^2 r^2\, dr = \frac{r_1^2}{2}\,[J_{l+3/2}(a_n)]^2, \tag{5}$$

by equation (9.32). The R_{nl}'s are orthogonal, by the previous reasoning.

Thus far, we have stated, without proof, the nature of the normalizing factor in each case. The form of the expression is determined by the condition that the square of the amplitude, integrated over the volume, is equal to unity. Since the volume element is $d\tau = r^2 \sin\theta\, dr\, d\theta\, d\phi$, we have

$$\int \psi\psi^*\, d\tau = \int_0^{2\pi} \Phi_m\Phi_m^*\, d\phi \int_0^{\pi} \Theta^2 \sin\theta\, d\theta \int_0^{r_1} R_{nl}^2 r^2\, dr = 1. \tag{6}$$

From equations (10.4), (10.27), and (5), we write down the complete eigenfunction:

$$\psi_{mln} = \frac{1}{\sqrt{2\pi}}e^{im\phi}\sqrt{\frac{(l-m)!(2l+1)}{2(l+m)!}}\,P_l^m(\cos\theta)\,\frac{1}{r_1}\sqrt{\frac{2}{r}}\,\frac{J_{l+1/2}[a_n(r/r_1)]}{J_{l+3/2}(a_n)}. \tag{7}$$

12. The significance of wave functions. As far as the present volume is concerned, we shall not go into details concerning the relationship between ordinary vibrations and the atomic problems of wave mechanics. Nevertheless, the reader who may be familiar with the elements of the quantum theory cannot fail to see an analogy between the foregoing discussion and that of the classical methods of quantization. In the problem of the hydrogen atom, as originally discussed by Bohr and Sommerfeld, arbitrary rules were employed to select, from a manifold of mechanically possible electron orbits, a few stationary states of special significance. The older atomic model postulated no vibrating medium capable of sending out light waves. The model was definitely incomplete. The mechanical problem of selecting certain stable modes from all possible modes of the vibrating sphere has many features in common with quantization. The parameters m, l, and n, are analogous to quantum numbers. The distinguishing difference is that in the mechanical problems the quantum numbers appeared naturally. We needed no special assumption to introduce them, aside from the physical limitation that the amplitudes fulfill the boundary conditions and that they be everywhere finite and continuous.

Schroedinger, in 1925, succeeded in applying the wave equation to atomic problems. Although he needed certain additional assumptions to adapt the fundamental equation to wave mechanics, the solutions follow in a manner similar to that already employed. The letters m, l, and n, as introduced in the problem of the vibrating sphere, have been adopted because of their relation to the atomic quantum numbers. Indeed, the functions Φ and Θ for the hydrogen atom are identical with those for the vibrating sphere.

For the present we need not concern ourselves with details. We should merely consider the general method of attacking wave problems. We see that the wave equation is a partial differential equation, involving ψ and the various coordinates. By considering periodic solutions, we may eliminate the time factor. Some assumption is necessary concerning the form of ψ, e.g., we may substitute $\psi = R\Theta\Phi$ in the wave equation. The substitution is, in effect, a transformation of variable, made with the hope of separating the portions of the equation that depend on only a single coordinate from those depending on other coordinates. To effect the separation, we may try various coordinate systems or various assumptions concerning the nature of the wave functions.

When we find that we can separate out all the terms that depend on some single coordinate, we may set those terms of the equation equal to a constant and solve the resulting ordinary differential equation by standard methods. The boundary conditions together with the required finiteness of the amplitude serve to define the eigenfunctions. The resulting wave functions, which are usually orthogonal, are then normalized. This procedure, generally employed in the vibration problems of classical mechanics, is taken over almost without change into wave mechanics.

13. The vibrating string with complex wave velocity. In the previous vibration problems we have considered the velocity u as a constant. When we remove this restriction many interesting possibilities arise. For example, let us consider the problem of a vibrating string, equation (6.1), where

$$u^2 = (a' - b'x^2)^{-1}. \tag{1}$$

We shall suppose, further, that the string extends from $-\infty$ to $+\infty$. Thus the velocity will be real over the region $-\sqrt{a'/b'} \leq x \leq \sqrt{a'/b'}$, and imaginary over the rest of the string. Even though an imaginary velocity is difficult to picture physically, we obtain a real solution from the equation, which takes the form

$$\frac{d^2\psi}{dx^2} + (a - bx^2)\psi = 0, \tag{2}$$

where a and b are new constants, which bear a single relationship to a' and b'.

For large values of x, the equation becomes, asymptotically,

$$\frac{d^2\psi}{dx^2} \sim bx^2\psi, \tag{3}$$

which is satisfied in this region by the function, as follows:

$$\psi = e^{\pm(\sqrt{b}/2)x^2}, \quad \frac{d^2\psi}{dx^2} = (\pm b + bx^2)e^{\pm(\sqrt{b}/2)x^2} \sim bx^2\psi. \tag{4}$$

Of the two possible values for ψ, only the one with the minus sign is acceptable as a wave function. With the other, ψ tends to infinity for large values of x. Accordingly we shall look for a solution of the form

$$\psi = e^{-(\sqrt{b}/2)x^2} \sum_j A_j x^j. \tag{5}$$

Substituting this expression in (2), we find that

$$e^{-(\sqrt{b}/2)x^2} \sum_j A_j\{[\sqrt{b}\,(2j + 1) - a]x^j + j(j - 1)x^{j-2}\}. \tag{6}$$

Collecting coefficients of x^j, we have

$$e^{-(\sqrt{b}/2)x^2} \sum_j \{-A_j[\sqrt{b}\,(2j + 1) - a] + A_{j+2}(j + 2)(j + 1)\}x^j = 0,$$

or

$$A_j = A_{j+2}\frac{(j + 2)(j + 1)}{\sqrt{b}\,[2j + 1 - (a/\sqrt{b})]}. \tag{7}$$

If the series does not terminate at some value of j, say j', then the increasing powers of x, as $x \to \infty$, will tend to give an infinite value of ψ, which even the factor $e^{-\sqrt{b}\,x^2/2}$ cannot control. Thus if $A_{j'+2} = 0$, to keep $A_{j'}$ finite we must have

$$\left(2j' + 1 - \frac{a}{\sqrt{b}}\right) = 0, \quad \text{or} \quad 2j' + 1 = \frac{a}{\sqrt{b}}.$$

The quantity j' must also be integral, to eliminate negative powers of x.

Referring to equations (1) and (2), we have

$$a = 4\pi^2\nu^2 a', \quad b = 4\pi^2\nu^2 b', \quad \text{or} \quad \nu_{j'} = \frac{\sqrt{b'}}{2\pi a'}(2j' + 1). \tag{8}$$

The wave function, apart from a normalizing factor, takes the form

$$\psi_{j'} = e^{-y_{j'}/2}\left[y^{j'} - \frac{j'(j'-1)}{2^2 1!} y^{j'-2} + \frac{j'(j'-1)(j'-2)(j'-3)}{2^4 2!} y^{j'-4} + \dots\right], \tag{9}$$

where

$$y_{j'} = \sqrt[4]{b}\, x = \sqrt{2\pi\nu_{j'}}\, \sqrt[4]{b'}\, x. \tag{10}$$

If we multiply (9) through by $2^{j'}$, we may write the wave equation as follows:

$$\psi_{j'} = e^{-y_{j'}/2} H_{j'}(y_{j'}), \tag{11}$$

where $H_{j'}$ stands for the summation, the well-known Hermite polynomials,

$$H_j(x) = (2x)^j - \frac{j(j-1)}{1!}(2x)^{j-2} + \frac{j(j-1)(j-2)(j-3)}{2!}(2x)^{j-4}$$

$$= (-1)^j e^{x^2} \frac{d^j}{dx^j}(e^{-x^2}). \tag{12}$$

Also

$$\frac{dH_j(x)}{dx} = 2jH_{j-1}(x). \tag{13}$$

By means of these relations we may establish both the orthogonality and the normalization of the functions, through successive integration by parts. The final wave function is as follows:

$$\psi_j = \frac{1}{(2^j j!)^{1/2}}\left(\frac{b'}{\pi}\right)^{1/8} e^{-y^2/2} H_j(y), \tag{14}$$

where we have now written j for j'. Note that no special boundary conditions are necessary. The required continuity and finiteness of ψ, suffice to determine the wave functions.

The Effects of Damping in Vibrating Systems

14. Damped oscillator. The time variation of the vibrating string led to the well-known differential equation of harmonic motion, (4.3),

$$\frac{d^2S}{dt^2} + a^2u^2S = 0, \tag{1}$$

whose solution is the real part of

$$S = e^{2\pi i\nu_0(t-t_0)}, \tag{2}$$

where

$$au = 2\pi\nu_0. \tag{3}$$

Let us now introduce into (1) a damping resistance that is proportional to the velocity, or

$$\frac{d^2S}{dt^2} + \gamma\frac{dS}{dt} + a^2u^2S = 0. \tag{4}$$

The substitution of

$$S = e^{mt}$$

into (4) yields the equation of condition

$$m^2 + \gamma m + a^2u^2 = 0, \tag{5}$$

or

$$m = \frac{-\gamma \pm \sqrt{\gamma^2 - 4a^2u^2}}{2}. \tag{6}$$

The nature of the solution depends upon the nature of the radical. If

$$\gamma^2 > 4a^2u^2, \tag{7}$$

m is real and the motion is

$$S = Ae^{m_1t} + Be^{m_2t}, \tag{8}$$

where m_1 and m_2 denote the two roots of (5). The resulting motion is non-oscillatory. Such a mechanical system is said to be *overdamped*. When

$$\gamma^2 < 4a^2u^2, \tag{9}$$

we may set

$$\frac{1}{2}\sqrt{4a^2u^2 - \gamma^2} = 2\pi\nu_0', \tag{10}$$

and the solution is

$$\begin{aligned} S &= Ae^{2\pi i\nu_0't-(\gamma/2)t} \\ &= e^{-(\gamma/2)t}(A_r \cos 2\pi\nu_0't + A_i \sin 2\pi\nu_0't) \\ &= A'e^{-(\gamma/2)t} \sin(2\pi\nu_0't - \alpha). \end{aligned} \tag{11}$$

The factor $e^{2\pi i\nu_0't}$ represents a sinusoidal wave, of frequency ν_0', but the amplitude of the wave diminishes with the time because of the factor $e^{-\gamma t/2}$. The motion is said to be *underdamped*.

When

$$\gamma^2 = 4a^2u^2 = 4\pi^2\nu_0^2, \tag{12}$$

we have a case of *critical damping*.

The resulting motion for the underdamped example is not a pure sine wave. A sine wave must be of infinite extent and must have constant amplitude. We suspect that the motion may be represented as a sum of sine waves of various amplitudes and frequencies, superposed to give the effect of a damped sinusoidal wave. The problem is not dissimilar to that already discussed in § (7), where we represented an arbitrary displacement in terms of a Fourier series. Consider a displacement of the form

$$\psi = 0; \quad t < 0. \quad \psi = Ae^{-(\gamma/2)t} \sin 2\pi\nu_0' t; \quad t \geq 0, \tag{13}$$

which represents a disturbance that starts initially at $t = 0$ and is sinusoidally damped thereafter.

Since this function extends from zero to infinity, we can represent it in terms of a Fourier integral, not a Fourier series. Substituting $\psi(t)$ for $f(x)$ in (7.36), we get

$$\psi(t) = \frac{A}{2\pi} \int_{-\infty}^{\infty} \int_{0}^{\infty} e^{-\gamma\xi/2} \sin 2\pi\nu_0'\xi e^{im(t-\xi)} \, d\xi \, dm. \tag{14}$$

The lower limit in the second integration is zero because $\psi(t) = 0$ over the range $-\infty < t \leq 0$. Replace $\sin 2\pi\nu_0' t$ by its equivalent, in expanded form, from (4.8),

$$2 \cos \theta = e^{i\theta} + e^{-i\theta}. \quad 2i \sin \theta = e^{i\theta} - e^{-i\theta}. \tag{15}$$

Then (14) becomes

$$\psi(t) = \frac{A}{4\pi i} \int_{-\infty}^{\infty} \int_{0}^{\infty} [e^{-(\gamma/2)\xi + 2\pi i\nu_0'\xi + imt - im\xi} - e^{-(\gamma/2)\xi - 2\pi i\nu_0'\xi + imt - im\xi}] \, d\xi \, dm. \tag{16}$$

Integrating with respect to ξ, we get

$$\psi(t) = \frac{A}{4\pi i} \int_{-\infty}^{\infty} e^{imt} \left[\frac{1}{\gamma/2 - 2\pi i\nu_0' + im} - \frac{1}{\gamma/2 + 2\pi i\nu_0' + im} \right] dm. \tag{17}$$

Break up the integration into ranges, from $-\infty$ to 0, and from 0 to ∞. Over these ranges make the respective substitutions:

$$m = -2\pi\nu, \quad dm = -2\pi \, d\nu, \quad \text{and} \quad m = 2\pi\nu, \quad dm = 2\pi \, d\nu.$$

We can then write the integral as

$$\psi(t) = \frac{A}{4\pi i} \int_{0}^{\infty} \left\{ e^{2\pi i\nu t} \left[\frac{1}{\gamma/4\pi - i\nu_0' + i\nu} - \frac{1}{\gamma/4\pi + i\nu_0' + i\nu} \right] + e^{-2\pi i\nu t} \left[\frac{1}{\gamma/4\pi - i\nu_0' - i\nu} - \frac{1}{\gamma/4\pi + i\nu_0' - i\nu} \right] \right\} d\nu. \tag{18}$$

Combining the first and fourth terms, and similarly the second and third, and rationalizing the denominators, we get

$$\psi(t) = \frac{A}{2\pi}\int_0^\infty \left[\frac{(\gamma/4\pi)\sin 2\pi\nu t + (\nu_0' - \nu)\cos 2\pi\nu t}{(\gamma/4\pi)^2 + (\nu_0' - \nu)^2} - \frac{(\gamma/4\pi)\sin 2\pi\nu t - (\nu_0' + \nu)\cos 2\pi\nu t}{(\gamma/4\pi)^2 + (\nu_0' + \nu)^2}\right] d\nu. \tag{19}$$

Now, clearing fractions and introducing the expression, equation (10),

$$\nu_0'^2 = \nu_0^2 - \left(\frac{\gamma}{4\pi}\right)^2, \tag{20}$$

we obtain

$$\psi(t) = \frac{A[\nu_0^2 - (\gamma/4\pi)^2]^{1/2}}{\pi}\int_0^\infty \frac{(\gamma\nu/2\pi)\sin 2\pi\nu t + (\nu_0^2 - \nu^2)\cos 2\pi\nu t}{(\nu^2 - \nu_0^2)^2 + (\gamma\nu/2\pi)^2}\, d\nu$$

$$= \int_0^\infty F_\nu \cos(2\pi\nu t - \alpha_\nu)\, d\nu$$

$$= \int_0^\infty F_\nu(\sin\alpha_\nu \sin 2\pi\nu t + \cos\alpha_\nu \cos 2\pi\nu t)\, d\nu, \tag{21}$$

which equation defines F_ν.

Let us set

$$F_\nu \sin\alpha_\nu = \frac{A[\nu_0^2 - (\gamma/4\pi)^2]^{1/2}}{\pi}\,\frac{\gamma\nu/2\pi}{(\nu^2 - \nu_0^2)^2 + (\gamma\nu/2\pi)^2}, \tag{22}$$

and

$$F_\nu \cos\alpha_\nu = \frac{A[\nu_0^2 - (\gamma/4\pi)^2]^{1/2}}{\pi}\,\frac{(\nu_0^2 - \nu^2)}{(\nu^2 - \nu_0^2)^2 + (\gamma\nu/2\pi)^2}. \tag{23}$$

Therefore

$$\tan\alpha_\nu = \gamma\nu/2\pi(\nu_0^2 - \nu^2). \tag{24}$$

$$F_\nu = \frac{A[\nu_0^2 - (\gamma/4\pi)^2]^{1/2}}{\pi}\,\frac{1}{[(\nu^2 - \nu_0^2)^2 + (\gamma\nu/2\pi)^2]^{1/2}}. \tag{25}$$

The phase factor, α_ν, depends on the frequency. Collecting these significant expressions, we write finally,

$$\psi(t) = \frac{A[\nu_0^2 - (\gamma/4\pi)^2]^{1/2}}{\pi}\int_0^\infty \frac{\cos(2\pi\nu t - \alpha_\nu)\, d\nu}{[(\nu^2 - \nu_0^2)^2 + (\gamma\nu/2\pi)^2]^{1/2}}. \tag{26}$$

Equation (26) represents a "spectrum analysis" of the damped oscillation. If the oscillator is, for example, an electron which radiates electromagnetic energy, (26) gives the distribution function of that radiation, with respect to frequency, ν. A spectrograph, for example, would show the radiation as having a peak for $\nu = \nu_0$ and rapidly decaying for wavelengths on both sides of the true resonant frequency. The radiation of a damped oscillator is not monochromatic, but covers a range from $\nu = 0$ to $\nu = \infty$. The sharpness of the peak depends on the magnitude of the damping constant, γ.

The student may recognize that the square of the coefficient, F_ν, is of the same form, functionally, as the expression for the coefficients of line absorption and emission. The correspondence is no accident. A direct physical connection exists between the equations of the present section and the form of the atomic absorption coefficient, as we shall demonstrate in Part IV, § (31).

15. Forced vibrations. Let us consider the problem of a string (or of any oscillator in general), subjected to a periodic force of frequency ν. Let ν_0 be the fundamental frequency to which the oscillator is resonant. Assume that the impressed force, F, is periodic with the time,

$$F = F_0 \cos 2\pi\nu t. \tag{1}$$

Introducing this expression into the right-hand side of (14.4), we obtain the equation

$$\frac{d^2S}{dt^2} + \gamma \frac{dS}{dt} + a^2u^2S = F_0 \cos 2\pi\nu t. \tag{2}$$

In solving differential equations of this type, we first set the right-hand part equal to zero, to obtain the general solution. To the resultant equation, which contains the arbitrary constants, we add a particular solution that satisfies the equation. The general solution is already available (14.11), where ν_0' is given, as before, by

$$2\pi\nu_0' = \frac{1}{2}\sqrt{4a^2u^2 - \gamma^2} = \frac{1}{2}\sqrt{4\pi^2\nu_0^2 - \gamma^2}. \tag{3}$$

To obtain the particular solution, make the substitution

$$S = C \cos (2\pi\nu t - \alpha), \tag{4}$$

where C and α are quantities to be determined. The result is

$$\begin{aligned} C[4\pi^2(\nu_0^2 - \nu^2) \cos (2\pi\nu t - \alpha) - 2\pi\nu\gamma \sin (2\pi\nu t - \alpha)] \\ = C[4\pi^2(\nu_0^2 - \nu^2) \cos \alpha + 2\pi\nu\gamma \sin \alpha] \cos 2\pi\nu t \\ + C[4\pi^2(\nu_0^2 - \nu^2) \sin \alpha - 2\pi\nu\gamma \cos \alpha] \sin 2\pi\nu t \\ = F_0 \cos 2\pi\nu t. \end{aligned} \tag{5}$$

The coefficient of the cosine term on the left-hand side must be equal to F_0 and that of the sine term must be zero, for the equation to be satisfied identically. Hence

$$\tan \alpha = \frac{\nu}{(\nu_0^2 - \nu^2)} \left(\frac{\gamma}{2\pi}\right). \tag{6}$$

From this equation we may evaluate $\sin \alpha$ and $\cos \alpha$ as well. The amplitude C may then be determined directly:

$$C = \frac{F_0}{4\pi^2} \frac{1}{[(\nu_0^2 - \nu^2)^2 + (\gamma\nu/2\pi)^2]^{1/2}}. \tag{7}$$

The complete solution of (2) is

$$S = [B \cos 2\pi\nu_0' t + A \sin 2\pi\nu_0' t]e^{-(\gamma/2)t} + C \cos (2\pi\nu t - \alpha). \tag{8}$$

The term in brackets represents a *transient* disturbance, which decays exponentially with the time. The second term represents a so-called *forced oscillation,* since its periodicity agrees with that of the impressed force. The oscillation, however, will be out of phase with the applied harmonic force. If $\nu_0 > \nu$, $\tan \alpha$ is positive and the displacement will lag behind the force by less than one-quarter period. If $\nu = \nu_0$, $\tan \alpha = \infty$, and the lag becomes exactly one-quarter period. When $\nu_0 < \nu$, $\tan \alpha$ is negative. Since $\sin \alpha$ is positive, α is in the second quadrant and the lag lies between one-quarter and one-half period.

Of special interest are the identities of (6) with (14.24) and (7) with (14.25). For the latter pair of equations, the constant factor is, of course, different. But the variation of the amplitude with frequency is identical. The analysis of the previous section applies only to the term $A \sin 2\pi\nu_0' t e^{-\gamma t/2}$ in (8). A similar investigation gives the same result for the cosine term, with a different α_ν, however. The spectral distribution of the radiation from the transient term of (8) is the same as that of the forced oscillations.

From (7), we see that the maximum amplitude occurs for $\nu = \nu_0$, when, if it were not for the damping term, the amplitude would be infinite. This result is one illustration of the phenomenon of resonance.

16. General wave equation for a damped oscillator. The complete wave equation for the damped one-dimensional oscillator becomes

$$\frac{\partial^2 \psi}{\partial x^2} = \frac{1}{u^2}\left(\frac{\partial^2 \psi}{\partial t^2} + \gamma \frac{\partial \psi}{\partial t}\right). \tag{1}$$

To solve this equation we write, as before,

$$\psi = SR. \tag{2}$$

Then we have

$$\frac{1}{R}\frac{d^2R}{dx^2} = \frac{1}{Su^2}\left(\frac{d^2S}{dt^2} + \gamma \frac{dS}{dt}\right) = -a^2. \tag{3}$$

We have already solved these two equations. The function for the x-coordinate is unaltered by the presence of damping. Its normalized form is given by (8.5). The time variation is most simply represented by (14.21), where the ψ refers to the variation for a given value of k.

The complete eigenfunction with the time factor, therefore, is

$$\psi = \sqrt{\frac{2}{X}} \sin k\pi \frac{x}{X} \int_0^\infty F_\nu \cos (2\pi\nu t - \alpha_\nu)\, d\nu, \tag{4}$$

where F_ν and α_ν are given by (14.25) and (14.24), respectively. In these expressions we must set ν_0 equal to ν_k, the normal vibration frequency of the kth mode of the undamped oscillator. Equation (4), because of the appearance of the extra amplitude factor, F_ν, is no longer normalized. The F_ν factor must now represent the initial amplitude. We have, in effect, split up our original wave function ψ_k, into an infinite number of wave functions, each characterized by a frequency ν. These individual subfunctions are not orthogonal to one another. Indeed, one may legitimately question whether "sub" wave functions exist. The individual amplitudes, F_ν, ascribed to various frequencies, cannot be added, because the phases are different.

The general form of the wave equation, when a damping resistance proportional to the velocity is present, is the extension of (1) to three coordinates, or

$$\nabla^2\psi = \frac{1}{u^2}\left(\frac{\partial^2\psi}{\partial t^2} + \gamma \frac{\partial\psi}{\partial t}\right). \tag{5}$$

17. Dissipation of energy by a damped linear oscillator. The discussion of these sections on damped vibrations is leading up to the classical problem of the interaction of atoms and radiation. Before the days of quantum theory, physicists generally attacked such problems on the basis of a very simplified atomic model. The emitting electrons were supposed to be oscillating about a position of equilibrium. Physicists assumed the existence of a restoring force proportional to the displacement and a damping force proportional to the velocity. The resulting differential equations of motion are identical with those already discussed.

The motion of an oscillator subject to a damping force is given in equation (14.11):

$$S = Ae^{-(\gamma/2)t} \sin (2\pi\nu_0't - \alpha). \tag{1}$$

A represents a linear amplitude and S measures the displacement as a function of t. The velocity of the oscillating particle will be

$$\frac{dS}{dt} = Ae^{-(\gamma/2)t}\left[2\pi\nu_0' \cos (2\pi\nu_0't - \alpha) - \frac{\gamma}{2} \sin (2\pi\nu_0't - \alpha)\right]. \tag{2}$$

As a result of damping, the average kinetic energy exhibits a slow decrease from cycle to cycle. To eliminate the variations that occur in any one cycle, we shall calculate the average value $\overline{T}$ of the kinetic energy. We suppose the damping to be small, so that we may neglect the change in the exponential term during a single oscillation. Since

$$\frac{1}{2\pi}\int_0^{2\pi} \sin^2\theta\, d\theta = \frac{1}{2\pi}\int_0^{2\pi} \cos^2\theta\, d\theta = \frac{1}{2}, \tag{3}$$

and

$$\frac{1}{2\pi}\int_0^{2\pi} \cos\theta \sin\theta\, d\theta = 0, \tag{4}$$

we have, if m is the mass of the oscillator,

$$\overline{T} = \frac{1}{2}\, m\overline{\left(\frac{dS}{dt}\right)^2} = \frac{\oint T\, dt}{\oint dt} \sim \frac{A^2 m e^{-(\gamma/2)t}}{4}\left(4\pi^2\nu_0'^2 + \frac{\gamma^2}{4}\right)$$

$$= \pi^2\nu_0^2 A^2 m e^{-\gamma t}, \tag{5}$$

by (14.20). The symbol $\oint$ signifies that the integral is to be taken over one complete cycle.

Since, on the average, the energy is half kinetic and half potential, the rate at which energy disappears from the system is

$$\frac{d\overline{E}}{dt} = 2\,\frac{d\overline{T}}{dt} = -2\pi^2\nu_0^2 A^2 m\gamma e^{-\gamma t}. \tag{6}$$

The logarithmic decrement is

$$\frac{d \ln 2\overline{T}}{dt} = -\gamma. \tag{7}$$

We should note two points in particular: the fact that the damping factor, γ, directly determines the rate of energy loss, and that the energy is proportional to the square of the amplitude.

The forces, of course, are not conservative and no general potential function exists. The rate at which the forces do work against the resistance, $\gamma(dS/dt)$, is determined by the integral of the force times the displacement ds, averaged over a given cycle. The length of a cycle is equal to $1/\nu_0'$, and

$$\frac{dE}{dt} = m\gamma\nu_0' \oint \frac{dS}{dt}\, ds = m\gamma \int_{t_1}^{t_1+(1/\nu_0')} \left(\frac{ds}{dt}\right)^2 dt, \tag{8}$$

because $dS = ds$, as shown by equation (1). This result proves to be twice the average value of the rate of change of kinetic energy, as previously given.

Let us now discuss the problem of dissipation of energy in a system subject to forced oscillations. If the frequency of the forced oscillation is ν, the solution of the equations of motion, after the transient term has died away, is given by (15.8):

$$S = C \cos(2\pi\nu t - \alpha). \tag{9}$$

C and α are given by (15.6) and (15.7). The velocity is

$$\frac{dS}{dt} = -2\pi\nu C \sin(2\pi\nu t - \alpha). \tag{10}$$

The rate of energy loss from the system is still to be calculated from (8), where ν_0' is set equal to ν. We find that

$$\frac{dE}{dt} = -2\pi m\gamma\nu C^2 = -F_0^2\left\{\frac{m}{2\pi}\,\frac{(\gamma/4\pi)\nu^2}{[(\nu_0^2 - \nu^2)^2 + (\gamma\nu/2\pi)^2]}\right\}. \tag{11}$$

The quantity in braces is proportional to the absorption coefficient, for upon it depends the rate at which the oscillator subtracts energy from the external field and dissipates the energy as friction, radiation, or in some other fashion. The absorption coefficient shows a sharp maximum for energies near or equal to the fundamental frequency.

In the neighborhood of this maximum we may set, without appreciable error, $\nu \sim \nu_0$ and $\nu + \nu_0 \sim 2\nu_0$. We shall still have to retain the factor $\nu - \nu_0$. Let us now suppose that the driving force consists of a continuous band of frequencies, instead of a single sharp frequency. To obtain the total rate of energy absorption over all frequencies, we shall have to multiply (11) by $d\nu$ and integrate over all ν. With the foregoing approximations this equation becomes

$$\frac{dE}{dt} = \frac{d}{dt}\int E_\nu\, d\nu = -F_{0\nu}^2\frac{m}{8\pi}\int_0^\infty \frac{\gamma/4\pi}{[(\nu - \nu_0)^2 + (\gamma/4\pi)^2]}\, d\nu$$

$$= -F_{0\nu}\frac{m}{8\pi}\left[\operatorname{arc\,tan}\frac{4\pi(\nu - \nu_0)}{\gamma}\right]_0^\infty = -F_{0\nu}\frac{m}{8}. \tag{12}$$

At the lower limit, we have used the approximation

$$\operatorname{arc\,tan}\left(-\frac{4\pi\nu_0}{\gamma}\right) \sim \operatorname{arc\,tan}(-\infty) = -\frac{\pi}{2}, \tag{13}$$

in accord with the assumption that

$$\gamma/4\pi \ll \nu_0. \tag{14}$$

The foregoing problem is closely related to the question of the formation of an absorption line, when radiation passes through a layer of atoms. The factor $F_{0\nu}$ measures the amplitude of the radiation; its square times $d\nu$ is proportional to the intensity of radiation within the frequency interval $d\nu$. $F_{0\nu}$ must also contain a factor specifying the nature of the interaction between the incident energy and the atomic oscillator. This latter factor might conceivably be a function of the frequency in certain types of problems.

Boundary Value Problems in General

18. Flow of heat. The differential equation for vibrations and wave motion is closely related to that for heat conduction. We shall, therefore, find it interesting and instructive to derive the equation for heat transfer and solve it in several representative examples.

Consider some medium, solid, liquid, or gaseous, through which the temperature, T, is not constant. We shall suppose that T is a function of the coordinates and the time. At any given instant we may define, in the medium, surfaces over which the temperature is constant. The equation,

$$T = C, \tag{1}$$

is analogous to the equation defining a surface of constant potential. Such a surface, or any line drawn upon it, is called an *isotherm* or *isothermal*. The force, $\mathbf{F}$, a vector normal to the surface $V = C$, is given by $\mathbf{F} = -\nabla V$. Similarly we represent the normal to the isothermal surface by a vector $\mathbf{g}$,

$$\mathbf{g} = -\nabla T = -\frac{\partial T}{\partial n}\mathbf{n}, \tag{2}$$

where n, as in previous sections, denotes a coordinate normal to the isothermal surface. The factor $\partial T/\partial n$ is the temperature gradient along the coordinate n.

We assume, in accord with experiment, that heat will tend to flow from the hotter portions of the medium toward the cooler. This flow takes place in a direction opposite to the temperature gradient. For example, if the temperature increases along the x-axis, the vector representing the flow of heat will be in the negative direction, because heat flows from hotter to colder regions. The rate of flow is proportional to the instantaneous temperature gradient and to the specific conductivity, κ, of the material. Hence, we set the vector $\mathbf{f}$, representing the vector rate of flow equal to

$$\mathbf{f} = \kappa\mathbf{g} = -\kappa\nabla T. \tag{3}$$

Consider, now, some definite volume of the medium. The net rate of heat flow out of this volume must be

$$\frac{dE}{dt} = \int \mathbf{f} \cdot d\mathbf{S}, \tag{4}$$

where $d\mathbf{S}$ is an element of the boundary surface. Let c be the specific heat and ρ the density of the material. According to the definition of specific heat, we associate the amount of heat dE with a temperature change dT, through the equation

$$dE = -\int (c\rho\, dT)\, d\tau, \tag{5}$$

where $d\tau$ is an element of volume. The negative sign indicates a temperature drop because dE represents a negative gain. Hence, the rate of heat loss depends upon the rate of temperature change, at each point of the medium, as follows:

$$\frac{dE}{dt} = -\int c\rho \frac{\partial T}{\partial t}\, d\tau. \tag{6}$$

Equating (4) and (6), we have

$$\int \mathbf{f} \cdot d\mathbf{S} = -\int c\rho \frac{\partial T}{\partial t}\, d\tau. \tag{7}$$

If any heat energy is generated within the volume, we must add a term $\int \varepsilon\, d\tau$ to the right-hand side of (7), where ε is the rate of heat generation per unit volume. By Gauss' theorem, we transform the surface integral of (7) into a volume integral:

$$\int \mathbf{f} \cdot d\mathbf{S} = \int \nabla \cdot \mathbf{f}\, d\tau = -\int \nabla \cdot (\kappa \nabla T)\, d\tau, \tag{8}$$

by (3). Then if κ is a constant over the medium (κ may still vary with the time),

$$\nabla \cdot (\kappa \nabla T) = \kappa \nabla^2 T, \tag{9}$$

and (7) becomes

$$\int \kappa \nabla^2 T\, d\tau = \int c\rho \frac{\partial T}{\partial t}\, d\tau. \tag{10}$$

Hence, for any volume element, we must have

$$\nabla^2 T = \frac{c\rho}{\kappa} \frac{\partial T}{\partial t}. \tag{11}$$

This differential equation applies only to the transfer of heat by conduction. It fails in any medium where the transfer occurs by convective or radiative processes. Equation (11) has much in common with the wave equation involving the time. We effect the solution most simply by writing ∇^2 in whatever form the geometry of the medium suggests and then attempting to separate variables by means of a product function. For example, in spherical coordinates, we try

$$T = SR\Theta\Phi, \tag{12}$$

where S is a function only of the time, etc. One can show that the solution of (11) is unique, when the proper initial and boundary conditions are specified.

19. Steady flow of heat. We shall discuss first the case of steady flow. Then

$$\partial T/\partial t = 0, \tag{1}$$

and (11) takes the form of Laplace's equation,

$$\nabla^2 T = 0. \tag{2}$$

Consider a medium bounded by the two infinite planes $x = 0$ and $x = a$. Further assume that the temperatures of the surfaces are constant and equal to T_0 and T_a, respectively. The symmetry shows that the isotherms are planes parallel to the boundary surfaces; hence we must regard T as a function of x alone. Equation (2) becomes

$$\frac{\partial^2 T}{\partial x^2} = \frac{d^2 T}{dx^2} = 0, \tag{3}$$

the integral of which is

$$T = Ax + B. \tag{4}$$

The constants are easily evaluated from the boundary conditions. Thus

$$T = \frac{T_a - T_0}{a} x + T_0. \tag{5}$$

Similarly, if the medium is in the form of a spherical shell, bounded by two spheres of radii r_a and r_b, with the boundary temperatures maintained constant at T_a and T_b, the temperature distribution will depend only on r. Therefore the conduction equation becomes

$$\frac{\partial}{\partial r} r^2 \frac{\partial T}{\partial r} = 0. \tag{6}$$

Hence
$$T = -\frac{A}{r} + B, \tag{7}$$

which gives, with the assumed boundary values ($r_a < r_b$),

$$T = \frac{T_a - T_b}{r} \frac{r_b r_a}{r_b - r_a} + \frac{T_b r_b - T_a r_a}{r_b - r_a}. \tag{8}$$

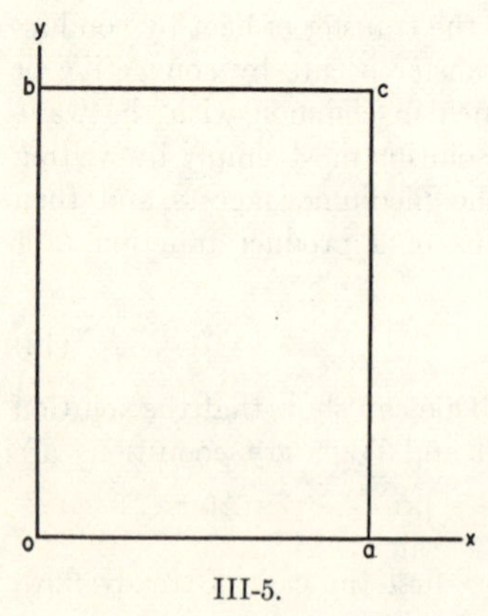

III-5.

We shall now treat flow in two dimensions. Consider a slab in the form of a long rectangular bar bounded by the surfaces $x = 0, x = a, y = 0, y = b$, and extending to $+\infty$ along z. A cross section of this slab by a plane perpendicular to the z-axis appears in Fig. 5. We further assume that the temperatures along ob, oa, and oc, are constant and equal, and of magnitude T_0. Along bc we take the temperature to be variable, a function of x alone, i.e., we set $T = T(x)$. Because of symmetry, the temperature distribution is independent of z. The conduction equation becomes

$$\frac{\partial^2 T}{\partial x^2} + \frac{\partial^2 T}{\partial y^2} = 0, \tag{9}$$

Laplace's equation in two dimensions.

To solve this equation we try the solution

$$T = XY + K, \tag{10}$$

where K is a constant. Then

$$\frac{1}{X}\frac{d^2X}{dx^2} = -\frac{1}{Y}\frac{d^2Y}{dy^2} = -\alpha^2. \tag{11}$$

The justification for the negative sign of the constant, to which we equate the independent sides of the equation will appear later. The solutions are

$$X = Ae^{i\alpha x} + Be^{-i\alpha y}, \quad Y = Ce^{\alpha y} - De^{-\alpha y}. \tag{12}$$

The temperature along the face $x = 0$, i.e., along ob, must be independent of y, and equal to T_0:

$$T = (A + B)(Ce^{\alpha y} + De^{-\alpha y}) + K = T_0. \tag{13}$$

We can fulfill this condition only if

$$B = -A \quad \text{and} \quad K = T_0. \tag{14}$$

Similarly, along oa, $y = 0$, we must have

$$T = A(e^{i\alpha x} - e^{-i\alpha x})(C + D) + T_0 = T_0. \tag{15}$$

Hence

$$D = -C. \tag{16}$$

We also note that

$$e^{i\alpha x} - e^{-i\alpha x} = 2i \sin \alpha x. \tag{17}$$

We now introduce the third condition, that along $x = a$,

$$T = G \sin \alpha a(e^{\alpha y} - e^{-\alpha y}) + T_0 = T_0, \tag{18}$$

where G is a new constant defined in terms of the old by

$$G = 2ACi. \tag{19}$$

To satisfy (18), we must have

$$\begin{aligned} G \sin \alpha a &= 0, \\ \text{or} \qquad \alpha a &= k\pi, \quad k = 0, 1, 2, \ldots. \end{aligned} \tag{20}$$

We now point out that had we adopted a positive sign for α^2 in (11), instead of a sine term for x we should have obtained the factor $e^{\alpha x} - e^{-\alpha x}$, which does not vanish for $x = a$. Since any one (or all) of the values of k in (20) will satisfy (18), we must write the solution in the general form

$$T = \sum_{k=1}^{\infty} G_k \sin \frac{k\pi}{a} x(e^{k\pi y/a} - e^{-k\pi y/a}) + T_0. \tag{21}$$

We must determine the coefficients G_k of the Fourier series from the fourth condition, that for $y = b$,

$$T = T(x). \tag{22}$$

thus

$$T = \sum_{k=1}^{\infty} H_k \sin \frac{k\pi}{a} x + T_0 = T(x), \tag{23}$$

where

$$H_k = G_k(e^{k\pi b/a} - e^{-k\pi b/a}). \tag{24}$$

We find the coefficients, H_k, as before, by integration of the half-range Fourier series, as in (7.17):

$$H_k = \frac{2}{a} \int_0^a (T_x - T_0) \sin \frac{k\pi}{a} x \, dx. \tag{25}$$

To proceed further, we must know the form of $T(x)$. If we set

$$T(x) = T_0 + \frac{2\beta x}{a} \qquad 0 \leq x \leq a/2,$$

and

$$T(x) = T_0 + \frac{2\beta(a - x)}{a}, \qquad a/2 \leq x \leq a, \tag{26}$$

so that the temperature increases linearly from T_0 at each edge to the value $T_0 + \beta$ at the center of the face, the analytical problem becomes identical with that of § (7), with the solution as given in (7.23). If we introduce the notation,

$$e^{k\pi y/a} - e^{-k\pi y/a} = 2 \sinh \frac{k\pi y}{a}, \tag{27}$$

the final expression for the temperature distribution in the interior of the bar becomes

$$T = T_0 + \sum_{k=0}^{\infty} \frac{8\beta}{\pi} \frac{\sin (2k + 1)\pi x/a}{(2k + 1)^2} \frac{\sinh (2k + 1)\pi y/a}{\sinh (2k + 1)\pi b/a}. \tag{28}$$

Fourier first introduced series of this character in connection with problems of heat flow.

20. Variable heat flow. We may divide problems of heat conduction that depend on the time into two rough classifications. The first and simpler arises when we permit a body with a known initial temperature distribution to come to equilibrium, either with itself or with external sources of constant temperature. The second deals with the variations in temperature that contact with a non-constant (often periodic) source of heat may induce in the given body.

As an example of the former, we consider the flow of heat in a long uniform rod of length a. We suppose the rod to be covered with perfect insulation so that heat can enter or escape only from the ends. At time $t = 0$, we suppose that the temperature distribution is

$$T = T(x, 0). \tag{1}$$

We plan to maintain each end of the rod at a constant temperature T_0. We are required to find the temperature at any point for all later times.

Our transfer equation is

$$\frac{\partial^2 T}{\partial x^2} = \frac{c\rho}{\kappa}\frac{\partial T}{\partial t}. \tag{2}$$

We try the solution

$$T = XS + K, \tag{3}$$

and further suppose that c, ρ, and κ are constant with x and t. Then (2) becomes

$$\frac{1}{X}\frac{d^2X}{dt^2} = \frac{c\rho}{\kappa}\frac{1}{S}\frac{dS}{dt} = -\alpha^2, \tag{4}$$

which yields the solutions,

$$S = Ce^{-\kappa\alpha t/c\rho}, \qquad X = Ae^{i\alpha x} + Be^{-i\alpha x}. \tag{5}$$

When $t = 0$,

$$T(x, 0) = Ae^{i\alpha x} + Be^{-i\alpha x} + K = T_0, \tag{6}$$

for $x = 0$, and $x = a$. As before, this condition requires that

$$B = -A, \qquad K = T_0, \qquad \alpha a = k\pi. \tag{7}$$

Thus the general solution becomes

$$T(x, t) = T_0 + \sum_1^\infty H_k \sin\frac{k\pi x}{a} e^{-k^2\kappa\pi^2 t/c\rho a^2}. \tag{8}$$

We are to obtain the coefficients H_k, as usual, by means of Fourier series. Note, especially, that the higher modes are damped out more quickly than the lower ones, because of the k^2 factor in the exponential. The final state, as $t \to \infty$, is that of a uniform temperature, as expected. The last remaining trace of a non-uniform distribution will come from the sine term for $k = 1$. If we had adopted a positive sign for α^2, in (4), the temperature would have increased without limit, in direct contradiction to experiment. Further, we could not have satisfied our boundary conditions.

As an example of the second type of problem, we consider the temperature variations in a semi-infinite rod, one face of which is in contact with a surface whose temperature varies sinusoidally with the time. We take this face to be the plane $x = 0$, and suppose the rod to extend indefinitely toward $x = \infty$. We further suppose that the rod is covered, except for the front surface, with perfect insulation. The equation to be solved is the same as (2), whose solution is

$$T(x, t) = (Ae^{i\alpha x} + Be^{-i\alpha x})e^{-\kappa\alpha^2 t/c\rho} + K. \tag{9}$$

The constant C in (5) is superfluous, since we may combine it with A and B.

Let ν be the frequency of the temperature variation on the face $x = 0$, so that

$$T(0, t) = T_0 + \beta \sin 2\pi\nu t; \quad 2\pi\nu = \kappa\gamma/c\rho, \tag{10}$$

where β denotes the magnitude of the extreme temperature variation. Then we must have

$$T_0 + \beta \sin 2\pi\nu t = (A + B)e^{-\kappa\alpha^2 t/c\rho} + K, \tag{11}$$

which requires that, for $t = 0$,

$$K = T_0. \tag{12}$$

This equation, however, leads us to an impasse, for the sine term on one side and the exponential function on the other are certainly not compatible. This example illustrates the difficulties that may arise when we inadvertently adopt the wrong form for the constant α^2. The discordance disappears if we set

$$\alpha^2 = -\gamma i, \tag{13}$$

and regard the constants A and B as complex, and of the form

$$A = A_r + iA_i, \qquad B = B_r + iB_i, \tag{14}$$

for then the complex exponential is equivalent to a sine or cosine function. Taking the real part of the result, we have

$$\beta \sin 2\pi\nu t = (A_r + B_r) \cos \frac{\kappa\gamma t}{c\rho} - (A_i + B_i) \sin \frac{\kappa\gamma t}{c\rho}, \tag{15}$$

whence

$$A_r + B_r = 0, \quad A_i + B_i = \beta, \quad \gamma = 2\pi\nu c\rho/\kappa. \tag{16}$$

We must make similar changes in the first factor of (9). From (13),

$$\alpha = i\sqrt{\gamma i}. \tag{17}$$

Since

$$(1 + i)^2 = 2i, \qquad \sqrt{i} = (1 + i)/\sqrt{2}, \tag{18}$$

$$i\alpha = -\sqrt{\gamma/2}\,(1 + i). \tag{19}$$

Then (9) becomes

$$T(x, t) = [(A_r + iA_i)e^{-\sqrt{\gamma/2}\,(1+i)x} + (B_r + iB_i)e^{\sqrt{\gamma/2}\,(1+i)x}]e^{i\kappa\gamma t/c\rho} + T_0. \tag{20}$$

As here defined, γ is a real positive quantity. Therefore if we wish to exclude the possibility of infinite temperatures for $x = \infty$, we must take

$$B_r = B_i = 0, \qquad A_r = 0, \qquad A_i = \beta, \tag{21}$$

by (16). Hence (20) becomes

$$\begin{aligned} T(x, t) &= -\beta e^{-\sqrt{\gamma/2}\,x} \Re[ie^{i\kappa\gamma t/c\ -\sqrt{\gamma/2}\,x}] \\ &= \beta e^{-\sqrt{\gamma/2}\,x} \sin(\kappa\gamma t/c\rho - \sqrt{\gamma/2}\,x) \\ &= \beta e^{-\sqrt{\pi\nu c\rho/\kappa}\,x} \sin(2\pi\nu t - \sqrt{\pi\nu c\rho/\kappa}\,x), \end{aligned} \tag{22}$$

which is the general solution of the problem. The temperature variation shows three characteristics. The amplitude of the variation diminishes exponentially with depth. The variation below the surface is sinusoidal with frequency ν, but the phase is shifted by the term $\sqrt{\pi\nu c\rho/\kappa}\,x$, i.e., the variation in the interior lags behind that for the surface. This shift is due to the finite velocity, v, of the temperature wave. When

$$\sqrt{\pi\nu c\rho/\kappa}\,x_0 = 2\pi, \tag{23}$$

the lag amounts to a full period. The wave has taken the time ν^{-1} to penetrate a distance x_0. Therefore the velocity is

$$v = x_0/\nu^{-1} = 2\sqrt{\pi\nu\kappa/c\rho}, \tag{24}$$

a function of the frequency.

SELECTED PROBLEMS FOR PART III

1. Prove that

$$e^\theta = \frac{e^\pi - e^{-\pi}}{\pi}\left[\frac{1}{2} + \frac{1}{2}\cos\theta + \frac{1}{5}\cos 2\theta - \frac{1}{10}\cos 3\theta + \frac{1}{17}\cos 4\theta - \ldots \right.$$
$$\left. + \frac{1}{2}\sin\theta - \frac{2}{5}\sin 2\theta + \frac{3}{10}\sin 3\theta - \frac{4}{17}\sin 4\theta + \ldots\right].$$

Hint: In the integration use the sine and cosine in complex form. $-\pi \leq \theta \leq \pi$.

2. Let $f(\theta) = \sin\frac{2}{3}\theta,\quad 0 \leq \theta \leq 2\pi$. Express $f(\theta)$ as a Fourier series.

3. Use Maclaurin's expansion to derive the series expansion for e^x.

4. Derive the expression for the form of a vibrating string at time $t = 1$ sec. Assume $\psi = XT$ and expand in orthogonal cosine functions. Assume that $\psi = f(x)$ when $t = 0$. Let l be the length of the string.

5. Prove that

$$1 + \cos\theta + \cos 2\theta + \ldots \infty = 1/2, \qquad 0 < \theta < 2\pi.$$

$$\sin\theta + \sin 2\theta + \sin 3\theta + \ldots \infty = (1/2)\cot\theta/2, \qquad 0 < \theta < 2\pi.$$

$$\cos\theta - \cos 2\theta + \cos 3\theta - \ldots \infty = 1/2, \qquad -\pi < \theta < \pi.$$

$$\sin\theta - \sin 2\theta + \sin 3\theta - \ldots \infty = (1/2)\tan\theta/2, \qquad -\pi < \theta < \pi.$$

6. Prove that

$$\frac{\pi}{2}e^{\theta} = \frac{1}{2}(1 + e^{\pi})\sin\theta + \frac{2}{5}(1 - e^{\pi})\sin 2\theta + \frac{3}{10}(1 + e^{\pi})\sin 3\theta$$

$$+ \frac{4}{17}(1 - e^{\pi})\sin 4\theta + \ldots \infty.$$

7. A vibrating string, extending from $x = -l/2$ to $x = +l/2$, has a varying density along its length so that the wave velocity varies as $u^{-2} = u_0^{-2}(1 - a \sin \pi x/l)$, where $a \gg 1$. Obtain the first approximation according to the assumption that $a = 0$. Then use this first approximation, $\psi = \cos \pi m x/l$ to get the equation

$$\frac{d^2\psi}{dx^2} + \frac{1}{u_0^2}\psi = \frac{a}{u_0^2}\psi \sin\frac{\pi x}{l}\cdot$$

Substitute for ψ on the right-hand side to express it as a function of x only. Express the product $(\cos \pi m x/l)(\sin \pi x/l)$ in terms of exponential equivalents and solve for the particular integral.

8. Suppose that a vibrating string of length l is subject to a damping resistance of magnitude $-a\,|\,v\,|$, where $v = \partial\psi/\partial t$. Determine the motion as a function of the time, and discuss the rates of damping of the various overtones. If the string has initially the V-shaped form of equation (7.20), discuss the subsequent motion as a function of the time. At what rate will the string lose energy?

9. For the vibrating circular plate, show that eigenfunctions are independent of the origin chosen for the angular coordinate.

10. Discuss the vibrating rectangular plate, the rectangular box, and the circular cylinder. Write down the normalized wave functions for each, in terms of solutions already derived.

11. The equation of motion of a solid, vibrating bar is

$$\frac{\partial^4\psi}{\partial x^4} = -\frac{\rho}{Q\kappa}\frac{\partial^2\psi}{\partial t^2}\cdot$$

where ρ is the density, κ the so-called "radius of gyration," which depends on the shape of the bar, and Q a constant of the material, known as "Young's modulus." For a circular bar of radius a, $k = a/2$. Discuss the motion of a cylindrical rod and show that $\psi = f(x \pm ut)$ is not a solution. Find the harmonic solution. Show that, if we try to define a wave velocity, it depends on the frequency, whereas the ordinary vibrating string has a wave velocity independent of frequency.

12. The Bessel functions whose order is half-odd integral are series of sines and cosines. If

$$J_{1/2}(z) = \left(\frac{2}{\pi z}\right)^{1/2} \sin z, \quad J_{3/2}(z) = \left(\frac{2}{\pi z}\right)\left(\frac{\sin z}{z} - \cos z\right),$$

etc., show that $J_{l+(1/2)}(z)$ is finite at the origin. Show that the alternative solutions, which satisfy Bessel's equation,

$$J_{-1/2}(z) = \left(\frac{2}{\pi z}\right)^{1/2} \cos z, \quad J_{-3/2}(z) = \left(\frac{2}{\pi z}\right)^{1/2}\left(-\frac{\cos z}{z} - \sin z\right),$$

are infinite at the origin and, therefore, not satisfactory as eigenfunctions in the vicinity of the origin.

13. A particle moves with constant velocity parallel to the axis of x back and forth between two points located at $x = 0$ and $x = l$. At $t = 0$, the particle is at the origin. Its motion is subject to the law: $x = vt$ for $l/2v$; $x = l - vt$ for $l/2v \leq t \leq 3l/2v$,

$$x = \begin{cases} vt, & 0 \leq t \leq \dot{x}_1/v, \\ 2x_1 - vt, & x_1/v \leq t \leq 2x_1/v. \end{cases}$$

If the moving particle is an electron, its radiation will be on certain characteristic frequencies, of which $v/2l$ is the fundamental. Assuming that the damping by loss of radiation is negligible, express the motion in terms of its representative Fourier series. If we assume that the rate of radiation is proportional to the square of the amplitude and to the fourth power of the frequency, discuss the type of radiation to be expected from such a system. Compare the result with that of the vibrating string.

14. Show that the eigenfunctions of the vibrating sphere are independent of the choice of axes.

Problems 15–23 are representative of the use of eigenfunctions in wave and matrix mechanics.

15. Prove the following expansions:

$$\frac{\partial}{\partial \theta}\Theta(l, m) = \frac{1}{2}\sqrt{(l-m)(l+m+1)}\,\Theta(l, m+1) - \frac{1}{2}\sqrt{(l+m)(l-m+1)}\,\Theta(l, m-1);$$

$$(m \cot \theta)\Theta(l, m) = -\frac{1}{2}\sqrt{(l-m)(l+m+1)}\,\Theta(l, m+1) - \frac{1}{2}\sqrt{(l+m)(l-m+1)}\,\Theta(l, m-1).$$

Hint: Use Rodrigues' formula.

16. Show that

$$(\cos\theta)\Theta(l, m) = \Theta(l+1, m)\sqrt{\frac{(l-m+1)(l+m+1)}{(2l+1)(2l+3)}} + \Theta(l-1, m)\sqrt{\frac{(l-m)(l+m)}{(2l-1)(2l+1)}};$$

$$(\sin\theta)\Theta(l, m) = -\Theta(l+1, m+1)\sqrt{\frac{(l+m+1)(l+m+2)}{(2l+1)(2l+3)}} + \Theta(l-1, m+1)\sqrt{\frac{(l-m)(l-m+1)}{(2l-1)(2l+1)}}$$

$$= \Theta(l+1, m-1)\sqrt{\frac{(l-m+1)(l-m+2)}{(2l+1)(2l+3)}} - \Theta(l-1, m-1)\sqrt{\frac{(l+m-1)(l+m)}{(2l-1)(2l+1)}}.$$

17. Express the vector $\mathbf{r} = \mathbf{i}x + \mathbf{j}y + \mathbf{k}z$ as a bivector, in terms of the coordinates r, θ, ϕ.

Ans. $\mathbf{r} = r[(\boldsymbol{\alpha}(e^{-i\phi}/\sqrt{2})\sin\theta + \boldsymbol{\beta}(e^{i\phi}/\sqrt{2})\sin\theta + \mathbf{k}\cos\theta)]$,

where $\boldsymbol{\alpha} = (\mathbf{i} + i\mathbf{j})/\sqrt{2}$; $\boldsymbol{\beta} = (\mathbf{i} - i\mathbf{j})/\sqrt{2}$. Note that $\boldsymbol{\alpha}$ and $\boldsymbol{\beta}$ play the roles of unit vectors, in the sense that

$$\boldsymbol{\alpha}\cdot\boldsymbol{\alpha}^* = \boldsymbol{\alpha}\cdot\boldsymbol{\beta} = 1, \quad \boldsymbol{\alpha}\cdot\boldsymbol{\beta}^* = \boldsymbol{\alpha}\cdot\boldsymbol{\alpha} = 0, \quad \text{ETC.}$$

Show that $\mathbf{r}\cdot\mathbf{r} = r^2$. For further details cf. IV-28.

18. We have defined angular momentum as $\mathbf{L} = \mathbf{r}\times\mathbf{p}$, cf. equation (II-22.4). In rectangular components,

$$\mathbf{L} = \mathbf{i}(yp_z - zp_y) + \mathbf{j}(zp_x - xp_z) + \mathbf{k}(xp_y - yp_x) = \mathbf{i}L_x + \mathbf{j}L_y + \mathbf{k}L_z.$$

Show that in terms of the bivectors of the preceding problem,

$$\mathbf{L} = \frac{\boldsymbol{\alpha}(L_x - iL_y)}{\sqrt{2}} + \frac{\boldsymbol{\beta}(L_x + iL_y)}{\sqrt{2}} + \mathbf{k}L_z.$$

19. Now, with the standard transformation,

$$x = r\sin\theta\cos\phi, \quad y = r\sin\theta\sin\phi, \quad z = r\cos\theta,$$

and with the arbitrary definitions,

$$p_x = \frac{h}{2\pi i}\frac{\partial}{\partial x}, \quad p_y = \frac{h}{2\pi i}\frac{\partial}{\partial y}, \quad p_z = \frac{h}{2\pi i}\frac{\partial}{\partial z},$$

where h is Planck's constant, show that $\mathbf{L}$ takes the form

$$\mathbf{L} = \frac{h}{2\pi}\left[\frac{\boldsymbol{\alpha}}{\sqrt{2}}e^{-i\phi}\left(-\frac{\partial}{\partial\theta} + i\cot\theta\frac{\partial}{\partial\phi}\right) + \frac{\boldsymbol{\beta}}{2}e^{i\phi}\left(\frac{\partial}{\partial\theta} + i\cot\theta\frac{\partial}{\partial\phi}\right) - i\mathbf{k}\frac{\partial}{\partial\phi}\right].$$

20. From (19) and (15) show that

$$\mathbf{L}[\Theta(l, m)\Phi(m)] = \frac{h}{2\pi}\left[\frac{\alpha}{\sqrt{2}}\sqrt{(l + m)(l - m + 1)}\,\Theta(l, m - 1)\,\Phi(m - 1) + \frac{\beta}{\sqrt{2}}\sqrt{(l - m)(l + m + 1)}\,\Theta(l, m + 1)\Phi(m + 1) + \mathbf{k}m\Theta(l, m)\Phi(m)\right].$$

21. From the preceding problem and considerations of orthonormality of the wave functions, calculate

$$\overline{\mathbf{L}} = \int_0^\pi \int_0^{2\pi} \Theta(l' m')\Phi(m')\mathbf{L}[\Theta(l, m)\Phi(m)] \sin\theta \, d\theta \, d\phi.$$

Ans.
$\overline{\mathbf{L}} = (h/2\pi)(\alpha/\sqrt{2})\sqrt{(l + m)(l - m + 1)}\;\delta(l', l)\,\delta(m', m - 1);$ ETC.

22. Show that the operator

$$\mathbf{L}\cdot\mathbf{L} = -\left(\frac{h}{2\pi}\right)^2\left[\frac{1}{\sin\theta}\frac{\partial}{\partial\theta}\left(\sin\theta\frac{\partial}{\partial\theta}\right) + \frac{1}{\sin^2\theta}\frac{\partial^2}{\partial\phi^2}\right].$$

Note the similarity of this expression to the angular portion of the operator ∇^2, (10.1).

23. Apply the operator **L** to the result of problem (20), to show that

$$\mathbf{L}\cdot[\mathbf{L}\Theta(lm)\Phi(m)] = \left(\frac{h}{2\pi}\right)^2 l(l + 1)\Theta(l, m)\Phi(m).$$

Then show that

$$\overline{\mathbf{L}\cdot\mathbf{L}} = \int_0^\pi \int_0^{2\pi} \Theta(l', m')\Phi(m')\mathbf{L}\cdot\mathbf{L}\Theta(l, m)\Phi(m) \sin\theta \, d\theta \, d\phi$$

$$= \left(\frac{h}{2\pi}\right)^2 l(l + 1)\,\delta(l', l)\,\delta(m', m).$$

24. A cylindrical wire of radius a, heated electrically, along the axis of a cylinder, maintains the surface $r = a$ at constant temperature T_a. The cylinder itself, of radius b, is immersed in an oil bath of temperature T_b. Calculate the distribution of temperature within the cylinder. Assume that the two plane boundary faces are heavily insulated.

25. Consider a source of heat whose intensity fluctuates cyclically as follows:

$$T = T_0 \cos 2\pi\nu_0 t, \quad -\nu_0/4 \le t \le \nu_0/4; \quad 3\nu_0/4 \le t \le 5\nu_0/4; \quad \text{ETC.};$$

$$T = 0, \quad \nu_0/4 \le t \le 3\nu_0/4; \quad 5\nu_0/4 \le t \le 7\nu_0/4; \quad \text{ETC.}$$

Express T as a Fourier series.

26. Consider a uniform solid, bounded by the plane $z = 0$, and extending to infinity in the direction of $z = \infty$. Assume that a controlled source (or sink) of energy maintains the temperature over the entire plane $z = 0$ at the value given by problem (25). Calculate the distribution of temperature within the solid as a function of z and t. If ν_0 corresponds to a period of one day ($\nu_0 = 1/86{,}400$), and if $c\rho/\kappa = 200$, calculate the time required for the wave to penetrate to a depth of one meter. At what depth will the temperature be exactly one year out of phase with the impressed temperature at the boundary? If $T_0 = 300°$, what will be the amplitude of the fluctuation at the two depths referred to above?

PART IV

Classical Electromagnetic Theory

Electrostatics and Magnetostatics

1. Electric and magnetic fields. The fundamental equation for two point-masses attracting one another gravitationally is

$$F = -G\frac{m_1 m_2}{r^2}, \tag{1}$$

Newton's well-known law, where F is the force. The minus sign (sometimes omitted for convenience) indicates that the force is one of attraction, i.e., its action tends to diminish the distance r between the bodies. If equation (1) were not of further use, we should never have to worry about the factor G. Equation (1) is important, however, in that it relates quantities of a purely physical nature, such as mass and distance, to the dynamical quantity F. When we introduce the dimensional constant G, whose value depends on experiment, the equation is satisfied dimensionally as well as numerically. We may calculate the resulting trajectories from the fundamental differential equation.

The laws governing electric and magnetic fields are of similar character to those governing gravitational fields. As we demonstrated in Part I, § (4), we may write expressions analogous to (1) for the respective fields. We must allow for one marked difference, however. Electric and magnetic forces, unlike gravitational force, depend on the nature of the intervening medium. Accordingly we must introduce the factors μ and κ, defined, respectively, as the magnetic permeability and dielectric constant, which are properties of the media and not of the force centers. The respective forces are (I-4.1 and I-4.2)

$$F = C_1\frac{q_1 q_2}{\kappa r^2} \tag{2}$$

and

$$F = C_2\frac{p_1 p_2}{\mu r^2}, \tag{3}$$

where q_1 and q_2 refer to the electric charge, and p_1 and p_2 to the magnetic pole strengths. When the charges or the poles are of opposite sign (one positive and the other negative) the forces are attractive.

We must choose the various parameters so that the forces appear in ordinary dynamical units. In equation (1), we might have adopted a special unit of mass so as to make G unity, had not other physical concepts proved to be more convenient for defining the mass. For electrical and magnetic quantities, several systems of units are available. For the treatment of dynamical problems, when the force is to be expressed in cgs units, we may define the unit of q or p in such a way as to make the constants C_1 and C_2 equal unity. Similarly, we may set both κ and μ equal to unity for empty space.

We now restate the above definitions in quantitative form. Two equal electric charges (or magnetic poles) of the same sign possess unit charge (unit quantity of magnetism), if they repel one another with a force equal to one dyne, when placed one centimeter apart in a vacuum.

The two systems, thus established, are called, respectively, the electrostatic (es) and the electromagnetic (em) systems of units. They cannot be completely independent. A circulating electric current of given magnitude produces an electromagnetic field of a definite intensity. We shall determine the relationship presently. The physical dimensions of q and p are, from Part I,

$$q = [M^{1/2}L^{3/2}T^{-1}\kappa^{1/2}]. \qquad \text{es} \tag{4}$$

$$p = [M^{1/2}L^{3/2}T^{-1}\mu^{1/2}]. \qquad \text{em} \tag{5}$$

We have also seen, from I-4, that force in the MKS system becomes

$$F = \mu\bar{p}_1\bar{p}_2/4\pi r^2 = B\bar{p}_2, \tag{3a}$$

where $\bar{p}$ represents the induced pole strengths of magnetic charge and B the magnetic induction.

To extend the usefulness of this section we give the formulas in duplicate. The original form of each equation is Gaussian but its MKS counterpart follows directly, with the distinguishing symbol, a, in the equation number. Equations that appear only once possess identical form in both systems. In the MKS portion, some abbreviations occur. For example, when relativly long identical brackets are common to both, the internal factor is indicated by [ ], { }, or analogous symbol.

2. Electrostatic potential. Electric potential is analogous to gravitational potential. For example, a spherically symmetrical distribution of charge acts like a point charge at the center. As for gravitation, a scalar potential function V exists, such that the components of the electric force vector $\mathbf{E}$ are

$$E_x = -\frac{\partial V}{\partial x}, \quad E_y = -\frac{\partial V}{\partial y}, \quad E_z = -\frac{\partial V}{\partial z}, \tag{1}$$

(equations II-8.4), or in vector notation (II-9.4),

$$\mathbf{E} = -\nabla V. \tag{2}$$

The quantities E_x, etc., are *potential gradients*, since they determine the rate of change of V with distance. They are also termed the components of electric intensity, because they represent the force acting on a unit charge.

Let Q be the number of units of free electric charge symmetrically distributed through a spherical volume of radius r_0. The radial force acting on a positive unit charge at distance r from the center, where $r > r_0$, is

$$E_r = \frac{Q}{\kappa r^2} = -\frac{\partial V}{\partial r}. \tag{3}$$

$$E_r = \frac{Q}{4\pi\epsilon r^2} = -\frac{\partial V}{\partial r}. \tag{3a}$$

We shall provisionally adopt the gravitational convention of setting the potential energy of a particle, moving in the electrostatic field, equal to zero for $r = \infty$. Hence, integrating (3), and setting the constant of integration equal to zero, we have

$$V = \frac{Q}{\kappa r}. \tag{4}$$

$$V = \frac{Q}{4\pi\epsilon r}. \tag{4a}$$

The sign of Q determines the sign of V.

The electrostatic problem, compared with the gravitational, is complicated by the existence of repulsive as well as attractive forces. However, most of the ordinary mechanical formulae relating to the gravitational fields apply without substantial change. This similarity of procedure results from the fact that formulae already derived depend on an assumed inverse-square law for the force field and are otherwise independent of the nature of the bodies involved.

Gauss' law, (II-15.4), relating the total charge q contained within a volume to the total flux ϕ, i.e., to the integral of the normal component of $\mathbf{E}$ over the boundary surface, becomes

$$\phi = \iint \mathbf{E} \cdot d\mathbf{S} = 4\pi q = \iiint \nabla \cdot \mathbf{E}\, d\tau = -\iiint \nabla^2 V\, d\tau, \tag{5}$$

$$\phi = \iint \mathbf{E} \cdot d\mathbf{S} = q/\epsilon, \tag{5a}$$

by Gauss' theorem and (2). And since

$$q = \int \rho\, d\tau, \tag{6}$$

where ρ is the density of charge, we immediately obtain Poisson's law (II-16.3)

$$\nabla^2 V = -4\pi\rho \tag{7}$$

$$\nabla^2 V = -\rho/\epsilon \tag{7a}$$

and Laplace's law

$$\nabla^2 V = 0, \tag{8}$$

the latter for a region devoid of charge.

From the standpoint of electrical properties, we may divide matter roughly into two classes: conductors and insulators. Conductors will convey away the charge from an electrified body whereas insulators will allow it to remain. Physically, the distinction between a conductor and an insulator is that in the former the outer atomic electrons are only loosely bound, if at all, to the core of the atom. The charges are free to move and, if possible, occupy an equilibrium position under the forces of their mutual attraction and repulsion and of the external fields. In an insulator, however, the electrons are tightly bound.

Let $\mathbf{E}_0$ be the force intensity resulting from all other charges acting on an elementary charge q_0, in the conductor. Since, by hypothesis, q_0 is in a position of equilibrium, $\mathbf{E}_0$ must vanish throughout the conductor, i.e.,

$$\begin{aligned} \mathbf{E}_0 &= -\nabla V = 0. \\ V &= \text{CONST.} \end{aligned} \tag{9}$$

The potential is constant in the entire conducting medium. Since the earth is a conductor, one often chooses the arbitrary constant in the potential to give zero for the ground potential.

In electrostatics, the charges must lie entirely on the surface of a conductor. If we draw any closed surface entirely within the conductor, $\mathbf{E}$

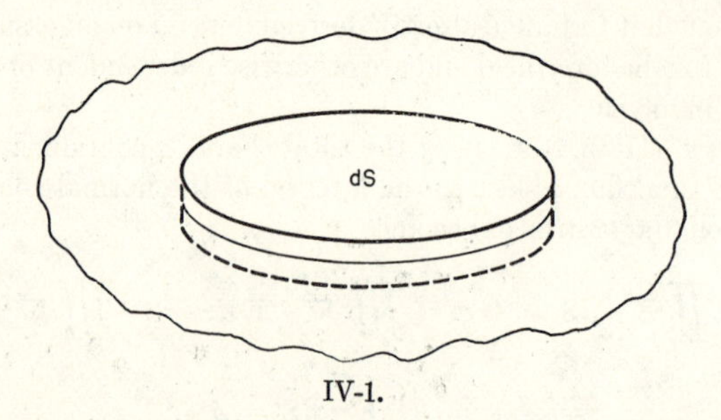

IV-1.

must vanish over it, so that $\phi = 0$, and hence, by (5), the charge within the surface must be zero. In consequence, for conductors we deal with a surface density σ, rather than with volume density, and we must modify our formulae accordingly. Outside the conductor the vector $\mathbf{E}$ will be normal to the boundary because the latter is a surface of constant potential. Let its value be E_n. Consider, now, a small volume, shaped like a pillbox, with its lower surface imbedded in the conductor and the upper surface, of area dS, outside and parallel to that of the conductor. The total charge inside the box is $\sigma\, dS$. The scalar product $\mathbf{E} \cdot d\mathbf{S}$ vanishes over the lower boundary because $\mathbf{E} = 0$ and over the sides because $\mathbf{E}$ and $d\mathbf{S}$ are perpendicular. Hence the total flux (5), is

$$\phi = \mathbf{E} \cdot d\mathbf{S} = E_n\, dS = 4\pi\sigma\, dS \tag{10}$$

$$\phi = \sigma\, dS/\epsilon. \tag{10a}$$

and

$$E_n = 4\pi\sigma. \tag{11}$$

$$E_n = \sigma/\epsilon. \tag{11a}$$

This equation relates the normal component of electric intensity to the surface charge density.

In the calculation of electric fields between conductors, the mobility of the charge introduces a new complication that did not concern us in the gravitational problem. For example, consider a grounded conducting sphere brought in the neighborhood of a plate charged with positive electricity. Because matter is electric, there will be a tendency for the negative charge to flow from the ground to the surface of the sphere. These induced charges on the sphere act to produce second-order distortions of the charge distribution on the plate; these in turn react to give third-order induced charges on the sphere, etc. The phenomenon finds its closest analogue in ordinary mechanics in the realm of tidal perturbations.

To illustrate the problem we shall first consider a point charge of magnitude q, in the neighborhood of an infinite plane. We suppose the plane to be uncharged originally and connected with the ground, so that the electric potential in the plane is constant and equal to zero. We are required to find, at any point in space, the value of V resulting from the charge q and the induced charges on the plane.

We adopt a system of coordinates such that the charge lies at $(a,0,0)$, and take for the equation of the plane, $x = 0$, (Fig. 2). Let us now suppose that a charge $(-q)$ is placed at the point $(-a,0,0)$. The potential at any point b, in the yz-plane, resulting from the two symmetrically placed charges, is

$$V = \frac{q}{r} + -\frac{q}{r} = 0, \tag{12}$$

$$V = \frac{q}{4\pi\epsilon r} + -\frac{q}{4\pi\epsilon r} = 0, \tag{12a}$$

where $r = \overline{\text{ab}}$. This equation satisfies the condition $V = 0$. Hence the electric fields must be the same whether the conducting plate is in place or not. The positive point charge and the negative charge induced on the surface of the plane must produce a field on the positive side of the origin identical with that from the pair of point charges. The combined potential of these charges is

$$V = \frac{q}{[(x - a)^2 + y^2 + z^2]^{1/2}} + \frac{-q}{[(x + a)^2 + y^2 + z^2]^{1/2}}. \tag{13}$$

$$V = \frac{q}{4\pi\epsilon[(x - a)^2 + y^2 + z^2]^{1/2}} + \frac{-q}{4\pi\epsilon[(x + a)^2 + y^2 + z^2]^{1/2}}. \tag{13a}$$

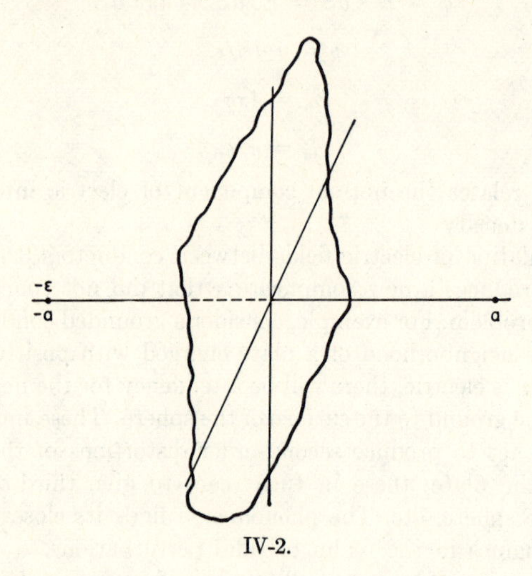

IV-2.

This equation satisfies (12) at the plane $x = 0$. The x-component of force normal to the plane is

$$E_x = -\frac{\partial V}{\partial x} = \frac{q(x - a)}{[(x - a)^2 + y^2 + z^2]^{3/2}} - \frac{q(x + a)}{[(x + a)^2 + y^2 + z^2]^{3/2}}. \tag{14}$$

$$E_x = -\frac{\partial V}{\partial x} = \frac{q(x - a)}{4\pi\epsilon[(x - a)^2 + y^2 + z^2]^{3/2}}$$

$$- \frac{q(x + a)}{4\pi\epsilon[(x + a)^2 + y^2 + z^2]^{3/2}}. \tag{14a}$$

In the plane $x = 0$,

$$E_x = E_n = -\frac{2qa}{(a^2 + y^2 + z^2)^{3/2}} = -\frac{2qa}{r^3} = 4\pi\sigma, \tag{15}$$

$$E_x = E_n = \frac{-2qa}{4\pi\epsilon(a^2 + y^2 + z^2)^{3/2}} = -\frac{2qa}{4\pi\epsilon r^3} = \frac{\sigma}{\epsilon}, \tag{15a}$$

by (11). We have written

$$r = (a^2 + y^2 + z^2)^{1/2}, \tag{16}$$

which is, as in (12), the distance from the external charge to the given point on the plate.

$$\sigma = -qa/2\pi r^3. \tag{17}$$

The induced charge is opposite in sign to the original, and σ has a maximum at the origin. The total induced charge is

$$\int_{-\infty}^{\infty}\int_{-\infty}^{\infty} \sigma \, dy \, dz = -q. \tag{18}$$

The method employed above is based on the so-called theory of images. We regard the plate as a plane mirror, in which the "image" of the charge, taken with opposite sign, makes the problem symmetrical. We have used images in similar fashion in II-§ 40, as an aid in calculating the motions of vortices near a plane boundary.

We next consider the problem of the charge induced on the surface of a grounded sphere centered at the origin, by a positive point charge q located at an external point $(a,0,0)$. Over the surface of the sphere $V = 0$. Let b be any point on the surface of the sphere, and let r be its radius. We are to find the magnitude, q_s, and location of an image point charge that will make the combined potential from this image and the original charge vanish over the sphere.

Considerations of symmetry dictate that the image must lie at some point

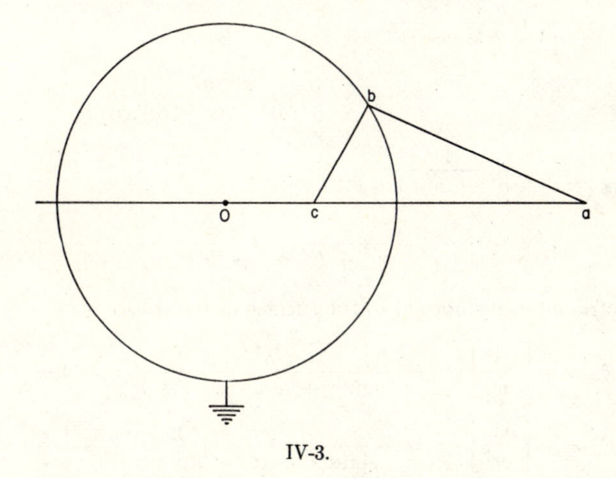

IV-3.

along the line Oa (Fig. 3). Take its coordinates as$(c,0,0)$ and its magnitude as $-qf$. Then we can determine c and f from the condition that

$$q/\overline{\text{ab}} - qf/\overline{\text{cb}} = 0 \tag{19}$$

for all positions of b. In terms of the coordinates, we have

$$\frac{q}{(a^2 + r^2 - 2ax)^{1/2}} = \frac{qf}{(c^2 + r^2 - 2cx)^{1/2}}, \tag{20}$$

which gives the condition

$$f^2a^2 + f^2r^2 - 2f^2ax = c^2 + r^2 - 2cx. \tag{21}$$

The coefficients of x on both sides of this equation must be equal; hence

$$f^2 = c/a. \tag{22}$$

Equation (19) becomes

$$c^2 - c(a^2 + r^2)/a + r^2 = 0, \tag{23}$$

the solution of which gives

$$c = a, \quad f = 1 \quad \text{or} \quad c = r^2/a, \quad f = r/a. \tag{24}$$

The first of these solutions is trivial because it corresponds to an equal negative charge superposed on the original. The second leads to the desired result: the image charge is one of magnitude $(-\varepsilon r/a)$ located at $(r^2/a,0,0)$. Using the law of cosines we prove that the potential at some point in space indicated by the spherical coordinates (r',θ,ϕ), with θ measured from the axis Oa, becomes

$$V = \frac{q}{(r'^2 + a^2 - 2r'a\cos\theta)^{1/2}} - \frac{q(r/a)}{[r'^2 + (r^2/a)^2 - 2r'(r^2/a)\cos\theta]^{1/2}}. \tag{25}$$

$$V = \frac{q}{4\pi\epsilon}\left[\frac{1}{(r'^2 + a^2 - 2r'a\cos\theta)^{1/2}} - \frac{(r/a)}{[r'^2 + (r^2/a)^2 - 2r'(r^2/a)\cos\theta]^{1/2}}\right]. \tag{25a}$$

The electric intensity normal to the surface of the sphere is

$$E_n = -\left[\frac{\partial V}{\partial r'}\right]_{r'=r} = -\frac{q(a^2 - r^2)}{a(r^2 + a^2 - 2ra\cos\theta)^{3/2}} = 4\pi\sigma. \tag{26}$$

$$E_n = -\left[\frac{\partial V}{\partial r'}\right]_{r'=r} = -\frac{q(a^2 - r^2)}{4\pi\epsilon a(r^2 + a^2 - 2ra\cos\theta)^{3/2}} = \frac{\sigma}{\epsilon}. \tag{26a}$$

The factor in the denominator is simply the cube of the distance, R, from the point a to the surface of the sphere. Therefore

$$\sigma = -q(a^2 - r^2)/aR^3. \tag{27}$$

The quantity σ varies inversely as R^3, as before. We find that

$$q_s = \int_0^{\pi} \int_0^{2\pi} \sigma r^2 \sin \theta \, d\theta \, d\phi = -\frac{qr}{a}, \tag{28}$$

as we might have inferred from the fact that the image charge possessed this value.

Suppose now that the conducting sphere is insulated from the ground, and originally uncharged. The image problem is the same as before, but the negative charges must come from the side of the sphere opposite to a. Since the total charge on the sphere is zero, instead of $(-qr/a)$, as in (28), we must introduce into the image field a compensating positive charge of magnitude $(+qr/a)$. This charge excess we place at the center of the sphere, in order to give a constant potential over the boundary surface. We obtain the potential by adding a term $(+qr/ar')$ to the right-hand side of (25). In MKS units the added term is $(qr/4\pi\epsilon ar')$.

Of special interest is the case when a becomes very large, with the value of q sufficiently large so that the field at the origin is finite:

$$E = -q/a^2. \tag{29}$$

$$E = -q/4\pi\epsilon a^2. \tag{29a}$$

Here the minus sign arises from the nature of the coordinate system, which is not centered at the charged body. Because of the repulsion, a positive charge tends to move in the negative direction of the x-axis. As $a \to \infty$, the field near the origin becomes parallel to the x-axis. Hence we may investigate by this procedure the charge induced on a sphere by a uniform field. In the limit, the last term in (25) gives rise to the potential

$$[-qr/ar' - (qr^3/a^2r'^2) \cos \theta].$$

For MKS units, multiply by $1/4\pi\epsilon$. This excess potential, added to that of the central image, gives the combined potential of the induced charge:

$$V_i = -(Er^3/r'^2) \cos \theta, \tag{30}$$

by (29). Similarly expanding the first term in (25), we find the potential of the external field:

$$V_e = \frac{q}{a} + \frac{qr'}{a^2} \cos \theta = -Er' \cos \theta + C. \tag{31}$$

$$V_e = \frac{q}{4\pi\epsilon a} + \frac{qr'}{4\pi\epsilon a^2} \cos \theta = -Er' \cos \theta + C. \tag{31a}$$

Note that an infinitely distant charge sufficiently large to give a finite field at the origin, will give $C = \infty$. We here employ the concept merely as a device to reproduce the conditions of a uniform field. Hence C is not necessarily infinite in problems involving a uniform field. The total potential is

$$V = V_i + V_e = -E \frac{r'^3 - r^3}{r'^2} \cos \theta + C, \tag{32}$$

which reduces, as it must, to $V = C$ over the surface of the sphere $r' = r$.

The pair of infinitely close charges at the center of a sphere, producing a double source of potential, is called an electric dipole. Dipole moment, P, we define as the product of the charge by the distance between the charges. Therefore

$$P = \frac{qr}{a} \frac{r^2}{a} = Er^3, \tag{33}$$

$$P = 4\pi\epsilon Er^3, \tag{33a}$$

which informs us that a constant electric field of intensity E produces on the surface of a sphere, a charge distribution that acts *in the space exterior to the sphere* like a dipole source of moment P. P is a vector quantity. In the present example,

$$\mathbf{P} = \mathbf{E}r^3, \tag{34}$$

$$\mathbf{P} = 4\pi\epsilon\mathbf{E}r^3, \tag{34a}$$

since $\mathbf{E}$ and $\mathbf{P}$ are similarly directed along the axis of x. We shall discuss the dipole problem further in § (5), where we transform the formulae for the magnetic case to those for the electric problem through substitution of q for p, the magnetic pole strength.

We easily extend these examples to cover more complex cases. For example, we can reduce the distribution of σ over the surfaces of two charged spheres to an infinite series of images. We shall not follow the arguments further, but turn now to the problems presented by insulators in an electric field.

As previously mentioned, the distinction between conductors and insulators is chiefly one of the relative mobility of the electrons. In insulators the atomic electrons are tightly bound. Nevertheless, in an electric force field, there will still be a tendency for the positive charges to flow in one direction and the negative in the other. There results a slight net separation of charge, which, in a plate normal to $\mathbf{E}$, would give two external surfaces charged oppositely in sign. The phenomenon is called *polarization*. Insulators are often referred to as dielectrics because the induced charges occur in compensating pairs. The potential field in the dielectric, resulting from a charge q in the dielectric, is less than that in a vacuum, according to the formula

$$V = q/\kappa r. \tag{35}$$

$$V = q/4\pi\epsilon r. \qquad \kappa = \epsilon/\epsilon_0, \quad \epsilon_0 = 10^9/36\pi. \tag{35a}$$

Here κ is the dielectric constant (2.4) in Gaussian units.

Let us consider the field produced as the result of a positive charge q located at $(a,0,0)$, in the neighborhood of an infinite block of dielectric material $(x \leq 0)$ with a plane face at $x = 0$. As a result of electric polariza-

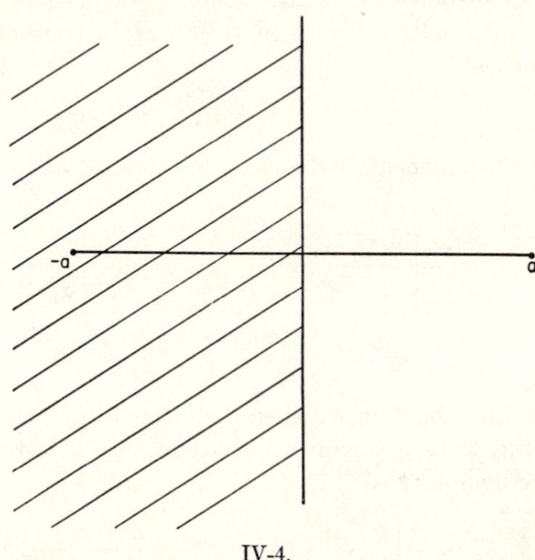

IV-4.

tion, a layer of negative charge will collect on this face. The compensating layer of positive charge will lie at $x = -\infty$ and can produce no effect at the origin. We cannot consider the plane a surface of constant potential, as for a conductor, since the charges are not free to move.

To solve the problem we apply the theory of images, supposing the dielectric to be removed, leaving only the surface layer of negative charge. Within the space previously occupied by the dielectric, the original potential, given by

$$V_1 = q/[(x - a)^2 + y^2 + z^2]^{1/2}, \tag{36}$$

$$V_1 = q/4\pi\epsilon[(x - a)^2 + y^2 + z^2]^{1/2}, \tag{36a}$$

will be partially "screened" by the layer. Let us assume that the potential, because of the screening, takes the reduced value

$$V_2 = fq/[(x - a)^2 + y^2 + z^2]^{1/2}, \tag{37}$$

$$V_2 = fq/4\pi\epsilon[(x - a)^2 + y^2 + z^2]^{1/2}, \tag{37a}$$

in the dielectric. Further, we assume that the "image" of a, as reflected in the surface charge, is equivalent to a charge of magnitude $-qf'$, located at $(-a,0,0)$ giving a potential

$$V_3 = -f'q/[(x + a)^2 + y^2 + z^2]^{1/2}. \tag{38}$$

$$V_3 = -f'q/4\pi\epsilon[(x - a)^2 + y^2 + z^2]^{1/2}. \tag{38a}$$

We must have

$$V_2 = V_1 + V_3. \tag{39}$$

Furthermore, these potentials satisfy Laplace's equation, except at the points $(a,0,0)$ and $(-a,0,0)$. Therefore, they must be continuous across the face $x = 0$, whence

$$f = 1 - f'. \tag{40}$$

The normal or x-component of the electric intensity, in free space, at the boundary is

$$E_x = E_n = -\left[\frac{\partial(V_1 + V_3)}{\partial x}\right]_{x=0} = -\frac{qa(1 + f')}{(a^2 + y^2 + z^2)^{3/2}}. \tag{41}$$

$$E_x = -\frac{qa(1 + f')}{4\pi\epsilon(a^2 + y^2 + z^2)^{3/2}}. \tag{41a}$$

This expression must hold for E_x when the image is replaced by the actual dielectric. In the space originally occupied by the dielectric, the electric intensity at the boundary is

$$E_x' = -\left[\frac{\partial V_2}{\partial x}\right]_{x=0} = -\frac{qaf}{(a^2 + y^2 + z^2)^{3/2}}. \tag{42}$$

$$E_x' = -\frac{qaf}{4\pi\epsilon(a^2 + y^2 + z^2)^{3/2}}. \tag{42a}$$

Now imagine the dielectric replaced. From (35), we see that the actual field E_x is less than E_x', as follows:

$$E_x' = E_x\kappa = \frac{E_x\epsilon}{\epsilon_0}, \tag{43}$$

which relation must be true for all values of y and z in the boundary (see equation 22.8). Therefore

$$(1 + f')\kappa = \frac{(1 + f')\epsilon}{\epsilon_0} = f = 1 - f', \tag{44}$$

from (41). Hence

$$f' = \frac{\kappa - 1}{\kappa + 1}, \quad f = \frac{2}{\kappa + 1}. \tag{45}$$

For MKS substitute ϵ/ϵ_0 for κ in (45) and (46). These conditions, in combination with (41) to (45), completely define the electric field both inside and outside the dielectric. The justification for our artificial assumptions lies in the fact that the potentials satisfy both Laplace's equation and the requisite boundary conditions.

Combining the foregoing approach with the image method used in deriving (32) we may prove that a homogeneous sphere of dielectric material, *in vacuo*, will give in a constant electric field, the potential outside the sphere:

$$V = -E\left[r' - \frac{r^3}{r'^2}\frac{\kappa - 1}{\kappa + 2}\right]\cos\theta + C. \tag{46}$$

Comparing this equation with (32), we note that a conductor behaves like a dielectric of infinite κ.

3. The charge on the surface of a star. There are some simple astrophysical applications of these theorems, first applied by Lindemann* to calculate the electrostatic potential at the surface of a star. For illustrative purposes we shall attempt to set an upper limit to the electric charge on the surface of the sun. The solar atmosphere consists of highly ionized gases, negative electrons, and positive ions. The former, being lighter, are the more likely to attain velocities in excess of the velocity of escape. The condition that a particle escape is simply that its kinetic energy be equal to or greater than its potential energy, i.e.,

$$\frac{1}{2}mv^2 \geq m\frac{MG}{R_0}, \tag{1}$$

where m is the mass of the particle, M the mass, and R_0 the radius of the sun. According to kinetic theory, the component of mean kinetic energy of an atomic constituent is independent of its mass, or

$$\overline{\frac{1}{2}mv^2} = \frac{1}{2}kT = 4.12 \times 10^{-13} \text{ erg}, \tag{2}$$

for the observed surface temperature, T, of 6000°K. Here k is Boltzmann's constant. The value of the potential energy of an electron in the gravitational field, obtained from the right-hand side of (1), is 1.71×10^{-12} erg.

Although the average kinetic energy is less than the potential energy, enough of the electrons possess kinetic energies greater than the critical value to make the escape appreciable. A star that was electrically neutral, initially, would rapidly acquire a positive charge through loss of negative electrons. The rate of escape, however, must diminish with the time, because the star's positive potential supplements the gravitational potential

Phil. Mag. 38, 674, 1919.

in retarding the escape. Theoretically, a finite fraction of the electrons always possesses velocities greater than any specified finite velocity of escape. Given an infinite time, a star could acquire an infinite positive charge if no other forces came into play.

Nature, however, sets an upper limit to the potential, for, when the positive charge reaches a certain value, positive protons will be repelled. From then on, the sun will expel protons and electrons at the same rate. Let Q_{max} be the maximum charge of the star. The condition that the positive potential energy of a proton in the electric field of the sun equal that of the negative gravitational energy is

$$q_1 \frac{Q_{max}}{R_0} = m_1 \frac{MG}{R_0}, \qquad Q_{max} = \frac{m_1}{\varepsilon_1} MG. \tag{3}$$

$$\frac{q_1 Q_{max}}{4\pi\epsilon_0 R_0} = m_1 \frac{MG}{R_0}. \tag{3a}$$

Therefore Q_{max} turns out to be independent of the star's radius. Putting in numerical values, we find that

$$Q_{max} = 4.61 \times 10^{11} \text{ esu} = 154 \text{ coulombs}. \tag{4}$$

Since a single electron has a charge, q_1, of 4.77×10^{-10} es unit or 1.6×10^{-19} coulomb, this value of Q_{max} results from the removal of a definite number, N, of electrons from the entire surface of the star.

$$N = \frac{Q_{max}}{q_1} = 9.7 \times 10^{20} \text{ electrons}. \tag{5}$$

The number is surprisingly small. The area of the sun is 6.1×10^{22} cm^2. At the maximum the excess of protons over electrons cannot be greater than one for every sixty square centimeters of solar surface.

The calculation illustrates how much more powerful electric forces are than gravitational; 10^{-5} gram of electrons or 2×10^{-2} gram of protons placed at the center of an otherwise empty sphere of radius equal to that of the sun, would produce an electrostatic attraction (or repulsion) upon a proton, equal to the the gravitational force of the entire solar mass. A stellar atmosphere must therefore be electrically neutral, to a very high degree of approximation.

The maximum electrical potential, V_{max} is

$$V_{max} = \frac{Q_{max}}{R_0} = 6.64 \text{ esu} \tag{6}$$

$$V_{max} = \frac{Q_{max}}{4\pi\epsilon_0 R_0} \sim 2000 \text{ volts}. \tag{6a}$$

Note that 1 esu = 300 volts.

The figure for V_{max}, representing an extreme upper limit, is not excessive.

The energy of cosmic rays, for example, which runs into billions of electron volts,* indicates that their origin is non-stellar, unless some very special processes are acting.

The actual value of the solar potential may be appreciably less than that given by (6). It is determined by a complicated balancing of the individual kinetic motions of electrons and atoms with electric, gravitational, and radiation-pressure forces.

4. Capacitance. The potential at the surface of the sun, or of any electrically charged sphere, by equation (3.6), is directly proportional to the quantity of electricity and inversely proportional to the radius. If the sphere is surrounded by a medium whose dielectric constant is κ, instead of unity, we have the relation, in accord with (2.4),

$$V = \frac{Q}{\kappa R_0}. \tag{1}$$

$$V = \frac{Q}{4\pi\epsilon R_0} \tag{1a}$$

If we have a series of spheres of various radii, each charged with the same amount of electricity, the potentials will vary inversely as κR_0. In electrical terminology, such a sphere or indeed any medium used for holding or storing an electrical charge is known as a capacitor.

When a capacitor attains a certain potential with respect to its surroundings as more and more electricity accumulates, electrical breakdown may very well occur. The capacitor will then discharge, much as an electrified cloud sends a flash of lightning to the earth. More electricity can be stored without danger of breakdown on a sphere of large than on one of small radius. In consequence, we often call the quantity κR_0 the *capacitance*, C, of the capacitor. Then

$$V = Q/C. \tag{2}$$

The physical dimensions of C thus become $[L\kappa]$. Although (1) holds explicitly only for a sphere, we may extend relation (2) to apply to a capacitor of any shape. The capacitance of the sun, with κ equal to unity, is equal to the solar radius in centimeters. The capacitance of a capacitor depends on its surroundings through the occurrence of the dielectric constant.

The electric intensity, i.e., the radial force acting on a unit electric charge outside of a charged sphere, is, as we have already seen,

$$E = \frac{Q}{\kappa r^2} = 4\pi \frac{\sigma}{\kappa}, \tag{3}$$

$$E = \frac{Q}{4\pi\epsilon r^2} = \frac{\sigma}{\epsilon}, \tag{3a}$$

*1 electron volt = $(1.601864 \pm 0.000024) \times 10^{-12}$ erg. See Table (I-12).

since $Q = 4\pi\sigma r^2$, where σ is the surface density of the charge. We shall calculate the force on a unit charge at distance r from a uniformly charged, circular plane sheet, of radius x_0. Then

$$\sigma = Q/A, \tag{4}$$

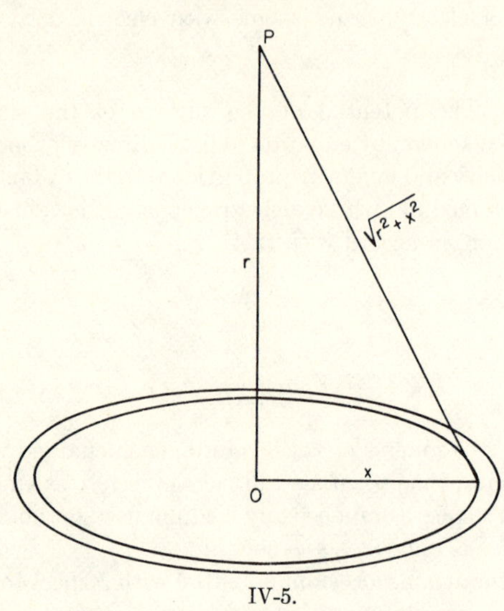

IV-5.

where A is the area of the plate. The potential at P resulting from the annulus between x and $x + dx$ is

$$dV = \frac{2\pi\sigma x\, dx}{\kappa(r^2 + x^2)^{1/2}}, \tag{5}$$

$$dV = \frac{\sigma x\, dx}{2\epsilon(r^2 + x^2)^{1/2}}, \tag{5a}$$

and

$$V = \int_0^{x_0} dV = \frac{2\pi\sigma}{\kappa}\,[(r^2 + x_0^2)^{1/2} - r]. \tag{6}$$

$$V = \frac{\sigma}{2\epsilon}\,[(r^2 + x_0^2)^{1/2} - r]. \tag{6a}$$

Then the electric intensity along r, is

$$E = -\frac{\partial V}{\partial r} = \frac{2\pi\sigma}{\kappa}\left[1 - \frac{r}{(r^2 + x_0^2)^{1/2}}\right]. \tag{7}$$

$$E = \frac{\sigma}{2\epsilon}\left[1 - \frac{r}{(r^2 + x_0^2)^{1/2}}\right]. \tag{7a}$$

When $x_0 \gg r$, we have the result that

$$E = \frac{2\pi\sigma}{\kappa} = \frac{2\pi Q}{\kappa A}. \tag{8}$$

$$E = \sigma/2\epsilon = Q/2\epsilon A. \tag{8a}$$

In a capacitor consisting of two parallel circular plates, carrying equal and opposite charges, each plate contributes an equal amount to the field. The field inside such a capacitor is uniform and of intensity

$$E = 4\pi Q/\kappa A. \tag{9}$$

$$E = Q/\epsilon A. \tag{9a}$$

For further discussion of these equations see § (20).

5. The magnetic dipole. The magnetostatic force-fields surrounding individual magnetic poles follow the laws we have derived for electrostatic fields, with κ replaced by μ and Q by p, the magnetic pole strength. The formulae for the single poles, however, are rarely applicable because, unlike electric charges, magnetic poles never occur singly. Each positive (north-seeking) pole is always accompanied by a negative (south-seeking) pole. As far as we know, no fundamental magnetic "stuff" exists. All magnetism, including that of permanent magnets, owes its existence to circulation of electricity, whether macroscopic in a coil of wire, or microscopic in a whirl about an atomic nucleus. The experimental data of magnetic phenomena can nevertheless be graphically and quantitatively represented on the supposition that fundamental magnetic poles of both signs actually exist, but that their occurrence is always in pairs. With some limitations we may describe the resulting force field in terms of a distribution of fictitious magnetic dipoles. Nevertheless, the distribution is not unique. If we wish to describe the field at the position we adopted for our original fictitious magnets, we have to employ some alternative fictitious distribution for that purpose. In brief, we can describe magnetic phenomena in terms of fictitious dipoles; we cannot always say that poles actually exist at some definite location in space.

Laboratory study of the force fields arising from a uniformly magnetized needle, e.g., mapping of the "lines of force" by means of iron filings, shows that the effective magnetic poles reside near either end of the needle. In a magnetized needle of infinitesimal cross section and infinite length, the poles would act like the theoretical pole of given sign and the force field would obey the inverse-square law assumed above. The presence of magnetic material introduces complications that we shall discuss later on. The potential at distance r from a single pole of magnetic charge $\bar{p}$ is, analogous to (2.4) for the electrostatic case,

$$V = \bar{p}/r. \tag{1}$$

$$V = \bar{p}/4\pi r. \tag{1a}$$

The potential, V, at a given point is the work required to bring a unit positive pole from infinity up to that point. The force vector, $\mathbf{H}$, at the point, is, as before,

$$\mathbf{H} = -\nabla V. \tag{2}$$

Let us calculate the magnetic potential at a point, C, at distance r from a small bar magnet, i.e., from a dipole of length a and pole strength p. Suppose the magnet to be placed along the x-axis, with the center at the

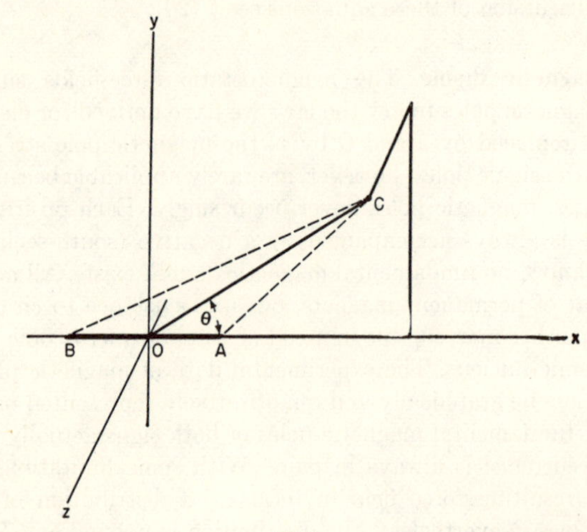

IV-6.

origin, the positive pole at A and the negative pole at B (Fig. 6). Let the coordinates of C be x, y, z. Then

$$\mathrm{OC} = r = [x^2 + y^2 + z^2]^{1/2},$$

$$\mathrm{AC} = r_1 = \left[\left(x - \frac{a}{2}\right)^2 + y^2 + z^2\right]^{1/2}$$

$$\sim [x^2 + y^2 + z^2 - ax]^{1/2} = r\left(1 - \frac{ax}{r^2}\right)^{1/2}, \tag{3}$$

$$\mathrm{BC} = r_2 = \left[\left(x + \frac{a}{2}\right)^2 + y^2 + z^2\right]^{1/2} \sim r\left(1 + \frac{ax}{r^2}\right)^{1/2},$$

where the expansion and approximations depend on the supposition that $r \gg a$. To get the potential at C we add the potentials of the respective poles. Or, from (1),

$$V = \frac{\bar{p}}{r_1} - \frac{\bar{p}}{r_2} = \frac{\bar{p}}{r}\left[\left(1 - \frac{ax}{r^2}\right)^{1/2} - \left(1 + \frac{ax}{r^2}\right)^{1/2}\right] \sim \frac{\bar{p}a}{r^2}\frac{x}{r} = \frac{\bar{p}a}{r^2}\cos\theta. \tag{4}$$

$$V = \frac{\bar{p}}{4\pi r_1} - \frac{\bar{p}}{4\pi r_2} = \ldots = \frac{\bar{p}a}{4\pi r^2}\frac{x}{r} = \frac{\bar{p}a}{4\pi r^2}\cos\theta. \tag{4a}$$

We have obtained (4) by expanding the parentheses in accord with the binomial theorem and by discarding second-order terms. The direction cosine of the line OC with respect to the x-axis is $x/r = \cos\theta$; θ is the angle between OC and the positive direction of the axis of the magnet. The product, $\bar{p}a$, of the pole strength by the length of the magnet, is the *magnetic moment*, M. Hence

$$V = \frac{M}{r^2}\cos\theta. \tag{5}$$

$$V = \frac{M}{4\pi r^2}\cos\theta. \tag{5a}$$

To evaluate the force at the point r, θ, ϕ, we express ∇ in polar coordinates (II-35.24). Then by (2),

$$\mathbf{H} = M\left(\mathbf{i}_r\frac{2\cos\theta}{r^3} + \mathbf{i}_\theta\frac{\sin\theta}{r^3}\right). \tag{6}$$

$$\mathbf{H} = \frac{M}{4\pi}\left(\mathbf{i}_r\frac{2\cos\theta}{r^3} + \mathbf{i}_\theta\frac{\sin\theta}{r^3}\right). \tag{6a}$$

The magnetic moment is really a vector, and $M\cos\theta$ is the projection of the vector M upon the vector r. Hence in vector notation, we may write, instead of (5),

$$V = \frac{\mathbf{M}\cdot\mathbf{r}}{r^3}. \tag{7}$$

$$V = \frac{\mathbf{M}\cdot\mathbf{r}}{4\pi r^3}. \tag{7a}$$

C. The potential of a uniform magnetic shell. Consider a thin shell and suppose that n magnetic dipoles per unit area, each of magnetic moment M, are distributed uniformly over the surface, with their positive poles coinciding with one face and the negative poles with the other face of the shell. We need to know the potential at the point P (Fig. 7). Let dS be an element of area of the shell at distance r from P. The magnetic moment of the area dS is $Mn\,dS$. Hence, by (5.5), the contribution to the potential from the area element is

$$dV = \frac{Mn\,dS\cos\theta}{r^2}. \tag{1}$$

$$dV_2 = \frac{Mn\,dS\cos\theta}{4\pi r^2}. \tag{1a}$$

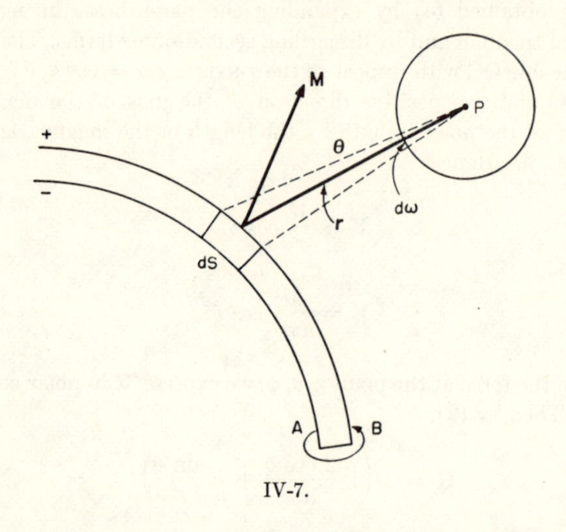

IV-7.

About P draw a sphere of unit radius. From P as a vertex draw the elementary cone having dS as its base. The projection of dS upon the normal to the axis of the cone is $dS\cos\theta$. Furthermore,

$$\frac{dS\cos\theta}{r^2} = d\omega, \tag{2}$$

where $d\omega$ is the area of the unit sphere intercepted by the cone. Hence $d\omega$ is the *solid angle* of the cone. For a complete sphere, $\int d\omega = 4\pi$.

$$V_+ = \int dV = Mn\int d\omega = Mn\Omega = \Phi\Omega \tag{3}$$

$$V_+ = \frac{Mn}{4\pi}\int d\omega = \frac{Mn\Omega}{4\pi} = \frac{\Phi\Omega}{4\pi}, \tag{3a}$$

where V_+ denotes the potential when P is on the positive side of the shell, Ω is the solid angle subtended by the shell at P, and Φ is the magnetic moment per unit area. The potential, V_-, at a point on the negative side of the shell, is

$$V_- = -\Phi\Omega. \tag{4}$$

$$V_- = -\frac{\Phi\Omega}{4\pi}. \tag{4a}$$

The work, W, necessary to carry a positive unit induced magnetic charge $\bar{p}$ from a point A in the negative surface to a point B in the positive surface is, by (2.3a) or (I-4.15).

$$\begin{aligned} W &= \int_A^B \mathbf{F} \cdot d\mathbf{s} = \int_A^B \mathbf{B} \cdot d\mathbf{s} \\ &= \mu \int_A^B \mathbf{H} \cdot d\mathbf{s} = \mu(V_B - V_A) = \mu\Phi(\Omega_B + \Omega_A), \end{aligned} \tag{5}$$

$$W = \frac{\mu\Phi(\Omega_B + \Omega_A)}{4\pi}, \tag{5a}$$

where **H** is the magnetic field. **B**, the magnetic induction, is equal to $\mu\mathbf{H}$. Equation (5) is a line integral. The reversal of sign from the original equation (II-8.6), relating work to potential difference, comes in because two unit positive poles repel one another, i.e., work must be done on the system to bring the poles together, whereas in the gravitational problem, the forces are attractive and negative work is done when the particles are made to approach one another.

The solid angles Ω_A and Ω_B depend only on the shape of the periphery of

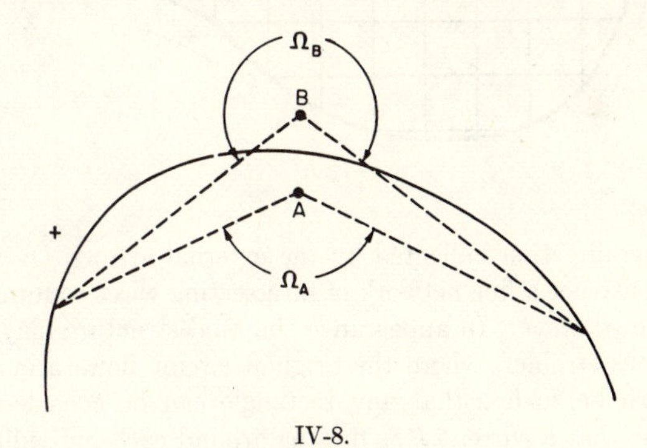

IV-8.

the shell. A schematic representation appears in Fig. 8. When the shell is very thin, so that A and B are almost coincident,

$$\Omega_A + \Omega_B \sim 4\pi \tag{6}$$

and

$$W = 4\pi\mu\Phi. \tag{7}$$

$$W = \mu\Phi. \tag{7a}$$

Electromagnetism

7. Ampere's theorem. Thus far we have considered electric and magnetic fields separately. To relate them we must either make some assumption or appeal to some experiment. Since the fate of any assumption ultimately rests in the laboratory, we shall resort to the experimental data at once. In 1820 Oersted noted that a wire carrying an electric current was surrounded by a magnetic field. Three years later Ampere announced the fundamental theorem concerning the nature of the relationship. Ampere discovered that, as far as magnetic effects are concerned, an electric current, I', flowing through a wire bent to form a closed circuit, produces the same effect as a thin magnetic shell bounded by the periphery of the circuit and of strength proportional to I'.

In Fig. 9, we suppose the current to be flowing through the closed circuit

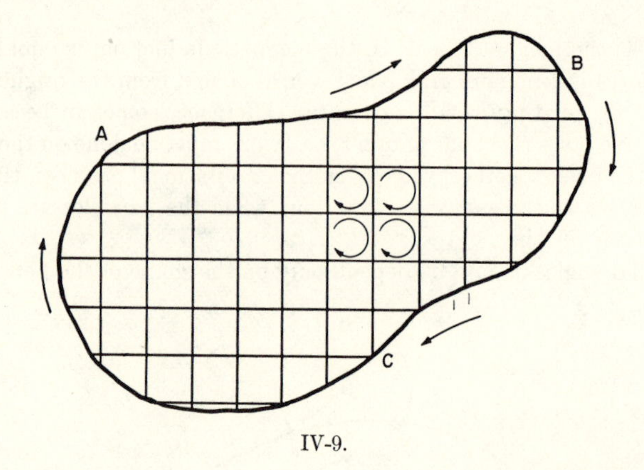

IV-9.

(ABC) in the direction indicated by the external arrows. Over ABC as a boundary construct a fine network of intersecting wires to form a covering of any shape whatever. In appearance the final structure may resemble a badly-bent tea-strainer, where the original circuit flows around the rim. Let the mesh be so fine that any rectangle can be considered a plane. Now suppose that a current I' is flowing around each individual mesh, as indicated by the small arrows. Throughout the network, each current is balanced by a current in the opposite direction, except at the boundary. But each individual network produces a magnetic moment perpendicular to its plane. Hence, as far as magnetic effect is concerned, we may replace the current by a magnetic shell of shape coinciding with the network. The theorem holds for a network of any shape. We may, therefore, take any shell we choose, bounded by the circuit, as the equivalent magnetic shell.

The reader will recognize the similarity of the above demonstration to that used previously in the derivation of Stokes' law (II-32). We shall see that the resemblance has actual physical significance.

To remember the direction of the magnetism in the equivalent shell, follow the simple rule: if the face of a clock dial represents the positive poles and the back of the dial the negative, the direction of flow of the positive current is counterclockwise.

8. The electromagnetic and electrostatic systems. In accordance with Ampere's theorem, we may define the current I' in terms of the equivalent magnetic shell, by the equation

$$\Phi = kI', \tag{1}$$

$$\Phi = kI, \tag{1a}$$

where k is a constant of proportionality. Since equation (1) amounts to a definition, we may set k equal to unity, and write

$$I' = \Phi. \tag{2}$$

$$I = \Phi. \tag{2a}$$

In MKS units, the dimensions of all other quantities appear in terms of M, L, T, and Q. Thus $[I] = [QT^{-1}]$.

The current, defined as in (2), is said to be in the electromagnetic system. We shall expect to find that it differs, both numerically and dimensionally, from its value in the electrostatic system. The current, I, in electrostatic units, possesses physical dimensions of electric charge divided by the time, or

$$[I] = [M^{1/2}L^{3/2}T^{-2}\kappa^{1/2}], \tag{3}$$

where κ is set equal to unity for free space. The physical dimensions of I' are

$$[I'] = [M^{1/2}L^{1/2}T^{-1}\mu^{-1/2}], \tag{4}$$

where μ is to be set equal to unity for a vacuum. If, now, we let

$$I = c'I', \tag{5}$$

where c' is some constant, we note that the physical dimensions of c' are given by

$$[c'] = [LT^{-1}], \tag{6}$$

i.e., c' has the dimensions of a velocity. Had we not suppressed the dimensions κ and μ, by calling them unity, we should have been able to allow for the difference between the two sets of units by setting

$$c' = (\mu\kappa)^{-1/2}, \tag{7}$$

where c' is a dimensional constant, whose value is to be determined by experiment.

By reference to Chapter I, §§ (10) and (11), we ascertain the following relationships between given physical quantities measured in es and em units. In (8) we have purposely written c for c'.

$$\begin{aligned} \text{Quantity of electricity: } 1 \text{ em unit} &= c \text{ es units.} \\ \text{Current: } 1 \text{ em unit} &= c \text{ es units.} \\ \text{Electric Potential: } 1 \text{ em unit} &= 1/c \text{ es units.} \\ \text{Capacitance: } 1 \text{ em unit} &= c^2 \text{ es units.} \end{aligned} \tag{8}$$

The tabulation may be extended if desired.

To evaluate c' we must measure, in the laboratory, any one of the physical quantities, first in one system and then in the other system of units. For example we might collect a certain amount of electricity on a sphere, measure the quantity by electrostatic means, and then discharge the sphere over a known interval of time through a loop of wire and measure the resultant magnetic field. This procedure, although theoretically sound, meets with practical difficulties because of the great difference in magnitude of the units. An enormous quantity of electricity, as measured in es units, might still be insufficient to produce a perceptible magnetic field. We can compare units more conveniently through the medium of capacitance, because we are able to calculate the capacitance of a capacitor, in es units, from simple geometry. The student will recall that the es capacitance of a spherical conductor is equal to its radius in centimeters.

Equation (4.2) holds for either the es or em system, as may be proved from the conversion factors of (8). That is,

$$V = \frac{Q}{C}, \quad V' = \frac{Q'}{C'} = \frac{\bar{I}' \, \Delta t}{C'}, \tag{9}$$

where the primed quantities refer to the em system and Δt is the time that the average current, $\bar{I}'$, flows to charge a capacitor. With proper procedure, V', $\bar{I}'$, and Δt can be measured and C' calculated from experiment. Then, C is derived geometrically. The mean of the best experimental results is

$$c' = \sqrt{C/C'} = (2.9979 \pm 0.0001) \times 10^{10} \text{ cm sec}^{-1}. \tag{10}$$

(We have already seen that c' represents a velocity.) Equation (10) agrees within its experimental error with the direct determinations of the velocity of light:

$$c = (2.99796 \pm 0.00004) \times 10^{10} \text{ cm sec}^{-1}.$$

This agreement led Maxwell to conclude that light is an electromagnetic phenomenon. Henceforth we shall assume that $c' = c$.

9. Practical units. As we have previously stated, the es and em systems are specially adapted for calculations of a dynamical nature. The numbers that result from their use are, however, frequently too large or too small for general purposes. Consequently a "practical" system has been devised as follows, founded primarily upon the em system:

Name	Symbol	Practical or MKS unit	em units	es units
Quantity of electricity	Q	1 coulomb	10^{-1}	3×10^9
Current	I	1 ampere	10^{-1}	3×10^9
Electric potential / Electromagnetic force	V	1 volt	10^8	1/300
Capacitance	C	1 farad	10^{-9}	9×10^{11}
		1 microfarad	10^{-15}	9×10^5
		1 micromicrofarad	10^{-21}	9×10^{-1}
Resistance	R	1 ohm	10^9	$1/(9 \times 10^{11})$
Conductivity	σ	1 mho/meter	10^{-11}	9×10^9
Electric field	$\mathbf{E}$	1 volt/meter	10^6	$1/(3 \times 10^4)$
Electric displacement	$\mathbf{D}$	1 coulomb/meter^2	$4\pi \times 10^{-3}$	$12\pi \times 10^3$
Flux		1 weber	10^8 maxwells	1/300
Magnetic induction	$\mathbf{B}$	1 weber/meter	10^4 gausses	$1/(3 \times 10^6)$
Magnetic field	$\mathbf{H}$	1 ampere turn/meter	$4\pi \times 10^{-3}$ oersted	$12\pi \times 10^7$
Inductance		1 henry	10^9	$1/(9 \times 10^{11})$
Permittivity of free space	ϵ_0	$(1/36\pi) \times 10^9$ farad/meter	1	$1/(9 \times 10^{20})$
Permeability of free space	μ_0	$4\pi \times 10^{-7}$ henry/meter	$1/(9 \times 10^{20})$	1
Length	L	1 meter	10^2 cm	10^2 cm
Mass	m	1 kg	10^3 g	10^3 g
Time	t	1 sec	1 sec	1 sec
Force	F	1 newton	10^5 dynes	10^5 dynes
Work	W	1 joule	10^7 ergs	10^7 ergs
Energy	E	1 joule	10^7 ergs	10^7 ergs
Power	P	1 watt	10^7 ergs/sec	10^7 ergs/sec

The figures in the last two columns are based on $c = 3 \times 10^{10}$ cm/sec.

To obtain greater accuracy, use the exact value of the velocity of light. The resistance, R, of an electric circuit is

$$R = V/I. \tag{1}$$

The practical units, defined above, are said to be on the absolute scale. We may thus speak of the abs volt or the abs amp, often written as one word, e.g., abamp. For commercial purposes, alternative definitions are employed by international agreement.

In certain problems we must know the work required to move an electron through a difference of potential V. The charge, ε, of an electron is 4.770 $\times 10^{-10}$ es unit. The potential is the work per unit charge. In falling through a potential difference of one volt = 1/300 es unit, the electron acquires energy $\varepsilon V = \varepsilon/300$. In falling through V_p abs volts, therefore, the electron acquires energy:

$$W_e = 1.5910 \times 10^{-12} V_p \text{ erg}. \tag{2}$$

This equation is particularly important for relating the ionization potential of an atom to the energy of ionization.

10. The circuital theorem. In Figure 10, the path abc represents the projection of a circular loop of wire carrying a current I'. Let the plane, circular area bounded by this wire represent the equivalent magnetic shell.

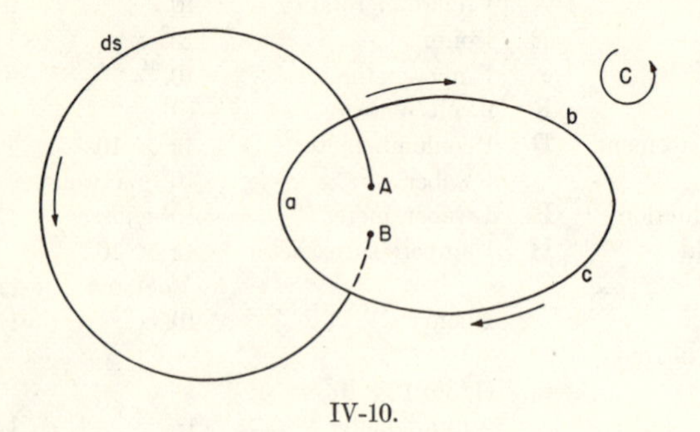

IV-10.

By the rule of § (7), the negative poles are on the upper surface. The work required to carry a unit positive magnetic pole around the circuit from A to B is given by (6.7), or

$$W = \int_A^B \mathbf{B} \cdot d\mathbf{s} = 4\pi\mu\Phi = 4\pi\mu I' = \frac{4\pi\mu I}{c}, \tag{1}$$

$$W = \int_A^B \mathbf{B} \cdot d\mathbf{s} = \mu \int_A^B \mathbf{H} \cdot d\mathbf{s} = \mu\Phi = \mu I, \tag{1a}$$

where $d\mathbf{s}$ is an element of path. This law holds only for steady currents. Otherwise a displacement current enters. For the present we shall consider only steady currents.

Ampere's theorem would not apply over the remaining distance BA, if a physical magnetic shell composed of real magnetic dipoles existed. In traversing the distance BA, the pole would regain the energy lost and the work done would be zero. When the magnetic field results from an electric current, however, the shell has only a fictitious existence. As in § (7), we may represent the magnetic field by any one of an indefinite number of possible shells. Therefore, when the particle arrives at B, let us adopt some other shell to represent the field, which will be undisturbed by this change. Then the unit pole may continue on to A, in air. The distance BA is so small that the work done can be neglected. Equation (1) therefore represents the work done for the complete circuit.

The potential at A is multivalued, depending on the number of times the unit pole traverses the circuit. If the circuit does not enclose part of the current, as for the path C, the work done is zero and we may employ the magnetostatic potential.

If J represents the current density, i.e., the current per unit area,

$$I = \int J \, dS, \tag{2}$$

where dS is an element of surface. Equation (2) holds in the above form only when the current is flowing in a direction normal to dS. In general, we shall have to write

$$I = \int J \cos \theta \, dS = \int \mathbf{J} \cdot d\mathbf{S}, \tag{3}$$

since $\mathbf{J}$ and $d\mathbf{S}$ are vectors. From (1) and (3), with intermediate application of Stokes' theorem (II-34.8), we obtain the result for a circuit over a closed path, γ,

$$W = \mu \int_\gamma \mathbf{H} \cdot d\mathbf{s} = \mu \int \nabla \times \mathbf{H} \cdot d\mathbf{S} = \frac{4\pi\mu}{c} \int \mathbf{J} \cdot d\mathbf{S}. \tag{4}$$

$$W = \mu \int \mathbf{J} \cdot d\mathbf{S}. \tag{4a}$$

This equation must be independent of the shape of the boundary or surface. Hence it must hold for each differential element. Therefore

$$\nabla \times \mathbf{H} = \frac{4\pi\mathbf{J}}{c}, \tag{5}$$

$$\nabla \times \mathbf{H} = \mathbf{J}, \tag{5a}$$

which immediately gives the result that

$$\nabla \cdot \mathbf{J} = \frac{c}{4\pi} \nabla \cdot \nabla \times \mathbf{H} = 0, \tag{6}$$

$$\nabla \cdot \mathbf{J} = \nabla \cdot \nabla \times \mathbf{H} = 0, \tag{6a}$$

by (II-23.13).

The magnetic field produced by a circulating current is clearly dissimilar to ordinary gravitational, electrostatic, or magnetostatic fields. No scalar potential V exists that will enable us to evaluate $\mathbf{H}$ *at every point* of a closed path encircling the current. We cannot define **H** by

$$\mathbf{H} = -\nabla V, \tag{7}$$

because we should then have

$$\nabla \times \mathbf{H} = -\nabla \times \nabla V = 0, \tag{8}$$

by (II-23.14), in contradiction to (5).

The reader will have already noted the analogy of the behavior of the vector **H** to the behavior of the velocity vector **v**, in a vortex. Since we cannot define them in terms of a *scalar* potential, we may introduce a *vector* potential **A**, as in (II-35.8). Then we make the following definition:

$$\mathbf{B} = \mu\mathbf{H} = \nabla \times \mathbf{A}. \tag{9}$$

For MKS units, in free space, $\mu = \mu_0 = 10^{-7}$ weber/ampere meter. In Gaussian units, **H** and **B** are identical in free space. In MKS units they are entirely different, dimensionally as well as numerically. Here **A** is provisionally assumed to be a solenoidal vector, i.e., $\nabla \cdot \mathbf{A} = 0$. In fact, the descriptive adjective "solenoidal" results from this application, since a circuit through which current flows in a circular path is a type of *solenoid*. Then

$$4\pi\mathbf{J}/c = -\nabla^2\mathbf{A}, \tag{10}$$

$$\mu\mathbf{J} = -\nabla^2\mathbf{A}, \tag{10a}$$

by (II-35.10). The components of **J** are given by (II-23.9). Also

$$\nabla \cdot \mathbf{H} = 0. \tag{11}$$

$$\nabla \cdot \mathbf{B} = 0. \tag{11a}$$

The condition expressed by equation (6) is equivalent to that of (II-34.7), viz., that there are no sources or sinks of electricity in the volume and that electricity behaves like an incompressible fluid. An incompressible fluid, however, is free to move, in the absence of sources or sinks, only in closed circuits. Analogously, the fundamental equation (5) is directly applicable only to motion in a closed circuit. Since transient currents may exist in an open circuit, as when a capacitor is being charged, we shall later extend the equations to allow for the existence of *displacement currents*, in contrast with *continuous currents*.

We shall define **A** as follows:

$$\mathbf{A} = \frac{\mu I}{c} \int \left(\frac{1}{r}\right) d\mathbf{s}, \tag{12}$$

$$\mathbf{A} = \frac{\mu I}{4\pi} \int \left(\frac{1}{r}\right) d\mathbf{s}, \tag{12a}$$

where $d\mathbf{s}$ is now an element of the length along the circuit, parallel to the direction of current flow. We shall show later on in this section that (12) leads to the correct expression for $\mathbf{H}$. From (9) and (12) we have

$$\mathbf{B} = \nabla \times \mathbf{A} = \frac{\mu I}{c} \int \nabla \times \frac{d\mathbf{s}}{r}, \tag{13}$$

$$\mathbf{B} = \nabla \times \mathbf{A} = \frac{\mu I}{4\pi} \int \nabla \times \frac{d\mathbf{s}}{r}; \quad \mathbf{H} = \frac{I}{4\pi} \int \nabla \times \frac{d\mathbf{s}}{r}, \tag{13a}$$

where the derivatives are taken with respect to the field point at which we wish to measure $\mathbf{B}$ and not with respect to parts of the circuit element. We shall define this problem in greater detail in equations (24) to (32). Since

$$\nabla \times a\mathbf{C} = a\nabla \times \mathbf{C} - \mathbf{C} \times \nabla a, \tag{14}$$

where $\mathbf{C}$ is any vector and a a scalar, (see problem II-59, p. 181), we have

$$\mathbf{B} = -\frac{\mu I}{c} \int d\mathbf{s} \times \nabla\left(\frac{1}{r}\right) = -\frac{\mu I}{c} \int \frac{\mathbf{r}}{r^3} \times d\mathbf{s}. \tag{15}$$

$$\mathbf{B} = \frac{\mu I}{4\pi} \int \left(\nabla \frac{1}{r}\right) \times d\mathbf{s} = \frac{\mu I}{4\pi} \int \frac{d\mathbf{s} \times \mathbf{r}}{r^3}. \tag{15a}$$

Here $\nabla \times d\mathbf{s}$ vanishes, since $d\mathbf{s}$ is a constant of the circuit and hence does not depend on the coordinates of r; $\mathbf{r}$ is a vector from $d\mathbf{s}$ to the field point.

This result is the Biot-Savart law, which becomes, in its differential form,

$$d\mathbf{H} = -\frac{I}{c} \frac{\mathbf{r}}{r^3} \times d\mathbf{s}. \tag{16}$$

$$d\mathbf{H} = \frac{I}{4\pi} \frac{d\mathbf{s} \times \mathbf{r}}{r^3}. \tag{16a}$$

This expression gives the component of the magnetic field arising from the current flowing in the circuit element $d\mathbf{s}$. In general, $d\mathbf{H}$ is a vector normal to $\mathbf{r}$ and $d\mathbf{s}$. Its absolute magnitude is

$$|\, d\mathbf{H} \,| = I \sin \psi \frac{ds}{cr^2}, \tag{17}$$

$$|\, d\mathbf{H} \,| = \frac{I}{4\pi} \sin \psi \frac{ds}{r^2}, \tag{17a}$$

where ψ is the angle between $d\mathbf{s}$ and $\mathbf{r}$.

Consider a current flowing in the positive direction of the x-axis. The contribution, $d\mathbf{H}$, to the magnetic field at $\mathbf{r} = \mathbf{j}y$, from an element $d\mathbf{s} = \mathbf{i}\, dx$, at the origin, is

$$d\mathbf{H} = \mathbf{k} \frac{I}{c} \frac{dx}{y^2}. \tag{18}$$

$$d\mathbf{H} = \mathbf{k} \frac{I}{4\pi} \frac{dx}{y^2}. \tag{18a}$$

Thus a positive magnetic pole at y is urged to rotate about the wire carrying the current. The rotation, as viewed from a point on the positive portion of the x-axis, is counterclockwise. Upon this simple fundamental theorem, with its important consequences, rests the industrial development of the electric motor.

The quantity I must necessarily be the magnitude of a vector parallel to the circuit element, $d\mathbf{s}$. Thus we may replace $I\, d\mathbf{s}$ by $\mathbf{I}\, ds$ in the foregoing equations. Also, we may express $\mathbf{I}$ in terms of the current density vector, $\mathbf{J}$, by (3). We take dS everywhere as a surface element normal to $d\mathbf{s}$, the direction of flow. Then

$$\mathbf{I} = \int \mathbf{J}\, dS. \tag{19}$$

Since $dS\, ds = d\tau$, an element of volume, we find in place of (15),

$$\mathbf{H} = -\int \frac{\mathbf{r} \times \mathbf{J}}{cr^3}\, d\tau. \tag{20}$$

$$\mathbf{H} = -\int \frac{\mathbf{r} \times \mathbf{J}}{4\pi r^3}\, d\tau. \tag{20a}$$

We have still to prove that (12) and (13) give the correct magnetic field, which by Ampere's law we compute from the potential of any equivalent magnetic shell. In equation (6.1), Mn, i.e., the magnetic moment per unit area, and dS are really vector quantities parallel to one another. Further, θ is the angle between $d\mathbf{S}$ and the vector $\mathbf{r}$ to the point where the potential is being measured. Then $r\, dS \cos\theta = \mathbf{r} \cdot d\mathbf{S}$. Finally, in accord with Ampere's law, (8.2), we may set

$$\Phi = I' = I/c. \tag{21}$$

$$\Phi = I. \tag{21a}$$

Thus the entire potential of the fictitious magnetic shell becomes

$$V = \frac{I}{c}\int \frac{\mathbf{r} \cdot d\mathbf{S}}{r^3}, \tag{22}$$

$$V = \frac{I}{4\pi}\int \frac{\mathbf{r} \cdot d\mathbf{S}}{r^3}, \tag{22a}$$

and we may calculate $\mathbf{H}$ *for any point on any path that does not cut through the given fictitious shell*, by

$$\mathbf{H} = -\nabla V = -\frac{I}{c}\int \nabla\left(\frac{\mathbf{r} \cdot d\mathbf{S}}{r^3}\right). \tag{23}$$

$$\mathbf{H} = -\frac{I}{4\pi}\int \nabla\left(\frac{\mathbf{r} \cdot d\mathbf{S}}{r^3}\right). \tag{23a}$$

We must show the equivalence of (23) and (13).

First take the scalar product of **A**, equation (12), with some constant vector **a**. Then

$$\mathbf{A} \cdot \mathbf{a} = \frac{\mu I}{c} \int \left(\frac{\mathbf{a}}{r}\right) \cdot d\mathbf{s} = \frac{\mu I}{c} \int \nabla' \times \left(\frac{\mathbf{a}}{r}\right) \cdot d\mathbf{S}, \tag{24}$$

$$\mathbf{A} \cdot \mathbf{a} = \frac{\mu I}{4\pi} \int \left(\frac{\mathbf{a}}{r}\right) \cdot d\mathbf{s} = \frac{\mu I}{4\pi} \int \nabla' \times \left(\frac{\mathbf{a}}{r}\right) \cdot d\mathbf{S}, \tag{24a}$$

by Stokes' theorem. Let x, y, z denote the coordinates of the field point for which we are to calculate the vector potential, and x', y', z' the coordinates of a point in the shell. In performing the integration, we must hold x, y, z as fixed. Thus the differentiation must be carried out with respect to the primed coordinates, whereas in (23) the differentiation refers to the coordinates of the field point. For this reason we must distinguish between the operators ∇ and ∇'.

We have the following relation for the curl of the product of a scalar b and a vector (see problem II-59, p. 181):

$$\nabla' \times b\mathbf{a} = b\nabla' \times \mathbf{a} - \mathbf{a} \times \nabla' b. \tag{25}$$

Therefore, since **a** is constant,

$$\mathbf{A} \cdot \mathbf{a} = -\frac{\mu I}{c} \int \mathbf{a} \times \nabla\left(\frac{1}{r}\right) \cdot d\mathbf{S} = -\frac{\mu I}{c} \mathbf{a} \cdot \int \nabla'\left(\frac{1}{r}\right) \times d\mathbf{S}, \tag{26}$$

$$\mathbf{A} \cdot \mathbf{a} = -\frac{\mu I}{4\pi} \int \mathbf{a} \times \nabla\left(\frac{1}{r}\right) \cdot d\mathbf{S} = -\frac{\mu I}{4\pi} \mathbf{a} \cdot \int \nabla'\left(\frac{1}{r}\right) \times d\mathbf{S}, \tag{26a}$$

wherein we have interchanged the dot and cross in the triple scalar product. This relation must hold whatever the orientation and magnitude of **a**. Consequently the vectors into which **a** is multiplied must also be equal. Further, since

$$r^2 = (x - x')^2 + (y - y')^2 + (z - z')^2, \tag{27}$$

we readily prove that

$$\nabla'\left(\frac{1}{r}\right) = -\nabla\left(\frac{1}{r}\right). \tag{28}$$

Therefore

$$\mathbf{A} = \frac{\mu I}{c} \int \nabla\left(\frac{1}{r}\right) \times d\mathbf{S}. \tag{29}$$

$$\mathbf{A} = \frac{\mu I}{4\pi} \int \nabla\left(\frac{1}{r}\right) \times d\mathbf{S}. \tag{29a}$$

From equation (25), we see that

$$\nabla\left(\frac{1}{r}\right) \times d\mathbf{S} = \nabla \times \frac{d\mathbf{S}}{r} - \frac{1}{r}\nabla \times d\mathbf{S}, \tag{30}$$

the differentiation is now carried out with respect to the field point, so that the vector $d\mathbf{S}$ is held constant during the partial differentiation. Hence

$$\mathbf{A} = \frac{\mu I}{c}\int \nabla \times \frac{d\mathbf{S}}{r}. \tag{31}$$

$$\mathbf{A} = \frac{\mu I}{4\pi}\int \nabla \times \frac{d\mathbf{S}}{r}. \tag{31a}$$

And now, applying (13), we have

$$\mathbf{B} = \nabla \times \mathbf{A} = \frac{\mu I}{c}\int \nabla \times \nabla \times \frac{d\mathbf{S}}{r}. \tag{32}$$

$$\mathbf{B} = \nabla \times \mathbf{A} = \frac{\mu I}{4\pi}\int \nabla \times \nabla \times \frac{d\mathbf{S}}{r}. \tag{32a}$$

But

$$\nabla \times \nabla \times \frac{d\mathbf{S}}{r} = \nabla\nabla \cdot \frac{d\mathbf{S}}{r} - \nabla^2 \frac{d\mathbf{S}}{r}. \tag{33}$$

$$\nabla^2 \frac{d\mathbf{S}}{r} = d\mathbf{S}\,\nabla^2 \frac{1}{r} = 0. \tag{34}$$

$$\nabla \cdot \frac{d\mathbf{S}}{r} = \frac{1}{r}\nabla \cdot d\mathbf{S} + d\mathbf{S} \cdot \nabla \frac{1}{r} = -\frac{\mathbf{r}}{r^3} \cdot d\mathbf{S}. \tag{35}$$

Thus we get the relation

$$\mathbf{H} = -\frac{I}{c}\int \nabla \frac{\mathbf{r} \cdot d\mathbf{S}}{r^3}, \tag{36}$$

$$\mathbf{H} = -\frac{I}{4\pi}\int \nabla \frac{\mathbf{r} \cdot d\mathbf{S}}{r^3}, \tag{36a}$$

which agrees with (23). The use of equations (12) and (13) is therefore justified. Equation (13) is to be distinguished from (23) in that it does not depend on the introduction of a fictitious magnetic shell. Therefore we may use (13) to calculate the force for any path in the field, whereas (23) has the limitations previously specified. Note that we may employ the vector potential to represent the field resulting from any distribution of double or dipole sources.

The questions arise, if we provisionally accept equation (16), giving $d\mathbf{H}$, produced by the current in the element $d\mathbf{s}$, what sort of elementary magnetic shell are we entitled to associate with this current element, and what is its equivalent magnetic moment? The problem, of course, is to find a shell that gives the correct value for $d\mathbf{H}$. Let $d\mathbf{s}$, in Fig. 11, represent an element of a circuit carrying current I, and let P be the external point at which we are to calculate the force. In the plane containing $d\mathbf{s}$ and P construct an infinite sector as shown and suppose it to be bounded by a wire carrying current I, as indicated by the arrows. This circuit is now closed and we may regard the plane surface thus bounded as the equivalent shell, whose magnetic moment per unit area is $\mathbf{\Phi}$. By the rules previously dis-

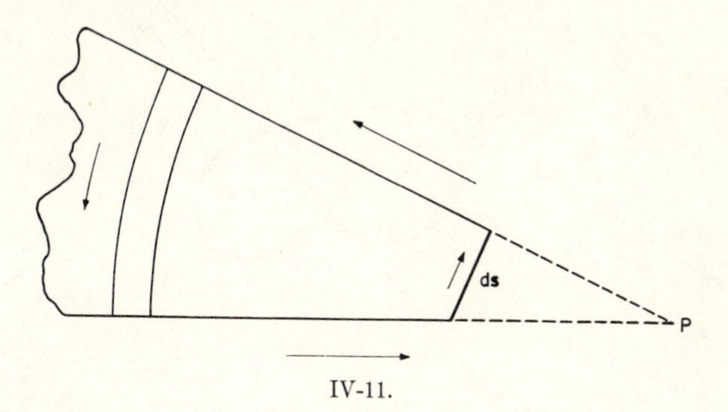

IV-11.

cussed, we see that the positive poles lie an infinitesimal distance above, and the negative poles an infinitesimal distance below the plane of the paper. If $\mathbf{r}_0$ is any vector from P to $d\mathbf{s}$, we see that $\mathbf{r}_0 \times d\mathbf{s}$ is perpendicular to the plane of the paper, and thus points in the direction of $\mathbf{\Phi}$.

The force from the area $r\alpha\, dr$ of the shell between r and $r + dr$ is obtained from (5.6). Here $\theta = \pi/2$, and the radial component of force vanishes. In the present problem, $\mathbf{i}_\theta$ is perpendicular *inwards* to the plane of the paper. Hence

$$d\mathbf{H} = \Phi \mathbf{i}_\theta \alpha \int_{r_0}^{\infty} \frac{dr}{r^2} = \Phi \mathbf{i}_\theta \frac{\alpha}{r_0}, \tag{37}$$

$$d\mathbf{H} = \frac{\Phi}{4\pi} \mathbf{i}_\theta \alpha \int_{r_0}^{\infty} \frac{dr}{r^2} = \frac{\Phi}{4\pi} \mathbf{i}_\theta \frac{\alpha}{r_0}, \tag{37a}$$

wherein we have taken $d\mathbf{s}$ and therefore also α to be very small. Let ψ be the angle between $\mathbf{r}$ and $d\mathbf{s}$. Then

$$ds \sin \psi = r\alpha. \tag{38}$$

$$|\, d\mathbf{H}\,| = \Phi \frac{\sin \psi}{r^2}\, ds = I \sin \psi \frac{ds}{cr^2}, \tag{39}$$

$$|\, d\mathbf{H}\,| = \frac{\Phi}{4\pi} \frac{\sin \psi}{r^2}\, ds = \frac{I}{4\pi} \frac{\sin \psi}{r^2}\, ds, \tag{39a}$$

in agreement with (17). Thus the shell we are considering gives the proper magnitude and direction for $d\mathbf{H}$, and hence may be adopted as the equivalent magnetic shell.

Now construct an infinite cone, with its vertex at P and directrices cutting all elements of the closed surface (Fig. 12), and imagine the cone

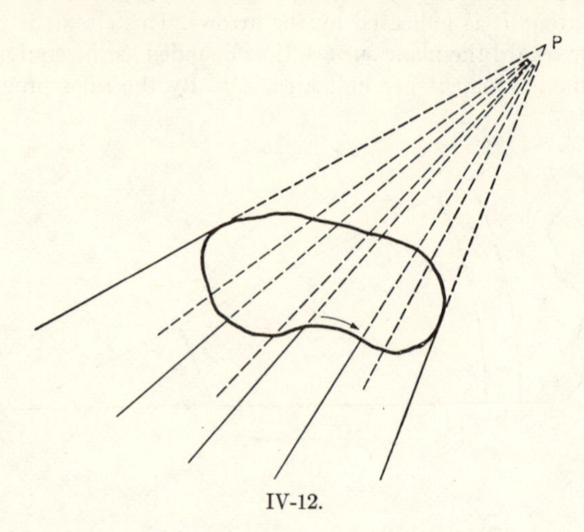

IV-12.

to be closed at infinity by a spherical cap centered at P. The potential of the cap is, from (6.4),

$$V = -\Phi\Omega = \text{CONST}, \tag{40}$$

$$V = -\Phi\Omega/4\pi = \text{CONST}, \tag{40a}$$

because Ω, the solid angle of the cone, is independent of r. Therefore

$$\mathbf{H} = -\nabla V = 0, \tag{41}$$

and the cap contributes nothing to the magnetic field at P. We thus find it convenient to regard as the equivalent magnetic shell that portion of the cone and spherical cap cut off by the circuit boundary.

If we imagine the cone to be complete, the surface integral, M, of the magnetic moment vector $\mathbf{\Phi}$ over the closed surface must vanish, i.e.,

$$M = \int \mathbf{\Phi} \cdot d\mathbf{S} = \int \nabla \cdot \mathbf{\Phi}\, d\tau = 0, \tag{42}$$

because $\mathbf{\Phi}$, like $\mathbf{I}$, is a solenoidal vector. Therefore, if we divide M into two portions M_c, the part arising from the conical shell under consideration, and M_c' the part from the vertex down to the circuit, we have

$$M_c = -M_c'. \tag{43}$$

The latter expression is easily evaluated. The area of any one of the infinitesimal triangles subtended by $d\mathbf{s}$ from the point P, is $\frac{1}{2}\mathbf{r} \times d\mathbf{s}$ by (II-26.19). Thus

$$dM_c = -\frac{1}{2}\mathbf{\Phi} \cdot \mathbf{r} \times d\mathbf{s}. \tag{44}$$

We have chosen $\mathbf{r}$ as the vector from $d\mathbf{s}$ to P, hence $d\mathbf{s} \times \mathbf{r}$ is parallel to $\mathbf{\Phi}$. In our present integration we shall regard P as fixed. It is more convenient to replace $\mathbf{r}$ by $-\mathbf{r}$, and reinterpret $\mathbf{r}$ as the vector from P to $d\mathbf{s}$. We shall denote by $d\mathbf{M}$ the resultant total magnetic moment arising from the current element in $d\mathbf{s}$, and drop the subscript c as no longer necessary. We may write, replacing $\mathbf{\Phi}$ by its equivalent in terms of the current,

$$d\mathbf{M} = \frac{I}{2c}\mathbf{r} \times d\mathbf{s} = \frac{1}{2c}\mathbf{r} \times \mathbf{I}\, ds. \tag{45}$$

$$d\mathbf{M} = \frac{I}{2}\mathbf{r} \times d\mathbf{s} = \frac{1}{2}\mathbf{r} \times \mathbf{I}\, ds. \tag{45a}$$

Further, we express $\mathbf{I}$ in terms of $\mathbf{J}\, dS$, the current density, and consider $dS\, ds = d\tau$, an element of volume, as in (19) and (20). Then we have, for the resultant total magnetic moment of any current distribution at the point P,

$$\mathbf{M} = \frac{1}{2c}\int \mathbf{r} \times \mathbf{J}\, d\tau, \tag{46}$$

$$\mathbf{M} = \frac{1}{2}\int \mathbf{r} \times \mathbf{J}\, d\tau, \tag{46a}$$

a formula of great importance in radiation problems. The vector potential consistent with the foregoing demonstrations is, from (12) and (19),

$$\mathbf{A} = \frac{\mu}{c}\int \frac{\mathbf{J}}{r}\, d\tau. \tag{47}$$

$$\mathbf{A} = \frac{\mu}{4\pi}\int \frac{\mathbf{J}}{r}\, d\tau. \tag{47a}$$

11. Magnetic field of a circular loop. In a magnetic field produced by a current, no scalar potential exists for the purpose of calculating the work done by a pole that traverses a closed path encircling the current. As we have seen, the difficulty arises because the potential is multivalued. Nevertheless, we can calculate the force at any point in the field by means of a scalar potential, referred to some equivalent magnetic shell.

We shall now calculate the magnetic field resulting from a current I flowing in a circular path of radius r. Let the circle be centered at the origin, its plane corresponding with the xz plane (Fig. 13). We shall restrict our calculations to points along the y-axis.

From equations (6.4) and (8.2) the potential at the point P,

$$V = \Phi\Omega = \frac{I}{c}\,\Omega, \tag{1}$$

$$V = \frac{\Phi\Omega}{4\pi} = \frac{I\Omega}{4\pi}, \tag{1a}$$

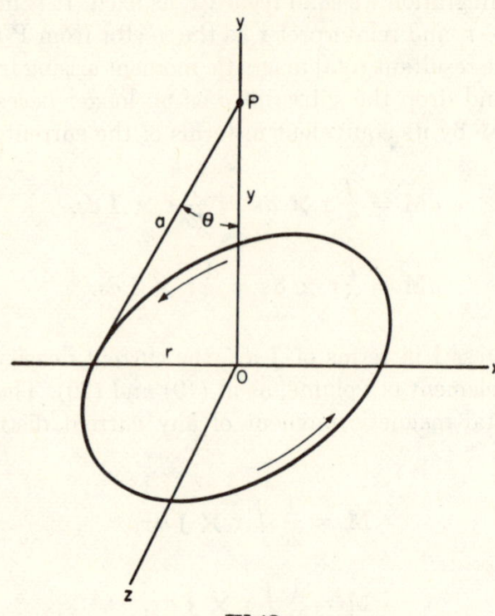

IV-13.

where Ω is the solid angle subtended at the point by the circle. Let a be the distance from P to the circumference of the circle. Then

$$a^2 = r^2 + y^2. \tag{2}$$

We easily find that

$$\Omega = 2\pi \int_0^{\text{arc cos } y/a} \sin\theta \, d\theta = 2\pi\left[1 - \frac{y}{(r^2 + y^2)^{1/2}}\right], \tag{3}$$

whence
$$H_y = -\frac{\partial V}{\partial y} = \frac{2\pi I}{c}\frac{r^2}{(r^2 + y^2)^{3/2}}. \tag{4}$$

$$H_y = \frac{I}{2}\frac{r^2}{(r^2 + y^2)^{3/2}}, \tag{4a}$$

The field at the center of the circle and, indeed, for any point in the xz-plane inside the circle, is

$$H = 2\pi I/cr. \tag{5}$$

$$H = I/2r. \tag{5a}$$

H is normal to the xz-plane. I is to be measured in es units and r in centimeters. The unit of magnetic field in the em system is the *gauss*.

12. The magnetic field of a rotating charged disk. Consider the field H produced by a rotating disk carrying a uniform charge of n electrons, each of charge ε, per unit area. Any small ring located between r and $r + dr$ will

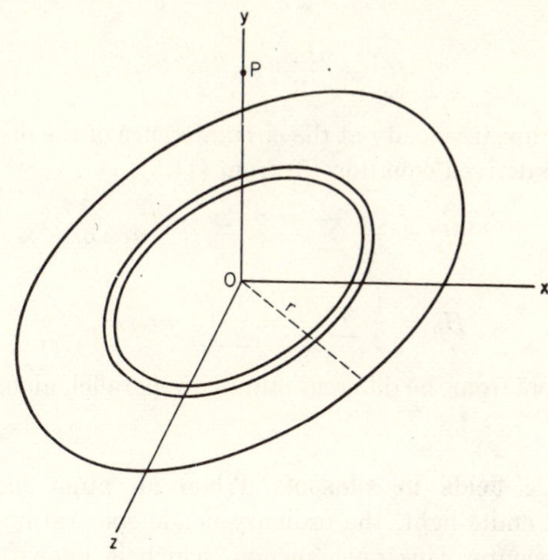

IV-14.

behave like the loop of the previous section. If ω is the angular velocity of the disk and v the linear velocity at distance r from the center,

$$v = r\omega \tag{1}$$

and the electric current in the annulus is

$$dI = n\varepsilon v\,dr = n\varepsilon\omega r\,dr. \tag{2}$$

The magnetic potential at P resulting from current flowing in the ring is given by (11.1) and (11.3):

$$dV = \frac{2\pi n\varepsilon\omega}{c}\left[1 - \frac{y}{(r^2 + y^2)^{1/2}}\right] r\,dr. \tag{3}$$

$$dV = \frac{n\varepsilon\omega}{2}\left[1 - \frac{y}{(r^2 + y^2)^{1/2}}\right] r\,dr. \tag{3a}$$

Then

$$V = \int_0^{r_0} dV = \frac{2\pi n\varepsilon\omega}{c}\left[\frac{r_0^2}{2} - y(r_0^2 + y^2)^{1/2} + y^2\right], \tag{4}$$

$$V = \frac{n\varepsilon\omega}{2}\left[\frac{r_0^2}{2} - y(r_0^2 + y^2)^{1/2} + y^2\right], \tag{4a}$$

$$H_y = -\frac{\partial V}{\partial y} = \frac{2\pi n\varepsilon\omega}{c}\left[\frac{r_0^2 + 2y^2}{(r_0^2 + y^2)^{1/2}} - 2y\right], \tag{5}$$

$$H_y = \frac{n\varepsilon\omega}{2}\left[\frac{r_0^2 + 2y^2}{(r_0^2 + y^2)^{1/2}} - 2y\right], \tag{5a}$$

and

$$H_0 = \frac{2\pi n\varepsilon\omega r_0}{c} = \frac{2\pi n\varepsilon v_0}{c}, \tag{6}$$

$$H_0 = \frac{n\varepsilon\omega r_0}{2} = \frac{n\varepsilon v_0}{2}, \tag{6a}$$

where v_0 is the linear velocity at the circumference of the disk. We note that we could have derived equation (6) from (11.5),

$$H_0 = \frac{2\pi}{c}\sum\frac{di}{r} = \frac{2\pi}{c}\int_0^{r_0} n\varepsilon\omega\, dr, \tag{7}$$

$$H_0 = \frac{1}{2}\sum\frac{di}{r} = \frac{1}{2}\int_0^{r_0} n\varepsilon\omega\, dr, \tag{7a}$$

since the vectors from the different annuli are parallel, along the axis of the disk.

13. Magnetic fields in sunspots. When an atom located within a magnetic field emits light, the ordinary single spectral lines split up into several components. This phenomenon, which is known as the Zeeman effect, will be discussed in detail later on. We may, in the meanwhile, employ the observational result that spectral lines in sunspots exhibit an unquestioned magnetic splitting. The fields responsible for the solar Zeeman effect occasionally attain, in large spots, values as high as 4000 gausses.

The origin of these intense magnetic fields is one of the outstanding astrophysical problems. We easily dispose of the more obvious hypotheses. In the first place, we cannot appeal to permanent magnetism. No alignment of the individual atom magnets could possibly be maintained in the presence of the turbulent motion and high temperatures existing on and in the sun. The effect is undoubtedly electrical and we may calculate the magnitude of the electric currents and electric charges involved.

A simple model, based on the formulae of the proceeding section will indicate the nature of the difficulties. Observations suggest that sunspots are vortices. We shall assume, for purposes of demonstration, that the spot contains a disk of n electrons per cm^2, which rotates with uniform angular velocity. Then, from equation (12.6),

$$n = cH_0/2\pi\varepsilon v_0, \tag{1}$$

$$n = 2H_0/\varepsilon v_0, \tag{1a}$$

where v_0 represents the tangential velocity at the periphery. Although the vortical speeds of spots are unknown, we may take $v_0 = 300 \text{ km sec}^{-1} =$

3×10^7 cm sec^{-1}, as an upper limit to the value. Also set $H_0 = 4000$ gausses, to agree with observation. Then we find that n, at the very least, will be

$$n \sim 10^{15} \text{ electrons cm}^{-2}. \tag{2}$$

The electric currents are truly enormous.

$$I = n\varepsilon \int_0^{r_0} v\, dr = n\varepsilon v_0 r_0/2. \tag{3}$$

We shall take r_0, the radius of the spot, to be 10^9 cm. Then

$$I \sim 10^{22} \text{ es units} \sim 3 \times 10^{12} \text{ amperes}. \tag{4}$$

A change of model will alter these figures by at most an order of magnitude or two. We may safely draw certain general conclusions from the result. We have previously found (§ 3) that an excess of one proton per sixty square centimeters of solar surface would produce sufficient positive potential just to overcome the solar attraction by electrostatic repulsion of an electron. An excess of protons, as given by equation (2), would produce forces 6×10^{15} times as great. A sunspot having such an excess would break up with explosive violence. We have been forced to the conclusion that the sun is practically neutral electrically. Circulation of the minute residual charges would produce fields of negligible strength. Currents of the order of 10^{12} amperes are required, but they must be *galvanic* in nature. The charges of one sign must drift with respect to those of the other sign, while the medium remains macroscopically neutral. The forces that produce the galvanic current still remain to be identified.

14. Derivation of Maxwell's equations. We may summarize the fundamental electromagnetic relationships, derived up to the present point. We have the circuital theorem, (10.6) or (10.8),

$$\nabla \times \mathbf{H} = 4\pi\mathbf{J}/c. \tag{1}$$

$$\nabla \times \mathbf{H} = \mathbf{J}. \tag{1a}$$

Also, from (10.14),

$$\nabla \cdot \mathbf{B} = 0, \tag{2}$$

$$\nabla \cdot \mathbf{B} = 0, \tag{2a}$$

and from (10.9),

$$\nabla \cdot \mathbf{J} = 0. \tag{3}$$

$$\nabla \cdot \mathbf{J} = 0. \tag{3a}$$

These are not the most general equations that could be derived. They hold for the steady state, when the current is flowing through a closed circuit.

When the currents or the magnetic fields are fluctuating, either in magnitude or position, with the time, we shall have to re-analyze the problem.

Consider the electric circuit shown in Fig. 15, where A indicates the circular plates of a capacitor and B a source of electric potential, e.g., batteries, connected across the wires leading to A. The electric intensity E, between the plates, where we suppose the separation to be negligible compared with the linear dimensions, is given by (4.9),

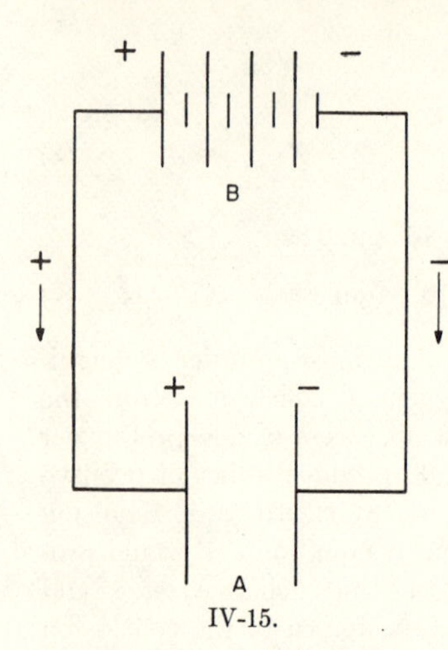

IV-15.

$$E = 4\pi Q/\kappa A, \tag{4}$$

$$E = Q/\epsilon A, \tag{4a}$$

where Q is the quantity of electricity on the plates and A is the area. As long as the voltage applied to the capacitor remains constant, both Q and E are constant and no current flows in the leads.

If the voltage increases with the time, instead of remaining constant, electricity will flow from the batteries along the leads, negative charges on the left and positive along the right as indicated by the arrows. The resulting flow of electricity is analogous to that of a closed circuit, as if an actual instead of virtual flow took place, directly through the medium between the plates. Differentiating (4), we have

$$\frac{d}{dt}(\kappa E) = \frac{4\pi}{A}\frac{dQ}{dt}, \tag{5}$$

$$\frac{d}{dt}(\epsilon E) = \frac{1}{A}\frac{dQ}{dt}, \tag{5a}$$

which relates the rate of change of E to the "virtual" current across the plates. The virtual current density, J_v, across the capacitor is $(dQ/dt)/A$. Hence

$$\frac{d}{dt}(\kappa E) = \frac{dD}{dt} = 4\pi J_v, \tag{6}$$

$$\frac{d}{dt}(\epsilon E) = \frac{dD}{dt} = J_v, \tag{6a}$$

where we have set

$$D = \kappa E. \tag{7}$$

$$D = \epsilon E. \tag{7a}$$

Although this equation applies specifically for a field in a one-dimensional coordinate system, we readily deduce the generalization. First, we note that E indicates the magnitude of the vector defining the electric intensity at a given point of space.

Ordinarily, E will vary from point to point, hence the time differentiation must be a partial one. In an isotropic medium the vectors **D** and **E** are parallel. Then for the three vector components we have the relation

$$\frac{\partial}{\partial t}(\mathbf{i}D_x + \mathbf{j}D_y + \mathbf{k}D_z) = 4\pi(\mathbf{i}J_{vx} + \mathbf{j}J_{vy} + \mathbf{k}J_{vz}),$$

or

$$\frac{\partial \mathbf{D}}{\partial t} = 4\pi \mathbf{J}_v. \tag{8}$$

$$\frac{\partial \mathbf{D}}{\partial t} = \mathbf{J}_v. \tag{8a}$$

Maxwell termed the quantity **D** the "displacement current" or the "electric displacement." He regarded the true currents as being composed of steady currents and virtual currents and made the additional supposition, now fully justified by experiment, that the magnetic effects of the two were identical. Accordingly, we have only to replace **J** in (1) by $\mathbf{J} + \mathbf{J}_v$, where **J** represents as before the constant component. Then we have

$$\nabla \times \mathbf{H} = \frac{1}{c}\frac{\partial \mathbf{D}}{\partial t} + \frac{4\pi}{c}\mathbf{J}. \tag{9}$$

$$\nabla \times \mathbf{H} = \frac{\partial \mathbf{D}}{\partial t} + \mathbf{J}. \tag{9a}$$

This is one of the famous Maxwell equations.

If we take the divergence of both sides of (9) we have

$$\nabla \cdot \nabla \times \mathbf{H} = \nabla \cdot \left(\frac{1}{c}\frac{\partial \mathbf{D}}{\partial t} + \frac{4\pi}{c}\mathbf{J}\right) = 0. \tag{10}$$

$$\nabla \cdot \nabla \times \mathbf{H} = \nabla \cdot \left(\frac{\partial \mathbf{D}}{\partial t} + \mathbf{J}\right) = 0. \tag{10a}$$

Therefore

$$\nabla \cdot \mathbf{J} + \frac{1}{4\pi}\frac{\partial}{\partial t}(\nabla \cdot \mathbf{D}) = 0. \tag{11}$$

$$\nabla \cdot \mathbf{J} + \frac{\partial}{\partial t}(\nabla \cdot \mathbf{D}) = 0. \tag{11a}$$

The equation of continuity (II-34.4) is, if the total charge is conserved (no sources or sinks),

$$\frac{\partial \rho}{\partial t} + \nabla \cdot (\rho \mathbf{v}_0) = 0. \tag{12}$$

If we interpret ρ as the density of electric charge with $\mathbf{v}_0$ the vector velocity of flow, the current density

$$\mathbf{J} = \rho\mathbf{v} \tag{13}$$

and

$$\frac{\partial \rho}{\partial t} + \nabla \cdot \mathbf{J} = 0. \tag{14}$$

Equations (11) and (14) require, then, that

$$\nabla \cdot \mathbf{D} = 4\pi\rho, \tag{15}$$

$$\nabla \cdot \mathbf{D} = \rho, \tag{15a}$$

which replaces (3) in the more general relationships. Equation (2) is taken over unchanged:

$$\nabla \cdot \mathbf{B} = 0. \tag{16}$$

since there can be no divergence of magnetic poles of one sign into a given volume, magnetic poles always occurring in pairs.

The fourth Maxwell equation defines the current produced by a magnetic field, whereas (9) defines the magnetic field resulting from an electric current. The relationships are not quite reciprocal. A steady current will give rise to a magnetic field. But, as the experiments have shown, electric currents are induced only by changes in a magnetic field. Consider some surface bounded by any closed curve, γ. The flux of magnetic intensity through a surface is given by the expression (II-13.1), which becomes

$$\phi = \mu \int \mathbf{H} \cdot d\mathbf{S}. \tag{17}$$

The symbol ϕ represents the "number of magnetic lines of force" passing through the surface. The work required to carry a unit electric charge around the closed curve γ is the electromotive force, emf:

$$\text{emf} = \int_\gamma \mathbf{E} \cdot d\mathbf{s} = \int \nabla \times \mathbf{E} \cdot d\mathbf{S}. \tag{18}$$

($\mathbf{E}$ represents the electric field.) Stokes' theorem (II-32.8) has been employed. The law found experimentally relates the current to the rate of change of flux, as follows:

$$\text{emf} = -\frac{1}{c}\frac{d\phi}{dt}. \tag{19}$$

$$\text{emf} = -\frac{d\phi}{dt}. \tag{19a}$$

The factor c, in (19) converts emf from the em to the es system. Then

$$\int \nabla \times \mathbf{E} \cdot d\mathbf{S} = -\frac{1}{c}\frac{d}{dt}\mu \int \mathbf{H} \cdot d\mathbf{S} = -\frac{1}{c}\int \frac{\partial}{\partial t}\mu\mathbf{H} \cdot d\mathbf{S}. \tag{20}$$

$$\int \nabla \times \mathbf{E} \cdot d\mathbf{S} = -\frac{d}{dt}\mu \int \mathbf{H} \cdot d\mathbf{S} = -\int \frac{\partial}{\partial t}\mu\mathbf{H} \cdot d\mathbf{S}. \tag{20a}$$

Note that we require a partial differentiation when we carry out the differentiation under the integral sign. This equation must hold for each differential element, hence

$$\nabla \times \mathbf{E} = -\frac{1}{c}\frac{\partial \mathbf{B}}{\partial t}, \tag{21}$$

$$\nabla \times \mathbf{E} = -\frac{\partial \mathbf{B}}{\partial t}, \tag{21a}$$

where
$$\mathbf{B} = \mu \mathbf{H}, \tag{22}$$

the *magnetic induction*. Equation (21) is the fourth Maxwell equation, which, in conjunction with (9), (15), and (16), forms the basis of electromagnetic theory.

The four field quantities **E**, **D**, **H**, and **B** are fundamental to electromagnetic problems. We might have introduced **D** and **B** earlier (cf. equation 10.1a). They are essentially more basic than the parameters μ and κ, by which we have defined them. The simpler representation above given holds only for isotropic media. In general, we should regard both μ and κ as tensors or dyadics: $\boldsymbol{\mu}$ and $\boldsymbol{\kappa}$. Then the equations

$$\mathbf{D} = \boldsymbol{\kappa} \cdot \mathbf{E} \quad \text{and} \quad \mathbf{B} = \boldsymbol{\mu} \cdot \mathbf{H} \tag{23}$$

$$\mathbf{D} = \boldsymbol{\varepsilon} \cdot \mathbf{E} \quad \text{and} \quad \mathbf{B} = \boldsymbol{\mu} \cdot \mathbf{H} \tag{23a}$$

replace (7) and (22), respectively. When equations (23) apply, **D** and **E** are proportional only along the principal axes of the tensor.

The quantities κ and μ take values differing from unity because the respective fields induce polarizations within the media. We define two quantities χ_e and χ_m, respectively the electric and magnetic susceptibilities, as

$$4\pi\chi_e = \kappa - 1 \quad \text{and} \quad 4\pi\chi_m = \mu - 1. \tag{24}$$

$$\chi_e = \frac{\epsilon}{\epsilon_0} - 1 \quad \text{and} \quad \chi_m = \frac{\mu}{\mu_0} - 1. \tag{24a}$$

Experiment shows that $\chi_e \geq 0$, whereas χ_m may be of either sign. We distinguish between the two types of magnetic susceptibilities as follows: $\chi_m > 0$ for paramagnetic substances and $\chi_m < 0$ for diamagnetic materials.

For most materials χ_m is nearly zero and, in practice, we may usually set $\chi_m = 0$. But a certain group of materials, iron, cobalt, nickel, and various alloys of these substances exhibit extreme positive susceptibility. Equation (22) does not apply to ferromagnetic media of this type, since **B** does not necessarily vanish with **H**. Although the phenomenon will not concern us for much of the following, we shall digress for the moment with a brief discussion of hysteresis.

For ordinary material a plot of B vs. H defines a straight line whose slope is μ. For ferromagnetic materials, however, such a diagram produces a curve known as a hysteresis loop. Consider the behavior of a bar of iron inside a coil of wire through which an alternating current runs. As the current, measured by H, drops to zero, the magnetism B of the bar does not vanish. To reduce B to zero we must apply some current in the opposite direction. The fact that B lags behind H constitutes the phenomenon of hysteresis. The area of the loop measures the rate of energy dissipation within the material.

We define the polarization **P** and magnetization **M** by the equations, cf. (31.25),

$$\mathbf{P} = (\mathbf{D} - \mathbf{E})/4\pi \quad \text{and} \quad \mathbf{M} = (\mathbf{B} - \mathbf{H})/4\pi, \tag{25}$$

$$\mathbf{P} = \mathbf{D} - \epsilon_0\mathbf{E} \quad \text{and} \quad \mathbf{M} = \mathbf{B}/\mu_0 - \mathbf{H}, \tag{25a}$$

so that the χ_e and χ_m of (24) are

$$\chi_e = |\,\mathbf{P}\,|/|\,\mathbf{E}_0\,| \quad \text{and} \quad \chi_m = |\,\mathbf{M}\,|/|\,\mathbf{H}\,|. \tag{26}$$

$$\chi_e = \frac{|\,\mathbf{P}\,|}{\epsilon_0\,|\,\mathbf{E}\,|}. \tag{26a}$$

Electromagnetic Radiation

15. The electromagnetic theory of light. For a medium where no conduction current flows and where κ and μ are independent both of the time and the coordinates, we may rewrite Maxwell's equation in a simpler form. Let **E** and **H** be vectors representing the electric and magnetic fields, with respective components E_x, E_y, and E_z and H_x, H_y, H_z. Then the equations become

$$\nabla \times \mathbf{H} = \frac{\kappa}{c}\frac{\partial \mathbf{E}}{\partial t}, \tag{1}$$

$$\nabla \times \mathbf{H} = \frac{\partial \mathbf{D}}{\partial t} = \epsilon\frac{\partial \mathbf{E}}{\partial t}, \tag{1a}$$

$$\nabla \times \mathbf{E} = -\frac{\mu}{c}\frac{\partial \mathbf{H}}{\partial t}, \tag{2}$$

$$\nabla \times \mathbf{E} = -\frac{\partial \mathbf{B}}{\partial t} = -\mu\frac{\partial \mathbf{H}}{\partial t}, \tag{2a}$$

$$\nabla \cdot \mathbf{H} = 0, \tag{3}$$

$$\nabla \cdot \mathbf{H} = 0, \tag{3a}$$

$$\nabla \cdot \mathbf{E} = 4\pi\rho/\kappa. \tag{4}$$

$$\nabla \cdot \mathbf{D} = \rho. \tag{4a}$$

As an illustration of the superiority of vector methods, and also as an example, we shall carry out the solution of these equations, first in the notation of differential calculus, and second in the vector notation. Rewriting equations (1) to (4) in terms of their vector components, we have

$$\frac{\partial H_z}{\partial y} - \frac{\partial H_y}{\partial z} = \frac{\kappa}{c}\frac{\partial E_x}{\partial t}, \quad \frac{\partial H_x}{\partial z} - \frac{\partial H_z}{\partial x} = \frac{\kappa}{c}\frac{\partial E_y}{\partial t}, \qquad (5)$$

$$\frac{\partial H_y}{\partial x} - \frac{\partial H_x}{\partial y} = \frac{\kappa}{c}\frac{\partial E_z}{\partial t}.$$

$$\frac{\partial H_z}{\partial y} - \frac{\partial H_y}{\partial z} = \epsilon\frac{\partial E_x}{\partial t}, \quad \text{ETC.} \qquad (5a)$$

$$\frac{\partial E_z}{\partial y} - \frac{\partial E_y}{\partial z} = -\frac{\mu}{c}\frac{\partial H_x}{\partial t}, \quad \frac{\partial E_x}{\partial z} - \frac{\partial E_z}{\partial x} = -\frac{\mu}{c}\frac{\partial H_y}{\partial t}, \qquad (6)$$

$$\frac{\partial E_y}{\partial x} - \frac{\partial E_x}{\partial y} = -\frac{\mu}{c}\frac{\partial H_z}{\partial t}.$$

$$\frac{\partial E_z}{\partial y} - \frac{\partial E_y}{\partial z} = -\mu\frac{\partial H_x}{\partial t}, \quad \text{ETC.} \qquad (6a)$$

$$\frac{\partial H_x}{\partial x} + \frac{\partial H_y}{\partial y} + \frac{\partial H_z}{\partial z} = 0. \qquad (7)$$

$$\frac{\partial E_x}{\partial x} + \frac{\partial E_y}{\partial y} + \frac{\partial E_z}{\partial z} = \frac{4\pi\rho}{\kappa}. \qquad (8)$$

$$\frac{\partial E_x}{\partial x} + \frac{\partial E_y}{\partial y} + \frac{\partial E_z}{\partial z} = \frac{\rho}{\epsilon}. \qquad (8a)$$

Differentiate partially the first equation of (5) with respect to t, the second of (6) with respect to z, and the third of (6) with respect to y. Use the two last equations to eliminate $\partial^2 H_y/\partial y\, \partial t$ and $\partial^2 H_y/\partial z\, \partial t$ from the first. The result is

$$\frac{\mu\kappa}{c^2}\frac{\partial^2 E_x}{\partial t^2} = \frac{\partial^2 E_x}{\partial x^2} + \frac{\partial^2 E_x}{\partial y^2} + \frac{\partial^2 E_x}{\partial z^2} - \frac{\partial}{\partial x}\left(\frac{\partial E_x}{\partial x} + \frac{\partial E_y}{\partial y} + \frac{\partial E_z}{\partial z}\right), \qquad (9)$$

$$\mu\epsilon\frac{\partial^2 E_x}{\partial t^2} = \frac{\partial^2 E_x}{\partial x^2} + \frac{\partial^2 E_x}{\partial y^2} + \frac{\partial^2 E_x}{\partial z^2} - \frac{\partial}{\partial x}\left(\frac{\partial E_x}{\partial x} + \frac{\partial E_y}{\partial y} + \frac{\partial E_z}{\partial z}\right), \qquad (9a)$$

where we have added and subtracted the term $\partial^2 E_x/\partial x^2$ on the right-hand side. In a region where no free electric charges exist, $\rho = 0$, and the second member vanishes, by (8). Hence

$$\frac{\mu\kappa}{c^2}\frac{\partial^2 E_x}{\partial t^2} = \frac{\partial^2 E_x}{\partial x^2} + \frac{\partial^2 E_x}{\partial y^2} + \frac{\partial^2 E_x}{\partial z^2} = \nabla^2 E_x. \qquad (10)$$

$$\mu\epsilon\frac{\partial^2 E}{\partial t^2} = \nabla^2 E_x. \qquad (10a)$$

Let

$$v^2 = c^2/\mu\kappa. \tag{11}$$

$$v^2 = 1/\mu\epsilon. \tag{11a}$$

We treat each independent equation of (5) and (6) similarly, to give six fundamental equations of the form

$$\nabla^2 E_x = \frac{1}{v^2}\frac{\partial^2 E_x}{\partial t^2}, \quad \nabla^2 E_y = \frac{1}{v^2}\frac{\partial^2 E_y}{\partial t^2}, \quad \nabla^2 E_z = \frac{1}{v^2}\frac{\partial^2 E_z}{\partial t^2}. \tag{12}$$

$$\nabla^2 H_x = \frac{1}{v^2}\frac{\partial^2 H_x}{\partial t^2}, \quad \nabla^2 H_y = \frac{1}{v^2}\frac{\partial^2 H_y}{\partial t^2}, \quad \nabla^2 H_z = \frac{1}{v^2}\frac{\partial^2 H_z}{\partial t^2}. \tag{13}$$

The vector analogue of the foregoing transformations is as follows. Differentiate (1) partially with respect to t, substitute from (2), and reduce further by means of (II-23.16).

$$\begin{aligned}\frac{\partial}{\partial t}\nabla \times \mathbf{H} = \nabla \times \frac{\partial \mathbf{H}}{\partial t} &= -\frac{c}{\mu}\nabla \times \nabla \times \mathbf{E} \\ &= -\frac{c}{\mu}(\nabla\nabla \cdot \mathbf{E} - \nabla^2\mathbf{E}) = \frac{\kappa}{c}\frac{\partial^2 \mathbf{E}}{\partial t^2}.\end{aligned} \tag{14}$$

$$\begin{aligned}\frac{\partial}{\partial t}(\nabla \times \mathbf{H}) &= -\frac{1}{\mu}\nabla \times \nabla \times \mathbf{E} \\ &= -\frac{1}{\mu}(\nabla\nabla \cdot \mathbf{E} - \nabla^2\mathbf{E}) = \epsilon\frac{\partial^2 \mathbf{E}}{\partial t^2}.\end{aligned} \tag{14a}$$

Equation (2) follows a similar transformation. When $\rho = 0$, $\nabla \cdot \mathbf{E} = 0$. Then the results are

$$\nabla^2\mathbf{E} = \frac{1}{v^2}\frac{\partial^2 \mathbf{E}}{\partial t^2} \quad \text{and} \quad \nabla^2\mathbf{H} = \frac{1}{v^2}\frac{\partial^2 \mathbf{H}}{\partial t^2}, \tag{15}$$

the vector components of which agree with (12) and (13).

Equations of the type (12) and (13) have already been studied in detail. The reader will already have recognized them to be of the form of the wave equation, (III-1.15). The general solution of (12), which is representative, is

$$E_x = f_1(x + vt) + f_2(x - vt), \tag{16}$$

by (III-2.6). The f's are arbitrary functions, which define the form of a disturbance traveling through space with a velocity v. For free space, $\kappa = \mu = 1$, in Gaussian units. In the MKS system, for free space, $\epsilon = \epsilon_0$, $\mu = \mu_0$. Then

$$v = c, \tag{17}$$

$$v = 1/\sqrt{\epsilon_0\mu_0} = 3 \times 10^8\ M/\text{sec} = c, \tag{17a}$$

the velocity of light. Maxwell thus recognized the electromagnetic character of radiation.

We have already seen that we cannot make further progress in solving the wave equation, unless the original conditions or certain boundary conditions are specified. A light wave, presumably, has its origin in some atom. More precisely, the electromagnetic radiation is emitted as the result of oscillations of some atomic electron (or electrons). If the electronic motion is periodic the light wave will also be periodic and the frequency of the wave will be equal to that of the atomic oscillation. If the electron's vibrations are aperiodic, the light wave will also exhibit no characteristic frequency. As a next step, we shall require the equation of motion of an electron in a composite field involving electric and magnetic forces.

16. The electromagnetic force field. We begin by calculating the electromagnetic forces produced by a moving point charge, of magnitude q. The first term of the force, that arising from the electrostatic field, is merely Coulomb's law,

$$E = q/\kappa r^2, \tag{1}$$

$$E = q/4\pi\epsilon r^2, \tag{1a}$$

where r is the distance from the charge to the point where we measure the field, E.

The second term, arising from the action of magnetic forces, is more difficult to calculate. First of all, we note that only charges in motion exert magnetic forces. The electrostatic field is symmetrical about the charge in a coordinate system moving with the charge. Transforming this moving electrostatic field to coordinates at rest with respect to the observer introduces distortions, which we shall interpret as a magnetic field.

Suppose that the charge moves parallel to the axis of x, that its velocity

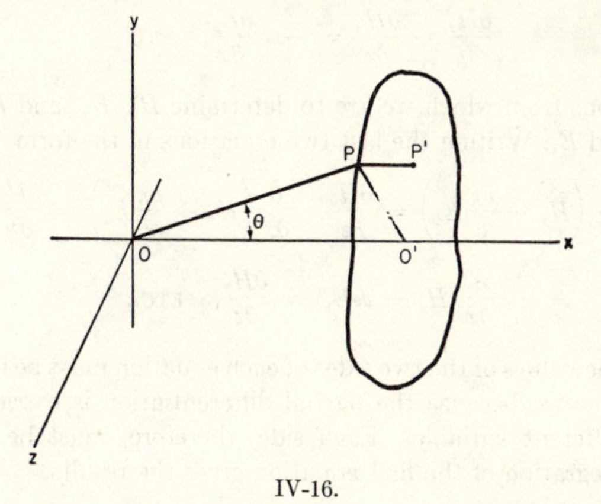

IV-16.

is uniform, and that at time t its position coincides with the origin. We are required to calculate the magnetic force produced by the motion at some point P, at distance r from the origin (Fig. 16). At time $t + dt$ the charge will have moved to a position O′, and its x-coordinate will be $v\,dt$. From P draw the vector PP′ parallel to and equal to OO′. The electric field about the charge follows the charge in its motion. Hence the field at P' at time $t + dt$ is equal to the field at P at time t. Since changes of the intensity of the electric field at any point are propagated with the velocity of light, we suppose that the speed of the moving charge is much less than that of light. If X now denotes the electric intensity at P, we have

$$\frac{\partial E_x}{\partial t} = -v\frac{\partial E_x}{\partial x}. \tag{2}$$

This result follows from the equation of continuity (14.12), if we write E_x for ρ. It follows more directly from the fact that we obtain the same change in field in an instantaneous traversal of a distance $-dx$, as we do by remaining stationary for a time dt, during which interval the field attached to the electron advances through a distance $dx = v\,dt$. We also have the equations

$$\frac{\partial E_y}{\partial t} = -v\frac{\partial E_y}{\partial x}, \quad \frac{\partial E_z}{\partial t} = -v\frac{\partial E_z}{\partial x}. \tag{3}$$

We may substitute these values into the Maxwell equations (15.5), whence

$$\frac{\partial H_z}{\partial y} - \frac{\partial H_y}{\partial z} = -\frac{v\kappa}{c}\frac{\partial E_x}{\partial x}, \quad \frac{\partial H_x}{\partial z} - \frac{\partial H_z}{\partial x} = -\frac{v\kappa}{c}\frac{\partial E_y}{\partial x}, \quad \frac{\partial H_y}{\partial x} - \frac{\partial H_x}{\partial y} = -\frac{v\kappa}{c}\frac{\partial E_z}{\partial x}, \tag{4}$$

$$\frac{\partial H_z}{\partial y} - \frac{\partial H_y}{\partial z} = -v\epsilon\frac{\partial E_x}{\partial x}, \quad \text{ETC.}, \tag{4a}$$

three equations from which we are to determine H_x, H_y, and H_z, in terms of E_x, E_y, and E_z. Writing the last two equations in the form

$$\frac{\partial}{\partial x}\left(H_z - \frac{v\kappa}{c}E_y\right) = \frac{\partial H_x}{\partial z}, \quad \frac{\partial}{\partial x}\left(H_y + \frac{v\kappa}{c}E_z\right) = \frac{\partial H_x}{\partial y}, \tag{5}$$

$$\frac{\partial}{\partial x}(H_z - v\epsilon E_y) = \frac{\partial H_x}{\partial z}, \quad \text{ETC.}, \tag{5a}$$

we see that the values of the two sides of each equation must be independent of the coordinates, because the partial differentiation is carried out with respect to different variables. Each side, therefore, must be equal to a constant. Integration of the first equation gives the results:

$$H_z - \frac{v\kappa}{c}E_y = C_1x + C_2, \quad H_x = C_1z + C_3, \tag{6}$$

$$H_z - v\epsilon E_y = C_1x + C_2, \quad \text{ETC.}, \tag{6a}$$

where the "constant" of integration, C_2, is independent of x but may still be a function of y and z. Similarly, C_3 may be a function of x and y. We know, however, that the magnetic field, and hence H_x, H_y, and H_z, must vanish when $v = 0$, whatever may be the coordinates of the charge. Therefore all the integration "constants" must disappear; and we have the result:

$$H_x = 0, \quad H_y = -v\kappa E_z/c, \quad H_z = v\kappa E_y/c. \tag{7}$$

$$H_x = 0, \quad H_y = -v\epsilon E_z, \quad H_z = v\epsilon E_y. \tag{7a}$$

From (15.4),

$$\nabla \cdot \mathbf{E} = 0,$$

for the region devoid of charge, we prove that the solution also satisfies equation (4a).

The components of electric intensity at P, (coordinates x, y, z), produced by a point charge, result when we multiply the normal component $q/\kappa r^2$, by the respective direction cosines; thus

$$E_x = \frac{q}{\kappa r^2}\frac{x}{r}, \quad E_y = \frac{q}{\kappa r^2}\frac{y}{r},$$

$$E_x = \frac{q}{4\pi\epsilon r^2}\frac{x}{r}, \quad \text{ETC.} \tag{a}$$

Hence

$$H_x = 0, \quad H_y = -vqz/cr^3, \quad H_z = vqy/cr^3. \tag{8}$$

$$H_x = 0, \quad H_y = -vqz/4\pi r^3, \quad H_z = vqy/4\pi r^3. \tag{8a}$$

In vector form, the magnetic field at P is

$$\mathbf{H} = \frac{v}{c}\frac{q}{r^2}\left(-\mathbf{j}\frac{z}{r} + \mathbf{k}\frac{y}{r}\right). \tag{9}$$

$$\mathbf{H} = \frac{vq}{4\pi r^2}\left(-\mathbf{j}\frac{z}{r} + \mathbf{k}\frac{y}{r}\right). \tag{9a}$$

There is no x-component of magnetic force in the present example. The absolute magnitude of $\mathbf{H}$ is

$$H = |\mathbf{H}| = \frac{v}{c}\frac{q}{r^2}\left(\frac{y^2 + z^2}{r^2}\right)^{1/2} = \frac{v}{c}\frac{q}{r^2}\sin\theta, \tag{10}$$

$$H = |\mathbf{H}| = \frac{vq}{4\pi r^2}\left(\frac{y^2 + z^2}{r^2}\right)^{1/2} = \frac{vq}{4\pi r^2}\sin\theta, \tag{10a}$$

where θ is the angle between r and the axis of x. $(y^2 + z^2)^{1/2}$ is the radius

PQ of the circle drawn in Fig. 17, centered about the x-axis and normal to it. Equation (10) is essentially equivalent to the Biot-Savart law (10.15), the application of which is now justified for open circuits.

We determine the direction in which the force acts most readily from consideration of the point $z = 0$, where

$$\mathbf{H} = \mathbf{k}y/r. \tag{11}$$

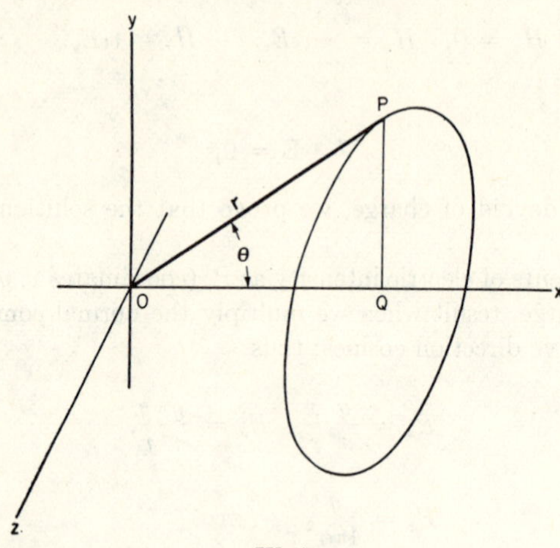

IV-17.

The force is normal to the xy-plane, i.e., to the plane defined by the direction of the charge's motion (the vector $\mathbf{v}$) and the point P (the vector $\mathbf{r}$).

Thus far we have calculated the magnetic field intensity at P produced by the charge in motion. The force upon a positive magnetic charge of strength $\bar{p}$ is, by (1.3a),

$$\mathbf{F} = \mathbf{B}\bar{p} = \mu\mathbf{H}\bar{p} = \mu\frac{vq}{c}\frac{\bar{p}}{r^2}\left(-\mathbf{j}\frac{z}{r} + \mathbf{k}\frac{y}{r}\right). \tag{12}$$

$$\mathbf{F} = \frac{\mu vq\bar{p}}{4\pi r^2}\left(-\mathbf{j}\frac{z}{r} + \mathbf{k}\frac{y}{r}\right). \tag{12a}$$

As a result of the presence, at P, of the magnetic pole, the moving charge at the origin will experience an equal and opposite force,

$$\mathbf{F}_m = \frac{vq}{c}\,\mu H'\left(\mathbf{j}\frac{z}{r} - \mathbf{k}\frac{y}{r}\right), \tag{13}$$

$$\mathbf{F}_m = \mu vqH'\left(\mathbf{j}\frac{z}{r} - \mathbf{k}\frac{y}{r}\right), \tag{13a}$$

since

$$H' = \bar{p}/r^2, \tag{14}$$

$$H' = \bar{p}/4\pi r^2, \tag{14a}$$

the intensity of the magnetic field at the origin resulting from the pole at P. In (13) z and y refer to the coordinates of the pole, P. As in (10),

$$| \mathbf{F}_m | = \frac{vq}{c} \mu H' \sin \theta = \frac{vq}{c} B' \sin \theta, \tag{15}$$

$$| F_m | = \mu vqH' \sin \theta = vqB' \sin \theta, \tag{15a}$$

where B' is given by (14.22).

Equation (13) is, of course, not the most general equation, since the vector velocity will not necessarily be parallel to the x-axis. We may remove this restriction, however.

The vector, $\mathbf{F}_m$ is perpendicular to the plane defined by the vectors $\mathbf{v}$ and $\mathbf{H}$, as we prove by setting $z = 0$ in (13). Furthermore, the magnitude of the resultant, as in (15), contains the factor $\sin \theta$. The procedure of evaluating $\mathbf{F}_m$ corresponds to that of taking the vector product, as defined in (II-22). Hence

$$\mathbf{F}_m = \frac{q}{c} \mathbf{v} \times \mathbf{B}, \tag{16}$$

$$\mathbf{F}_m = q\mathbf{v} \times \mathbf{E}, \tag{16a}$$

or, since we may regard $q\mathbf{v}$ as the current vector $\mathbf{I}$,

$$\mathbf{F}_m = \frac{I}{c} d\mathbf{s} \times \mathbf{B}. \tag{17}$$

$$\mathbf{F}_m = I\, d\mathbf{s} \times \mathbf{B}. \tag{17a}$$

That $\mathbf{v} \times \mathbf{B}$ and not $\mathbf{B} \times \mathbf{v}$ is required we see from the vector diagram shown in Fig. 17. Since $\mathbf{B}$ represents a repulsive force emanating from P, the vector $\mathbf{B}$ is opposite (anti-parallel) to the vector $\mathbf{r}$, which we suppose to be in the xy-plane ($z = 0$). We may set

$$\mathbf{B} = -\mathbf{i}B_x - \mathbf{j}B_y, \quad \mathbf{v} = \mathbf{i}v, \tag{18}$$

where B_x and B_y are positive magnitudes. The vector product,

$$\mathbf{v} \times \mathbf{B} = -\mathbf{k}vB_y, \tag{19}$$

is compatible with the sign of (13), with $z = 0$. Had we adopted $\mathbf{B} \times \mathbf{v}$, the product would have had reversed sign.

We note that, for free space,

$$\mathbf{B} = \mathbf{H}. \tag{20}$$

$$\mathbf{B} = \mu_0 \mathbf{H}. \tag{20a}$$

If $\mathbf{E}$ is the electric intensity at the origin, the complete force vector becomes

$$\mathbf{F} = \mathbf{F}_e + \mathbf{F}_m = q\left(\mathbf{E} + \frac{1}{c} \mathbf{v} \times \mathbf{B}\right). \tag{21}$$

$$\mathbf{F} = q(\mathbf{E} + \mathbf{v} \times \mathbf{B}). \tag{21a}$$

Equation (19) gives the true force; to derive the force per unit charge, we must divide by q. In some problems, we may replace q by ρ, the density of electric charge, to obtain the force acting on a unit volume.

The foregoing equations are fundamental and apply to many physical problems. Our derivations may imply that the velocities of the charges must be small compared with that of light. We have assumed that the forces exerted at P at a given instant by the charge are determined by the instantaneous position of the charge on the x-axis. If the velocities are high, this assumption will no longer be true, for the force fields are propagated with the velocity of light and not with infinite velocity. To allow for high velocities we should have to re-derive the equations on the basis of restricted relativity theory. However, a detailed analysis shows that the field equations, given above, are invariant irrespective of the velocity. Nevertheless we shall have to make use of the fact that the electric potential at P is not determined by the instantaneous position of the charge, by introducing the so-called "retarded potential."

17. The energy of an electromagnetic field. We have seen, equation (4.9), that the electrostatic field between two parallel capacitor plates is constant and of magnitude

$$E = 4\pi Q/\kappa A, \tag{1}$$

$$E = Q/\epsilon A, \tag{1a}$$

along the normal to the plates. The work done by a charge dQ that moves from one plate to the other, then, is simply

$$dW = Ea\, dQ, \tag{2}$$

where a is the distance between the plates. The charge may move either through the intervening medium, or, by means of a battery, along the leads. An increment of charge, dQ, changes the electric intensity by an amount

$$dE = \frac{4\pi}{\kappa A}\, dQ. \tag{3}$$

$$dE = dQ/A\epsilon. \tag{3a}$$

Hence the total work done, as the field changes from an initial value of zero to some value E, is

$$W = \frac{\kappa}{4\pi}\, aA \int_0^E E\, dE = \frac{\kappa}{8\pi}\, aAE^2. \tag{4}$$

$$W = A\epsilon a \int_0^E E\, dE = \frac{A\epsilon a}{2}\, E^2. \tag{4a}$$

W represents the potential energy stored in the capacitor. Experiment shows that the energy is actually distributed through the volume between the capacitor plates. The volume is aA; hence the electric energy per unit volume is

$$\kappa E^2/8\pi. \tag{5}$$

$$\epsilon E^2/2. \tag{5a}$$

The total electric energy, in general, when the vector $\mathbf{E}$ is not necessarily always parallel to one axis, is

$$V = \frac{\kappa}{8\pi}\int \mathbf{E}\cdot\mathbf{E}\,d\tau = \frac{\kappa}{8\pi}\int (E_x^2 + E_y^2 + E_z^2)\,d\tau, \tag{6}$$

$$V = \frac{\epsilon}{2}\int \mathbf{E}\cdot\mathbf{E}\,d\tau, \tag{6a}$$

where $d\tau$ is an element of volume.

The magnetic energy per unit volume, analogously, proves to be

$$\mu H^2/8\pi,$$

$$\mu H^2/2, \tag{a}$$

and the total magnetic energy, T,

$$T = \frac{\mu}{8\pi}\int \mathbf{H}\cdot\mathbf{H}\,d\tau = \frac{\mu}{8\pi}\int (H_x^2 + H_y^2 + H_z^2)\,d\tau. \tag{7}$$

$$T = \frac{\mu}{2}\int \mathbf{H}\cdot\mathbf{H}\,d\tau. \tag{7a}$$

We determine the total energy of the medium by integrating the sum of the two expressions over the volume.

18. The Poynting vector. The work done by electromagnetic forces that displace a unit charge of electricity through a distance $d\mathbf{s}$ becomes

$$dW = \mathbf{F}\cdot d\mathbf{s}. \tag{1}$$

If the displacement occurs in time dt, the rate of doing work is

$$\frac{dW}{dt} = \mathbf{F}\cdot\frac{d\mathbf{s}}{dt} = \mathbf{F}\cdot\mathbf{v}, \tag{2}$$

where $\mathbf{v}$ is the velocity of the charge. If ρ is the volume density of the electric charge, we have, from (16.21), the force per unit volume:

$$\mathbf{F} = \rho\left(\mathbf{E} + \frac{1}{c}\mathbf{v}\times\mathbf{B}\right). \tag{3}$$

$$\mathbf{F} = \rho[\mathbf{E} + \mathbf{v}\times\mathbf{B}]. \tag{3a}$$

Then the total work done in time dt over a given volume, results from an integration over the volume:

$$\frac{dW}{dt} = \int \rho\left(\mathbf{E} \cdot \mathbf{v} + \frac{1}{c}\,\mathbf{v} \times \mathbf{B} \cdot \mathbf{v}\right) d\tau. \tag{4}$$

$$\frac{dW}{dt} = \int \rho(\mathbf{E} \cdot \mathbf{v} + \mathbf{v} \times \mathbf{B} \cdot \mathbf{v})\, d\tau. \tag{4a}$$

By the definition of the vector product, the vector $\mathbf{v} \times \mathbf{B}$ is perpendicular to $\mathbf{v}$. The scalar product of one vector and another perpendicular to it is zero. Hence the second term of (4) vanishes. The magnetic forces do no work because the motion is always at right angles to the field. Since

$$\rho\mathbf{v} = \mathbf{J}, \tag{5}$$

the current density,

$$\frac{dW}{dt} = \int \mathbf{E} \cdot \mathbf{J}\, d\tau = \frac{c}{4\pi}\int \left(\mathbf{E} \cdot \nabla \times \mathbf{H} - \frac{1}{c}\,\mathbf{E} \cdot \frac{\partial \mathbf{D}}{\partial t}\right) d\tau, \tag{6}$$

$$\frac{dW}{dt} = \int \left(\mathbf{E} \cdot \nabla \times \mathbf{H} - \mathbf{E} \cdot \frac{\partial \mathbf{D}}{\partial t}\right) d\tau, \tag{6a}$$

where we have used one of Maxwell's equations, (14.9). Let

$$\mathbf{E} = \mathbf{i}E_x + \mathbf{j}E_y + \mathbf{k}E_z$$

and

$$\mathbf{H} = \mathbf{i}H_x + \mathbf{j}H_y + \mathbf{k}H_z, \tag{7}$$

as before. Then, by performing the elementary vector multiplication, we obtain the identity

$$\mathbf{E} \cdot \nabla \times \mathbf{H}$$
$$= E_x\left(\frac{\partial H_z}{\partial y} - \frac{\partial H_y}{\partial z}\right) + E_y\left(\frac{\partial H_x}{\partial z} - \frac{\partial H_z}{\partial x}\right) + E_z\left(\frac{\partial H_y}{\partial x} - \frac{\partial H_x}{\partial y}\right). \tag{8}$$

At this point we shall make a simple vector transformation. Analogous to (8), we have

$$\mathbf{H} \cdot \nabla \times \mathbf{E}$$
$$= H_x\left(\frac{\partial E_z}{\partial y} - \frac{\partial E_y}{\partial z}\right) + H_y\left(\frac{\partial E_x}{\partial z} - \frac{\partial E_z}{\partial x}\right) + H_z\left(\frac{\partial E_y}{\partial x} - \frac{\partial E_x}{\partial y}\right). \tag{9}$$

Now consider the vector $\mathbf{N}$, defined by

$$\frac{4\pi}{c}\,\mathbf{N} = \mathbf{E} \times \mathbf{H}$$
$$= \mathbf{i}(E_yH_z - E_zH_y) + \mathbf{j}(E_zH_x - E_xH_z) + \mathbf{k}(E_xH_y - E_yH_x), \tag{10}$$

$$\mathbf{N} = \mathbf{E} \times \mathbf{H} = \ldots, \tag{10a}$$

and its divergence,

$$\frac{4\pi}{c}\nabla\cdot\mathbf{N} = -\left[E_x\left(\frac{\partial H_z}{\partial y} - \frac{\partial H_y}{\partial z}\right) + E_y\left(\frac{\partial H_x}{\partial z} - \frac{\partial H_z}{\partial x}\right) + E_z\left(\frac{\partial H_y}{\partial x} - \frac{\partial H_x}{\partial y}\right)\right]$$
$$+\left[H_x\left(\frac{\partial E_z}{\partial y} - \frac{\partial E_y}{\partial z}\right) + H_y\left(\frac{\partial E_x}{\partial z} - \frac{\partial E_z}{\partial x}\right) + H_z\left(\frac{\partial E_y}{\partial x} - \frac{\partial E_x}{\partial y}\right)\right.$$
$$= -\mathbf{E}\cdot\nabla\times\mathbf{H} + \mathbf{H}\cdot\nabla\times\mathbf{E}. \tag{11}$$

$$\nabla\cdot\mathbf{N} = -\mathbf{E}\cdot\nabla\times\mathbf{H} + \mathbf{H}\cdot\nabla\times\mathbf{E}. \tag{11a}$$

Replacing $\mathbf{E}\cdot\nabla\times\mathbf{H}$ by its equivalent from (11), we obtain instead of (9),

$$\frac{dW}{dt} = \frac{c}{4\pi}\int\left(\mathbf{H}\cdot\nabla\times\mathbf{E} - \frac{1}{c}\mathbf{E}\cdot\frac{\partial\mathbf{D}}{\partial t} - \frac{4\pi}{c}\nabla\cdot\mathbf{N}\right)d\tau$$
$$= -\frac{1}{4\pi}\int\left(\mathbf{H}\cdot\frac{\partial\mathbf{B}}{\partial t} + \mathbf{E}\cdot\frac{\partial\mathbf{D}}{\partial t}\right)d\tau - \int\nabla\cdot\mathbf{N}\,d\tau, \tag{12}$$

$$\frac{dW}{dt} = \int\left(\mathbf{H}\cdot\nabla\times\mathbf{E} - \mathbf{E}\cdot\frac{\partial\mathbf{D}}{\partial t} - \nabla\cdot\mathbf{N}\right)d\tau$$
$$= -\int\left(\mathbf{H}\cdot\frac{\partial\mathbf{B}}{\partial t} + \mathbf{E}\cdot\frac{\partial\mathbf{D}}{\partial t}\right)d\tau - \int\nabla\cdot\mathbf{N}\,d\tau, \tag{12a}$$

by Maxwell's fourth equation, (14.21). But

$$\mathbf{E} = \mu\mathbf{H}, \qquad \mathbf{D} = \kappa\mathbf{E}, \tag{13}$$

$$\mathbf{B} = \mu\mathbf{H}, \qquad \mathbf{D} = \epsilon\mathbf{E}, \tag{13a}$$

by (14.7) and (14.22).

Then, if κ, (ϵ), and μ are independent of the time and coordinates,

$$\frac{\kappa}{4\pi}\int\mathbf{E}\cdot\frac{\partial\mathbf{E}}{\partial t}\,d\tau = \frac{\kappa}{4\pi}\int\frac{1}{2}\frac{\partial}{\partial t}(\mathbf{E}\cdot\mathbf{E})\,d\tau = \frac{\kappa}{8\pi}\frac{d}{dt}\int E^2\,d\tau, \tag{14}$$

$$\epsilon\int\mathbf{E}\cdot\frac{\partial\mathbf{E}}{\partial t}\,d\tau = \epsilon\int\frac{1}{2}\frac{\partial}{\partial t}(\mathbf{E}\cdot\mathbf{E})\,d\tau = \frac{\epsilon}{2}\frac{d}{dt}\int E^2\,d\tau, \tag{14a}$$

since the scalar product of any vector by itself is equal to the square of the absolute value of the vector. A similar relation holds for **H**. By Gauss' theorem (II-14.5) or (II-23.19), we transform the second volume integral into one over the boundary surface; thus

$$\int\nabla\cdot\mathbf{N}\,d\tau = \int\mathbf{N}\cdot d\mathbf{S}, \tag{15}$$

Thus, finally,

$$\frac{dW}{dt} = -\frac{d}{dt}\int\frac{1}{8\pi}(\kappa E^2 + \mu H^2)\,d\tau - \int\mathbf{N}\cdot d\mathbf{S}. \tag{16}$$

$$\frac{dW}{dt} = -\frac{d}{dt}\int\frac{1}{2}(\epsilon E^2 + \mu H^2)\,d\tau - \int\mathbf{N}\cdot d\mathbf{S}. \tag{16a}$$

We are now in a position to interpret the result physically. The sum of the two integrals represents the rate at which the electromagnetic forces do work on the charges. The first term, we note by comparison with (17.9), we may interpret as the rate of change of energy stored in the medium. When the second term is not zero, we see that it represents a flux over the boundary surface. We must therefore identify this term with radiation escaping from the volume. The vector

$$\mathbf{N} = \frac{c}{4\pi} \mathbf{E} \times \mathbf{H}, \tag{17}$$

$$\mathbf{N} = \mathbf{E} \times \mathbf{H}, \tag{17a}$$

which is perpendicular to both **E** and **H** by definition of the vector product, is known as the Poynting vector. Its magnitude and direction at any point of space determine the flow of electromagnetic energy at the point.

19. Propagation of a plane electromagnetic wave. The equations,

$$E_x = 0, \quad E_y = B \cos\left[\frac{2\pi}{\lambda}(x - vt)\right], \quad E_z = 0, \tag{1}$$

represent a continuous wave, of electric amplitude B and wavelength λ, progressing with velocity v in the direction of the axis of x (Fig. 18). We

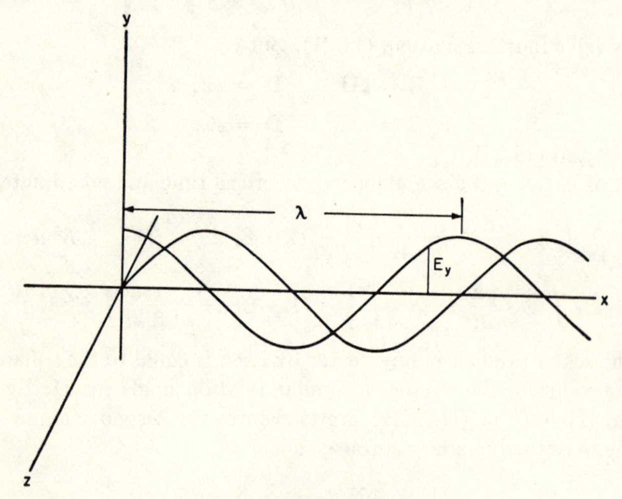

IV-18.

shall suppose that E_y, at any given x-coordinate, represents the electric intensity of the electromagnetic wave. We are to calculate the direction and magnitude of the magnetic wave that must be associated with an electric field of varying intensity. (Do not confuse B with **B**.) We have

$$\frac{\partial E_y}{\partial x} = -\frac{2\pi}{\lambda} B \sin \frac{2\pi}{\lambda}(x - vt), \tag{2}$$

all the other partial derivatives being zero. Substituting these results in the Maxwell equations for free space (15.6), we obtain the conditions

$$-\frac{\mu}{c}\frac{\partial H_x}{\partial t} = 0, \quad -\frac{\mu}{c}\frac{\partial H_y}{\partial t} = 0,$$
$$-\frac{\mu}{c}\frac{\partial H_z}{\partial t} = \frac{\partial E_y}{\partial x} = -\frac{2\pi}{\lambda} B \sin\frac{2\pi}{\lambda}(x - vt). \tag{3}$$

$$-\mu\frac{\partial H_x}{\partial t} = 0, \quad -\mu\frac{\partial H_y}{\partial t} = 0,$$
$$-\mu\frac{\partial H_z}{\partial t} = \frac{\partial E_y}{\partial t} = -\frac{2\pi}{\lambda} B \sin\frac{2\pi}{\lambda}(x - vt). \tag{3a}$$

The integrals of these equations are

$$H_x = H_y = 0, \quad H_z = \frac{c}{\mu v} B \cos\frac{2\pi}{\lambda}(x - vt), \tag{4}$$

$$H_z = \frac{1}{\mu v} B \cos\frac{2\pi}{\lambda}(x - vt) = \frac{\sqrt{\epsilon}}{\sqrt{\mu}} B \cos\frac{2\pi}{\lambda}(x - vt)$$
$$= \epsilon v B \cos\frac{2\pi}{\lambda}(x - vt), \tag{4a}$$

the constants of integration vanishing because the magnetic field must go to zero when $B \to 0$.

Since H_z is the z-component of the magnetic field, we see that the electric

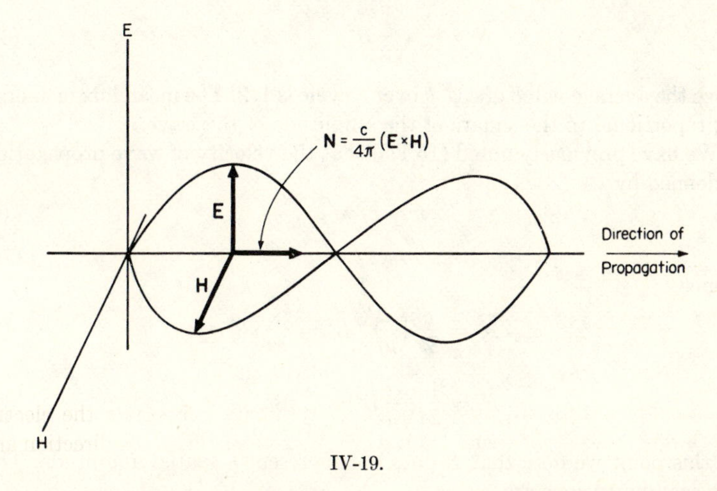

IV-19.

and magnetic fields are perpendicular to one another. They are of the same phase. Both go to zero simultaneously. A representation of a plane wave of this variety appears in Fig. 19. Here E_y and H_z comprise, respectively, the electric and magnetic vectors. A wave of this variety is said to be "plane polarized," but since the xy- and the xz-planes are equally involved, we shall have to define polarization in some arbitrary manner. The older convention of physical optics specified polarization in terms of the plane of the magnetic vector. There is an increasing tendency among physicists, however, to choose the electric vector as defining the fundamental plane. We shall adopt the latter procedure in this book.

The Poynting vector, **N**, which represents the direction of propagation and the magnitude of the energy transferred across a given plane, lies along the positive direction of the x-axis. Its magnitude follows from the equation

$$\left.\begin{aligned} \frac{4\pi}{c}\mathbf{N} &= \mathbf{E}\times\mathbf{H} = (E_y H_z)\mathbf{j}\times\mathbf{k} = \mathbf{i}E_y H_z. \\ N &= |\,\mathbf{N}\,| = \frac{c^2}{4\pi\mu v}B^2\cos^2\frac{2\pi}{\lambda}(x - vt). \end{aligned}\right\} \tag{5}$$

$$\left.\begin{aligned} \mathbf{N} &= \mathbf{E}\times\mathbf{H}. \\ N &= |\,\mathbf{N}\,| = \sqrt{\frac{\epsilon}{\mu}}B^2\cos^2\frac{2\pi}{\lambda}(x - vt). \end{aligned}\right\} \tag{5a}$$

The value of the energy transferred across a given plane is, therefore, continually fluctuating. The average value of N during a cycle is

$$\overline{N} = \frac{\int_0^{\lambda/v} N\,dt}{\int_0^{\lambda/v} dt} = \frac{c^2}{8\pi\mu v}B^2, \tag{6}$$

$$\overline{N} = \frac{1}{2}\sqrt{\frac{\epsilon}{\mu}}B^2, \tag{6a}$$

since the average value of $\cos^2\theta$ over a cycle is 1/2. The mean flux of energy is proportional to the square of the amplitude of the wave.

We have previously noted (15.11), that the velocity of wave propagation is defined by

$$v = c/\sqrt{\mu\kappa}. \tag{7}$$

$$v = 1/\sqrt{\epsilon\mu}. \tag{7a}$$

Hence

$$\overline{N} = \frac{c}{8\pi}\left(\frac{\kappa}{\mu}\right)^{1/2}B^2 = \frac{\kappa}{8\pi}vB^2. \tag{8}$$

$$\overline{N} = \frac{1}{2}\sqrt{\frac{\epsilon}{\mu}}B^2 = \frac{\epsilon}{2}vB^2. \tag{8a}$$

At this point we note that B does not represent a spatial amplitude. The physical dimensions are

$$[B] = [\mathrm{flux}/v\kappa]^{1/2} = [\mathrm{M}^{1/2}\mathrm{L}^{-1/2}\mathrm{T}^{-1}\kappa^{-1/2}]. \tag{9}$$

$$[B] = [\mathrm{force}/Qv] = \mathrm{MT}^{-1}\mathrm{Q}^{-1}. \tag{9a}$$

By (I-10.2) the dimensions of (9) are equivalent to those of electric intensity, as equation (1) requires.

20. Radiation pressure. Let us calculate the pressure exerted by a beam of radiation falling normally on one side of a perfectly absorbing slab of unit area. The energy per unit volume on the side of the surface from which the beam is coming is, by equations (17.6) and (17.7),

$$W = \frac{1}{8\pi}(\kappa E^2 + \mu H^2). \tag{1}$$

$$W = \frac{1}{2}(\epsilon E^2 + \mu H^2). \tag{1a}$$

Let us consider the field in the neighborhood of a hollow, electrically

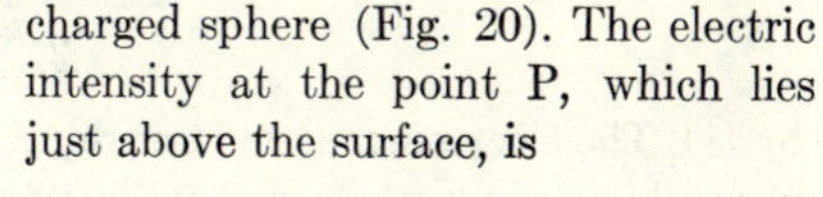

charged sphere (Fig. 20). The electric intensity at the point P, which lies just above the surface, is

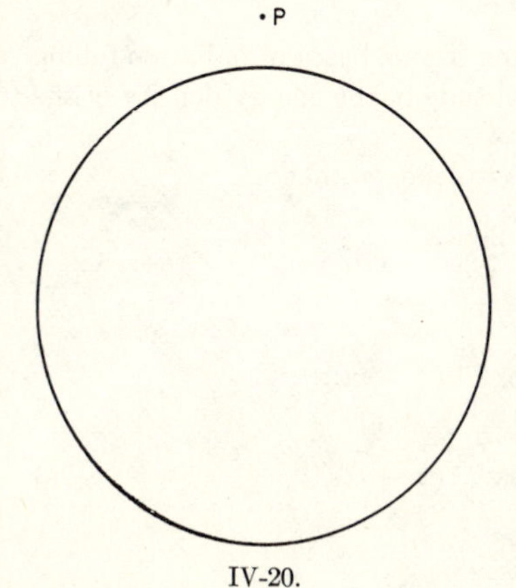

IV-20.

$$E = 4\pi\sigma/\kappa, \tag{2}$$

$$E = \sigma/\epsilon, \tag{2a}$$

where σ is the surface density of the electric charge, equation (4.3). We shall consider the electric intensity at P to consist of two parts, E_1, that arising from charges in the immediate vicinity of P, and E_2, that from all other charges, so that $E = E_1 + E_2$. Now, when P approaches indefinitely close to the surface of the sphere, the charges in the immediate neighborhood of P act as if they were distributed over an infinite plane and therefore produce an electric intensity,

$$E_1 = 2\pi\sigma/\kappa = E/2 = E_2, \tag{3}$$

$$E_1 = \sigma/2\epsilon = E/2 = E_2, \tag{3a}$$

equation (4.8) for a plane capacitor. The force acting upon unit area, resulting from charges in the vicinity of P is therefore

$$F = E_1\sigma = 2\pi\sigma^2/\kappa = \kappa E^2/8\pi. \tag{4}$$

$$F = E_1\sigma = \sigma^2/2\epsilon = \epsilon E^2/2. \tag{4a}$$

Equation (2), an expression of Gauss' law, holds whatever may be the form of the surface. The work done when a unit area moves through a unit distance in a direction parallel to the force vector is

$$V = \int_0^1 F\,ds = \frac{\kappa E^2}{8\pi}, \tag{5}$$

$$V = \int_0^1 F\,ds = \frac{\epsilon E^2}{2}, \tag{5a}$$

which agrees with equation (17.5).

If the electric shell is replaced by a magnetic shell, we may reproduce the equivalent of (17.7). We see then that the total electromagnetic force per unit area is

$$F = \frac{\kappa E^2 + \mu H^2}{8\pi} = W, \tag{6}$$

$$F = \frac{\epsilon E^2 + \mu H^2}{2} = W, \tag{6a}$$

by (1). The electromagnetic force resulting from a beam of radiation falling normally upon a surface is equal in magnitude to the energy density of the radiation.

If the beam is plane polarized in the xz-plane, so that

$$E = E_y = B\cos\frac{2\pi}{\lambda}(x - vt), \tag{7}$$

$$H = H_z = \frac{c}{\mu v}B\cos\frac{2\pi}{\lambda}(x - vt),$$

$$H = H_z = \frac{1}{\mu v}B\cos\frac{2\pi}{\lambda}(x - vt), \tag{7a}$$

equations (19.1) and (19.4), then

$$W = \frac{\kappa}{8\pi}B^2\left(1 + \frac{c^2}{\mu\kappa v}\right)\cos^2\frac{2\pi}{\lambda}(x - vt) = \frac{\kappa}{4\pi}B^2\cos^2\frac{2\pi}{\lambda}(x - vt), \tag{8}$$

$$W = \frac{1}{2}B^2\left(\epsilon + \frac{1}{2\mu v^2}\right)\cos^2\frac{2\pi}{\lambda}(x - vt) = \epsilon B^2\cos^2\frac{2\pi}{\lambda}(x - vt), \tag{8a}$$

by (19.7). The energy W is continually fluctuating, but the variations are so rapid that we observe only the average value, $\overline{E^2}$. The $\cos^2$ term, averaged over a cycle, is equal to $1/2$, as determined in the previous section. Hence the mean values of the density of electromagnetic energy and of the forces, are

$$\overline{W} = \frac{\kappa\overline{E^2}}{4\pi} = \frac{\kappa B^2}{8\pi} = \overline{F}. \tag{9}$$

$$\overline{W} = \epsilon\overline{E^2} = \frac{\epsilon B^2}{2} = \overline{F}. \tag{9a}$$

We shall now derive, by induction, the more general forms of the equations (6) to (9), when the radiation is moving in random directions, by considering the summed effect of all the rectangular components.

When the radiation was flowing along the x-axis and polarized in the xy plane, we found that the Poynting vector was $\mathbf{N} = \mathbf{i}E_y H_z$. If the radiation had been polarized in some other plane, but still traveling in the same direction, we should have had two component vectors instead of one, thus:

$$\mathbf{N} = \mathbf{i}(E_y^x H_z^x + E_z^x H_y^x),$$

wherein the superscript represents the *direction* of flow of the given component, as indicated by the subscript. In general, we should have to write

$$\mathbf{N} = \mathbf{i}(E_y^x H_z^x + E_z^x H_y^x) + \mathbf{j}(E_z^y H_x^y + E_x^y H_z^y) + \mathbf{k}(E_x^z H_y^z + E_y^z H_x^z), \tag{10}$$

a vector that we break up into six superposed components, since E_y^x may not be equivalent to E_y^z, etc. The six components arise because we have three independent directions, with two polarization planes for each direction.

We must write the general expression for the density of radiation flowing in random directions as follows:

$$W = \kappa \frac{(E_x^y)^2 + (E_x^z)^2 + (E_y^x)^2 + (E_y^z)^2 + (E_z^x)^2 + (E_z^y)^2}{8\pi} + \mu \frac{(H_x^y)^2 + (H_x^z)^2 + (H_y^x)^2 + (H_y^z)^2 + (H_y^x)^2 + (H_z^y)^2}{8\pi}$$

$$= \frac{\kappa E^2 + \mu H^2}{8\pi} \tag{11}$$

$$W = \frac{\epsilon E^2 + \mu H^2}{2}. \tag{11a}$$

The force per area normal to any one of the directions, say that of the axis of x, will be

$$F = \kappa \frac{(E_y^x)^2 + (E_z^x)^2}{8\pi} + \mu \frac{(H_y^x)^2 + (H_z^x)^2}{8\pi} \tag{12}$$

$$F = \epsilon \frac{(E_y^x)^2 + (E_z^x)^2}{2} + \mu \frac{(H_y^x)^2 + (H_z^x)^2}{2} \tag{12a}$$

We can say nothing regarding the instantaneous values of the various components. But, if the medium is homogeneous and isotropic and if the beam is unpolarized, the average values of the various components must be very nearly the same, i.e.,

$$\begin{aligned} (E_x^y)^2 &= (E_x^z)^2 = (E_y^x)^2 = \cdots = \frac{E^2}{6}, \\ (H_x^y)^2 &= (H_x^z)^2 = (H_y^x)^2 = \cdots = \frac{H^2}{6}, \end{aligned} \tag{13}$$

from (1). Hence

$$F = W/3, \tag{14}$$

the force of radiation pressure is equal to one-third of the energy density. For each of the components the time variation follows an equation of the form of (7),

$$E_{xy} = A_y \cos \frac{2\pi}{\lambda}(x - vt), \quad H_{xz} = \frac{c}{\mu v} A_y \cos \frac{2\pi}{\lambda}(x - vt), \quad \text{ETC.} \tag{15}$$

$$H_{xz} = \frac{1}{\mu v} A_y \cos \frac{2\pi}{\lambda}(x - vt), \quad \text{ETC.} \tag{15a}$$

The average value of W, obtained as before, is

$$W = \kappa \frac{A_y^2 + A_z^2 + B_z^2 + B_z^2 + C_z^2 + C_y^2}{8\pi}. \tag{16}$$

$$W = \epsilon \frac{A_y^2 + A_z^2 + B_z^2 + B_y^2 + C_z^2 + C_y^2}{2}. \tag{16a}$$

But we may regard the sum

$$A_y^2 + A_z^2 = A^2, \tag{17}$$

where A is the amplitude of a vector defining the actual magnitude and instantaneous direction of the polarization along the x-axis. Similar expressions hold for B and C. Likewise, we write

$$E_0^2 = A^2 + B^2 + C^2, \tag{18}$$

where E_0 denotes the amplitude of the actual wave. Then

$$\overline{F} = \overline{W}/3, \tag{19}$$

$$\overline{W} = \kappa E_0^2/8\pi. \tag{20}$$

$$\overline{W} = \epsilon E_0^2/2. \tag{20a}$$

One should note at this point that when the medium is not isotropic, as, e.g., in the case of a crystal of Iceland spar, the various components of (13) will differ. A tensor replaces the vector of (10) and the phenomenon of double refraction results. We shall not concern ourselves with these complexities, however.

21. Momentum of radiation. The pressure of a beam of radiation incident normally upon a surface is

$$\overline{F} = \overline{W} = \kappa B^2/8\pi, \tag{1}$$

$$\overline{F} = \overline{W} = \epsilon B^2/2, \tag{1a}$$

equation (20.9). This equation represents the momentum received per second by the surface, because we define force as the rate of change of momentum. This momentum must be associated with the radiation itself.

Energy incident per unit time on the surface is given by the Poynting flux, (19.8):

$$\overline{N} = \kappa v B^2/8\pi. \tag{2}$$

$$\overline{N} = \epsilon v B^2/2. \tag{2a}$$

Hence we may write

$$F = \overline{N}/v. \tag{3}$$

For problems that involve light transmission in a vacuum, we have

$$\kappa = 1, \qquad v = c. \tag{4}$$

$$\epsilon = \epsilon_0, \qquad v = c. \tag{4a}$$

To determine the radiation pressure, we first calculate the average value of the normal component of the radiation, i.e., the flux through the surface element, and divide by the velocity of the radiation. This procedure gives the radiation pressure on the area when all the energy is absorbed. When a fraction is transmitted by the surface the radiation pressure is diminished. If a fraction is reflected backwards, the radiation pressure is increased. Let f_1 and f_2 denote the respective fractions. Thus the general formula for the radiation pressure is

$$\overline{F} = \overline{N}(1 - f_1 + f_2). \tag{5}$$

For many purposes we may employ different notation. We shall call the upward flux F_+, the downward flux F_-, and let p_r and ρ be the respective radiation pressure and density. Then equation (5) takes the form

$$p_r = (F_+ - F_-)/v. \tag{6}$$

For an enclosure, wherein the radiation is isotropic,

$$p_r = \rho/3. \tag{7}$$

22. Conditions to be satisfied at a boundary surface of a medium. When a light beam crosses a surface of discontinuity, from one medium into another, the various conditions that must be satisfied follow directly from the Maxwell equations. Let us integrate the Maxwell equation (14.9), over

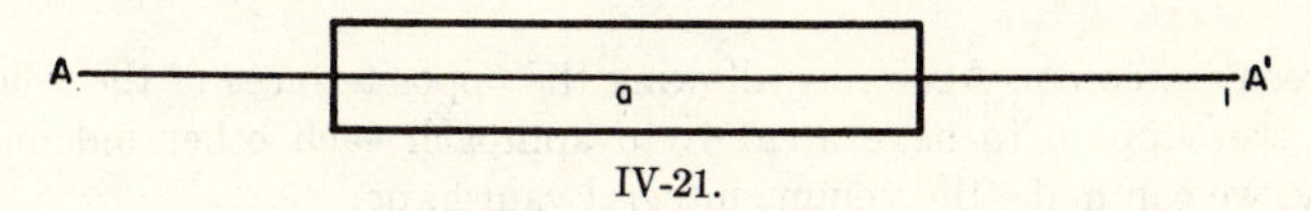

IV-21.

a long rectangular surface like that shown in (a), Fig. 21, set across the boundary, AA′. The condition is

$$\int \nabla \times \mathbf{H} \cdot d\mathbf{S} = \int \mathbf{H} \cdot d\mathbf{s} = \int \left(\frac{1}{c}\frac{\partial \mathbf{D}}{\partial t} + \frac{4\pi \mathbf{J}}{c}\right) \cdot d\mathbf{S}, \tag{1}$$

$$\int \nabla \times \mathbf{H} \cdot d\mathbf{S} = \int \left(\frac{\partial \mathbf{D}}{\partial t} + \mathbf{J}\right) \cdot d\mathbf{S}, \tag{1a}$$

by Stokes' theorem (II-34.8). By allowing the two sides of the rectangle normal to the boundary to decrease indefinitely, but still keeping the two opposite sides in the different media, we may make the surface integral vanish. Whence

$$\int \mathbf{H} \cdot d\mathbf{s} = 0. \tag{2}$$

We may ignore the contribution to this contour integral, of the infinitesimal sides. Let the length of the side tangent to the surface be l, and let the components of $\mathbf{H}$ *tangent* to the surface in the respective media be H_{t_1} and H_{t_2}. Then

$$\int \mathbf{H} \cdot d\mathbf{s} = l(H_{t_1} - H_{t_2}) = 0$$

and

$$H_{t_2} = H_{t_1}. \tag{3}$$

By a similar treatment of the equation (14.21),

$$\int \nabla \times \mathbf{E} \cdot d\mathbf{S} = \int \mathbf{E} \cdot d\mathbf{s} = -\frac{1}{c} \int \frac{\partial \mathbf{B}}{\partial t} \cdot d\mathbf{S} = 0, \tag{4}$$

$$\int \nabla \times \mathbf{E} \cdot d\mathbf{S} = -\int \frac{\partial \mathbf{B}}{\partial t} \quad d\mathbf{S} = 0, \tag{4a}$$

we may show that the tangential components of the electric vector are also continuous on both sides of the medium.

We derive the third boundary condition by integrating the Maxwell equation (14.15) over a volume such as would be produced when the thin rectangle (a) moves normal to its own surface and parallel to the surface of the medium. Then

$$\int \nabla \cdot \mathbf{D} \, d\tau = \int \mathbf{D} \cdot d\mathbf{S} = \int 4\pi\rho \, d\tau, \tag{5}$$

$$\int \nabla \cdot \mathbf{D} \, d\tau = \int \rho \, d\tau, \tag{5a}$$

by Green's theorem. Again, by allowing the opposite faces of the volume, which we suppose to have area A, to approach each other indefinitely closely, we can make the volume integral vanish, or

$$\int \mathbf{D} \cdot d\mathbf{S} = 0. \tag{6}$$

The vector $d\mathbf{S}$ is normal to the surface. Hence, to evaluate the integral we take the respective normal components of $\mathbf{D}$, D_{n_1}, and D_{n_2}, on opposite sides of the media, and write

$$A(D_{n_1} - D_{n_2}) = 0, \qquad D_{n_1} = D_{n_2}, \tag{7}$$

or, by (14.7),

$$\kappa_1 E_{n_1} = \kappa_2 E_{n_2}, \tag{8}$$

$$\epsilon_1 E_{n_1} = \epsilon_2 E_{n_2}, \tag{8a}$$

which is analogous to (2.43). Similarly, from (14.6), we may show that

$$B_{n_1} = B_{n_2} \tag{9}$$

or

$$\mu_1 H_{n_1} = \mu_2 H_{n_2}, \tag{10}$$

by (14.22).

Thus the four boundary conditions require, (a) that the tangential components of the magnetic field intensity (and likewise of the electric intensity) be equal on opposite sides of the boundary surface, and (b) that the normal components of the electric displacement **D** (and also of the magnetic induction **B**) balance.

23. Refraction and reflection of light waves. We have seen that for light traveling along the x-axis, with its polarization (electric vector) in the xy plane, the electric and magnetic vectors follow the relationships

$$\begin{aligned} \mathbf{E}' &= \mathbf{E}'_y = \mathbf{j}' E'_y \cos \frac{2\pi}{\lambda'} (x' - v't), \\ \mathbf{H}' &= \mathbf{H}'_z = \mathbf{k}' H'_z = \mathbf{k}' \frac{c}{\mu' v'} E'_y \cos \frac{2\pi}{\lambda'} (x' - v't). \end{aligned} \tag{1}$$

$$\mathbf{H}' = \mathbf{H}'_z = \mathbf{k}' H'_z = \mathbf{k}' \frac{1}{\mu' v'} E'_y \cos \frac{2\pi}{\lambda'} (x' - v't). \tag{1a}$$

These equations are the vector equivalents of (20.7). We have introduced the primed notation for later convenience. Similarly, light polarized in the xz-plane will be defined by the relationships

$$\begin{aligned} \mathbf{E} &= \mathbf{E}'_z = \mathbf{k}' E'_z \cos \frac{2\pi}{\lambda'} (x' - v't), \\ \mathbf{H} &= H'_y = \mathbf{j}' H'_y = \mathbf{j}' \frac{c}{\mu' v'} E'_z \cos \frac{2\pi}{\lambda'} (x' - v't). \end{aligned} \tag{2}$$

$$\mathbf{H} = \mathbf{j}' \frac{1}{\mu' v'} E'_z \cos \frac{2\pi}{\lambda'} (x' - v't). \tag{2a}$$

Now, when a light ray traveling along the x-axis in one medium meets the boundary of another medium of different optical properties, the light beam splits into two components. Part of the ray is reflected, say, along x'' and part of it is refracted into the second medium, along x'''. We shall suppose that the normal to the boundary surface lies in the $x'y'$-plane. θ_1, θ_2, and θ_3,

(Fig. 22), are then the respective angles of incidence, of refraction, and of reflection.

The experimental results are well known. We have

$$\theta_1 = \theta_2 \tag{3}$$

and

$$\frac{\sin \theta_1}{\sin \theta_3} = \frac{n_3}{n_1} = \text{CONST}, \tag{4}$$

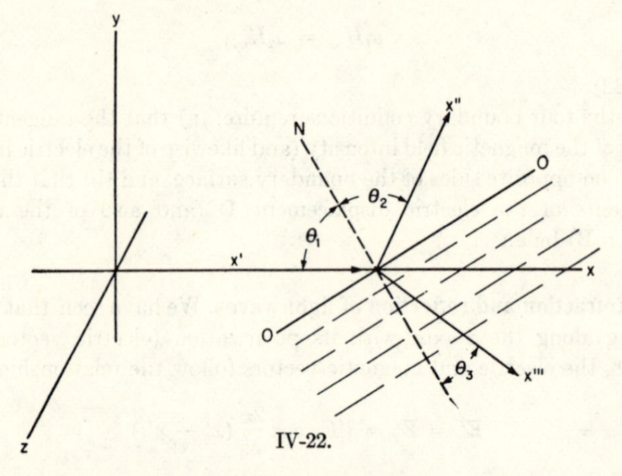

IV-22.

where n_1 and n_3 are the respective refractive indices of the first and second media. We know the latter relation as Snell's law. We wish to see whether the electromagnetic theory can account for the observed phenomena of physical optics, including the quantitative test of predicting the relative intensities of the reflected and refracted beams.

We shall distinguish two cases: (a) when the incident beam is polarized in the $x'y'$-plane (perpendicular to the plane of incidence) equations (1), and (b) when it is polarized in the $x'y'$-plane (in the plane of incidence) equations (2). The two solutions will enable us to derive the general result, since the vectors associated with light polarized at any other angle may be resolved into the two components of the solution here given.

We employ, for sake of convenience, different coordinate systems for each of the three rays, with the x-axis along the respective beams, and with the origin at the point where the rays meet the boundary surface. The following equations represent the electric and magnetic vectors of the reflected beam:

$$\left.\begin{aligned} \mathbf{E}'' &= (\mathbf{j}''E_y + \mathbf{k}''E_z) \cos \frac{2\pi}{\lambda''}(x'' - v''t), \\ \mathbf{H}'' &= (\mathbf{j}''H_y'' + \mathbf{k}''H_z'') \cos \frac{2\pi}{\lambda''}(x'' - v''t) \\ &= \frac{c}{\mu''v''}(\mathbf{j}''E_z'' + \mathbf{k}E_y'') \cos \frac{2\pi}{\lambda''}(x'' - v''t). \end{aligned}\right\} \tag{5}$$

$$\mathbf{H}'' = \frac{1}{\mu''v''}(\mathbf{j}''E_z'' + \mathbf{k}E_y'')\cos\frac{2\pi}{\lambda''}(x'' - v''t). \tag{5a}$$

The equations for the refracted wave, in its own system of coordinates, are identical with those of (5), with triple instead of double primes.

We now reduce the vector equations to a coordinate system relative to the boundary surface. We need no formal proof to show that the axes of x', x'', and x''' all lie in the same plane. The forces acting to deflect the initial beam are those associated with the boundary surface, the orientation of which is specified by the normal. Since the forces acting are normal to the boundary surface, the deflections must take place in the plane defined by the axis of x' and the normal. We have already chosen our coordinate system in accord with this condition.

In our new coordinate system we shall take an axis of y along the normal to the boundary surface and an axis of x along the surface, but in the original $x'y'$-plane. We thus have four coordinate systems, related to one another by simple rotation about the z-axis, which is common to all systems. The unit vector, $\mathbf{k}$, is common to all. Hence

$$\mathbf{k}' = \mathbf{k}'' = \mathbf{k}''' = \mathbf{k}, \tag{6}$$

where the unprimed system refers to the new coordinates. Referring to Fig. 23a, which depicts the relationship of the unit vector $\mathbf{j}'$ to the new coordinate system (cf. Fig. 22), we see that

$$\mathbf{j}' = \mathbf{i}\cos\theta_1 + \mathbf{j}\sin\theta_1. \tag{7}$$

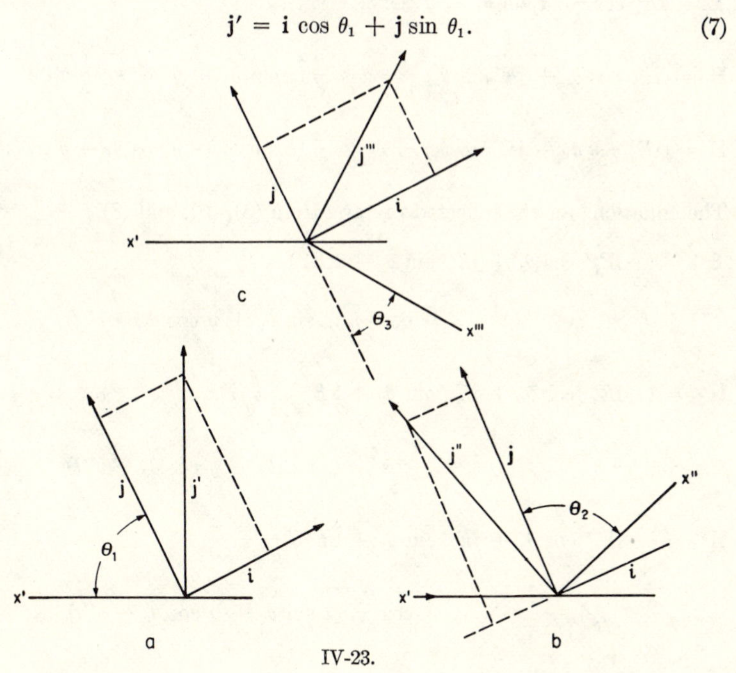

IV-23.

Similarly, for the other systems, 23b and 23c,

$$\mathbf{j}'' = -\mathbf{i} \cos \theta_2 + \mathbf{j} \sin \theta_2 \tag{8}$$

and

$$\mathbf{j}''' = \mathbf{i} \cos \theta_3 + \mathbf{j} \sin \theta_3. \tag{9}$$

Also

$$\begin{aligned} x' &= x \sin \theta_1 - y \cos \theta_1. \\ x'' &= x \sin \theta_2 + y \cos \theta_2. \\ x''' &= x \sin \theta_3 - y \cos \theta_3. \end{aligned} \tag{10}$$

Equations (1) become

$$\begin{aligned} \mathbf{E}' &= (\mathbf{i}E'_y \cos \theta_1 + \mathbf{j}E'_y \sin \theta_1) \cos \frac{2\pi}{\lambda'} (x \sin \theta_1 - y \cos \theta_1 - v't), \\ \mathbf{H}' &= \mathbf{k} \frac{c}{\mu' v'} E'_y \cos \frac{2\pi}{\lambda'} (x \sin \theta_1 - y \cos \theta_1 - v't), \end{aligned} \tag{11}$$

$$\mathbf{H}' = \mathbf{k} \frac{1}{\mu' v'} E'_y \cos \frac{2\pi}{\lambda'} (x \sin \theta_1 - y \cos \theta_1 - v't), \tag{11a}$$

for light polarized in the $x'y'$-plane. Similarly, for light polarized in the $x'z'$-plane, equations (2), we have

$$\begin{aligned} \mathbf{E}' &= \mathbf{k}E'_z \cos \frac{2\pi}{\lambda'} (x \sin \theta_1 - y \cos \theta_1 - v't). \\ \mathbf{H}' &= (\mathbf{i}E'_z \cos \theta_1 + \mathbf{j}E'_z \sin \theta_1) \frac{c}{\mu' v'} \cos \frac{2\pi}{\lambda'} (x \sin \theta_1 - y \cos \theta_1 - v't). \end{aligned} \tag{12}$$

$$\mathbf{H}' = (\mathbf{i}E'_z \cos \theta_1 + \mathbf{j}E'_z \sin \theta_1) \frac{1}{\mu' v'} \cos \frac{2\pi}{\lambda'} (x \sin \theta_1 - y \cos \theta_1 - v't). \tag{12a}$$

The equations for the reflected ray are, from (5), (6), and (8),

$$\begin{aligned} \mathbf{E}'' &= (-\mathbf{i}E''_y \cos \theta_2 + \mathbf{j}E''_y \sin \theta_2 + \mathbf{k}E''_z) \\ &\qquad \cdot \cos \frac{2\pi}{\lambda''} (x \sin \theta_2 + y \cos \theta_2 - v''t), \\ \mathbf{H}'' &= (-\mathbf{i}E''_z \cos \theta_2 + \mathbf{j}E''_z \sin \theta_2 + \mathbf{k}E''_y) \frac{c}{\mu'' v''} \\ &\qquad \cdot \cos \frac{2\pi}{\lambda''} (x \sin \theta_2 + y \cos \theta_2 - v''t), \end{aligned} \tag{13}$$

$$\begin{aligned} \mathbf{H}'' &= (-\mathbf{i}E''_z \cos \theta_2 + \mathbf{j}E''_z \sin \theta_2 + \mathbf{k}E''_y) \frac{1}{\mu'' v''} \\ &\qquad \cdot \cos \frac{2\pi}{\lambda''} (x \sin \theta_2 + y \cos \theta_2 - v''t), \end{aligned} \tag{13a}$$

and for the refracted ray,

$$\mathbf{E}''' = (\mathbf{i}E_y''' \cos\theta_3 + \mathbf{j}E_y''' \sin\theta_3 + \mathbf{k}E_z''') \cdot \cos\frac{2\pi}{\lambda'''}(x\sin\theta_3 - y\cos\theta_3 - v'''t). \tag{14}$$

$$\mathbf{H}''' = (\mathbf{i}E_z''' \cos\theta_3 + \mathbf{j}E_z''' \sin\theta_3 + \mathbf{k}E_y''') \frac{c}{\mu'''v'''} \cdot \cos\frac{2\pi}{\lambda'''}(x\sin\theta_3 - y\cos\theta_3 - v'''t).$$

$$\mathbf{H}''' = (\mathbf{i}E_z''' \cos\theta_3 + \mathbf{j}E_z''' \sin\theta_3 + \mathbf{k}E_y''') \frac{1}{\mu'''v'''} \cdot \cos\frac{2\pi}{\lambda'''}(x\sin\theta_3 - y\cos\theta_3 - v'''t). \tag{14a}$$

By our boundary conditions, the resultants of each component of **E** and **H** on opposite sides of the boundary must be equal for all points on the boundary and at all times. The boundary lies at $y = 0$. From the cosine factors in (11) to (14) we have the conditions,

$$\frac{\sin\theta_1}{\lambda'} = \frac{\sin\theta_2}{\lambda''} = \frac{\sin\theta_3}{\lambda'''}, \tag{15}$$

and

$$v'/\lambda' = v''/\lambda'' = v'''/\lambda'''. \tag{16}$$

But since the media are the same for the incident and reflected ray, we must have

$$v' = v''. \tag{17}$$

We thus deduce the following equations:

$$\lambda' = \lambda'', \qquad \sin\theta_1 = \sin\theta_2, \tag{18}$$

$$\theta_1 = \theta_2, \tag{19}$$

$$\frac{\sin\theta_1}{\sin\theta_3} = \frac{\lambda'}{\lambda'''} = \frac{v'}{v'''} = \frac{n_3}{n_1}, \tag{20}$$

by (4). Equations (15) to (20) hold for both cases (a) and (b). For (a) we equate the sum of the vector components of (11) and (13) to those of (14), in accord with the boundary conditions of the previous section. Note that **i** and **k** are the unit vectors tangent to the surface and that **j** represents the normal component. For the tangential components mere balancing of field intensities is all that we require (22.3), but for the normal components, we must multiply the electric and magnetic intensities by κ and μ, respectively, for the different media, (22.8) and (22.10). The letters (t) and (n) after the following equations indicate the tangential and normal components.

$$E'_y \cos\theta_1 - E''_y \cos\theta_2 = E'''_y \cos\theta_3. \quad \text{(t)} \tag{21}$$

$$\kappa' E'_y \sin\theta_1 + \kappa'' E''_y \sin\theta_2 = \kappa''' E'''_y \sin\theta_3. \quad \text{(n)} \tag{22}$$

$$\epsilon' E'_y \sin\theta_1 + \epsilon'' E''_y \sin\theta_2 = \epsilon''' E'''_y \sin\theta_3. \quad \text{(n)} \tag{22a}$$

$$E''_z = E'''_z. \quad \text{(t)} \tag{23}$$

$$-\frac{\cos\theta_2}{\mu'' v''} E''_z = \frac{\cos\theta_3}{\mu''' v'''} E'''_z. \quad \text{(t)} \tag{24}$$

$$\frac{\sin\theta_2}{v''} E''_z = \frac{\sin\theta_3}{v'''} E'''_z. \quad \text{(n)} \tag{25}$$

$$\frac{E'_y}{\mu' v'} + \frac{E''_y}{\mu'' v''} = \frac{E'''_y}{\mu''' v'''}. \quad \text{(t)} \tag{26}$$

Equations (23), (24), and (25), are consistent, in view of (20), only with the solution

$$E''_z = E'''_z = 0. \tag{27}$$

Multiplying equation (22) by (26) and making use of (19) and (20), we have

$$(E'_y + E''_y) = \sqrt{\frac{\mu' \kappa'''}{\mu''' \kappa'}}\, E'''_y. \tag{28}$$

$$E'_y + E''_y = \sqrt{\frac{\mu' \epsilon'''}{\mu''' \epsilon'}}\, E'''_y. \tag{28a}$$

From (21) and (19),

$$(E'_y - E''_y) = \frac{\cos\theta_3}{\cos\theta_1} E'''_y. \tag{29}$$

If we set

$$u_1^2 = \frac{\mu''' \kappa'}{\mu' \kappa'''} \frac{\cos^2\theta_3}{\cos^2\theta_1}, \tag{30}$$

$$u_1^2 = \frac{\mu''' \epsilon'}{\mu' \epsilon'''} \frac{\cos^2\theta_3}{\cos^2\theta_1}, \tag{30a}$$

we find that

$$(E'_y - E''_y) = u_1(E'_y + E''_y). \tag{31}$$

The ratio of the amplitudes of the reflected and incident beams becomes

$$E''_y/E'_y = (1 - u_1)/(1 + u_1). \tag{32}$$

For all media capable of transmitting light, the magnetic permeability is equal to unity, to a high degree of approximation, as the experimental data show. Hence

$$u_1 \sim \sqrt{\frac{\kappa'}{\kappa'''}}\frac{\cos\theta_3}{\cos\theta_1} = \frac{\sin\theta_3\cos\theta_3}{\sin\theta_1\cos\theta_1}, \tag{33}$$

$$u_1 \sim \sqrt{\frac{\epsilon'}{\epsilon'''}}\frac{\cos\theta_3}{\cos\theta_1}, \tag{33a}$$

by (15.11) and (20).

After making a simple trigonometric transformation, we find that

$$\frac{E_y''}{E_y'} = \frac{\tan(\theta_3 - \theta_1)}{\tan(\theta_3 + \theta_1)}. \tag{34}$$

The analogous expression for the refracted beam is

$$\frac{E_y'''}{E_y'} = \frac{2}{1 + u_1} = \frac{2\sin\theta_1}{\sin 2\theta_1 + \sin 2\theta_3}, \tag{35}$$

when we set μ equal to unity for both media.

The reader should specifically note that no general solution of the equations (21) to (26) would result if we had originally set $E''_y = 0$, i.e., if we had assumed that there was no reflected beam.

Equations (21) to (34) refer to case (a). Turning now to case (b) and solving in an analogous manner, we find from (12), (13), and (14), that

$$\frac{E_z''}{E_z'} = \frac{1 - u_2^2}{1 + u_2^2} \quad \text{and} \quad \frac{E_z'''}{E_z'} = \frac{2}{1 + u_2}, \tag{36}$$

where

$$u_2^2 = \frac{\mu'}{\mu'''}\frac{\sin\theta_1\cos\theta_3}{\sin\theta_3\cos\theta_1} = \frac{\mu'\kappa'''}{\mu'''\kappa'}\frac{\cos^2\theta_3}{\cos^2\theta_1}. \tag{37}$$

$$u_2^2 = \frac{\mu'\epsilon'''}{\mu'''\epsilon'}\frac{\cos^2\theta_3}{\cos^2\theta_1}. \tag{37a}$$

When $\mu = 1$, we have

$$\frac{E_z''}{E_z'} = \frac{\tan\theta_3 - \tan\theta_1}{\tan\theta_3 + \tan\theta_1} = \frac{\sin(\theta_3 - \theta_1)}{\sin(\theta_3 + \theta_1)}. \tag{38}$$

We could have inferred these equations directly from those preceding. Since the electric and magnetic vectors of the initial wave are interchanged we have merely to interchange κ with μ in u_1, to obtain u_2. Otherwise the equations are identical.

Now consider an experiment where two beams of equal intensity, one polarized perpendicular to, and the other in, the plane of incidence, are allowed to fall on a surface and suffer reflection. We deduce the ratio of the amplitudes of the two reflected beams, for various angles of incidence, by setting $E'_y = E'_z$ in equations (34) and (37) and dividing the latter by the former.

$$\frac{\text{amplitude perpendicular}}{\text{amplitude parallel}} = \frac{E''_y}{E''_y} = \frac{\cos(\theta_3 - \theta_1)}{\cos(\theta_3 + \theta_1)}. \tag{39}$$

This ratio is always greater than unity for $\theta_1 > 0$. In other words, light polarized perpendicular to the plane of incidence suffers greater reflection and less transmission than light polarized in the plane of incidence. Experiment quantitatively verifies this theoretical result.

In equation (34), we see that when

$$\theta_3 + \theta_1 = \frac{\pi}{2}, \tag{40}$$

$$E''_y = 0, \tag{41}$$

and there is no reflected beam. For still greater angles, the tangent becomes negative and E''_y is opposite in sign to E'_y, i.e., the phase of the wave alters by 180°.

We have seen that

$$\sin\theta_3 = \frac{n_1}{n_3}\sin\theta_1, \tag{42}$$

equation (20). When the ratio n_1/n_3 is greater than unity, i.e., when the path of the beam is from the denser into the less dense medium, we can always find a critical angle of incidence, θ_κ, for which

$$\sin\theta_3 = \frac{n_1}{n_3}\sin\theta_\kappa = 1 \quad \text{or} \quad \theta_3 = \frac{\pi}{2}. \tag{43}$$

The refracted beam is parallel to the surface of the medium. For values of

$$\theta > \theta_\kappa, \tag{44}$$

we can find no real solution, since the right-hand side of (42) is greater than unity. A detailed solution, which we shall not reproduce here, confirms our suspicion that the refracted ray dies out and that only the reflected ray exists. The phenomenon, known as total reflection, is familiar to all students of optics. The totally reflecting prism is widely employed, since its efficiency appreciably exceeds that of the best mirror surface.

24. The dielectric constant and the refractive index. Media of the type we have been considering up to the present, in which no free electric charges exist, are termed *dielectrics*. All electrons are closely bound to the atomic nuclei and are not free to wander through the medium. As a preliminary to the more general problem of transmission of light in a medium containing electric charges, we may rewrite several of the previous equations in slightly different form.

Since we shall no longer be concerned with boundary surfaces, we may conveniently suppose the beam to travel along the axis of x. Then, dropping the primes, we have

$$E = E_y \cos \frac{2\pi}{\lambda}(x - vt), \qquad H = \frac{c}{\mu v} E_y \cos \frac{2\pi}{\lambda}(x - vt), \tag{1}$$

$$H = \frac{1}{\mu v} E_y \cos \frac{2\pi}{\lambda}(x - vt), \tag{1a}$$

for light polarized in the xy-plane, (23.1). There is no need to retain the vector form, because we are now dealing with single vector components. We must merely recall that E and H are perpendicular to one another. In these equations λ and v refer specifically to the wavelength and velocity in the medium under consideration; in vacuo the respective values would have been λ_0 and c. From (23.20), we have

$$\lambda_0/\lambda = c/v = n, \tag{2}$$

by the definition of refractive index. Also, by (15.11),

$$v^2 = c^2/\mu\kappa. \tag{3}$$

$$v^2 = c^2/\kappa_e\kappa_m, \quad \text{where} \quad \kappa_e = \epsilon/\epsilon_0, \quad \kappa_m = \mu/\mu_0. \tag{3a}$$

Hence

$$\mu\kappa = n^2, \tag{4}$$

or, since $\mu = 1$ for dielectrics, as experiment discloses,

$$\kappa = n^2. \tag{5}$$

$$n^2 = \kappa_e\kappa_m. \tag{5a}$$

These equations relate the dielectric constant to the refractive index of the medium.

The frequency, ν, associated with the wave is defined by

$$\nu = c/\lambda_0 = v/\lambda, \tag{6}$$

i.e., the frequency is independent of the velocity of propagation. The equations of (1) then become

$$E = E_y \cos 2\pi\nu\left(x\frac{n}{c} - t\right) = E_y e^{2\pi i\nu(xn/c - t)} \tag{7}$$

where we have introduced the complex exponential in place of the cosine term. We understand that the real part of (7) represents the solution of the problem. The procedure will be to determine the form of (7) in various types of media, evaluating n for each example.

25. Absorption of radiation in a charged medium. The wave equations as derived in § (15) are not completely general. We supposed that the medium was devoid of free electric charges. We also assumed that no electric currents could exist. For greater generality we shall have to replace (15.1) by (14.9),

$$\nabla \times \mathbf{H} = \frac{1}{c}\frac{\partial \mathbf{D}}{\partial t} + \frac{4\pi}{c}\mathbf{J}. \tag{1}$$

$$\nabla \times \mathbf{H} = \frac{\partial \mathbf{D}}{\partial t} + \mathbf{J}. \tag{1a}$$

Equations (15.2) to (15.4) remain unchanged. We shall suppose the electric current, **J**, to be that induced in the medium by the direct action of the light waves. The principal force acting on the electric charges is that of the electric intensity. If R is the specific resistance and σ the conductivity of the medium to the flow of the electric current, we shall have

$$\mathbf{J}' = \mathbf{E}'/R = \sigma\mathbf{E}', \tag{2}$$

by Ohm's law, (9.1).

In this equation $\mathbf{J}'$ and $\mathbf{E}'$ are expressed in either electromagnetic or MKS units. Converting them to es units by (8.8), we have

$$\mathbf{J} = c^2\mathbf{E}/R = c^2\sigma\mathbf{E}, \tag{3}$$

where R is still expressed in em units. Since **J** refers to the current per unit area, $1/R$ must also be the reciprocal resistance or the conductivity per unit area, per unit length. For the present we shall not specify the origin of the resistance. It may arise from a purely electric effect, as when the electrons fritter away their energy in collisions with atoms, or it may arise from the reaction of the radiation emitted from an accelerated charge.

Furthermore, if we assume κ to be independent of the time, we may, by (14.7), write (1) in the form

$$\nabla \times \mathbf{H} = \frac{\kappa}{c}\frac{\partial \mathbf{E}}{\partial t} + \frac{4\pi c}{R}\mathbf{E}. \tag{4}$$

$$\nabla \times \mathbf{H} = \epsilon\frac{\partial \mathbf{E}}{\partial t} + \sigma\mathbf{E}. \tag{4a}$$

Taking the divergence of both sides of (4), we obtain

$$0 = \frac{1}{c}\left(\kappa\frac{\partial}{\partial t} + \frac{4\pi c^2}{R}\right)\nabla \cdot \mathbf{E} = \frac{1}{c}\left(\kappa\frac{\partial}{\partial t} + \frac{4\pi c^2}{R}\right)\frac{4\pi\rho}{\kappa}, \tag{5}$$

$$0 = \left(\epsilon\frac{\partial}{\partial t} + \sigma\right)\nabla \cdot \mathbf{E} = \left(\epsilon\frac{\partial}{\partial t} + \sigma\right)\frac{\rho}{\epsilon}, \tag{5a}$$

by (15.4), since $\nabla \cdot \nabla \times \mathbf{H} = 0$. (6)

The equation

$$\frac{\partial \rho}{\partial t} + \frac{4\pi c^2}{R\kappa}\rho = 0 \tag{7}$$

$$\frac{\partial \rho}{\partial t} + \frac{\sigma}{\epsilon}\rho = 0 \tag{7a}$$

has the integral

$$\ln \rho = -\frac{4\pi c^2}{R\kappa} t + \text{const}, \tag{8}$$

$$\ln \rho = -\frac{\sigma t}{\epsilon} + \text{const}, \tag{8a}$$

or

$$\rho = \rho_0 e^{(-4\pi c^2/R\kappa)t}, \tag{9}$$

$$\rho = \rho_0 e^{-\sigma t/\epsilon}, \tag{9a}$$

where ρ_0, the constant of integration, the value of ρ at time $t = 0$, may be a function of x, y, and z, because (7) is only a partial differential equation.

Equation (9) shows that the effective charge of a medium, under action of a light beam, diminishes with the time, so that, in the limit,

$$\rho \to 0 \tag{10}$$

and

$$\nabla \cdot \mathbf{E} \to 0, \tag{11}$$

by (15.4). Differentiating (4) partially with respect to the time, we obtain

$$\nabla \times \frac{\partial \mathbf{H}}{\partial t} = \frac{\kappa}{c}\frac{\partial^2 \mathbf{E}}{\partial t^2} + \frac{4\pi c}{R}\frac{\partial \mathbf{E}}{\partial t}$$

$$= -\frac{c}{\mu} \nabla \times \nabla \times \mathbf{E} = -\frac{c}{\mu}(\nabla\nabla \cdot \mathbf{E} - \nabla^2\mathbf{E}),$$

$$\frac{\kappa}{c}\frac{\partial^2 \mathbf{E}}{\partial t^2} + \frac{4\pi c}{R}\frac{\partial \mathbf{E}}{\partial t} = \frac{c}{\mu}\nabla^2\mathbf{E}, \tag{12}$$

$$\nabla \times \frac{\partial \mathbf{H}}{\partial t} = \epsilon \frac{\partial^2 \mathbf{E}}{\partial t^2} + \sigma \frac{\partial \mathbf{E}}{\partial t}$$

$$= -\nabla \times \nabla \times \frac{\mathbf{E}}{\mu}$$

$$= -\frac{1}{\mu}(\nabla\nabla \cdot \mathbf{E} - \nabla^2\mathbf{E}) = \frac{1}{\mu}\nabla^2\mathbf{E}, \tag{12a}$$

by (II-23.16) and (11).

The general solution of this equation has no particular interest for us. Let us consider, as before, a plane wave, polarized in the xz-plane, and attempt to represent the solution of the form of (27),

$$E = E_y = E_{y0} e^{2\pi i\nu(x\mathfrak{n}/c - t)}. \tag{13}$$

We have written $\mathfrak{n}$ instead of n, to distinguish it from the true refractive index. We shall treat $\mathfrak{n}$ as an arbitrary parameter, whose value we are to determine from the differential equation. Then, since

$$E_y = E_z = 0, \tag{14}$$

(12) becomes

$$\frac{\mu\kappa}{c^2}\frac{\partial^2 E_y}{\partial t^2} + \frac{4\pi\mu}{R}\frac{\partial E_y}{\partial t} = \frac{\partial^2 E_y}{\partial x^2}. \tag{15}$$

$$\mu\epsilon\frac{\partial^2 E_y}{\partial t^2} + \mu\sigma\frac{\partial E_y}{\partial t} = \frac{\partial^2 E_y}{\partial x^2}. \tag{15a}$$

Performing the indicated differentiations, we have, from (13) and (15), that

$$\mathfrak{n}^2 = \mu\kappa + \frac{2\mu c^2 i}{\nu R}. \tag{16}$$

$$\mathfrak{n}^2 = \frac{c^2\mu\epsilon + ic^2\mu^2\sigma}{\omega}, \quad \text{where} \quad \omega = 2\pi\nu. \tag{16a}$$

This result is equivalent to setting a condition upon $\mathfrak{n}$ so that (13) will be satisfied. When R is infinite we recover our previous expression for the refractive index. When R is finite $\mathfrak{n}$ is complex.

Let us set

$$\mathfrak{n} = n + ib \tag{17}$$

where both n and b are real and positive. We have

$$\mathfrak{n}^2 = n^2 - b^2 + 2inb. \tag{18}$$

Then from the imaginary part of (16), we find that

$$b = \frac{\mu c^2}{\nu R n}. \tag{19}$$

$$b = \frac{c^2\mu\sigma}{2n\omega}. \tag{19a}$$

We determine n from the biquadratic:

$$n^2 - \frac{\mu^2 c^4}{\nu^2 R^2 n^2} = \mu\kappa. \tag{20}$$

$$n^2 - \frac{c^4\mu^2\sigma^2}{4n^2\omega^2} = c^2\epsilon\mu. \tag{20a}$$

We shall suppose that this equation has been solved for n. Substituting from (17) and (19) into (13), we obtain the result:

$$E = E_0 e^{2\pi i\nu(xn/c - t) - 2\pi\mu c x/Rn} = E_0 e^{-kx/2}\cos 2\pi\nu(xn/c - t), \tag{21}$$

$$E = E_0 e^{2\pi i\nu(xn/c - t) - c\mu\sigma x/2n} = E_0 e^{-kx/2}\cos 2\pi\nu(xn/c - t), \tag{21a}$$

where

$$k = \frac{4\pi\mu c}{Rn}. \tag{22}$$

$$k = \mu c\sigma/n. \tag{22a}$$

The quantity n determines, as before, the wave velocity in the medium and hence plays the role of a refractive index. Equation (22) shows that amplitude of the light beam decreases exponentially with the distance. The average value of the energy density is

$$\overline{W} = \frac{\kappa}{4\pi}\overline{E^2} = \frac{\kappa}{8\pi}E_0^2e^{-kx}. \tag{23}$$

$$\overline{W} = \epsilon\overline{E^2} = \frac{\epsilon E_0^2}{2}e^{-kx}, \tag{23a}$$

from (20.9). Here k is the absorption coefficient.

When R is sufficiently large that we may neglect the second term on the left-hand side of (20) in comparison with $\mu\kappa$, $n^2 \sim \mu\kappa$, (in MKS, $n^2 \sim c^2\epsilon\mu$) and we may write

$$n^2 \sim \mu\kappa\left(1 + \frac{c^4}{\nu^2R^2\kappa^2}\right). \tag{24}$$

$$n^2 = c^2\epsilon\mu\left(1 + \frac{\sigma^2}{4\epsilon^2\omega^2}\right). \tag{24a}$$

When R is small, we may neglect the term $\mu\kappa$. (In MKS, we neglect $c^2\epsilon\mu$). Then

$$n \sim c\left(\frac{\mu}{\nu R}\right)^{1/2}, \tag{25}$$

$$n \sim c\left(\frac{\mu\sigma}{2\omega}\right)^{1/2}, \tag{25a}$$

which becomes infinite for a perfect conductor. This result accords with the statement made in § (2), that conductors behave like dielectrics of infinite κ. We know from experience that all metals are poor transmitters of radiation, a fact in agreement with their low resistance. For a metal, from (22) and (25),

$$k \sim 4\pi\left(\frac{\mu\nu}{R}\right)^{1/2}.$$

For copper, $R = 1.7 \times 10^{-6}$ ohm $= 1.7 \times 10^3$ em units. $\mu = 1$, $\nu = 6 \times 10^{14}$, corresponding to a wavelength of 5000 Angstrom units. With these values,

$$k \sim 7 \times 10^6 \text{ cm}^{-1}.$$

A very thin sheet of metal, therefore, only 1.4×10^{-7} cm in thickness should be capable of reducing the intensity of an incident beam by a factor of $1/e$, ($e = 2.718$). This result does not agree well with experiment, but the metallic model is highly idealized. We have assumed that the electric resistance to the impressed high-frequency forces is equal to that for ordinary direct current. This assumption fails to allow for possible change of the resistance with frequency.

Before we can make further progress, we must examine the question whether the electron, subject to acceleration, resists the change of motion in accord with Newton's laws of motion. In other words, we are asking, "does electricity possess mass?" We know from experiment that an electron possesses mass, but is this mass that of a true material point, irrespective of charge? Or can we attribute the term, $m_e\, d^2x/dt^2$, in the equations of motion of an electron to an effective electromagnetic mass of magnitude m_e?

26. The scalar and vector potentials. In obtaining the necessary formulae for electromagnetic mass, we first return to § 10, and introduce the vector potential. The Maxwell equations are

$$\nabla \times \mathbf{H} = \frac{\kappa}{c}\frac{\partial \mathbf{E}}{\partial t} + \frac{4\pi}{c}\mathbf{J}, \tag{1}$$

$$\nabla \times \mathbf{H} = \epsilon \frac{\partial \mathbf{E}}{\partial t} + \mathbf{J}, \tag{1a}$$

$$\nabla \times \mathbf{E} = -\frac{\mu}{c}\frac{\partial \mathbf{H}}{\partial t}, \tag{2}$$

$$\nabla \times \mathbf{E} = -\mu \frac{\partial \mathbf{H}}{\partial t}, \tag{2a}$$

$$\nabla \cdot \mathbf{H} = 0, \tag{3}$$

$$\nabla \cdot \mathbf{E} = \frac{4\pi\rho}{\kappa}, \tag{4}$$

$$\nabla \cdot \mathbf{E} = \frac{\rho}{\epsilon}, \tag{4a}$$

for μ and κ constant with respect to the time.

Let $\mathbf{A}$ be a vector potential obeying the relations

$$\mathbf{B} = \nabla \times \mathbf{A}. \tag{5}$$

Then we have one of the Maxwell equations automatically satisfied since

$$\nabla \cdot \mathbf{B} = \mu \nabla \cdot \mathbf{H} = \nabla \cdot \nabla \times \mathbf{A} = 0. \tag{6}$$

In the electrostatic case we had

$$\mathbf{E} = -\nabla\phi, \tag{7}$$

where ϕ is a scalar potential. This expression cannot satisfy the conditions when the charges are moving, because it requires that

$$\nabla \times \mathbf{E} = -\nabla \times \nabla\varphi = 0, \tag{8}$$

in contradiction to equation (2). The question of electric potential must be approached via equation (2), which we write in the form

$$\nabla \times \mathbf{E} = -\frac{\mu}{c}\frac{\partial \mathbf{H}}{\partial t} = -\frac{\mu}{c}\frac{\partial}{\partial t}\nabla \times \mathbf{A} = -\frac{\mu}{c}\nabla \times \frac{\partial \mathbf{A}}{\partial t}. \tag{9}$$

This equation suggests our taking

$$\mathbf{E} = -\frac{1}{c}\frac{\partial \mathbf{A}}{\partial t}, \tag{10}$$

$$\mathbf{E} = -\frac{\partial \mathbf{A}}{\partial t}, \tag{10a}$$

although we note that it does not fulfill the limiting condition (7), when **A** is stationary. We may, however, fulfill both conditions simultaneously if we write

$$\mathbf{E} = -\nabla\varphi - \frac{1}{c}\frac{\partial \mathbf{A}}{\partial t}. \tag{11}$$

$$\mathbf{E} = -\nabla\varphi - \frac{\partial \mathbf{A}}{\partial t}. \tag{11a}$$

Then we have, by (4)

$$\nabla \cdot \mathbf{E} = -\nabla^2\varphi - \frac{1}{c}\frac{\partial}{\partial t}\nabla \cdot \mathbf{A} = \frac{4\pi\rho}{\kappa}. \tag{12}$$

$$\nabla \cdot \mathbf{E} = -\nabla^2\varphi - \frac{\partial}{\partial t}\nabla \cdot \mathbf{A} = \frac{\rho}{\epsilon}. \tag{12a}$$

The last Maxwell equation to be satisfied is

$$\begin{aligned}\nabla \times \mathbf{H} = \nabla \times \left(\frac{\nabla \times \mathbf{A}}{\mu}\right) &= \frac{1}{\mu}(\nabla\nabla \cdot \mathbf{A} - \nabla^2\mathbf{A}) = \frac{\kappa}{c}\frac{\partial \mathbf{E}}{\partial t} + \frac{4\pi}{c}\mathbf{J} \\ &= -\frac{\kappa}{c}\nabla\frac{\partial\varphi}{\partial t} - \frac{\kappa}{c^2}\frac{\partial^2\mathbf{A}}{\partial t^2} + \frac{4\pi}{c}\mathbf{J},\end{aligned} \tag{13}$$

$$\begin{aligned}\nabla \times \mathbf{H} = \nabla \times \left(\frac{\nabla \times \mathbf{A}}{\mu}\right) &= \frac{1}{\mu}(\nabla\nabla \cdot \mathbf{A} - \nabla^2\mathbf{A}) = \epsilon\frac{\partial \mathbf{E}}{\partial t} + \mathbf{J} \\ &= -\epsilon\nabla\frac{\partial\varphi}{\partial t} - \epsilon\frac{\partial^2\mathbf{A}}{\partial t^2} + \mathbf{J},\end{aligned} \tag{13a}$$

by (II-23.16) and (11). In order to determine a vector completely we must specify both its curl and its divergence. We have defined $\nabla \times \mathbf{A}$ by equation (5). We are at liberty to select the divergence in any manner we choose. Accordingly we shall set

$$\nabla \cdot \mathbf{A} + \frac{\kappa\mu}{c}\frac{\partial\phi}{\partial t} = 0, \tag{14}$$

$$\nabla \cdot \mathbf{A} + \epsilon\mu\frac{\partial\varphi}{\partial t} = 0, \tag{14a}$$

because it leads to the simplifying result that

$$\nabla\nabla \cdot \mathbf{A} + \frac{\kappa\mu}{c}\nabla\frac{\partial\phi}{\partial t} = 0. \tag{15}$$

$$\nabla\nabla \cdot \mathbf{A} + \epsilon\mu\nabla\frac{\partial\varphi}{\partial t} = 0. \tag{15a}$$

In passing we note that whereas **A** uniquely determines **B**, a knowledge of **B** alone will not enable us to evaluate **A**. In thus adopting a non-zero value for $\nabla \cdot \mathbf{A}$, we depart from the methods used in the analysis of vortices, where we assumed that the vector potential was strictly solenoidal.

Equations (13) and (12) become

$$\nabla^2\mathbf{A} = \frac{\mu\kappa}{c^2}\frac{\partial^2\mathbf{A}}{\partial t^2} - \frac{4\pi\mu\mathbf{J}}{c}, \tag{16}$$

$$\nabla^2\mathbf{A} = \epsilon\mu\frac{\partial^2\mathbf{A}}{\partial t^2} - \mu\mathbf{J}, \tag{16a}$$

$$\nabla^2\phi = \frac{\mu\kappa}{c^2}\frac{\partial^2\varphi}{\partial t^2} - \frac{4\pi\rho}{\kappa}. \tag{17}$$

$$\nabla^2\varphi = \epsilon\mu\frac{\partial^2\varphi}{\partial t^2} - \frac{\rho}{\epsilon}. \tag{17a}$$

These equations are closely related to Poisson's (II-16.3), to which they reduce when **A** and ϕ are constant with time. In regions devoid of currents or charges, the foregoing equations reduce to the wave equation, the potentials being propagated with velocity

$$v = c/(\mu\kappa)^{1/2}. \tag{18}$$

$$v = \sqrt{1/\epsilon\mu}. \tag{18a}$$

In a vacuum the potentials are propagated with a velocity equal to that of light.

If the velocity of propagation were infinite, instead of finite, the potentials would be directly determined by Poisson's equation. We could evaluate the electrostatic potential, as before, by means of the integral

$$\phi = \int \frac{\rho}{\kappa r}\,d\tau, \tag{19}$$

$$\phi = \int \frac{\rho}{4\pi\epsilon r}\,d\tau, \tag{19a}$$

where ρ is the charge density within the volume element $d\tau$ and r the distance from the element to the point at which the potential is being calculated. By analogy, we may also write, since the defining equations are identical in form,

$$\mathbf{A} = \int \frac{\mu \mathbf{J}}{cr}\, d\tau, \tag{20}$$

$$\mathbf{A} = \int \frac{\mu}{4\pi} \frac{\mathbf{J}}{r}\, d\tau, \tag{20a}$$

which breaks up into three scalar integrals, one for each vector component.

We must allow for the finite velocity of propagation of the potential wave. A potential so calculated is known as a "retarded" potential. The problem is analogous to that of making allowance for the finite speed of light, the so-called "light-time", in astronomical measurements. We regard the potential as a function of the time, as well as of the coordinates. Thus, with $\kappa = 1$,

$$\phi(x', y', z', t') = \int \frac{\rho(x_0, y_0, z_0, t_0)}{r_0}\, d\tau_0, \tag{21}$$

$$\phi(x', y', z', t') = \int \frac{\rho\, d\tau_0}{4\pi\epsilon r}, \tag{21a}$$

where x_0, y_0, and z_0 denote the coordinates of the volume element at time t_0. t_0 represents an instant earlier than t' by the "light-time" between the element and the point x', y', z'. Or,

$$t' - t_0 = r_0/c. \tag{22}$$

27. Electromagnetic mass; forces acting on an accelerated charge. We shall now be able to calculate the forces acting on an accelerated electric charge. Lorentz developed the fundamental theory. For the present we shall not attempt to define the shape of the charged volume. We shall, however, assume that all elements of the volume have the same velocity, acceleration, and rate of acceleration. Later on we shall attempt to identify the volume, containing total charge ε, with that of the electron.

Let O be a small volume element $dx\, dy\, dz$, containing charge of density ρ, (Fig. 24). This is the volume that we shall eventually associate with the electron. The discussion of this section applies to the potentials *within* the electron. At a certain time let the coordinates of O be x_0, y_0, z_0. If the charges are stationary, the electrostatic potential at some other point Q, located within the volume, at coordinates x', y', z', is

$$\phi = \int \frac{\rho}{r_0}\, d\tau_0, \tag{1}$$

$$\phi = \int \frac{\rho}{4\pi\epsilon r_0}\, d\tau_0, \tag{1a}$$

where $$r_0^2 = (x' - x_0)^2 + (y' - y_0)^2 + (z' - z_0)^2, \tag{2}$$

$$d\tau_0 = dx_0\, dy_0\, dz_0. \tag{3}$$

If O is moving, however, we shall have to make due allowance for the finite time required for propagation of the electrostatic field from O to Q, because the various elements do not lie all at the same distance from Q. The time, T, of propagation of the field, is

$$T = r_0/c, \tag{4}$$

where c is the velocity of light. While the field associated with the element at position O is travelling to Q, the element will have moved on to P at

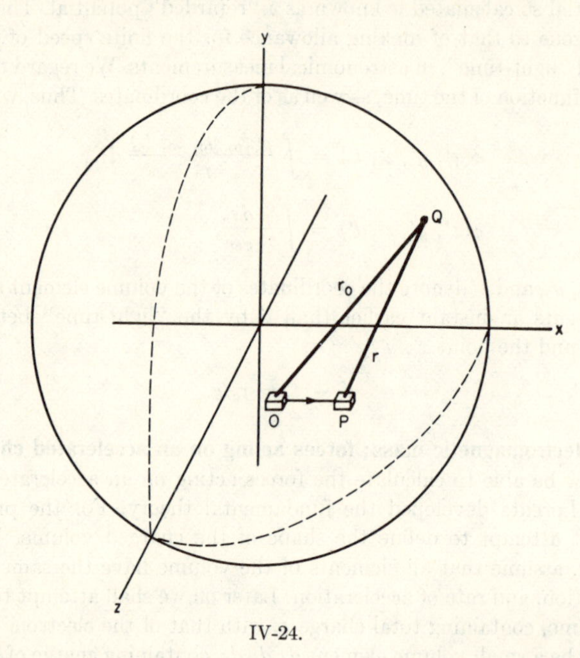

IV-24.

coordinates x, y, z, reaching P at time t, let us say. We shall suppose that the motion is parallel to the axis of x.

We wish to calculate the retarded potential at Q as if it had proceeded instantaneously from P, which means that we shall have to evaluate in terms of x, y, z, instead of in terms of x_0, y_0, z_0, making allowance for the retardation caused by the non-instantaneous time of propagation. Denote the distance PQ by r.

The position of the element along the axis of x is a function of the time; thus

$$x = f(t), \tag{5}$$

and the position x_0, occupied at time $t - T$, is

$$x_0 = f(t - T). \tag{6}$$

Let us expand this function by Taylor's theorem.

$$x_0 = f(t) - Tf'(t) + \frac{T^2}{2!} f''(t) - \frac{T^3}{3!} f'''(t) + \ldots$$

$$= x - T\frac{dx}{dt} + \frac{T^2}{2}\frac{d^2x}{dt^2} - \frac{T^3}{6}\frac{d^3x}{dt^3} + \ldots$$

$$= x - vT + \dot{v}\frac{T^2}{2} - \ddot{v}\frac{T^3}{6} + \ldots, \tag{7}$$

where v represents the velocity, $\dot{v}$ the acceleration, etc. In thus expanding our function, we have assumed that v, $\dot{v}$, and $\ddot{v}$ are of successively smaller orders, so that convergence is assured.

From (2) and (7), we have

$$r_0^2 = \left[\left(x' - x + vT - \frac{\dot{v}T^2}{2} + \frac{\ddot{v}T^3}{6}\right)^2 + (y' - y)^2 + (z' - z)^2\right], \tag{8}$$

since

$$y = y_0 \quad \text{and} \quad z = z_0. \tag{9}$$

We have, from (4),

$$T = r_0/c \sim r/c. \tag{10}$$

This value of T is sufficiently accurate to be entered in the right-hand side of (8), where it multiplies second-order terms. Inserting (10) into (8), expanding the result by means of the binomial theorem, and dropping terms of higher order, we find

$$\frac{1}{r_0} \sim \frac{1}{r}\left[1 + \frac{2(x' - x)}{r^2}\left(vT - \frac{\dot{v}T^2}{2} + \frac{\ddot{v}T^3}{6}\right)\right]^{-1/2} \tag{11}$$

$$\sim \frac{1}{r}\left[1 - \frac{x' - x}{r}\left(\frac{v}{c} - \frac{\dot{v}r}{2c^2} + \frac{\ddot{v}r^2}{6c^3}\right)\right]. \tag{12}$$

From (7) and (10),

$$x_0 = x - \frac{vr}{c} + \frac{\dot{v}r^2}{2c^2} - \frac{\ddot{v}r^3}{6c^3}, \tag{13}$$

and

$$dx_0 = \left[1 + \frac{v}{c}\frac{(x' - x)}{r} - \frac{\dot{v}}{c^2}(x' - x) + \frac{\ddot{v}r}{2c^3}(x' - x)\right]dx, \tag{14}$$

since

$$\frac{\partial r}{\partial x} = -\frac{x' - x}{r}, \tag{15}$$

by (11). Substituting from (12) and (14) into (1), we find that

$$\varphi = \int \frac{\rho}{r}\left[1 - \frac{\dot{v}}{2c^2}(x' - x) + \frac{\ddot{v}r}{3c^3}(x' - x)\right]dx\,dy\,dz. \tag{16}$$

$$\varphi = \int \frac{\rho}{4\pi\epsilon r}\left[\quad\right]dx\,dy\,dz. \tag{16a}$$

To obtain (16), we have multiplied together (12) and (14) and dropped the higher order products. We shall require the expression

$$-\frac{\partial\varphi}{\partial x'} = \int \rho\left\{\frac{x'-x}{r^3} + \frac{\dot{v}}{2c^2}\left[\frac{1}{r} - \frac{(x'-x)^2}{r^3}\right] - \frac{\ddot{v}}{3c^3}\right\} dx\, dy\, dz. \tag{17}$$

$$-\frac{\partial\varphi}{\partial x'} = \int \frac{\rho}{4\pi\epsilon}\left\{\quad\right\} dx\, dy\, dz. \tag{17a}$$

The differentiation may be performed under the integral sign, because x' remains constant as far as the integration is concerned.

We have now to calculate the retarded vector potential, by (26.20).

$$\mathbf{A} = \int \frac{\mathbf{J}}{cr_0}\, dx_0\, dy_0\, dz_0. \tag{18}$$

$$\mathbf{A} = \int \frac{\mu_0\mathbf{J}}{4\pi r}\, dx_0\, dy_0\, dz_0. \tag{18a}$$

The only current component is along the x-axis, and

$$\mathbf{J}_x = \mathbf{i}\rho v_0, \tag{19}$$

where v_0 is the velocity of the element when at O. Here $\mathbf{i}$, of course, indicates the unit vector.

Analogous to (13), we may write

$$v_0 = v - \frac{\dot{v}r}{c} + \frac{\ddot{v}r^2}{2c^2} - \dots. \tag{20}$$

Therefore, by (14), (19), and (20), equation (18) becomes

$$A = \frac{1}{c}\int \frac{\rho}{r}\left(v - \dot{v}\frac{r}{c} + \frac{\ddot{v}r^2}{2c^2}\right) dx\, dy\, dz. \tag{21}$$

$$A = \int \frac{\mu_0\rho}{4\pi r}\left(\quad\right) dx\, dy\, dz. \tag{21a}$$

The vector notation is no longer necessary, since we are dealing with the magnitude of a single component. Higher order terms have been excluded. We also have

$$-\frac{1}{c}\frac{\partial A}{\partial t} = -\frac{1}{c^2}\int \frac{\rho}{r}\left(\dot{v} - \ddot{v}\frac{r}{c}\right) dx\, dy\, dz, \tag{22}$$

$$-\frac{\partial A}{\partial t} = -\frac{\mu_0}{4\pi}\int \frac{\rho}{r}\left(\dot{v} - \ddot{v}\frac{r}{c}\right) dx\, dy\, dz, \tag{22a}$$

since

$$\frac{\partial v}{\partial t} = \dot{v}, \tag{23}$$

etc. From (26.11),

$$E_{x'} = -\frac{\partial \phi}{\partial x'} - \frac{1}{c}\frac{\partial A}{\partial t}, \tag{24}$$

$$E_{x'} = -\frac{\partial \varphi}{\partial x'} - \frac{\partial A}{\partial t}, \tag{24a}$$

which represents the instantaneous electric intensity at Q resulting from all the charges in the volume. Let ρ' be the density of electric charge at Q, within the volume element $dx'\, dy'\, dz'$. Then the x-component of force on the volume element is

$$E_{x'}\rho'\, dx'\, dy'\, dz'. \tag{25}$$

To obtain the total force acting on the total charged region, we must integrate over the volume a second time, because each element of charge interacts with every element. We thus have, for the force,

$$F_x = \int \rho'\, d\tau' \int \rho \frac{x' - x}{r^3}\, d\tau - \frac{\dot{v}}{2c^2} \int \rho'\, d\tau' \cdot \int \rho \left[\frac{1}{r} + \frac{(x' - x)^2}{r^3}\right] d\tau + \frac{2\ddot{v}}{3c^3} \int \rho'\, d\tau' \int \rho\, d\tau, \tag{26}$$

$$F_x = \frac{1}{4\pi\epsilon} \int \rho'\, d\tau' \int \rho \frac{x' - x}{r^3}\, d\tau - \frac{\mu_0 \dot{v}}{8\pi} \int \rho'\, d\tau \cdot \int \rho \left[\frac{1}{r} + \frac{(x' - x)^2}{r^3}\right] d\tau + \frac{\mu_0 \ddot{v}}{6\pi c} \int \rho'\, d\tau' \int \rho\, d\tau, \tag{26a}$$

where

$$d\tau = dx\, dy\, dz \quad \text{and} \quad d\tau' = dx'\, dy'\, dz'.$$

But

$$\int \rho\, d\tau = \int \rho'\, d\tau' = \varepsilon, \tag{27}$$

the total electronic charge. We may write the first integral of (26) in the form

$$\int \rho\, d\tau \int \rho' \frac{x'}{r^3}\, d\tau' - \int \rho'\, d\tau' \int \rho \frac{x}{r^3}\, d\tau = 0, \tag{28}$$

since the integrations with respect to τ' and τ cover the same volume.

The last term in (26) becomes

$$\frac{2\varepsilon^2}{3c^3}\ddot{v} = \frac{2\varepsilon^2}{3c^3}\frac{d^3x}{dt^3}, \tag{29}$$

$$\frac{\mu_0 \varepsilon^2}{6\pi c}\ddot{v} = \frac{\mu_0 \varepsilon^2}{6\pi c}\frac{d^3x}{dt^3} = \frac{\varepsilon^2}{6\pi\epsilon_0 c^3}\frac{d^3x}{dt^3}, \tag{29a}$$

by (27). These last two results, (28), and (29), are independent of the shape of the volume or of the distribution of the charge. To evaluate the middle term, however, we must know both these factors. Let us calculate the magnitude of the term on the assumption that the electron is a sphere of radius R_0 with its electric charge uniformly distributed over its surface. Assume the center of the electron to be at the origin.

Determine the value of the integral,

$$\iiint \rho\left[\frac{1}{r} + \frac{(x' - x)^2}{r^3}\right] dx\, dy\, dz, \tag{30}$$

at the point

$$x' = R_0, \qquad y' = z' = 0. \tag{31}$$

We have the relation

$$4\pi R_0^2\sigma = 4\pi \int_0^{R_0} \rho R^2\, dR = 4\pi R_0^2 \int_0^{R_0} \rho\, dR = \varepsilon, \tag{32}$$

where σ is the surface density of the charge. We have factored R from the integral sign since $\rho = 0$, except for $R = R_0$, by hypothesis.

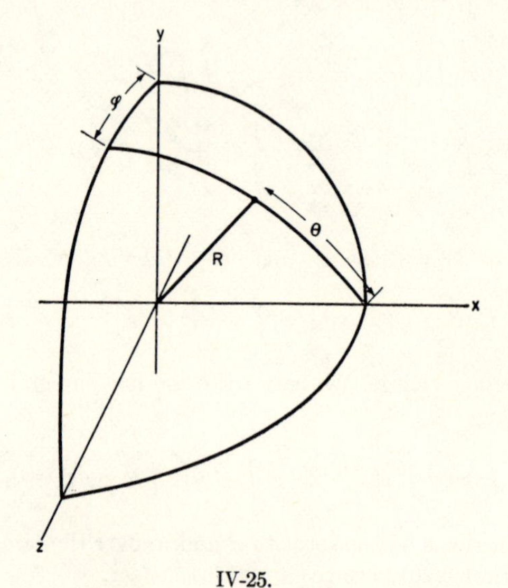

IV-25.

Take spherical coordinates as follows:

$$x = R\cos\theta, \quad y = R\sin\theta\cos\phi, \quad z = R\sin\theta\sin\varphi,$$
$$dx\, dy\, dz = R^2 \sin\theta\, dR\, d\theta\, d\varphi, \tag{33}$$

which differs from the conventional set of polar coordinates only by an axis of rotation. We find from (31) and (33) that

$$\begin{aligned} r^2 &= (x' - x)^2 + (y' - y)^2 + (z' - z)^2 \\ &= R_0^2 - 2RR_0 \cos\theta + R^2 = 2R_0^2(1 - \cos\theta), \end{aligned} \tag{34}$$

since the integrand is zero except when $R = R_0$. The integral becomes

$$\begin{aligned} &\frac{R_0}{\sqrt{2}} \int_0^{R_0} \int_0^{\pi} \int_0^{2\pi} \rho \left[\frac{1}{(1 - \cos\theta)^{1/2}} + (1 - \cos\theta)^{1/2} \right] \sin\theta \, dR \, d\theta \, d\varphi \\ &\qquad = \frac{2\pi\sigma R_0}{\sqrt{2}} \left[2(1 - \cos\theta)^{1/2} + \frac{2}{3}(1 - \cos\theta)^{3/2} \right]_0^{\pi} = \frac{4}{3} \frac{\varepsilon}{R_0}. \end{aligned} \tag{35}$$

Because of the spherical symmetry, the value of this integral is the same for any point on the surface. Carrying out second integration for this term of (28), we obtain the final result that

$$-\frac{2\varepsilon^2}{3R_0c^2}\dot{v} = -\frac{2\varepsilon^2}{3R_0c^2}\frac{d^2x}{dt^2}. \tag{36}$$

$$-\frac{\mu_0\varepsilon^2}{\partial\pi R_0}\dot{v} = -\frac{\mu_0\varepsilon^2}{6\pi R_0}\frac{d^2x}{dt^2} = -\frac{\varepsilon^2}{6\pi\epsilon_0 R_0 c^2}\frac{d^2x}{dt^2}. \tag{36a}$$

The quantity $[\varepsilon^2/R_0c^2]$ has the physical dimensions of mass, as it should, in order that the full term represent a force. If we set the coefficient of the differential term equal to the mass of the electron we obtain the result

$$R_0 = 2\varepsilon^2/3mc^2. \tag{37}$$

$$R_0 = 2\varepsilon^2/6\pi\epsilon_0 mc^2. \tag{37a}$$

This equation, of course, does not necessarily specify the radius of the electron. It merely states that a weightless sphere, of radius R_0, containing charge ε upon its surface, will behave mechanically as if it possessed a mass m. If we assume that the mass of the electron is all electromagnetic, then for the observed values of ε, m, and c, we find

$$R_0 = 1.874 \times 10^{-13} \text{ cm}. \tag{38}$$

The forces are calculated in the present section from the point of view of an observer within the electron. We shall not question the validity of the assumptions involved, bizarre as they appear in the light of modern atomic theory and wave mechanics. As a matter of fact, physicists often adopt certain of the relationships of classical electromagnetic theory as fundamental postulates of the newer mechanics. The ultimate test of the assumption lies, of course, in the experimental laboratory.

For the present we wish merely to transform our point of view to that of an observer external to the electron. According to the law of reaction, the external forces must be the negative of the internal forces. Thus, for the external observer, the electron is subject to the force,

$$F_x = m\frac{d^2x}{dt^2} - \frac{2\varepsilon^2}{3c^3}\frac{d^3x}{dt^3}, \tag{39}$$

$$F_x = m\frac{d^2x}{dt^2} - \frac{\mu_0\varepsilon^2}{6\pi c}\frac{d^3x}{dt^3}, \tag{39a}$$

from (26), (29), (36), and (37). We note, in passing, that the y and z force components are zero, as a result of our assumption that the acceleration components are zero along these axes.

The work done by an electron that moves from x_1 to x_2 is

$$W = \int_{x_1}^{x_2} F_x\,dx = m\int_{x_1}^{x_2}\frac{d^2x}{dt^2}\,dx - \frac{2\varepsilon^2}{3c^3}\int_{x_1}^{x_2}\frac{d^3x}{dt^3}\,dx$$

$$= m\int_{t_1}^{t_2}\frac{d^2x}{dt^2}\frac{dx}{dt}\,dt - \frac{2\varepsilon^2}{3c^3}\int_{t_1}^{t_2}\frac{d^3x}{dt^3}\frac{dx}{dt}\,dt. \tag{40}$$

$$W = m\int_{t_1}^{t_2}\frac{d^2x}{dt^2}\frac{dx}{dt}\,dt - \frac{\mu_0\varepsilon^2}{6\pi c}\int_{t_1}^{t_2}\frac{d^3x}{dt^3}\frac{dx}{dt}\,dt. \tag{40a}$$

We readily integrate the first term. For the second we carry out an integration by parts:

$$W = \left[\frac{1}{2}mv^2 - \frac{2\varepsilon^2}{3c^3}v\dot{v}\right]_{t_1}^{t_2} + \frac{2\varepsilon^2}{3c^3}\int_{t_1}^{t_2}\dot{v}^2\,dt. \tag{41}$$

$$W = \left[\frac{1}{2}mv^2 - \frac{\mu_0\varepsilon^2}{6\pi c}v\dot{v}\right]_{t_1}^{t_2} + \frac{\mu_0\varepsilon^2}{6\pi c}\int_{t_1}^{t_2}\dot{v}^2\,dt. \tag{41a}$$

The term, $mv^2/2$, gives the change of kinetic energy from time t_1 to t_2. The second term is more difficult to interpret, but an intricate analysis shows that it actually represents changes in the electromagnetic energy stored within the electron. The presence of such a term is consistent with the theory of relativity, which requires a variation of the effective mass with velocity.

If the electron under consideration is vibrating in an atomic linear oscillator, and if we set for t_1 and t_2 the moments when the oscillation has reached maximum amplitude at the extremities of the swing, $v_1 = v_2 = 0$. The bracketed terms vanish and we have left only the relationship

$$W = \frac{2\varepsilon^2}{3c^3}\int_{t_1}^{t_2}\dot{v}^2\,dt. \tag{42}$$

$$W = \frac{\mu_0\varepsilon^2}{6\pi c}\int_{t_1}^{t_2}\dot{v}^2\,dt. \tag{42a}$$

This last term represents the loss of energy by a linear oscillator. The instantaneous rate of loss is

$$\frac{dW}{dt} = \frac{2\varepsilon^2\dot{v}^2}{3c^3}. \tag{43}$$

$$\frac{dW}{dt} = \frac{\mu\varepsilon^2}{6\pi c}\dot{v}^2 = \frac{\varepsilon^2}{6\pi\epsilon c^3}\dot{v}^2. \tag{43a}$$

28. Bivectors. Consider the following expression for the electric intensity of a light wave of frequency ν, moving in the positive direction along the z-axis, cf. (23.1):

$$\mathbf{E} = \mathbf{i}E_x \cos 2\pi\nu\left(t - \frac{z}{c}\right) + \mathbf{j}E_y \cos 2\pi\nu\left(t - \frac{z}{c}\right). \tag{1}$$

$$\nu = c/\lambda. \tag{2}$$

Let us examine the time variation of $\mathbf{E}$ at $z = 0$. The vector components represent plane waves, polarized, respectively, in the xz- and yz-planes. The two components are "in phase," i.e., both components attain their maximum and minimum values simultaneously in the same period. Thus the two vectors add to give a resultant plane wave, polarized in some intermediate plane. A simple rotation of axes will give an equation of the form

$$\mathbf{E}_0 = \mathbf{i}'A \cos 2\pi\nu t, \tag{3}$$

where the new amplitude is related to the old ones by

$$A^2 = E_x^2 + E_y^2, \tag{4}$$

because $\mathbf{E}_0 \cdot \mathbf{E}_0$ must give identical results for both (1) and (4). Thus $\mathbf{E}_0$ changes in magnitude but not in direction. Introducing complex notation, we may represent (1) as the real part of

$$\mathbf{E}_{0c} = (\mathbf{i}E_x + \mathbf{j}E_y)e^{2\pi i\nu t}. \tag{5}$$

Now consider the expression

$$\mathbf{E}_0 = \mathbf{i}E_x \cos 2\pi\nu t + \mathbf{j}E_y \sin 2\pi\nu t. \tag{6}$$

This time the two components are exactly out of phase. When $t = 0$, $\mathbf{E} = \mathbf{i}E_x$ and the intensity lies along x. One-fourth of a period later, i.e., for $t = \frac{1}{4}\nu$, $\mathbf{E} = \mathbf{j}E_y$, where $\mathbf{E}$ is a vector, rotating about the origin and constantly changing in magnitude. Its tip describes an ellipse. We may express $\mathbf{E}_0$ as the real part of the expression

$$\mathbf{E}_{0c} = (\mathbf{i}E_x - i\mathbf{j}E_y)e^{2\pi i\nu t} = [\mathbf{i}E_x \cos 2\pi\nu t + \mathbf{j}E_y \sin 2\pi\nu t] + i[\mathbf{i}E_x \sin 2\pi\nu t - \mathbf{j}E_y \cos 2\pi\nu t]. \tag{7}$$

This quantity is called a *bivector*, because it consists of one vector in *real* space and another in *imaginary* space. The magnitudes of each vector must be treated separately.

Now let us examine the more general vector

$$\mathbf{E}_0 = \mathbf{i}E_x \cos(2\pi\nu t + \alpha) + \mathbf{j}E_y \sin(2\pi\nu t + \beta), \tag{8}$$

where α and β are *phase angles*. Equation (1) corresponds to $\alpha = 0, \beta = \pi/2$, and (2) to $\alpha = 0$, $\beta = 0$. Equation (8) thus represents a variable vector rotating in counterclockwise fashion about the origin, in an ellipse (Fig. 26).

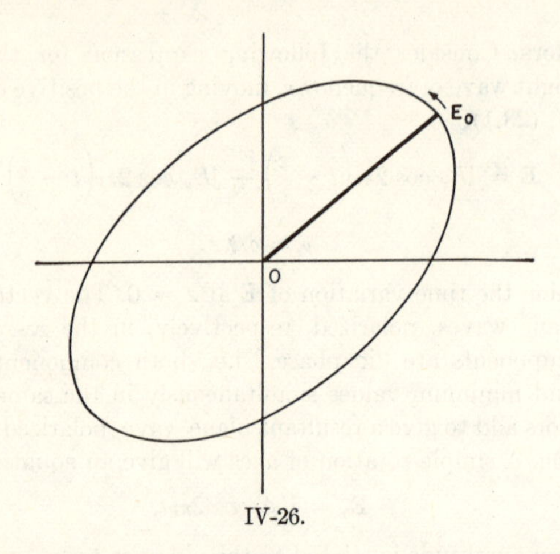

IV-26.

With a rotation of axes, we may express (8) in the form of (6) or (7). This axis rotation is not always convenient, however. For example, if we wish to follow the variation of $\mathbf{E}_0$ at z we should have to adopt an axis system rotating as though attached to a screw that advances as it turns. Hence we seek another way of expressing $\mathbf{E}_0$ in complex form. We assume that $\mathbf{E}_0$ is the real part of

$$\mathbf{E}_{0c} = (\mathbf{i}E_x e^{i\alpha} - i\mathbf{j}E_y e^{i\beta})e^{2\pi i\nu t}. \tag{9}$$

We may write this equation as follows:

$$\begin{aligned}\mathbf{E}_{0c} &= [\mathbf{i}E_x(\cos\alpha + i\sin\alpha) - i\mathbf{j}E_y(\cos\beta + i\sin\beta)]e^{2\pi i\nu t} \\ &= (\mathbf{i}E_{xc} - i\mathbf{j}E_{yc})e^{2\pi i\nu t},\end{aligned} \tag{10}$$

where

$$E_{xc} = E_x(\cos\alpha + i\sin\alpha) \quad \text{and} \quad E_{yc} = E_y(\cos\beta + i\sin\beta). \tag{11}$$

More generally, for any point on the z-axis, we may write

$$\mathbf{E}_c = (\mathbf{i}E_{xc} - i\mathbf{j}E_{yc})e^{2\pi i\nu(t-z/c)}. \tag{12}$$

The symbols E_{xc} and E_{yc} represent what are called *complex amplitudes*. Both components of the bivector, $\mathbf{i}E_{xc} - i\mathbf{j}E_{yc}$, now have physical significance, since they are multiplied by the complex factor $e^{2\pi i\nu t}$.

At this point we may introduce a mathematical simplification. In most problems of electromagnetic theory, the vector **E** is important chiefly as an intermediary. Our ultimate interest usually lies in values of products like $\mathbf{E} \cdot \mathbf{E}$, $\mathbf{E} \times \mathbf{H}$, etc. We evaluate such quantities by taking the real part of (12), performing the scalar or vector product, and finally averaging the time factors. The results of this somewhat long-winded process are as follows:

$$\overline{\mathbf{E} \cdot \mathbf{E}} = (E_x^2 \cos^2 \alpha + E_y^2 \sin^2 \beta)\, \overline{\cos^2 2\pi\nu(t - z/c)}$$
$$+ (E_x^2 \sin^2 \alpha + E_y^2 \cos^2 \beta)\, \overline{\sin^2 2\pi\nu(t - z/z)} = (E_x^2 + E_y^2)/2. \qquad (13)$$

With $\mathbf{E}_c$ defined as above, the associated $\mathbf{H}_c$ vector becomes

$$\mathbf{H}_c = \frac{c}{\mu v}(\mathbf{j}E_{xc} + i\mathbf{i}E_{yc})e^{2\pi i\nu(t-z/c)}, \qquad (14)$$

$$\mathbf{H}_c = \frac{1}{\mu v}(\mathbf{j}E_{xc} + i\mathbf{i}E_{yc})e^{2\pi i\nu(t-z/c)}, \qquad (14a)$$

$$\overline{\mathbf{E} \times \mathbf{H}} = \frac{c}{\mu v}\mathbf{k}\frac{E_x^2 + E_y^2}{2}. \qquad (15)$$

$$\overline{\mathbf{E} \times \mathbf{H}} = \frac{1}{\mu v}\mathbf{k}\frac{E_x^2 + E_y^2}{2}. \qquad (15a)$$

These results prove to be independent of the actual magnitudes of α and β.

It is interesting to note that we can obtain identical results by a much shorter calculation of a purely formal nature. We make direct use of the bivectors. Perform as follows the symbolic multiplication of the vectors and complex conjugates, replacing (i) every place it occurs by $(-i)$, noting that

$$E_{xc}E_{xc}^* = E_x^2 \quad \text{and} \quad E_{yc}E_{yc}^* = E_y^2, \qquad (16)$$

$$\overline{\mathbf{E} \times \mathbf{H}} = \frac{1}{2}(\mathbf{E}_c \times \mathbf{H}_c^*) = \frac{c}{\mu v}\mathbf{k}\frac{E_x^2 + E_y^2}{2}, \qquad (17)$$

$$\overline{\mathbf{E} \times \mathbf{H}} = \frac{1}{\mu v}\mathbf{k}\frac{E_x^2 + E_y^2}{2}, \qquad (17a)$$

and

$$\mathbf{E} \cdot \mathbf{E} = \frac{1}{2}\mathbf{E}_c \cdot \mathbf{E}_c^* = \frac{E_x^2 + E_y^2}{2}. \qquad (18)$$

When $E_x = E_y = A$ and $\alpha = \beta$, the tip of the vector **E** describes a circle. Under these circumstances equation (12) takes the simple form

$$\mathbf{E}_c = A_c(\mathbf{i} \mp i\mathbf{j})e^{2\pi i\nu(t-z/c)}. \qquad (19)$$

The upper sign corresponds to counterclockwise and the lower to clockwise rotation. When $\beta = \alpha + \pi/2$, we see from (11) and (12) that

$$\mathbf{E}_c = (\mathbf{i}E_x + \mathbf{j}E_y)e^{2\pi i\nu(t-z/c)+i\alpha}. \qquad (20)$$

The vector, being real, does not rotate when we come back to the case of plane-polarized light. Equation (19) refers to circular polarization. These formulae represent the two extremes of the general expression (12), which refers to elliptical polarization.

The foregoing notation is easily adapted to scalar quantities. For example, the sinusoidal variation of E_x, expressed by

$$E_x = E_{x0} \cos (2\pi\nu t + \alpha), \tag{21}$$

where E_{x0} is the amplitude of the oscillation, may be written as

$$E_x = E_{xc} e^{2\pi i \nu t}. \tag{22}$$

with the phase constant absorbed in the complex coefficient. We may disentangle α, if necessary, by taking the ratio of the imaginary and real coefficients of E_{xc}, since

$$\tan \alpha = E_{xi}/E_{xr}. \tag{23}$$

But the phase constants, important as they are when we set up an initial problem, are rarely of significance in the final results. Hence we find the formal shorthand method of calculating time averages extremely useful. We note that

$$\overline{E_x^2} = E_{xc} E_{xc}^*/2. \tag{24}$$

This formalized representation of a vector as the real part of a complex bivector is a natural extension of the methods we introduced for scalar quantities, cf. III-4. As one further example, note the existence of the identity

$$\begin{aligned} \mathbf{r} &= \mathbf{i}x + \mathbf{j}y + \mathbf{k}z \\ &= r\left[\frac{1}{2}(\mathbf{i} + i\mathbf{j}) \sin \theta \, e^{-i\varphi} + \frac{1}{2}(\mathbf{i} - i\mathbf{j}) \sin \theta \, e^{i\varphi} + \mathbf{k} \cos \theta\right]. \end{aligned} \tag{25}$$

We prove the above from the equations

$$x = r \sin \theta \cos \varphi, \quad y = r \sin \theta \sin \varphi, \quad z = r \cos \varphi,$$

$$e^{\pm i\varphi} = \cos \varphi \pm i \sin \varphi.$$

For bivectors, the quantities

$$\boldsymbol{\alpha} = \frac{1}{\sqrt{2}}(\mathbf{i} + i\mathbf{j}), \quad \boldsymbol{\beta} = \frac{1}{\sqrt{2}}(\mathbf{i} - i\mathbf{j}), \quad \boldsymbol{\gamma} = \mathbf{k}, \tag{26}$$

take the place of the unit vectors **i**, **j**, and **k**. They form an ortho-normal set in the sense that

$$\left.\begin{aligned} \boldsymbol{\alpha} \cdot \boldsymbol{\alpha}^* = 1, \quad \boldsymbol{\alpha} \cdot \boldsymbol{\beta}^* = 0, \quad \boldsymbol{\alpha} \cdot \boldsymbol{\gamma}^* = 0 \\ \boldsymbol{\beta} \cdot \boldsymbol{\alpha}^* = 0, \quad \boldsymbol{\beta} \cdot \boldsymbol{\beta}^* = 1, \quad \boldsymbol{\beta} \cdot \boldsymbol{\gamma}^* = 0 \\ \boldsymbol{\gamma} \cdot \boldsymbol{\alpha}^* = 0, \quad \boldsymbol{\gamma} \cdot \boldsymbol{\beta}^* = 0, \quad \boldsymbol{\gamma} \cdot \boldsymbol{\gamma}^* = 1 \end{aligned}\right\} \tag{27}$$

We can extend the concept to tensors or dyadics. For example, the dyadic **rr** becomes

$$\mathbf{rr} = r^2 \cdot \begin{bmatrix} \frac{1}{2}\,\boldsymbol{\alpha\alpha}\sin^2\theta\, e^{-2i\varphi} & \frac{1}{2}\,\boldsymbol{\alpha\beta}\sin^2\theta & \frac{1}{\sqrt{2}}\,\boldsymbol{\alpha\gamma}\sin\theta\cos\theta\, e^{-i\varphi} \\ \frac{1}{2}\,\boldsymbol{\beta\alpha}\sin^2\theta & \frac{1}{2}\,\boldsymbol{\beta\beta}\sin^2\theta\, e^{2i\varphi} & \frac{1}{\sqrt{2}}\,\boldsymbol{\beta\gamma}\sin\theta\cos\theta\, e^{i\varphi} \\ \frac{1}{\sqrt{2}}\,\boldsymbol{\gamma\alpha}\sin\theta\cos\theta\, e^{-i\varphi} & \frac{1}{\sqrt{2}}\,\boldsymbol{\gamma\beta}\sin\theta\cos\theta\, e^{i\varphi} & \boldsymbol{\gamma\gamma}\cos^2\theta \end{bmatrix}. \tag{28}$$

Both **r** and **rr** are real, despite the complex notation. However, the exponential form often simplifies the calculations.

Similarly we can write any real vector **A** in bivector notation as

$$\mathbf{A} = \mathbf{i}A_x + \mathbf{j}A_y + \mathbf{k}A_z = \frac{\boldsymbol{\alpha}}{\sqrt{2}}(A_x - iA_y) + \frac{\boldsymbol{\beta}}{\sqrt{2}}(A_x + iA_y) + \boldsymbol{\gamma}A_z \tag{29}$$

29. Radiation from a pulsating charge. Consider a volume containing an electric charge, whose density is fluctuating. In some volume element, suppose we express the density, ρ, as a function of the time, t', so that

$$\rho = \rho(t'). \tag{1}$$

Then the scalar potential, $d\phi$, at a distance r, arising from the charge within $d\tau$, is

$$d\phi = \left[\frac{\rho(t')}{r}\,d\tau\right]_{t=t'+r/c}. \tag{2}$$

$$d\varphi = \left[\frac{\rho(t')}{4\pi\epsilon r}\,d\tau\right]_{t=t'+r/c}. \tag{2a}$$

The brackets signify that we must calculate the potential at r for a time t, later than t', by the light-time, r/c. We assume that the potentials are propagated in a vacuum. We are concerned with the heterogeneous fluctuations in ρ only to the extent of evaluating the radiation that results from them. Hence we simplify the problem as follows. By means of a spectroscope at r, we resolve the radiation into its component frequencies. In this way we may confine our attention to a given frequency. We may express the fluctuations in density in a pseudo-Fourier series as follows:

$$\rho(t') = \rho_0 + \sum_n^\infty \rho_n \cos(2\pi\nu_n t' + \alpha), \tag{3}$$

where we use the index ν_n instead of n itself, because the frequencies involved may not be integral multiples of one another. Note in particular that since

$$\int \rho(t)\, d\tau = \rho_0, \tag{4}$$

the total charge in the volume, we must have

$$\int \rho_n\, d\tau = 0, \qquad n \geq 1. \tag{5}$$

If we substitute this expression into (2) and integrate over the volume, we see that the first term gives only the constant electrostatic potential, which is of no importance in the radiation problems. Further, if we restrict the problem to consideration of a given frequency, ν, ignoring possible fluctuations that give rise to radiation in another spectral region, we may study the series one term at a time.

The phase constant, α, is likely to prove troublesome, since different regions will be oscillating with diverse phases. We may at least postpone consideration of the problem, by means of the notation of the previous section (38.21) and (38.22). Introduce a complex density, ρ_c, such that

$$\rho(t') = \rho_c e^{2\pi i\nu t'}. \tag{6}$$

Then we may write the potential at r, a function of the time t, as the real part of

$$\phi_c = \int \frac{\rho_c e^{2\pi i\nu(t-r/c)}}{r}\, d\tau. \tag{7}$$

$$\phi_c = \int \frac{\rho_c e^{2\pi i\nu(t-r/c)}}{4\pi\epsilon r}\, d\tau. \tag{7a}$$

The waves from each volume element are propagated with spherical symmetry. Hence the elementary potential, resulting from the volume element $d\tau$,

$$d\varphi_c = \frac{\rho_c e^{2\pi i\nu(t-r/c)}}{r}\, d\tau, \tag{8}$$

$$d\varphi_c = \frac{\rho_c e^{2\pi i\nu(t-r/c)}}{4\pi\epsilon r}\, d\tau, \tag{8a}$$

must obey the wave equation for spherical symmetry,

$$\nabla^2\varphi_c = \frac{1}{r^2}\frac{\partial}{\partial r}\left(r^2\frac{\partial\varphi_c}{\partial r}\right) = \frac{1}{c^2}\frac{\partial^2\varphi_c}{\partial t^2}. \tag{9}$$

Direct substitution verifies that (8) satisfies (9). Hence we may proceed with (7).

We adopt spherical coordinates with an axis running from the origin to the external point, P, and measure θ from this axis. Then, if P lies at distance r_2 from the origin, and if the coordinates of $d\tau$ are r_2, θ, ϕ, we have by the law of cosines,

$$r = \sqrt{r_1^2 + r_2^2 - 2r_1r_2 \cos \theta}, \tag{10}$$

which we are to substitute into (5), and integrate over the volume

$$d\tau = r_1^2 \, dr_1 \, d\theta \, d\varphi. \tag{11}$$

Here r_2 is constant during the integration and r_1 the variable coordinate.

In evaluating φ, we make use of the condition, $r_2 \gg r_1$, to expand r in powers of r_1/r_2. The reader will recognize the identity of this procedure with that used earlier, in our development of spherical harmonics. From (II-17.7) and (II-17.8), we have, to the first approximation,

$$\frac{1}{r} = \frac{1}{r_2}\left[1 + \frac{r_1}{r_2} \cos \theta + \ldots\right]. \tag{12}$$

Inverting (12), by direct division or by the binomial theorem, we have

$$r = r_2\left[1 - \frac{r_1}{r_2} \cos \theta + \ldots\right], \tag{13}$$

and similarly

$$e^{-ikr} \sim e^{-ikr_2} e^{ikr_1\cos\theta} \sim e^{-ikr_2}\left[1 + ikr_1 \cos \theta - \frac{k^2r_1^2 \cos^2 \theta}{2_{,}} + \ldots\right], \tag{14}$$

where

$$k = 2\pi\nu/c = 2\pi/\lambda. \tag{15}$$

For visual light, $k \sim 10^5$. Since we shall later adopt r_1 as the radius of an atom, i.e., $r_1 \sim 10^{-8}$, the above expansion is legitimate. Then, with (12) and (14), (7) becomes

$$\varphi_c = \frac{e^{2\pi i\nu t - ikr_2}}{r_2} \int \rho_c\left(1 + ikr_1 \cos \theta - \frac{k^2r_1^2 \cos^2 \theta}{2}\right) d\tau. \tag{16}$$

$$\varphi_c = \frac{e^{2\pi i\nu t - ikr_2}}{4\pi\epsilon r_2} \int \rho_c\left(1 + ikr_1 \cos \theta - \frac{k^2r_1^2 \cos^2 \theta}{2}\right) d\tau. \tag{16a}$$

The neglected terms are all of powers (r_1/r_2) or higher, consistent with our setting $r_2 \gg r_1$.

We express the retarded potential $\mathbf{A}$, (10.47), as a complex bivector, again to avoid difficulties with phases. Hence,

$$\mathbf{J} = \mathbf{J}_c e^{2\pi i\nu t'} \tag{17}$$

and

$$\mathbf{A}_c = \frac{1}{c} \int \frac{\mathbf{J}_c}{r} e^{2\pi i\nu t'} \, d\tau = \frac{1}{c} \int \frac{\mathbf{J}_c}{r} e^{2\pi i\nu(t - r/c)} \, d\tau, \tag{18}$$

$$\mathbf{A}_c = \frac{\mu}{4\pi} \int \frac{\mathbf{J}_c}{r} e^{2\pi i\nu(t - r/c)} \, d\tau, \tag{18a}$$

where $\mathbf{J}_c$ is the complex current density. Since each vector component of $\mathbf{J}_c$ gives a scalar equation of the form of (7), the expansion proves to be identical with (7). Therefore

$$\mathbf{A}_c = \frac{e^{2\pi i\nu t - ikr_2}}{cr_2}\int \mathbf{J}_c(1 + ikr_1 \cos\theta)\, d\tau. \tag{19}$$

$$\mathbf{A}_c = \frac{\mu e^{2\pi i\nu t - ikr_2}}{4\pi r_2}\int \mathbf{J}_c(1 + ikr_1 \cos\theta)\, d\tau. \tag{19a}$$

Higher order terms are not necessary.

We shall now transform (16) and (19) to full vector form, and identify the successive terms. First, we note from (5) that the first term of (16) vanishes, unless $\nu = 0$, when φ reduces to the electrostatic potential. This does not interest us, however, because no radiation results. Let $\mathbf{i}_r$ be a unit vector along $\mathbf{r}_2$. Then

$$r_1 \cos\theta = \mathbf{r}_1 \cdot \mathbf{i}_r \tag{20}$$

and

$$(r_1 \cos\theta)^2 = (\mathbf{r}_1 \cdot \mathbf{i}_r)^2 = \mathbf{i}_r \cdot \mathbf{r}_1\mathbf{r}_1 \cdot \mathbf{i}_r. \tag{21}$$

The dyadic form is especially useful since we may factor the unit vectors, $\mathbf{i}_r$, from the sign of integration.

Let us set

$$\mathbf{P}_c = \int \rho_c \mathbf{r}_1\, d\tau \quad \text{and} \quad \mathfrak{R}_c = \int \rho_c \mathbf{r}_1\mathbf{r}_1\, d\tau. \tag{22}$$

Then

$$\varphi_c = \frac{e^{2\pi i\nu t - ikr}}{r}\left(ik\mathbf{i}_r \cdot \mathbf{P}_c - \frac{k^2}{2}\mathbf{i}_r \cdot \mathfrak{R}_c \cdot \mathbf{i}_r\right), \tag{23}$$

$$\varphi_c = \frac{e^{2\pi i\nu t - ikr}}{4\pi\epsilon r}\left(\quad\right), \tag{23a}$$

wherein we have dropped the subscript of r_2, as no longer needed. Here $\mathbf{P}_c$ is clearly the first-order moment and $\mathfrak{R}_c$ the second-order electric moment of the charge distribution. We have met vectors and dyadics of these types previously. The integral representative of $\mathfrak{R}_c$, apart from the complex notation, is similar to integrals we encountered in Part II, in connection with moments of inertia, which also are of the second order. Also we shall find the complex diadic form of (29.28) useful. Not knowing the phases in each elementary volume of the charge distribution, we still are unable to take the real part of these expressions. Until we specify the phases from physical considerations, these integrations are only symbolic. Note, however, that the phases from each charged element have reinforced or canceled one another so as to give some resultant phase for each of the integrals $\mathbf{P}$ and $\mathfrak{R}$.

We turn now to the vector potential. The equation of continuity (II-34.4), cf. also (14.12) to (14.14),

$$\frac{\partial \rho}{\partial t'} = -\nabla \cdot (\rho \mathbf{v}) = -\nabla \cdot \mathbf{J}, \tag{24}$$

takes the following form in complex notation:

$$\nabla \cdot \mathbf{J}_c = -2\pi i \nu \rho_c = -ikc\rho_c, \tag{25}$$

as we introduce (6) and (17). Consider the integral,

$$\int g\nabla \cdot \mathbf{J}\, d\tau = -ikc \int \rho g\, d\tau, \tag{26}$$

where g is any scalar quantity. Then, since

$$\nabla \cdot (g\mathbf{J}) = g\nabla \cdot \mathbf{J} + \mathbf{J} \cdot \nabla g, \tag{27}$$

(II, Problem 59), we may write

$$\int g\nabla \cdot \mathbf{J}\, d\tau = \int g\mathbf{J} \cdot d\mathbf{S} - \int \mathbf{J} \cdot \nabla g\, d\tau, \tag{28}$$

where the surface integral results from an application of Gauss' theorem. If we choose for our surface one completely enclosing and entirely outside the atom, $\mathbf{J}$ will vanish over the boundary; thus the surface integral will be equal to zero. Each component of a vector $\mathbf{r}$, e.g., $\mathbf{i}x$, $\mathbf{i}y$, . . . , or of a dyadic, $\mathbf{rr}$, e.g., $\mathbf{ii}x^2$, $\mathbf{ij}xy$, . . . , is a scalar. Hence, after performing the transformation (28), for each scalar component, we may multiply by the appropriate combination of unit vectors and sum. This procedure, however, is equivalent to setting $g = \mathbf{r}$ or $\mathbf{rr}$ originally, as the case may be. Now

$$\begin{aligned}\mathbf{J} \cdot \nabla\mathbf{r} &= \left(J_x \frac{\partial}{\partial x} + J_y \frac{\partial}{\partial y} + J_z \frac{\partial}{\partial z}\right)(\mathbf{i}x + \mathbf{j}y + \mathbf{k}z) \\ &= \mathbf{i}J_x + \mathbf{j}J_y + \mathbf{k}J_z = \mathbf{J},\end{aligned} \tag{29}$$

and

$$\mathbf{J} \cdot \nabla(\mathbf{rr}) = \left(J_x \frac{\partial}{\partial x} + J_y \frac{\partial}{\partial y} + J_z \frac{\partial}{\partial z}\right)\mathbf{rr}. \tag{30}$$

Expanding $\mathbf{rr}$ in terms of its nine dyad components, differentiating and collecting terms, we easily find that

$$\mathbf{J} \cdot \nabla(\mathbf{rr}) = \mathbf{Jr} + \mathbf{rJ}, \tag{31}$$

also a dyadic. Therefore

$$\begin{aligned}\int \mathbf{J}_c\, d\tau &= \int \mathbf{J}_c \cdot \nabla\mathbf{r}\, d\tau \\ &= -\int \mathbf{r}\nabla \cdot \mathbf{J}_c\, d\tau = ikc \int \rho_c \mathbf{r}\, d\tau = ikc\mathbf{P}_c,\end{aligned} \tag{32}$$

by (29), (28), (25), and (22). Similarly,

$$\int (\mathbf{J}_c\mathbf{r} + \mathbf{r}\mathbf{J}_c)\, d\tau = \int \mathbf{J}_c \cdot \nabla(\mathbf{r}\mathbf{r})\, d\tau$$

$$= -\int \mathbf{r}\mathbf{r}\nabla \cdot \mathbf{J}_c\, d\tau = ikc\mathfrak{N}_c. \tag{33}$$

Now return to the vector potential, (19). The first integral is simply $ikc\mathbf{P}_c$, by (30). The second integral becomes

$$ikc \int \mathbf{J}_c r_1 \cos\theta\, d\tau = ikc\mathbf{i}_r \cdot \int \mathbf{r}_1\mathbf{J}_c\, d\tau$$

$$= ikc\mathbf{i}_r \cdot \int \left[\frac{1}{2}(\mathbf{r}_1\mathbf{J}_c + \mathbf{J}_c\mathbf{r}_1) + \frac{1}{2}(\mathbf{r}_1\mathbf{J}_c - \mathbf{J}_c\mathbf{r}_1)\right] d\tau, \tag{34}$$

where we have expanded the unsymmetrical dyadic, $\mathbf{r}_1\mathbf{J}$, as the sum of a symmetrical and an anti-symmetrical dyadic, (II-27.14). Note the expansion of the triple vector product, (II-22.25)

$$\mathbf{i}_r \times (\mathbf{r}_1 \times \mathbf{J}_c) = -\mathbf{i}_r \cdot (\mathbf{r}_1\mathbf{J}_c - \mathbf{J}_c\mathbf{r}_1). \tag{35}$$

We have already shown, (10.46), that the total magnetic moment of any current distribution is

$$\mathbf{M}_c = \frac{1}{2c}\int \mathbf{r}_1 \times \mathbf{J}_c\, d\tau. \tag{36}$$

$$\mathbf{M}_c = \frac{1}{2}\int \mathbf{r}_1 \times \mathbf{J}_c\, d\tau. \tag{36a}$$

Thus, from the last group of equations, we finally obtain the result:

$$\mathbf{A}_c = \frac{e^{2\pi i\nu t - ikr}}{r}\, ik\left[\mathbf{P}_c + \frac{ik}{2}\,\mathbf{i}_r \cdot \mathfrak{N}_c - \mathbf{i}_r \times \mathbf{M}_c\right], \tag{37}$$

$$\mathbf{A}_c = \frac{\mu_0 e^{2\pi i\nu t - ikr}}{4\pi r}\, ik\left[\quad\right], \tag{37a}$$

which, together with (16) and equations (26.5) and (26.11),

$$\mathbf{H}_c = \nabla \times \mathbf{A}_c \quad \text{and} \quad \mathbf{E}_c = -\nabla\varphi_c - \frac{1}{c}\frac{\partial \mathbf{A}_c}{\partial t}, \tag{38}$$

$$\mathbf{B}_c = \nabla \times \mathbf{A}_c \quad \text{and} \quad \mathbf{E}_c = -\nabla\varphi_c - \frac{\partial \mathbf{A}_c}{\partial t}, \tag{38a}$$

serves to define completely the periodic complex magnetic and electric vectors at any point in space. Since $\mathbf{A}_c$ and φ_c are functions of r alone, we may apply ∇ in the form: $\nabla = \mathbf{i}_r\partial/\partial r$. Thus, keeping only terms of the first order in r^{-1}, we have

$$\mathbf{H}_c = \frac{e^{2\pi i\nu t - ikr}}{r} k^2 \left[\mathbf{i}_r \times \mathbf{P}_c + \frac{ik}{2} \mathbf{i}_r \times (\mathbf{i}_r \cdot \mathfrak{N}_c) - \mathbf{i}_r \times (\mathbf{i}_r \times \mathbf{M}_c) \right]. \quad (39)$$

$$\mathbf{H}_c = \frac{e^{2\pi i\nu t - ikr}}{4\pi r} k^2 c \left[\quad \right]. \quad (39a)$$

Employing the relations,

$$\mathbf{P} = \mathfrak{J} \cdot \mathbf{P} \quad \text{and} \quad \mathbf{i}_r \cdot \mathfrak{N} = \mathfrak{J} \cdot \mathfrak{N} \cdot \mathbf{i}_r, \quad (40)$$

where $\mathfrak{J}$ is the unit dyadic, we also find that

$$\mathbf{E}_c = \frac{e^{2\pi i\nu t - ikr}}{r} k^2 \left[(\mathfrak{J} - \mathbf{i}_r\mathbf{i}_r) \cdot \left(\mathbf{P}_c + \frac{ik}{2} \mathfrak{N}_c \cdot \mathbf{i}_r \right) - \mathbf{i}_r \times \mathbf{M}_c \right]. \quad (41)$$

$$\mathbf{E}_c = \frac{e^{2\pi i\nu t - ikr}}{4\pi\epsilon_0 r} k^2 \left[\quad \right]. \quad (41a)$$

The three terms of (39) and (41) are to be considered separately. Thus, from (28.18), we distinguish three cases:

$$\overline{\mathbf{N}}_1 = \frac{c}{4\pi} \overline{(\mathbf{E}_1 \times \mathbf{H}_1)} = \mathbf{i}_r \frac{ck^4}{8\pi r^2} [\mathbf{P}_c \cdot \mathbf{P}_c^* - (\mathbf{i}_r \cdot \mathbf{P}_c)(\mathbf{i}_r \cdot \mathbf{P}_c^*)], \quad (42)$$

$$\overline{\mathbf{N}}_1 = \overline{\mathbf{E}_1 \times \mathbf{H}_1} = \mathbf{i}_r \frac{ck^4}{32\pi^2 \epsilon_0 r^2} \left[\quad \right], \quad (42a)$$

$$\overline{\mathbf{N}}_2 = \mathbf{i}_r \frac{ck^6}{32\pi r^2} [\mathbf{i}_r \cdot \mathfrak{N}_c \cdot \mathfrak{N}_c^* \cdot \mathbf{i}_r - (\mathbf{i}_r \cdot \mathfrak{N}_c \cdot \mathbf{i}_r)(\mathbf{i}_r \cdot \mathfrak{N}_c^* \cdot \mathbf{i}_r)], \quad (43)$$

$$\overline{\mathbf{N}}_2 = \mathbf{i}_r \frac{ck^6}{128\pi^2 \epsilon_0 r^2} \left[\quad \right], \quad (43a)$$

$$\overline{\mathbf{N}}_3 = \mathbf{i}_r \frac{ck^4}{8\pi r^2} [\mathbf{M}_c \cdot \mathbf{M}_c^* - (\mathbf{i}_r \cdot \mathbf{M}_c)(\mathbf{i}_r \cdot \mathbf{M}_c^*)], \quad (44)$$

$$\overline{\mathbf{N}}_3 = \mathbf{i}_r \frac{ck^4}{32\pi^2 \epsilon_0 r^2} \left[\quad \right], \quad (44a)$$

which refer, respectively, to radiation resulting from the so-called electric dipole, electric quadrupole, and magnetic dipole moments of the fluctuating charge distribution. These expressions are by no means as complicated as they appear at first sight. They reduce to simple expressions when $\mathbf{P}_c$, $\mathfrak{N}_c$ and $\mathbf{M}_c$ are given. But we must have advance knowledge of the nature of the fluctuations. For example, if the electric moment is that of a linear oscillator, we may set

$$\mathbf{P}_c = \mathbf{k} P e^{2\pi i\nu t}, \quad \mathbf{P}_c \cdot \mathbf{P}_c^* = P^2. \quad (45)$$

By (28.25),

$$\mathbf{i}_r = \frac{\mathbf{r}}{r} = \mathbf{i} \sin \theta \cos \varphi + \mathbf{j} \sin \theta \sin \varphi + \mathbf{k} \cos \theta,$$

$$= \frac{1}{2}(\mathbf{i} + i\mathbf{j}) \sin \theta e^{-i\varphi} + \frac{1}{2}(\mathbf{i} - i\mathbf{j})e^{i\varphi} + \mathbf{k} \cos \theta. \tag{46}$$

We find that

$$\mathbf{N}_1 = \mathbf{i}_r \frac{ck^4}{8\pi r^2} P^2 \sin^2 \theta. \tag{47}$$

$$\mathbf{N}_1 = \mathbf{i}_r \frac{ck^4}{32\pi^2 \epsilon_0 r^2} P^2 \sin^2 \theta. \tag{47a}$$

If the electric moment arises from rotation of charges in a circle, we may set, as in (28.19),

$$\mathbf{P}_c = \frac{P}{\sqrt{2}} (\mathbf{i} \mp i\mathbf{j})e^{2\pi i\nu t}, \tag{48}$$

where the normalizing factor, $\sqrt{2}$, has been introduced in the denominator, so that we shall continue to have

$$\mathbf{P}_c \cdot \mathbf{P}_c^* = P^2, \tag{49}$$

as in (45). When the normalizing factor has been thus introduced, we must define P appropriately so that $\mathbf{P}$ will still be the real part of (48). For this case,

$$\mathbf{N}_1 = \mathbf{i}_r \frac{ck^4}{8\pi r^2} \frac{P^2}{2} (1 + \cos^2 \theta), \tag{50}$$

$$\mathbf{N}_1 = \mathbf{i}_r \frac{ck^4}{32\pi^2 r^2} \frac{P^2}{2} (1 + \cos^2 \theta), \tag{50a}$$

independent of the sign of rotation. The total flux over the surface of a sphere of radius r, is

$$\varphi = \int \mathbf{N}_1 \cdot d\mathbf{S} = \frac{16\pi^4 \nu^4}{3c^3} P^2, \tag{51}$$

$$\varphi = \frac{4\pi^3 \nu^4}{3\epsilon_0 c^3} P^2, \tag{51a}$$

where $d\mathbf{S} = \mathbf{i}_r r^2 \sin \theta \, d\theta \, d\phi$. Equation (50) holds for both expressions for $\mathbf{N}_1$. There are alternative forms for $\mathbf{P}$; e.g., we may consider the case of elliptic instead of circular or linear vibrations. But (45) and (48) represent the two extremes and we may express other cases in terms of these.

The magnetic dipole formulae become identical with those for electric dipole radiation, if we replace $\mathbf{P}$ by $\mathbf{M}$. The electric quadrupole problem is a little more complicated. We may, as for dipole radiation, distinguish several fundamental cases. We have defined $\mathfrak{N}$ as follows:

$$\mathfrak{N} = \int \rho \mathbf{rr} \, d\tau. \tag{52}$$

To write this expression in complex form, we set $\rho = \rho_c e^{2\pi i \nu t}$ as before, and build up the dyadic **rr** as in (28.28), from the bivectors of the dipole problem.

We may classify the elements of the dyadic **rr** (28.28) according to the factor $e^{im\phi}$, where $m = \pm 2, \pm 1$, or 0. When ρ_c contains the complex conjugate, $e^{-im\phi}$, as a factor, the integral over ϕ is finite. Otherwise it vanishes. Thus we shall usually find that $\mathfrak{N}$ falls into the following three classes, labeled according to the value of $(-m)$.

$$\mathfrak{N}(\pm 2) = \frac{1}{4} e^{2\pi i \nu t} (\mathbf{i} \pm i\mathbf{j})(\mathbf{i} \pm i\mathbf{j}) \int \rho_c \sin^2 \theta e^{\mp 2i\varphi} \, d\tau. \tag{53}$$

In this equation and in other analogous equations in this section we can advantageously employ the unit bivectors $\boldsymbol{\alpha}$ and $\boldsymbol{\beta}$ defined in (26.26):

$$\mathfrak{N}(\pm 1) = \frac{1}{2} e^{2\pi i \nu t} [(\mathbf{i} \pm i\mathbf{j})\mathbf{k} + \mathbf{k}(\mathbf{i} \pm i\mathbf{j})] \int \rho_c \sin \theta \cos \theta e^{\mp i\varphi} \, d\tau. \tag{54}$$

$$\mathfrak{N}(0) = e^{2\pi i \nu t} \left[\frac{1}{2} \mathfrak{J} \int \rho_c \sin^2 \theta \, d\tau + \mathbf{kk} \int \rho_c \left(\cos^2 \theta - \frac{\sin^2 \theta}{2} \right) d\tau \right], \tag{55}$$

where $\mathfrak{J}$ is the unit dyadic as before.

In actual practice, the significant quadrupole distributions defined by ρ_c, fall into several distinct classes. We may usually write

$$\mathfrak{N} = \mathfrak{R} e^{2\pi i \nu t} \mathfrak{N}_c, \tag{56}$$

where $\mathfrak{R}$ stands for any one of the following dyadics:

$$\mathfrak{R}(\pm 2) = \frac{1}{2} (\mathbf{i} \pm i\mathbf{j})(\mathbf{i} \pm i\mathbf{j}), \tag{57}$$

$$\mathfrak{R}(\pm 1) = \frac{1}{2} [(\mathbf{i} \pm i\mathbf{j})\mathbf{k} + \mathbf{k}(\mathbf{i} \pm i\mathbf{j})], \tag{58}$$

$$\mathfrak{R}(0) = \sqrt{\frac{2}{3}} \left[\mathbf{kk} - \frac{1}{4} (\mathbf{i} + i\mathbf{j})(\mathbf{i} - i\mathbf{j}) - \frac{1}{4} (\mathbf{i} - i\mathbf{j})(\mathbf{i} + i\mathbf{j}) \right]. \tag{59}$$

These brackets have been normalized to unity, so that

$$\mathfrak{R} : \mathfrak{R}^* = 1 \quad \text{or} \quad \mathfrak{N}_c : \mathfrak{N}_c^* = \mathfrak{N}^2. \tag{60}$$

Quadrupole radiation has special significance in connection with atomic problems. We should point out that the three equations refer to quadrupole oscillations of three distinctive types. Whereas dipole moment results from the presence of one plus and one minus charge in close proximity, quadrupole moment requires the presence of a pair of such dipoles, which execute linear or circular oscillations simultaneously and with specified phases. We find it easier to interpret these dyadics from the vector $\mathbf{i}_r \cdot \mathfrak{N}_c$ than from $\mathfrak{N}_c$ itself. Thus we prove, with the aid of (46),

$$\mathbf{i}_r \cdot \mathfrak{K}(\pm 2) = \frac{1}{2}(\mathbf{i} \pm i\mathbf{j}) \sin \theta e^{\pm i\phi}, \tag{61}$$

$$\mathbf{i}_r \cdot \mathfrak{K}(\pm 1) = \frac{1}{2}[\mathbf{k} \sin \theta e^{\pm i\phi} + (\mathbf{i} \pm i\mathbf{j}) \cos \theta], \tag{62}$$

$$\mathbf{i}_r \cdot \mathfrak{K}(0) = \sqrt{\frac{2}{3}}\left[\mathbf{k} \cos \theta - \frac{1}{2}\mathbf{i} \sin \theta \cos \varphi - \frac{1}{2}\mathbf{j} \sin \theta \sin \varphi\right]. \tag{63}$$

These equations correspond respectively to a pair of rotating dipoles, to one dipole oscillating and the other rotating, and to both pairs oscillating.

From (43), (46), (52), and (60) to (63), we calculate the total flux.

$$\phi = \int \mathbf{N}_2 \cdot d\mathbf{S} = \frac{32\pi^6\nu^6}{5c^5}\mathfrak{N}^2. \tag{64}$$

$$\phi = \frac{8\pi^5\nu^6}{5\epsilon_0 c^5}\mathfrak{N}^2. \tag{64a}$$

30. Radiation from a linear oscillator; classical damping factor. An electric charge, oscillating with frequency ν_0 according to the law

$$x = A \cos 2\pi\nu_0 t, \tag{1}$$

$$v = -2\pi\nu_0 A \sin 2\pi\nu_0 t, \tag{2}$$

$$\dot{v} = -(2\pi\nu_0)^2 A \cos 2\pi\nu_0 t, \tag{3}$$

will radiate energy at the rate

$$\frac{dW}{dt} = -\frac{32\pi^4\nu_0^4\varepsilon^2 A^2}{3c^3}\cos^2(2\pi\nu_0 t), \tag{4}$$

$$\frac{dW}{dt} = -\frac{8\pi^3\nu_0^4\varepsilon^2 A^2}{3\epsilon_0 c^3}\cos^2(2\pi\nu_0 t), \tag{4a}$$

by equation (27.43). The minus sign indicates that the oscillator is losing energy. The average of the cosine-square term, over a complete period, is 1/2. Hence the theoretical mean rate of radiation from an oscillator is

$$\frac{dW}{dt} = -\frac{16\pi^4\nu_0^4\varepsilon^2 A^2}{3c^3} = -\frac{16\pi^4\nu_0^4 P^2}{3c^3}, \tag{5}$$

$$\frac{dW}{dt} = -\frac{4\pi^3\nu_0^4 P^2}{3\epsilon_0 c^3}, \tag{5a}$$

where

$$P = \varepsilon A, \tag{6}$$

the electric moment corresponding to the maximum displacement of the oscillator. Equation (5) agrees, as it should, with (29.51), thereby justifying our identification of the last term of (27.41) as the energy actually radiated by an accelerated charge.

The kinetic energy associated with the oscillator is

$$T = \frac{1}{2}mv^2 = 2\pi^2\nu_0^2 mA^2 \sin^2(2\pi\nu_0 t). \tag{7}$$

When

$$2\pi\nu_0 t = \pi/2, \qquad x = 0. \tag{8}$$

All the energy is kinetic and none is potential. Hence, for the total energy of the vibrating system, we may write

$$W = 2\pi^2\nu_0^2 A^2 m. \tag{9}$$

Eliminating A^2 from (5), we have

$$\frac{dW}{dt} = -\frac{8\pi^2\nu_0^2\varepsilon^2}{3mc^3}W = -\gamma W. \tag{10}$$

$$\frac{dW}{dt} = -\frac{2\pi\nu_0^2\varepsilon^2}{3\epsilon_0 mc^3}W = -\gamma W. \tag{10a}$$

We may write the expression as above only if the energy lost during a complete oscillation is small compared with the total energy of the oscillator. If this condition were not fulfilled, we should not be justified in making the step from (4) to (5), neglecting the variation in the amplitude between successive oscillations. If we interpret ε and m as the respective charge and mass of an electron, and if ν represents a frequency of the order of that of visible light, the coefficient of W, in (10), does prove to be small. Integrating (10), we find

$$W = W_0 e^{-\gamma t}, \tag{11}$$

where W_0 represents the initial energy at time $t = 0$. In an interval of time τ where

$$\tau = \frac{3mc^3}{8\pi^2\varepsilon^2\nu_0^2} = \frac{1}{\gamma}, \tag{12}$$

$$\tau = \frac{1}{\gamma} = \frac{3\epsilon_0 mc^3}{2\pi\varepsilon^2\nu_0^2}, \tag{12a}$$

the energy is depleted by a factor $1/e$. Here γ is the well-known classical "damping factor," so termed because it fixes the rate at which the oscillations are damped as a result of radiation processes. Numerically

$$\tau = 4.52\lambda^2 \tag{13}$$

where λ is the wavelength, in centimeters, of the light wave.

The foregoing solution represents in particular the case of an oscillator which, after initial excitation, executes its own damped oscillations. The problem is analogous to that of damped mechanical vibrations as discussed in Part III, §§ 14–17. We may analyze the exponentially-damped beam of radiation emitted by the oscillator, in terms of a Fourier integral. We shall find the spectrum to be composed of frequencies additional to ν_0, as in III-14.

We have previously noted, in the mechanical analogy referred to, that the spectral distribution of the damped and forced vibrations are identical. We shall discuss this question further in the following section.

31. Forced vibrations; electric polarization; refractive index; absorption coefficient. Ordinary, light-transparent media show no macroscopic electric charge. Each negative electron is balanced by the presence of an equal charge of opposite sign. Thus far we have specifically avoided discussing the problem of atomic oscillators. No simple electromechanical model will explain all the features of complex atomic spectra.

Various models of harmonic oscillators have been proposed. The simplest, perhaps, and at the same time most complete is that of Sir J. J. Thomson. Thomson's model of an atom consisted of a uniform spherical distribution of positive electricity in which a negative electron was imbedded. He supposed that the haze of positive electricity offered no resistance to the motion of the negative electron, which was free to oscilate about the position of equilibrium with some characteristic natural frequency ν. Let x be the distance of the electron from the center of the positive sphere. The total charge within the sphere of radius x varies as x^3 and the inverse-square law introduces a factor x^{-2}. Hence the total force acting on the charge is proportional to the displacement. If m is the mass of the electron,

$$m \frac{d^2x}{dt^2} = -ax. \tag{1}$$

The minus sign indicates that the force is attractive. The solution of this equation by (30.22), is

$$x = A \cos(\sqrt{a/m}\, t - \alpha), \tag{2}$$

where A and α are constants of integration. Obviously, we must have the relationship

$$\sqrt{a/m} = 2\pi\nu_0. \tag{3}$$

Equations (1) to (3) represent those of completely undamped oscillations. If now we introduce a resistance to the motion of the electron, such as that imposed by its own radiation, and if we assume a sinusoidal light wave of frequency ν to be falling on the oscillator, the entire equation of motion becomes

$$m \frac{d^2x}{dt^2} - \frac{2\varepsilon^2}{3c^3}\frac{d^3x}{dt^3} = -\alpha x + \varepsilon E_0 \cos 2\pi\nu t. \tag{4}$$

$$m\frac{d^2x}{dt^2} - \frac{\mu_0\varepsilon^2}{6\pi c}\frac{d^3x}{dt^3} = -ax + \varepsilon E_0 \cos 2\pi\nu t. \tag{4a}$$

The two terms on the left are those given by (27.39). The first term on the right is the normal restoring force, in which we may replace a by (3). Here E_0 represents the maximum amplitude of the electric intensity. The last term represents the periodic external driving force tending to set up the oscillations.

Consider, instead of (4), the equation

$$m\frac{d^2x}{dt^2} + b\frac{dx}{dt} = -4\pi^2\nu_0^2 mx + \varepsilon E_0 \cos 2\pi\nu t, \tag{5}$$

which is similar to (4), except that the damping force is proportional to the velocity instead of to the rate of change of acceleration. Equation (5) is identical in form with (III-15.2), which we have already solved. The solution, after the transient terms have died away, is

$$x = C\cos(2\pi\nu t - \alpha), \tag{6}$$

where

$$C = \frac{\varepsilon E_0}{4\pi^2 m}\frac{1}{[(\nu_0^2 - \nu^2)^2 + (b\nu/2\pi m)^2]^{1/2}}. \tag{7}$$

From this equation we have

$$\frac{dx}{dt} = -2\pi\nu C \sin(2\pi\nu t - \alpha), \tag{8}$$

$$\frac{d^3x}{dt^3} = -(2\pi\nu)^2\frac{dx}{dt}. \tag{9}$$

By setting

$$-\frac{2\varepsilon^2}{3c^3}\frac{d^3x}{dt^3} = \frac{8\pi^2\varepsilon^2\nu^2}{3c^3}\frac{dx}{dt} = b\frac{dx}{dt}, \tag{10}$$

$$-\frac{\mu_0\varepsilon^2}{6\pi c}\frac{d^3x}{dt^3} = \frac{2\pi\mu_0\varepsilon^2\nu^2}{3c}\frac{dx}{dt} = b\frac{dx}{dt}, \tag{10a}$$

from equations (4) and (5), we find that

$$b/m = \gamma, \tag{11}$$

the classical damping constant (30.12). Hence if we write (7) as follows:

$$C = \frac{\varepsilon E_0}{4\pi^2 m}\frac{1}{[(\nu_0^2 - \nu^2)^2 + (\gamma\nu/2\pi)^2]^{1/2}}, \tag{12}$$

we have solved equation (4).

The oscillator emits energy at the rate

$$\frac{dW}{dt} = -\frac{16\pi^4\nu^4\varepsilon^2C^2}{3c^3}, \tag{13}$$

$$\frac{dW}{dt} = \frac{4\pi^3\nu^4\varepsilon^2C^2}{3\epsilon_0c^3}, \tag{13a}$$

where the amplitude C takes the place of A in (30.5). The Poynting flux characterized by electric intensity $\mathbf{E}$ is

$$\mathbf{N} = \frac{c}{4\pi}\mathbf{E} \times \mathbf{H}, \tag{14}$$

$$\mathbf{N} = \mathbf{E} \times \mathbf{H}, \tag{14a}$$

Also, cf. (19.5) and (19.8), for free space,

$$|\mathbf{E}| = |\mathbf{H}|, \quad |\mathbf{N}| = \frac{c}{4\pi}E_0^2\cos^2(2\pi\nu t), \quad \overline{N} = \frac{cE_0^2}{8\pi}, \tag{15}$$

$$\sqrt{\epsilon_0}\,|\mathbf{E}| = \sqrt{\mu_0}\,|\mathbf{H}|, \quad |\mathbf{N}| = |\mathbf{E} \times \mathbf{H}| = \sqrt{\frac{\epsilon_0}{\mu_0}}\,E_0^2\cos^2(2\pi\nu t) \tag{15a}$$

$$\overline{N} = \frac{\epsilon_0cE_0^2}{2},$$

where we have averaged over the cycle. $\overline{N}$ represents the average radiation incident normally per sec upon an area of one cm^2. The oscillator presents a cross section for absorption of α_ν cm^2. The energy absorbed must be

$$\overline{N}\alpha_\nu = \frac{c}{8\pi}E_0^2\alpha_\nu = \frac{16\pi^4\nu^4\varepsilon^2C^2}{3c^3}, \tag{16}$$

$$\overline{N}\alpha_\nu = \frac{\epsilon_0cE_0^2\alpha\nu}{2} = \frac{4\pi^3\nu^4\varepsilon^2C^2}{3\epsilon_0c^3}, \tag{16a}$$

since the energy absorbed must be equal to that emitted, when a steady state has been set up.

Introducing C, from (12), we finally obtain the result:

$$\alpha_\nu = \frac{8\pi\varepsilon^4}{3m^2c^4}\frac{\nu^4}{[(\nu_0^2 - \nu^2)^2 + (\gamma\nu/2\pi)^2]}. \tag{17}$$

$$\alpha_\nu = \frac{\varepsilon^4}{6\pi\epsilon_0^2m^2c^4}\frac{\nu^4}{[(\nu^2 - \nu_0^2)^2 + (\gamma\nu/2\pi)^2]}. \tag{17a}$$

This formula is extremely important. We call α_ν the atomic absorption coefficient, as calculated by classical theory. We shall see in a moment that the above general result contains a number of special cases of scientific interest. We see that α_ν has the physical dimensions of an area.

There is an alternative derivation of equation (17) that illustrates still one further feature of the problem of radiation. Let us suppose that the medium contains N oscillators per unit volume. We have already seen that if the medium were subject to no outside disturbance, the amplitudes of the individual oscillators would rapidly decrease to zero. The electric moment would likewise vanish. We may consider that the light, traversing the medium, induces a certain electric moment per unit volume. The magnitude of the induced electric moment, often called the electric "polarization," is

$$P = N\varepsilon x. \tag{18}$$

We shall suppose the beam to be traveling along the z-axis. We shall suppose, also, that the beam consists of plane-polarized light, with the electric vector parallel to the axis of x. Then we may write, analogous to (24.7),

$$\mathbf{E} = \mathbf{i}E_0 \cos 2\pi\nu\left(\frac{zn}{c} - t\right) = \mathbf{i}E_0 e^{2\pi i\nu(zn/c - t)}, \tag{19}$$

where n is, by definition, the refractive index of the gas. Then c/n measures the wave velocity.

We are to determine n from the conditions of the problem. A medium that possesses neither free electric charges nor induced electric moment behaves electrically like a vacuum, with $\kappa = \epsilon/\epsilon_0$, the dielectric constant, equal to unity. If κ in the medium differs from unity, we must ascribe the variation to the presence of induced polarization.

Let us consider what forms the Maxwell equations assume for certain specified conditions. We restrict our attention to the equation whose general form is

$$\nabla \times \mathbf{H} = \frac{1}{c}\frac{\partial \mathbf{D}}{\partial t} + \frac{4\pi}{c}\mathbf{J}. \tag{20}$$

$$\nabla \times \mathbf{H} = \frac{\partial \mathbf{D}}{\partial t} + \mathbf{J}. \tag{20a}$$

In vacuo, (20) takes the form

$$\nabla \times \mathbf{H} = \frac{1}{c}\frac{\partial \mathbf{E}}{\partial t}, \tag{21}$$

$$\nabla \times \mathbf{H} = \epsilon_0 \frac{\partial \mathbf{E}}{\partial t}, \tag{21a}$$

and in a medium containing no free charges,

$$\nabla \times \mathbf{H} = \frac{1}{c}\frac{\partial \mathbf{D}}{\partial t} = \frac{\kappa}{c}\frac{\partial \mathbf{E}}{\partial t}. \tag{22}$$

$$\nabla \times \mathbf{H} = \epsilon \frac{\partial \mathbf{E}}{\partial t}. \tag{22a}$$

In Maxwell's equations, displacement currents are to be treated on a par with real currents. The polarization results in the production of a current density $\mathbf{J}$:

$$\mathbf{J} = \mathbf{i}N\varepsilon \frac{dx}{dt} = \mathbf{i}\frac{\partial P}{\partial t} = \frac{\partial \mathbf{P}}{\partial t}, \tag{23}$$

since $\mathbf{P}$ is itself a vector. Multiply this expression by $4\pi/c$ and add it to (21); thus

$$\nabla \times \mathbf{H} = \frac{1}{c}\frac{\partial}{\partial t}(\mathbf{E} + 4\pi\mathbf{P}), \tag{24}$$

$$\nabla \times \mathbf{H} = \frac{\partial}{\partial t}(\epsilon_0\mathbf{E} + \mathbf{P}), \tag{24a}$$

to allow for the effect of polarization in the dielectric, as we have previously termed a medium devoid of free charges, § (2). Equations (24) and (22) are consistent if we set, as in (14.25),

$$\mathbf{D} = \kappa\mathbf{E} = \mathbf{E} + 4\pi\mathbf{P}. \tag{25}$$

$$\epsilon\mathbf{E} = \epsilon_0\mathbf{E} + \mathbf{P}. \tag{25a}$$

Here $\mathbf{E}$ and $\mathbf{P}$ are parallel vectors because we assume the medium to be isotropic. Thus we have merely to consider their absolute values, and write

$$\kappa = 1 + \frac{4\pi P}{E}. \tag{26}$$

$$\kappa = \frac{\epsilon}{\epsilon_0} = 1 + \frac{P}{\epsilon_0 E}. \tag{26a}$$

We return, now, to equation (4), written in the form:

$$\frac{d^2x}{dt^2} + \gamma\frac{dx}{dt} + (2\pi\nu_0)^2x = \frac{\varepsilon E_0}{m}e^{2\pi i\nu(zn/c-t)}, \tag{27}$$

by (8), (9), and (19).

Although we may, if we choose, work entirely with real variables, employing a cosine instead of the imaginary-exponential factor, the algebra is considerably simplified by the present procedure. The reader is encouraged not to worry about obscure physical conceptions like complex amplitudes, complex refractive indices, etc. The appearance of the complex quantities is only temporary. To regain the physical meaning we must take the real part of the final solution as before.

Assume a solution of the form

$$x = Ae^{2\pi i\nu(zn/c-t)}, \tag{28}$$

where z, being the coordinate of the given oscillator, is a constant. Differentiating this expression, substituting the result into (27), and solving for the amplitude, we find that

$$A = \frac{\varepsilon E_0}{4\pi^2 m}\left[\frac{1}{(\nu^2 - \nu_0^2) - \gamma\nu i/2\pi}\right]. \tag{29}$$

From (18), (19), (28), and (29), we find that

$$\kappa = 1 + \frac{N\varepsilon^2}{\pi m}\frac{1}{(\nu_0^2 - \nu^2)^2 + (\gamma\nu/2\pi)^2}\left[(\nu_0^2 - \nu^2) + \frac{\gamma\nu i}{2\pi}\right], \tag{30}$$

$$\kappa = 1 + \frac{N\varepsilon^2}{4\pi^2\epsilon_0 m}\frac{1}{(\nu_0^2 - \nu^2)^2 + (\gamma\nu/2\pi)^2}\left[(\nu_0^2 - \nu^2) + \frac{\gamma\nu i}{2\pi}\right], \tag{30a}$$

where we have cleared the denominator of imaginary quantities.

The procedure we have followed up to the present bears a close resemblance to that followed in §§ 24 and 25. Our dielectric constant and hence the refractive index turns out to be complex.

Let $\mathfrak{n}$ be the complex index of refraction. Then, by (24.5),

$$\kappa = \mathfrak{n}^2. \tag{31}$$

Instead of setting $\mathfrak{n} = a + bi$, we shall anticipate the form of the final result and set the complex index

$$\mathfrak{n} = \frac{n + ikc}{4\pi\nu} = \kappa^{1/2},$$

where n and k are constants to be interpreted physically. For the physical conditions we shall ordinarily encounter, the second term of (30) proves to be small compared with unity. We may therefore expand $\kappa^{1/2}$ by means of the binomial theorem, and write

$$\mathfrak{n} = \frac{n + ikc}{4\pi\nu} = 1 + \frac{N\varepsilon^2}{2\pi m}\frac{1}{(\nu_0^2 - \nu^2)^2 + (\gamma\nu/2\pi)^2}\left[(\nu_0^2 - \nu^2) + \frac{\gamma\nu i}{2\pi}\right]. \tag{32}$$

$$\mathfrak{n} = 1 + \frac{N\varepsilon^2}{8\pi^2\epsilon_0 m}\frac{1}{(\nu_0^2 - \nu^2)^2 + (\gamma\nu/2\pi)^2}\left[(\nu_0^2 - \nu^2) + \frac{\gamma\nu i}{2\pi}\right]. \tag{32a}$$

The quantity written as n in equation (19) should now be interpreted as $\mathfrak{n}$. Dropping the vector form of (19), we have

$$\begin{aligned} E = E_0 e^{2\pi i\nu(z\mathfrak{n}/c - t)} &= E_0 e^{-kz/2} e^{2\pi i\nu(zn/c - t)} \\ &= E_0 e^{-kz/2}\cos 2\pi\nu(zn/c - t). \end{aligned} \tag{33}$$

We have finally taken the real part of the complex expression and are now in a position to interpret the result physically. The cosine term represents, as in (19), a wave progressing parallel to the axis of z. The induced polarization of the medium has had two effects on the incident light beam. The wave velocity has the value

$$v = c/n, \tag{34}$$

instead of c, as for a vacuum. n thus plays the role of the ordinary refractive index. Also the presence of the exponential term leads to actual absorption of the radiation. The amplitude of the beam is decreasing as $e^{-kz/2}$ and the energy, which is proportional to the square of the amplitude, varies as e^{-kz}. We may, therefore, interpret k as the absorption coefficient per unit path.

The magnitudes of n and k still remain to be fixed. Equating the real and imaginary parts on both sides of (32), we readily obtain the results

$$n - 1 = \frac{N\varepsilon^2}{2\pi m} \frac{\nu_0^2 - \nu^2}{(\nu_0^2 - \nu^2)^2 + (\gamma\nu/2\pi)^2}, \tag{35}$$

$$n - 1 = \frac{N\varepsilon^2}{8\pi^2\epsilon_0 m} \frac{\nu_0^2 - \nu^2}{(\nu_0^2 - \nu^2)^2 + (\gamma\nu/2\pi)^2}, \tag{35a}$$

$$k = N \frac{8\pi\varepsilon^4}{3m^2c^4} \frac{\nu^4}{(\nu_0^2 - \nu^2)^2 + (\gamma\nu/2\pi)^2}, \tag{36}$$

$$k = N \frac{\varepsilon^4}{6\pi\epsilon_0^2 m^2c^4} \frac{\nu^4}{(\nu_0^2 - \nu^2)^2 + (\gamma\nu/2\pi)^2}, \tag{36a}$$

where we have substituted for γ from (30.12). Equation (36) is consistent with (17) through the relationship

$$k = N\alpha_\nu. \tag{37}$$

32. Anomalous dispersion. According to equation (31.35), the index of refraction of most gases will lie in the vicinity of unity. For a frequency ν in the neighborhood of an absorption line that lies at frequency ν_0, we may write

$$\nu_0^2 - \nu^2 = (\nu_0 + \nu)(\nu_0 - \nu) \sim 2\nu_0(\nu_0 - \nu), \tag{1}$$

and (31.35) becomes

$$n - 1 \sim \frac{N\varepsilon^2}{4\pi m\nu_0} \frac{\nu_0 - \nu}{(\nu_0 - \nu)^2 + (\gamma/4\pi)^2}. \tag{2}$$

$$n - 1 = \frac{N\varepsilon^2}{16\pi^2\epsilon_0 m\nu_0} \frac{\nu_0 - \nu}{(\nu_0 - \nu)^2 + (\gamma/4\pi)^2}. \tag{2a}$$

When $\nu > \nu_0$, i.e., for frequencies greater than ν_0, $n - 1$ is negative. The refractive index is less than unity and the wave velocity is greater than that of light in vacuo.* For frequencies less than ν_0, the refractive index is greater than unity. The form of the curve of $n - 1$ is shown in Fig. 27. The explanation of the well-known phenomenon of anomalous dispersion or refraction lies in the above formula. Figure 28 illustrates a well-known laboratory experiment. Here A is a source, let us say, of sodium light; B is a spectrograph slit (horizontal); C is a collimator lens; D a prism; and F the

*The reader is assured that a wave velocity greater than c does not violate the law of relativity. As will be shown presently, the "signal velocity" is always less than c.

camera lens. If the second prism, E, were not present, the spectrum would extend in a horizontal plane and the lines would be vertical. Let GG′ represent the position of one of the *D* lines of sodium.

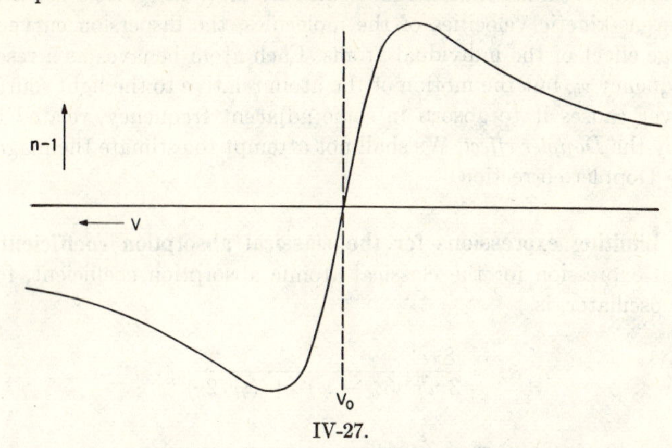

IV-27.

The prism E is merely a glass shell filled with sodium vapor. The base of the second prism is at right angles to that of the first. As a result of anomalous dispersion, the prism E bends light in the neighborhood of the

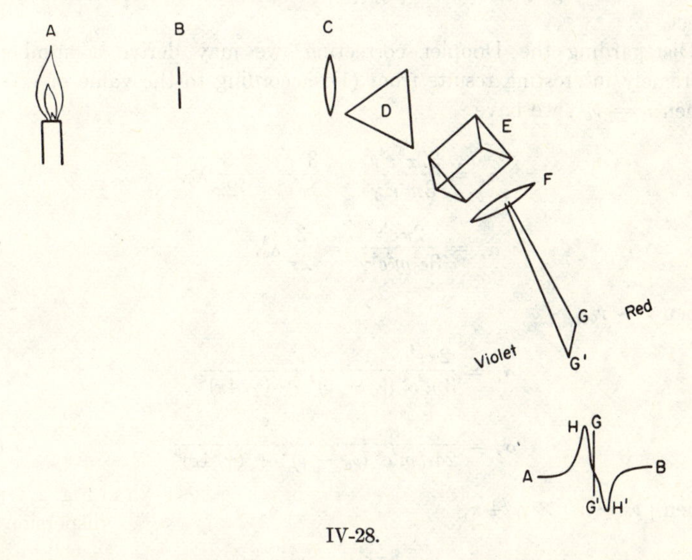

IV-28.

D lines into a special pattern. The light in the red wing bends in one direction, whereas that in the violet wing bends in the opposite. In consequence the light originally in GG′ assumes the pattern of an anomalous dispersion curve AHH′B′.

The observed magnitude of the phenomenon will not agree, however, with that predicted in equation (2). In deriving the foregoing formulae we have tacitly assumed that the molecules are motionless. As a consequence of the gas-kinetic velocities of the molecules, the dispersion curve is the average effect of the individual atoms. Each atom behaves as a resonator of frequency ν_0, but the motion of the atom relative to the light source and observer causes it to absorb in some adjacent frequency, related to the first by the *Doppler effect.* We shall not attempt to estimate the magnitude of the Doppler correction.

33. Limiting expressions for the classical absorption coefficient. The general expression for the classical atomic absorption coefficient, for the linear oscillator, is

$$\alpha_\nu = \frac{8\pi\varepsilon^4}{3m^2c^4} \frac{\nu^4}{(\nu_0^2 - \nu^2)^2 + (\gamma\nu/2\pi)^2}, \tag{1}$$

$$\alpha_\nu = \frac{\varepsilon^4}{6\pi\epsilon_0^2 m^2c^4} \frac{\nu^4}{(\nu_0^2 - \nu^2)^2 + (\gamma\nu/2\pi)^2}, \tag{1a}$$

(31.17). This formula, like that for anomalous dispersion, requires correction in the neighborhood of the spectral line to allow for the Doppler effect.

Disregarding the Doppler correction, we may derive a number of extremely interesting results from (1), according to the value of $\nu_0 - \nu$. When $\nu = \nu_0$, we have

$$\alpha_\nu = \frac{32\pi^3\varepsilon^4\nu^2}{3m^2c^4\gamma^2} = \frac{3}{2\pi}\frac{c^2}{\nu_0^2} = \frac{3}{2\pi}\lambda_0^2. \tag{2}$$

$$\alpha_\nu = \frac{2\pi\varepsilon^4\nu^2}{3\epsilon_0^2 m^2c^4\gamma^2} = \frac{3}{2\pi}\lambda_0^2. \tag{2a}$$

When $\nu \sim \nu_0$,

$$\alpha_\nu = \frac{2\pi\varepsilon^4}{3m^2c^4} \frac{\nu^2}{(\nu_0 - \nu)^2 + (\gamma/4\pi)^2}. \tag{3}$$

$$\alpha_\nu = \frac{\varepsilon^4}{24\epsilon_0^2 m^2c^4} \frac{\nu^2}{(\nu_0 - \nu)^2 + (\gamma/4\pi)^2}. \tag{3a}$$

When $|\nu_0 - \nu| \gg \gamma/4\pi$,

$$\alpha_\nu = \frac{8\pi\varepsilon^4}{3m^2c^4} \frac{\nu^4}{(\nu_0^2 - \nu^2)^2}. \tag{4}$$

$$\alpha_\nu = \frac{\varepsilon^4}{6\pi\epsilon_0^2 m^2c^4} \frac{\nu^4}{(\nu_0^2 - \nu^2)^2}. \tag{4a}$$

When $\nu_0 \ll \nu$, i.e., when the fundamental frequency lies far to the red,

$$\alpha_\nu = \frac{8\pi\varepsilon^4}{3m^2c^4}, \tag{5}$$

$$\alpha_\nu = \frac{\varepsilon^4}{6\pi\epsilon_0^2 m^2 c^4}, \tag{5a}$$

Thomson's well-known scattering formula for x rays. This equation also holds for free electrons, for which $\nu_0 \sim 0$.

Finally, when $\nu_0 \gg \nu$, i.e., when the fundamental frequency lies in the extreme ultraviolet,

$$\alpha_\nu = \frac{8\pi\varepsilon^4}{3m^2c^4}\frac{\nu^4}{\nu_0^4} = \frac{8\pi\varepsilon^4}{3m^2c^4}\frac{\lambda_0^4}{\lambda^4}, \tag{6}$$

$$\alpha_\nu = \frac{\varepsilon^4}{6\pi\epsilon_0^2 m^2 c^4}\frac{\nu^4}{\nu_0^4} = \frac{\varepsilon^4}{6\pi\epsilon_0^2 m^2 c^4}\frac{\lambda_0^4}{\lambda^4}, \tag{6a}$$

the well-known λ^{-4} law of Rayleigh scattering.

34. Wave and group velocity. We have already pointed out that no strictly monochromatic waves occur in nature. But, even if they did, we should find it impossible to measure their velocity of propagation. A pure sine wave extends from $-\infty$ to $+\infty$; its amplitude is uniform so that there is no distinguishing feature at any point of the beam. All experimental determinations of the velocity of light depend upon interrupting, chopping, or modulating the beam. The mechanical act of cutting off an otherwise pure sine wave will introduce, as we have seen, spurious frequencies differing from the natural one. The actual sharply terminated beam behaves like a Fourier integral of pure sine waves.

Suppose that we have a *group* of such sine waves, traveling in the positive direction of the z-axis. Let their amplitudes, A, all be identical and the frequency interval be narrow. As in (III-7.31), we shall express the individual frequencies as an integral multiple, k, of some very much lower fundamental frequency ν_f. Then we may express the amplitude resulting from the action of the entire group of waves, as

$$\psi = A \sum_{k_1}^{k_2} \cos 2\pi\nu_f k\left(\frac{zn_k}{c} - t\right), \tag{1}$$

where n_k is the refractive index for light of frequency $\nu_k = k\,\nu_f$. Each individual wave is assumed to travel with its own *wave velocity*, v_k, where

$$v_k = c/n_k \tag{2}$$

as before. Equation (1) follows directly from (31.19). In the limit (1) assumes the form of an integral.

Consider, now, any pair of these subwaves distinguished by the respective indices k and k'. The composite amplitude ψ_p, of the pair will be

$$\psi_p = A\left[\cos 2\pi\nu_f k\left(\frac{zn_k}{c} - t\right) + \cos 2\pi\nu_f k'\left(\frac{zn_{k'}}{c} - t\right)\right]. \tag{3}$$

Two harmonic interfering waves "heterodyne" with one another to give "beat" frequencies equal to the sum and difference of the original frequencies. This relation follows from the trigonometric formula:

$$\cos\alpha + \cos\beta = 2\cos\frac{1}{2}(\alpha + \beta)\cos\frac{1}{2}(\alpha - \beta). \tag{4}$$

For the present case the two frequencies are assumed to be very nearly equal. Hence we may, without appreciable error, set

$$k + k' \sim 2k, \quad \text{and} \quad kn_k + k'n_{k'} \sim 2kn_k. \tag{5}$$

For the difference $\alpha - \beta$, we may take

$$k - k' = \Delta k, \quad \text{and} \quad kn_k - k'n_{k'} = \Delta(kn_k), \tag{6}$$

since we cannot here ignore the difference between k and k'. Equation (3) becomes

$$\psi_p = 2A\left[\cos 2\pi\nu_f k\left(\frac{zn_k}{c} - t\right)\cos 2\pi\nu_f\,\Delta k\left(\frac{z}{c}\frac{\Delta(kn_k)}{\Delta k} - t\right)\right]. \tag{7}$$

The first of the cosine factors appears to be our original harmonic wave, with twice the amplitude. The second cosine factor possesses an angular argument that is very much less than that of the initial wave. Its effect

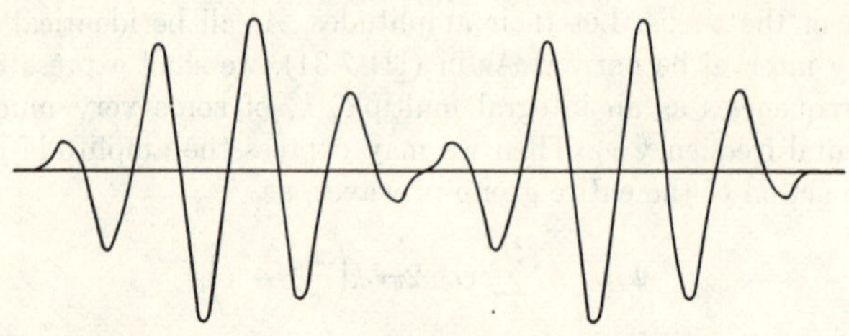

IV-29.

is to superpose on the initial wave a slowly varying amplitude. The resulting wave may be compared to a high-frequency radio wave "modulated" with a pure musical tone (Fig. 29).

Comparing the arguments of the two cosine terms, we note the correspondence between the factors

$$\frac{n_k}{c} = \frac{1}{v} \quad \text{and} \quad \frac{\Delta(kn_k)}{c\,\Delta k} = \frac{1}{u}. \tag{8}$$

Since v is the velocity at which the maxima of the individual wave progress, u will be the velocity of propagation for the maxima of the modulating wave. To the extent that $\Delta\,(kn_k)/\Delta k$ is a constant for the group, we may regard u as the velocity at which any modulated signal is propagated. It is u, therefore, the *group velocity*, and not v, the wave or phase velocity, that is found from experimental determinations of the velocity of light.

Since $\nu = k\,\nu_f$, we may pass to the differential form, and write, instead of (8),

$$u = \frac{c}{(d/d\nu)(\nu n)}. \tag{9}$$

We have previously noted in § 32, that the wave velocity may exceed that of light. But u, as may be seen from substitution of (32.2) in (9), is always less than c. Hence the theory of relativity, which specifies that no *signal* velocity may exceed that of light, is not violated by the case of anomalous dispersion.

35. Larmor's theorem and the Zeeman effect. We have seen that electrons vibrating harmonically with a given frequency will give rise to radiation of that frequency. If the oscillation occurs in the presence of a magnetic field, the character of the vibrations will be altered, with consequent change in the resulting radiation. We call the observed phenomenon, a splitting of the line into various components, the Zeeman effect.

The force on an electron of charge ε and mass m, moving with velocity $\mathbf{v}_0$ through a uniform magnetic field of intensity $\mathbf{B} = \mu\,\mathbf{H}$ is (16.21):

$$\mathbf{F} = \varepsilon\left(\mathbf{E} + \frac{\mu}{c}\,\mathbf{v}_0 \times \mathbf{H}\right) = m\mathbf{a}_0, \tag{1}$$

$$\mathbf{F} = \varepsilon(\mathbf{E} + \mu\mathbf{v}_0 \times \mathbf{H}) = m\mathbf{a}_0, \tag{1a}$$

where $\mathbf{a}_0$ is the acceleration. The last part of (1) is merely a statement of Newton's second law of motion.

Let $\mathbf{v}_0$ and $\mathbf{a}_0$ represent the velocity and acceleration with respect to fixed axes x_0, y_0, z_0, and $\mathbf{v}$ and $\mathbf{a}$ the same parameters with respect to axes x, y, z with origin common to the first set, but rotating with angular velocity $\boldsymbol{\omega}$. Then, by equations (II-25.14) and (II-25.17), we have

$$\begin{aligned} \mathbf{v}_0 &= \mathbf{v} + \boldsymbol{\omega} \times \mathbf{r}, \\ \mathbf{a}_0 &= \mathbf{a} + \boldsymbol{\omega} \times (\boldsymbol{\omega} \times \mathbf{r}) + 2(\boldsymbol{\omega} \times \mathbf{v}). \end{aligned} \tag{2}$$

Substituting these in (1), we get

$$ma - \varepsilon\mathbf{E} = \frac{\mu\varepsilon}{c}(\mathbf{v} + \boldsymbol{\omega} \times \mathbf{r}) \times \mathbf{H} - m\boldsymbol{\omega} \times (\boldsymbol{\omega} \times \mathbf{r}) - 2m(\boldsymbol{\omega} \times \mathbf{v})$$

$$= 2m\mathbf{v} \times \left(\boldsymbol{\omega} + \frac{\mu\varepsilon}{2mc}\mathbf{H}\right) + m(\boldsymbol{\omega} \times \mathbf{r}) \times \left(\boldsymbol{\omega} + \frac{\mu\varepsilon}{mc}\mathbf{H}\right). \tag{3}$$

$$ma - \varepsilon\mathbf{E} = \mu\varepsilon(\mathbf{v} + \boldsymbol{\omega} \times \mathbf{r}) \times \mathbf{H} - m\boldsymbol{\omega} \times (\boldsymbol{\omega} \times \mathbf{r}) - 2m'\boldsymbol{\omega} \times \mathbf{v})$$

$$= 2m\mathbf{v} \times \left(\boldsymbol{\omega} + \frac{\mu\varepsilon}{2m}\mathbf{H}\right) + m(\boldsymbol{\omega} \times \mathbf{r}) \times \left(\boldsymbol{\omega} + \frac{\mu\varepsilon}{m}\mathbf{H}\right). \tag{3a}$$

Thus far, $\boldsymbol{\omega}$ is an arbitrary vector introduced with the expectation of simplifying the equations. Let us try setting

$$\boldsymbol{\omega} = -\frac{\mu\varepsilon}{2mc}\mathbf{H}, \tag{4}$$

$$\boldsymbol{\omega} = -\frac{\mu\varepsilon}{2m}\mathbf{H}, \tag{4a}$$

which will cause the first parenthesis on the right to vanish. Then

$$ma = \varepsilon\mathbf{E} + \frac{\mu\varepsilon^2}{4mc}(\mathbf{r} \times \mathbf{H}) \times \mathbf{H}, \tag{5}$$

$$ma = \varepsilon\mathbf{E} + \frac{\mu\varepsilon^2}{4m}(\mathbf{r} \times \mathbf{H}) \times \mathbf{H}, \tag{5a}$$

which is the exact equation of motion of the electron referred to axes rotating with angular velocity, $\boldsymbol{\omega}$ as defined by (4).

Let us now investigate the relative magnitudes of the two terms on the right-hand side of (5). The electric field **E** arises from charges within the atom. If the effective charge of the nucleus and inner electrons in the atom is Z, where Z is some number generally greater than unity,

$$\mathbf{E} = Z\varepsilon\mathbf{r}/\kappa r^3, \quad E = |\,\mathbf{E}\,| = Z\varepsilon/\kappa r^2. \tag{6}$$

$$\mathbf{E} = Z\varepsilon\mathbf{r}/4\pi\epsilon r^3, \quad E = Z\varepsilon/4\pi\epsilon r^2. \tag{6a}$$

the last term of (5) attains its maximum value when **r** and **H** are perpendicular, so that its numerical value cannot exceed $\mu^2\varepsilon^2 rH^2/4mc^2$. For MKS, the value is $\mu^2\varepsilon^2 rH^2/4m$. Thus the order of magnitude of the ratio of the last to the first term is $\kappa\mu^2 r^3H^2/4mc^2Z$. For MKS, the ratio is $\pi\epsilon\mu^2 r^3H/mZ$. Here r will be of the order of an atomic radius, 10^{-8} cm. We may set κ and μ equal to unity, and we may also take $Z = 1$. The maximum value of H attainable in the laboratory is about 20,000 gausses. Therefore the ratio of the two terms in (5) will ordinarily not exceed 10^{-10}.

From this rather lengthy digression, we conclude that we may safely neglect the second term of (5). Hence we get for the equation of motion in a magnetic field,

$$m\mathbf{a} = \varepsilon\mathbf{E}. \tag{7}$$

But if the magnetic field were to drop to zero, we have from (1), that

$$m\mathbf{a}_0 = \varepsilon\mathbf{E}. \tag{8}$$

To a high degree of approximation, the equations of motion with respect to axes rotating with angular velocity $\boldsymbol{\omega}$ are the same as those in the non-rotating coordinate system when $\mathbf{H} = 0$. Thus to calculate the motion of a charge under the combined influence of electric and magnetic fields, we may first calculate the motion as if $\mathbf{H}$ were zero. And then, whatever this motion may be, we obtain the true motion by imposing, upon the initial trajectory, a rotation with angular velocity $\boldsymbol{\omega} = -\mu\varepsilon\mathbf{H}/2mc$. For MKS, $\boldsymbol{\omega} = -\mu\varepsilon\mathbf{H}/2m$. The axis of rotation is parallel to the field. This result is known as Larmor's theorem.

In the case of electrons moving in elliptic orbits, as for the Bohr atomic model, the motion in a magnetic field may still be described as elliptic with a superposed "precession." If the electron is merely oscillating linearly, a similar precession occurs. For example, consider, as in Fig. 30,

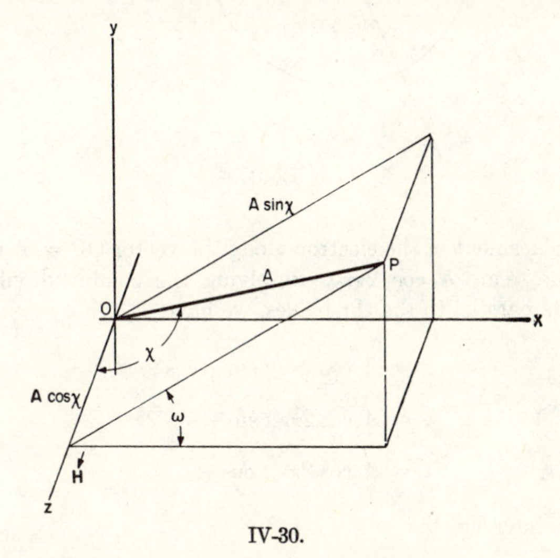

IV-30.

an electron oscillating along the line OP, with frequency ν_0 and amplitude A. Let OP make the angle χ with the positive z-direction, which we adopt as parallel to $\mathbf{H}$. Larmor's theorem then states that the electron will oscillate with frequency ν_0 and amplitude A along a line that precesses around the z-axis with angular velocity $\boldsymbol{\omega} = -\mu\varepsilon\mathbf{H}/2mc$, with the angle χ a constant. For MKS, cf. preceding paragraph. The line OP thus traces out a right circular cone, with vertex at O and generating angle χ. The direction of precession, for a *negative* electron, is shown in Fig. 31. From (4), we see that **H** and **ω** point in the same direction when the charge is negative. We may set κ and μ equal to unity because we consider the electron to move in a vacuum. For MKS, we thus set $\epsilon = \epsilon_0$ and $\mu = \mu_0$, to achieve the same purpose.

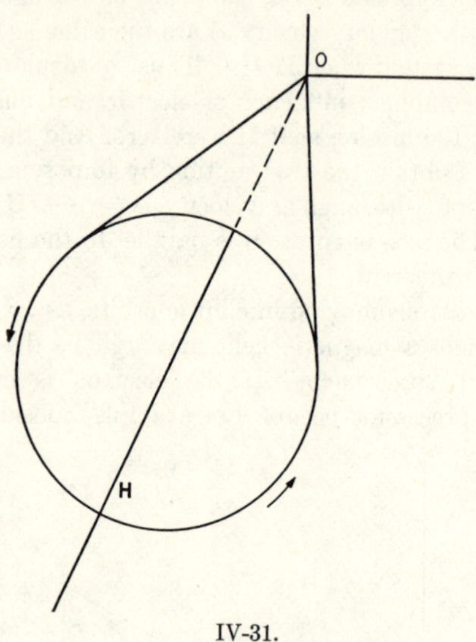

IV-31.

The displacement of the electron along the vector OP = **A** is a function of the time, e.g., **A** $\cos 2\pi\nu_0 t$. Resolving the combined vibration into components parallel to the three axes, we have

$$\begin{aligned} x &= A \cos 2\pi\nu_0 t \sin\chi \cos 2\pi\nu t, \\ y &= A \cos 2\pi\nu_0 t \sin\chi \sin 2\pi\nu t, \\ z &= A \cos 2\pi\nu_0 t \cos\chi, \end{aligned} \tag{9}$$

wherein we have written

$$\omega = 2\pi\nu, \quad \text{or} \quad \nu = \left|\frac{\mu\varepsilon H}{4\pi mc}\right|. \tag{10}$$

$$\nu = \left|\frac{\mu\varepsilon H}{4\pi m}\right|. \tag{10a}$$

Set $$A \sin\chi = 2A' \quad \text{and} \quad A \cos\chi = 2A''. \tag{11}$$

Then employ the trigonometric transformation (34.4) and the analogous expression for the sines. We thus obtain

$$
\begin{aligned}
x &= A'[\cos 2\pi(\nu_0 - \nu)t + \cos 2\pi(\nu_0 + \nu)t]. \\
y &= A'[-\sin 2\pi(\nu_0 - \nu)t + \sin 2\pi(\nu_0 + \nu)t]. \\
z &= A'' \cos 2\pi\nu_0 t.
\end{aligned}
\tag{12}
$$

The original displacement takes the form

$$
\begin{aligned}
\mathbf{r} = \mathbf{A} \cos 2\pi\nu_0 t = \mathbf{i}x + \mathbf{j}y + \mathbf{k}z = A''\mathbf{k} \cos 2\pi\nu_0 t \\
+ A'[\mathbf{i} \cos 2\pi(\nu_0 + \nu)t + \mathbf{j} \sin 2\pi(\nu_0 + \nu)t] \\
+ A'[\mathbf{i} \cos 2\pi(\nu_0 - \nu)t - \mathbf{j} \sin 2\pi(\nu_0 - \nu)t].
\end{aligned}
\tag{13}
$$

The expression $A'[\mathbf{i} \cos 2\pi(\nu_0 - \nu)t - \mathbf{j} \sin 2\pi(\nu_0 - \nu)t]$ is a vector of magnitude A', located in the xy-plane and rotating around the z-axis in a negative (clockwise) sense, with frequency $(\nu_0 - \nu)$. Similarly, the second bracket of (13) is a vector that rotates about z in a positive sense with frequency $(\nu_0 + \nu)$. An impressed magnetic field therefore resolves the

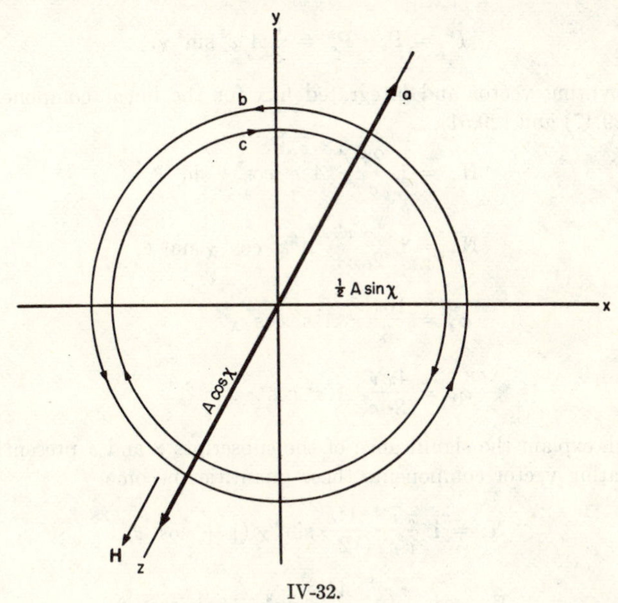

IV-32.

original linear oscillation into the three components, which we tabulate as follows:

	Component	*Direction*	*Amplitude*	*Frequency*
(a)	linear	parallel to **H**	$A \cos \chi$	ν_0
(b)	circular	positive rotation about **H**	$\frac{1}{2} A \sin \chi$	$\nu_0 + \nu$
(c)	circular	negative rotation about **H**	$\frac{1}{2} A \sin \chi$	$\nu_0 - \nu$

The component vectors are illustrated in Fig. 32.

Expressed as bivectors, the instantaneous electric moments, $\varepsilon\mathbf{r}$, for the three cases of (13) become

$$\text{(a)} \quad \mathbf{P}_c = \mathbf{k}A\varepsilon \cos \chi e^{2\pi i\nu_0 t}. \tag{14}$$

$$\text{(b)} \quad \mathbf{P}_c = \frac{(\mathbf{i} - i\mathbf{j})}{\sqrt{2}} \frac{A\varepsilon \sin \chi}{\sqrt{2}} e^{2\pi i(\nu_0+\nu)t}. \tag{15}$$

$$\text{(c)} \quad \mathbf{P}_c = \frac{(\mathbf{i} + i\mathbf{j})}{\sqrt{2}} \frac{A\varepsilon \sin \chi}{\sqrt{2}} e^{2\pi i(\nu_0-\nu)t}. \tag{16}$$

The three components of (13) are the real parts of these bivectors. The factor $(1/\sqrt{2})$, placed in front of the two rotating vectors, brings these equations into the form of (29.49). We are to compare equation (14) with (29.45). For this equation, we have

$$P^2 = \mathbf{P}_c \cdot \mathbf{P}_c^* = A^2\varepsilon^2 \cos^2 \chi, \tag{17}$$

whereas for (15) and (16) we have, as in (29.49),

$$P^2 = \mathbf{P}_c \cdot \mathbf{P}_c^* = \frac{1}{2} A^2\varepsilon^2 \sin^2 \chi. \tag{18}$$

The Poynting vector and integrated flux for the linear component are, from (29.47) and (29.51),

$$\mathbf{N}_\pi = \mathbf{i}_r \frac{2\pi^3\nu^4}{c^3r^2} A^2\varepsilon^2 \cos^2 \chi \sin^2 \theta, \tag{19}$$

$$\mathbf{N}_\pi = \mathbf{i}_r \frac{\pi^2\nu^4}{2\epsilon_0 c^3r^2} A^2\varepsilon^2 \cos^2 \chi \sin^2 \theta, \tag{19a}$$

$$\phi_\pi = \frac{16\pi^4\nu^4}{3c^3} A^2\varepsilon^2 \cos^2 \chi. \tag{20}$$

$$\phi_\pi = \frac{4\pi^3\nu^4}{3\epsilon_0 c^3} A^2\varepsilon^2 \cos^2 \chi. \tag{20a}$$

We shall explain the significance of the subscripts π and σ presently. For the rotating vector components these quantities become

$$\mathbf{N}_\sigma = \mathbf{i}_r \frac{\pi^3\nu'^4}{c^3r^2} \frac{A^2\varepsilon^2}{2} \sin^2 \chi\, (1 + \cos^2 \theta), \tag{21}$$

$$\mathbf{N}_\sigma = \mathbf{i}_r \frac{\pi^2\nu'^4}{4\pi\epsilon_0 c^3} \frac{A^2\varepsilon^2}{2} \sin^2 \chi\, (1 + \cos^2 \theta), \tag{21a}$$

and

$$\phi_\sigma = \frac{16\pi^4\nu'^4}{3c^3} \frac{A^2\varepsilon^2 \sin^2 \chi}{2}, \tag{22}$$

$$\phi_\sigma = \frac{4\pi^3\nu'^4}{3\epsilon_0 c^3} \frac{A^2\varepsilon^2 \sin^2 \chi}{2}, \tag{22a}$$

where

$$\nu' = \nu_0 \pm \nu. \tag{23}$$

For zero field, as $\nu' \to \nu_0$, the sum of the linear and the *two* circular components becomes

$$\phi_\pi + 2\phi_\sigma = \frac{16\pi^4\nu_0^4}{3c^3} A^2\varepsilon^2, \tag{24}$$

$$\phi_\pi + 2\phi_\sigma = \frac{4\pi^3\nu_0^4}{3\epsilon_0 c^3} A^2\varepsilon^2, \tag{24a}$$

which agrees, as it should, for the flux from the original linear oscillator, of moment $A\varepsilon$.

The foregoing equations refer to the energy emitted by a single atomic oscillator in a magnetic field. Ordinarily we shall have a large number of such oscillators, with their initial vibrations distributed at random. If all the oscillators have the same amplitude A, we may calculate the resulting intensities and fluxes by averaging $\sin^2 \chi$ and $\cos^2 \chi$ over a sphere. Thus

$$\overline{\cos^2 \chi} = \frac{\int_0^{2\pi} \int_0^{\pi} \cos^2 \chi \sin \chi \, d\chi \, d\varphi}{\int_0^{2\pi} \int_0^{\pi} \sin \chi \, d\chi \, d\varphi} = 1/3, \tag{25}$$

and

$$\overline{\sin^2 \chi} = 2/3. \tag{26}$$

Now set

$$A^2\varepsilon^2/3 = P^2. \tag{27}$$

Then (19) and (21) become, respectively,

$$\mathbf{N}_\pi = \mathbf{i}_r \frac{2\pi^3\nu^4}{c^3r^2} P^2 \sin^2 \theta, \quad \mathbf{N}_\sigma = \mathbf{i}_r \frac{\pi^3\nu'^4}{c^3r^2} P^2(1 + \cos^2 \theta), \tag{28}$$

and

$$\phi_\pi = \phi_\sigma = \frac{16\pi^4\nu^4}{3c^3} P^2,$$

$$\mathbf{N}_\pi = \mathbf{i}_r \frac{\pi^2\nu^4}{2\epsilon_0 c^3r^2} P^2 \sin^2 \theta, \quad \mathbf{N}_\sigma = \mathbf{i}_r \frac{\pi^2\nu'^4}{4\epsilon_0 c^3r^2} P^2(1 + \cos^2 \theta), \tag{28a}$$

$$\phi_\pi = \phi_\sigma = \frac{4\pi^3\nu^4}{3\epsilon_0 c^3} P^2,$$

as in (29.51). For zero field, the light in any given direction will be unpolarized.

From equations (28) we may now calculate the relative intensities of the three Zeeman components, as viewed from any angle. The electric vector, from (29.41), lies in the plane defined by the point of observation and the instantaneous position of P. Thus the linear component will give rise to linearly polarized light. When $\theta = 0$, the linear component vanishes and we view two lines circularly polarized in opposite directions and separated by frequency differences $\pm\mu\varepsilon H/4\pi mc$ from the central position

ν_0. In MKS the separation is $\pm\mu\varepsilon H/4\pi m$. This appearance of the spectrum, viewed along the field through an aperture drilled in the pole of the magnet, is called the longitudinal Zeeman pattern (Fig. 33).

If we view the oscillations from some point in the xy-plane, we observe the so-called transverse Zeeman pattern. Here the component parallel

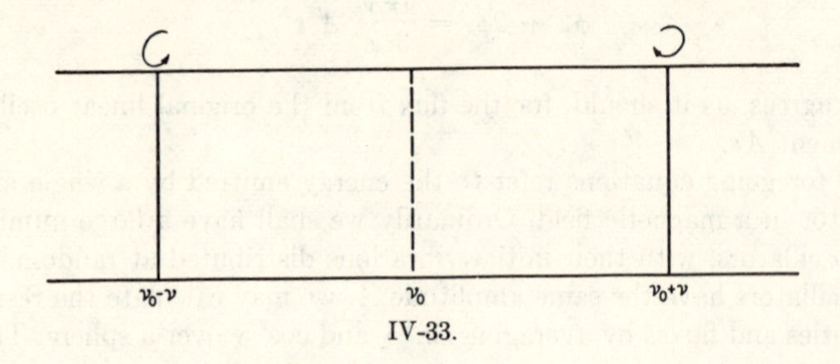

IV-33.

to the z-axis yields an undisplaced component linearly polarized. The two circular vibrations appear like linear oscillators, as seen from the transverse position. They thus give rise to light linearly polarized at right angles to the central component, with frequencies at $\nu_0 \pm \nu$. The two types of components are usually indicated by the letters π (parallel) and σ (perpendicular—German: *senkrecht*). Hence the significance of the notation in equations (19)-(22). Note that these letters refer to the direction of oscillation of the electron in the field and hence to the position of the *electric* vector. Hence a π component is polarized parallel and a σ component perpendicular to the field. The theoretical tranverse Zeeman

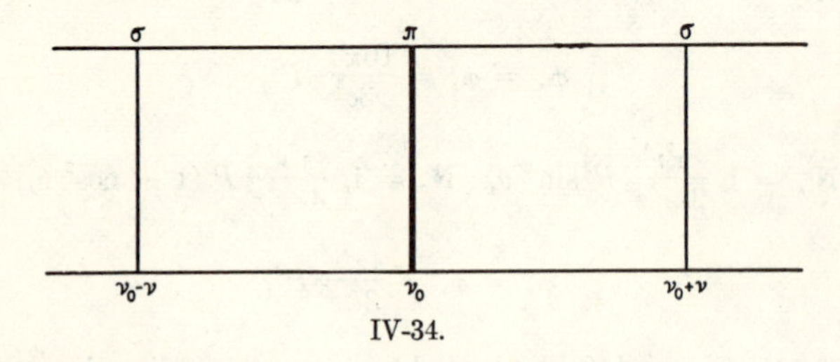

IV-34.

pattern is shown in Fig. 34. As viewed from intermediate positions, the pattern will consist of a linearly polarized central component and two displaced elliptically polarized components.

The Zeeman effect, as observed for actual atoms, is usually much more complex than the simple pattern predicted for the classical oscillator. Quantum theory completely removes the anomaly. Part of the difference results from the fact that the electron possesses an intrinsic magnetic moment in addition to a negative charge.

36. The Hamiltonian for the electromagnetic field. We have seen that the force acting upon a particle whose charge and mass are ε and m, respectively, conforms to the equation

$$\mathbf{F} = \varepsilon\left(\mathbf{E} + \frac{1}{c}\,\mathbf{v} \times \mathbf{B}\right), \tag{1}$$

$$\mathbf{F} = \varepsilon(\mathbf{E} + \mathbf{v} \times \mathbf{B}), \tag{1a}$$

(16.21). Furthermore, we have both **E** and **H** expressed in terms of the scalar and vector potentials ϕ and **A**, as follows:

$$\mathbf{E} = -\nabla\phi - \frac{1}{c}\frac{\partial\mathbf{A}}{\partial t}, \quad \mathbf{B} = \nabla \times \mathbf{A}. \tag{2}$$

$$\mathbf{E} = -\nabla\varphi - \frac{\partial\mathbf{A}}{\partial t}, \quad \mathbf{B} = \nabla \times \mathbf{A}. \tag{2a}$$

Let us write down in expanded form, the x-component of (1), in terms of (2). Then

$$F_x = \frac{d}{dt}(m\dot{x}) = -\varepsilon\frac{\partial\phi}{\partial x} - \frac{\varepsilon}{c}\frac{\partial A_x}{\partial t} + \frac{\varepsilon}{c}\left[\dot{y}\left(\frac{\partial A_y}{\partial x} - \frac{\partial A_x}{\partial y}\right) - \dot{z}\left(\frac{\partial A_x}{\partial z} - \frac{\partial A_z}{\partial x}\right)\right], \tag{3}$$

$$F_x = \frac{d}{dt}(m\dot{x}) = -\varepsilon\frac{\partial\phi}{\partial x} - \varepsilon\frac{\partial A_x}{\partial t} + \varepsilon\left[\dot{y}\left(\frac{\partial A_y}{\partial x} - \frac{\partial A_x}{\partial y}\right) - \dot{z}\left(\frac{\partial A_x}{\partial z} - \frac{\partial A_z}{\partial x}\right)\right], \tag{3a}$$

where the "dots" denote time derivatives as before. Now the total derivative

$$\frac{dA_x}{dt} = \frac{\partial A_x}{\partial x}\dot{x} + \frac{\partial A_x}{\partial y}\dot{y} + \frac{\partial A_x}{\partial z}\dot{z} + \frac{\partial A_x}{\partial t}. \tag{4}$$

Using this equation to eliminate $\partial A_x/\partial t$, we obtain

$$\frac{d}{dt}\left(m\dot{x} + \varepsilon\frac{A_x}{c}\right) + \varepsilon\frac{\partial\phi}{\partial x} - \frac{\varepsilon}{c}\left[\frac{\partial A_x}{\partial x}\dot{x} + \frac{\partial A_y}{\partial x}\dot{y} + \frac{\partial A_z}{\partial x}\dot{z}\right] = 0. \tag{5}$$

$$\frac{d}{dt}(m\dot{x} + \varepsilon A_x) + \varepsilon\frac{\partial\varphi}{\partial x} - \varepsilon\left[\frac{\partial A_x}{\partial x}\dot{x} + \frac{\partial A_y}{\partial x}\dot{y} + \frac{\partial A_z}{\partial x}\dot{z}\right] = 0. \tag{5a}$$

Similar equations hold for y and z.

Let us define a function L as follows:

$$L = \frac{1}{2}m(\dot{x}^2 + \dot{y}^2 + \dot{z}^2) - \varepsilon\phi + \frac{\varepsilon}{c}(A_x\dot{x} + A_y\dot{y} + A_z\dot{z}), \tag{6}$$

$$L = \frac{1}{2} m(\dot{x}^2 + \dot{y}^2 + \dot{z}^2) - \varepsilon\varphi + \varepsilon(A_x\dot{x} + A_y\dot{y} + A_z\dot{z}), \tag{6a}$$

so that

$$\frac{\partial L}{\partial \dot{x}} = m\dot{x} + \frac{\varepsilon A_x}{c} \tag{7}$$

$$\frac{\partial L}{\partial \dot{x}} = m\dot{x} + \varepsilon A_x \tag{7a}$$

and

$$\frac{\partial L}{\partial x} = -\varepsilon \frac{\partial \phi}{\partial x} + \frac{\varepsilon}{c}\left[\frac{\partial A_x}{\partial x}\dot{x} + \frac{\partial A_y}{\partial x}\dot{y} + \frac{\partial A_z}{\partial x}\dot{z}\right]. \tag{8}$$

$$\frac{\partial L}{\partial x} = -\varepsilon \frac{\partial \varphi}{\partial x} + \varepsilon\left[\frac{\partial A_x}{\partial x}\dot{x} + \frac{\partial A_y}{\partial x}\dot{y} + \frac{\partial A_z}{\partial x}\dot{z}\right]. \tag{8a}$$

Then we may write (5) in the form

$$\frac{d}{dt}\frac{\partial L}{\partial \dot{x}} - \frac{\partial L}{\partial x} = 0. \tag{9}$$

This equation is in the form of the Lagrangian equation of motion (II-39.22), if we interpret L as the Lagrangian function, $T - V$. (II-39.21). If conservative forces in addition to those of electrical origin are acting, we must replace the scalar electric potential energy, $\varepsilon\,\phi$, by V, the sum of all the scalar potentials of the system. Furthermore we must define the canonical momenta by means of (II-39.25). Thus

$$p_x = \frac{\partial L}{\partial \dot{x}}, \quad \text{ETC.} \tag{10}$$

The Hamiltonian function, H, is defined by (II-40.1) as

$$H = \sum p\dot{q} - L = \frac{1}{2} m(\dot{x}^2 + \dot{y}^2 + \dot{z}^2) + V. \tag{11}$$

Hence, in terms of the p's, we have, from (10) and (7), the following modified form of the Hamiltonian:

$$H = \frac{1}{2m}\left[\left(p_x - \frac{\varepsilon A_x}{c}\right)^2 + \left(p_y - \frac{\varepsilon A_y}{c}\right)^2 + \left(p_z - \frac{\varepsilon A_z}{c}\right)^2\right] + V. \tag{12}$$

$$H = \frac{1}{2m}[(p_x - \varepsilon A_x)^2 + (p_y - \varepsilon A_y)^2 + (p_z - \varepsilon A_z)^2] + V. \tag{12a}$$

The occurrence of a vector potential requires us to distinguish between the ordinary momentum mx of the charged particle and the canonical momentum p_x of the electromagnetic Hamiltonian. If $\mathbf{A}$ is zero, as for the electrostatic case, the expression reduces immediately to its usual form.

SELECTED PROBLEMS FOR PART IV*

1. By what factor would you have to increase the mass of two electrons in order to make their mutual gravitational attraction exactly counterbalance their mutual electric repulsion? Does this result indicate that we are justified in neglecting gravitational forces in most electrodynamical problems?

2. Two particles of mass m and electric charge $+q$ are suspended from strings of equal length, l, separated by a distance, d. Solve for the angle of deviation of the strings when the particles attain equilibrium in a gravitational field g cm/sec^{-2}.

3. Plot the field pattern and equipotentials for:

(a) two equal charges of the same sign;

(b) two equal charges of opposite sign. Let a be the distance of separation.

4. Plot the field and equipotentials of a charge $+q$ and $-2q$, separated by a distance d. Locate and identify the surface of zero potential. Hint: Consider the two charges to be images in a conducting grounded sphere.

5. Given a spherically symmetrical charge distribution, $\rho(r)$, prove that the field or potential at any point $r = r_0$, depends only on the charge in the region $r \leq r_0$, and that the effect of such a charge is equivalent to that of a charge concentrated at $r = 0$.

6. According to wave mechanics, a hydrogen atom in its normal state consists of an electric charge $+q$ located at $r = 0$, surrounded by a negative charge distribution whose density is $\rho(r) = -Ce^{-2r/a_0}$. Determine C from the condition that the total charge of the electron must be $-q$. This process we call "normalization." Then find the potential and electric field as a function of r.

7. Three particles, A, B, and C with respective electric charges q, $-qa/f$, and Va, lie in the same straight line such that AC $= f$, BC $= a^2/f$. Show that a spherical equipotential surface always exists. Discuss the position of the points of equilibrium on ABC when $V = q(f + a)/(f - a)^2$ or $V = q(f - a)/(f + a)^2$. (Cambridge exams.)

8. A and B are spherical conductors with respective charges $q + q'$ and $-q$. Prove that either a point or line of equilibrium exists, according to the relative radii and separation of the spheres and the ratio q'/q. (Cambridge exams.)

9. The charge density arising from the presence of electrons between the plates of a small, parallel-plate capacitor varies linearly as $\rho = \rho_0 x/d$, where d is the plate spacing and x represents the distance from one plate. One plate is grounded and the other held at potential V_0.

(a) What equation connects V_0, ρ_0, and d?

(b) If the field at any point between the plates reaches breakdown value E_b, (a function of the material between the plates), a spark discharge will occur. At what critical value of V_0 will this breakdown occur?

---

*Unless otherwise indicated, Gaussian units are used.

10. Solve for the potential at all points of space resulting from a charge q located at the point (x, y) in the plane $z = 0$, with the two planes $x = 0$ and $y = 0$ consisting of infinite grounded sheets of conducting metal. Use the method of images.

11. Solve for the potential at all points of space resulting from a charge q located on the x-axis at $x = a$, between two infinite conducting sheets at $x = 0$ and $x = b$, where $b > a$. Hint: This problem will require an infinite series of images and the solution will take the form of an infinite series.

12. Prove Earnshaw's theorem: A charged body placed in an electrostatic field cannot be maintained in a position of stable equilibrium under the influence of electric forces alone. Hint: Use Laplace's equation.

13. (a) Use Gauss' theorem to prove that the charge on a conductive body must reside at the surface.

(b) What can you conclude about the charge on the surface of a completely enclosed cavity within the conductor?

14. (a) Evaluate the integral of Gauss' theorem over a sphere of radius r whose center lies at the origin, where a charge q is located. Evaluate D at the surface.

(b) Now separate the charge into two parts. Let fq remain at $r = 0$ and put the remainder $(1 - f)q$ on the x-axis at the point $x = a$. Evaluate Gauss' integral over the sphere of radius $r = R > a$. Can you use this result to find D on the surface? If not, why not?

(c) What conditions must be satisfied so that we can apply Gauss' theorem to the evaluation of fields that arise from a given charge distribution? What method can you always apply, even when the Gaussian integral fails?

15. (a) The value of $\mathbf{E}$, in a perfect conductor, must be zero. In what type of medium will the polarization vector $\mathbf{P}$ behave analogously?

(b) Use the fact that $\mathbf{P}$ and the associated polarization charge follow from Gauss' law: $\int \mathbf{P} \cdot d\mathbf{S} = -q'$, to show that the surface charge density on the interface between a dielectric and a vacuum must be equal to the normal component of $\mathbf{P}$.

16. A charge q lies a distance d from the center of an insulated metal sphere of radius R, which carries a charge Q. What will be the force between the sphere and the charge when (a) $d > R$? (b) $d < R$? Show that the signs of Q and q are not the sole factors that fix the sign of the resultant force.

17. Assume that we can cut a small sphere of radius R out of a dielectric without affecting the polarization vector, $\mathbf{P}$. Calculate the field produced at the center of this sphere by the induced polarization charges on the surface of the cavity.

Ans. $E = 4\pi P/3$.

18. Calculate the work necessary to move a charge q to infinity from an initial point a distance d from an infinite conducting plane. Note that $W = q[V(d) - V(\infty)]$ is not the correct answer.

Ans. $W = q^2/4d$.

19. (a) Consider a plane sheet containing a surface charge of density σ. Is E discontinuous across the sheet? In what way does the discontinuity depend on σ? Is the potential, V, discontinuous across the sheet? What relation connects the discontinuity of V and σ?

(b) Consider a double layer of charge, e.g., the limiting case of a layer of surface charge density $+\sigma$ separated a distance l from one of $-\sigma$, in the limiting case where $l \to 0$, $\sigma \to \infty$, and $l\sigma$ = const = electric dipole moment per unit area. Discuss the possible discontinuities in E or V at the double layer and indicate how they are related to $l\sigma$.

20. Calculate the potential resulting from charges $+q$ at $x = a$ and $-q$ at $x = -a$, on the x-axis, at respective distances r_1 and r_2 from the respective points. Express the potential in polar coordinates (r, θ) with the simplifying assumption that $2a \ll r$. This expression represents the potential of a dipole of moment $M = 2aq$. Calculate the field intensity from such a dipole.

21. (a) Calculate the force and torque that a uniform field $\mathbf{E}_x$ exerts on an electric dipole of moment $\mathbf{M}$.

(b) Calculate the force and torque that an electric field whose gradient $d\mathbf{E}_x/dx$ is constant, exerts on a dipole.

Ans. (a) $\mathbf{F} = 0; \mathbf{T} = \mathbf{M} \times \mathbf{E}_x$.

(b) $\mathbf{F}_x = \mathbf{M} \cdot d\mathbf{E}_x/dx; \mathbf{T} = \mathbf{M} \times \mathbf{E}_x$.

(c) Show that, in general, $\mathbf{F} = \nabla(\mathbf{M} \cdot \mathbf{E})$ and $\mathbf{T} = \mathbf{M} \times \mathbf{E}$.

22. Two concentric spheres of radii r_1 and r_2 $(r_2 > r_1)$, with respective voltages $V_1 = 0$ and $V_2 = V$, are separated by a dielectric whose κ varies as $\kappa = (r_2/r)^n$, $n \geq 0 =$ const. Calculate the potential as a function of r. Use the relation $\nabla \cdot \mathbf{P} = -\rho_p$, where ρ_p is the polarization charge density, to solve for the distribution of polarization charge.

Ans.

$$V = V_0\left(\frac{r_1^{n-1} - r^{n-1}}{r_1^{n-1} - r_2^{n-1}}\right), \qquad n \neq 1;$$

$$V = V_0\frac{(\ln r/r_1)}{(\ln r_2/r_1)}, \qquad n = 1;$$

$$\rho_p = V_0\,\frac{n(n-1)}{4\pi}\,\frac{r^{n-3}}{r_1^{n-1} - r_2^{n-1}}, \qquad n \neq 1;$$

$$\rho_p = \frac{n V_0}{4\pi \ln r_1/r_2}\,\frac{r^{n-1}}{r_2^n}, \qquad n = 1.$$

23. Faraday performed his famous "ice pail experiment" to prove that Coulomb's law of force between two point charges varies inversely as the square of the distance between them. Maxwell and Cavendish later refined the experiment, substituting spheres for the original ice pails. Two concentric spherical conducting shells, completely insulated from one another, are temporarily connected with a wire. Now, in turn, place an electric charge on the outer shell, remove the connecting wire, discharge the outer sphere, and then test the inner sphere for charge. The result, perhaps surprising at first sight, is that the inner shell possesses no resultant charge. This condition will obtain only when the charges obey Coulomb's law. To show this theoretically, assume that the force between two charges ε, ε', varies as $\varepsilon\varepsilon'\phi(r)$, where r is the distance between them and ϕ some general function.

(a) Show that the potential at a point P within a uniformly charged sphere of radius a is

$$V = \int_0^{2\pi} \int_0^{\pi} \frac{Q}{4\pi a^2} \left(\int_r^{\infty} \phi(r)\, dr \right) a^2 \sin\theta\, d\theta\, d\phi,$$

where Q is the total charge and where r is the distance from P to the elemental area $a^2 \sin\theta\, d\theta\, d\phi$ on the surface.

(b) Suppose that P lies at $r = c$, $\theta = 0$. Show that, if $r^2 = a^2 + c^2 - 2ac\cos\theta$, then

$$V = \frac{1}{2} Q \int_{r=a-c}^{r=a+c} \left(\int_r^{\infty} \phi(r)\, dr \right) \frac{r\, dr}{ac}.$$

(c) Introduce the notation

$$f(r) = \int^r \left(\int_r^{\infty} \phi(r)\, dr \right) r\, dr,$$

and show that

$$V = \frac{Q}{2ac} [f(a + c) - f(a - c)].$$

This expression holds whether we regard c as point external to the smaller shell or internal to the larger one. During the time that the two spheres are in electrical contact their potentials must be the same. Further, this potential must be independent of c, and therefore constant.

(d) By successive differentiation with respect to c, show that

$$f''(a + c) = f''(a - c) = C$$

where C is a constant.

(e) Show that $f(r) = A + Br + Cr^2$ and $\phi(r) = B/r^2$. For further discussion, cf. Jeans, *Electricity and Magnetism*, Ch. II.

(f) If we write the law of force between two charges as $\phi(r) = 1/r^{2+p}$, prove that a charge Q_1 remains on the inner shell of the previously described Faraday "ice pail" experiment, so that, to the first order in p, we have

$$Q_1 = -pQ_2 \left[\frac{r_1}{r_2 - r_1} \ln 2r_2 - \frac{r_2 + r_1}{2(r_2 - r_1)} \ln (r_2 + r_1) + \frac{1}{2} \ln (r_2 - r_1) \right],$$

where Q_2 is the charge on the outer shell prior to discharge and r_2 and r_1 the respective radii of the outer and inner shells. Hint: Use Maclaurin's theorem to expand $\phi(r)$ as a power series, in p.

$$\phi(r) = \frac{1}{r^2} (1 + \alpha p + \ldots).$$

Evaluate V and then determine Q, by application of results from the previous problem. (Cf. Jeans, Ch. II.)

24. Calculate the capacitance of two concentric metallic spheres. How does the capacitance change when we move the inner sphere slightly off center? What is the capacitance of a single sphere? Hint: Let the radius of the inner sphere approach zero.

25. Two capacitors, of capacitances, C_1 and C_2, possess respective charges Q_1 and Q_2. What is the total energy of the pair? How does connecting the two capacitances in parallel affect the total energy? Account for the difference. Discuss the non-physical case when the connecting wire has zero resistance.

26. Cover the inner sphere of a capacitor consisting of two concentric spheres with a thin layer of thickness d, of dielectric whose constant is κ. What effect does this layer have on the capacitance of the capacitor?

27. A parallel-plate capacitor contains a dielectric whose κ increases linearly from one plate to the other. Calculate the capacitance of the capacitor in terms of κ_1 and κ_2, the values at each plate.

28. Consider a capacitor formed by two concentric spheres of radii $r = r_1$ and r_2. If the dielectric between them varies in some arbitrary fashion like $\kappa = f(\theta, \varphi)$, where θ and ϕ are polar coordinates in the usual sense, prove that

$$C = \frac{r_1 r_2}{4\pi(r_2 - r_1)} \int \kappa \sin\theta \, d\theta \, d\varphi.$$

29. Consider a perfect conductor, set into an electrostatic field (i.e., one produced by charges, not batteries), in such a way that it does not disturb the initial pattern of the equipotential surfaces, external to its boundaries. In other words, its boundaries must coincide exactly with some of the original equipotentials. How much work will you have to perform on the conductor to move it to infinity and thus restore the original pattern? Hint: Consider the field between the equipotentials bounding the volume previously occupied by the conductor.

30. A current flows in a straight wire of circular cross section, wherein the current density (in esu) is of the form: $i = f(rt)$. Neglecting the displacement current, show that

$$\frac{4\pi i}{c} r = \frac{\partial}{\partial r}(Hr) \quad \text{and} \quad \frac{\partial E}{\partial r} = \mu \frac{\partial H}{\partial t} \cdot$$

31. Use the Biot-Savart law (10.16), to solve for the field along the axis of a current-carrying wire loop. Compare with the discussion in § (11). Take care, before integrating, to get the proper directions for the d**H**'s and resolve into components along and perpendicular to the axis of symmetry.

32. (a) For a very long, closely wound solenoid (coil) the magnetic field just outside is very small. Symmetry considerations lead us to expect the internal field to be axial. Use these assumptions and the circuital theorem to calculate the field within the coil.

(b) The field along the coil axis also follows from the result of the previous problem. Perform the necessary integration and check the answer with that of (a). Note that the method (a) shows that the field anywhere inside the coil is equal to that along the axis, whereas method (b) gives the field on the axis for a solenoid of finite length.

33. Use the Biot-Savart law (10.16),

$$d\mathbf{H} = -\frac{I}{c}\frac{\mathbf{r}}{r^3} \times d\mathbf{S},$$

to solve for the magnetic field at a distance r from a straight wire of infinite length, carrying constant current.

Ans. $H = 2I/cr$.

34. From conditions of symmetry and the equation $\nabla \cdot \mathbf{H} = 0$, we should expect the magnetic field around a straight wire to possess cylindrical character, with $\mathbf{H}$ tangent to circles concentric with the wire. Use these considerations and Ampere's circuital theorem,

$$\int H_s \, ds = \frac{4\pi}{c} \iint \mathbf{J} \cdot d\mathbf{S}$$

to calculate the field from a straight wire of infinite length. Compare the answer with that of the previous problem.

35. Derive the equivalent wave equation for ϕ and $\mathbf{A}$, using Maxwell's equations without the displacement-current term. How could you test this equation experimentally? Do the solutions of these equations represent wave motion, such as light waves?

36. Calculate the magnetic moment at the center of a current-bearing loop.

(a) of circular shape, radius $= a$. (b) of square shape, side $= a$.

Use $\mathbf{M} = \dfrac{1}{2c} \iiint \mathbf{r} \times \mathbf{J} \, d\tau$.

(c) Note the difference in the form of the answer, when calculated in MKS units.

37. The intensity of the earth's magnetic field, at the pole, is about 2/3 gauss. Assume that this field results from a magnetic dipole. Calculate the magnetostatic potential V and the force field $\mathbf{H}$ in polar coordinates. Evaluate the magnetic moment M. Assume that the field results from a current loop circling the equator and calculate the total amount of current required to produce the field. The radius of the earth is 6.38×10^{10} cm. What is the differential equation of a line of force? Integrate this equation and check the fact that these lines of force are perpendicular to the equipotentials, V.

38. Prove that the area of a hysteresis loop measures the rate of energy dissipation (energy per cycle) within the magnetic material, as stated in § (14). Consider the work done by the agent that drives the current through the coil surrounding the core and relate the electrical parameters to $\mathbf{B}$ and $\mathbf{H}$.

39. What is the magnetic energy stored in a coil carrying current I and possessing a paramagnetic core of permeability μ?

40. Calculate the total energy, in ergs, stored up in the earth's magnetic field in the volume bounded by its surface, $r = R$, and $r = \infty$. Compare this amount with the quantity of solar energy incident on the surface of the earth in a second. cf. problem 51. If all this incident solar energy could be concentrated into the manufacture of this external field, how long would it take to manufacture the field?

Ans. Energy $= 8\pi R^3 H_0^2/3$, where H_0 is the value of the field at the pole, given in problem 37. $R = 6.37 \times 10^8$ cm.

41. Prove that the two Maxwell equations,

$$\nabla \times \mathbf{E} = -\frac{1}{c}\frac{\partial \mathbf{B}}{\partial t} \quad \text{and} \quad \nabla \times \mathbf{H} = \frac{1}{c}\frac{\partial \mathbf{D}}{\partial t} + \frac{4\pi}{c}\mathbf{J},$$

imply also the two divergence equations

$$\nabla \cdot \mathbf{D} = 4\pi\rho \quad \text{and} \quad \nabla \cdot \mathbf{B} = 0.$$

Hint: Take the divergence of the curl equations and apply the equation of continuity, $\nabla \cdot \mathbf{J} + \partial\rho/\partial t = 0$.

42. In a cyclotron, particles of charge q and mass m travel in circular paths under the influence of a magnetic field **B**. An alternating electric field is applied across the gap ab as shown in the diagram. **B** is perpendicular to the plane of the page. What is the relationship between the radius r of the orbit cd and the energy of the particle (non-relativistic) with no external voltage applied? If we wish to accelerate the particles by means of the alternating field, what frequency must we apply? Note that the frequency is independent of r and thus also independent of the energy.

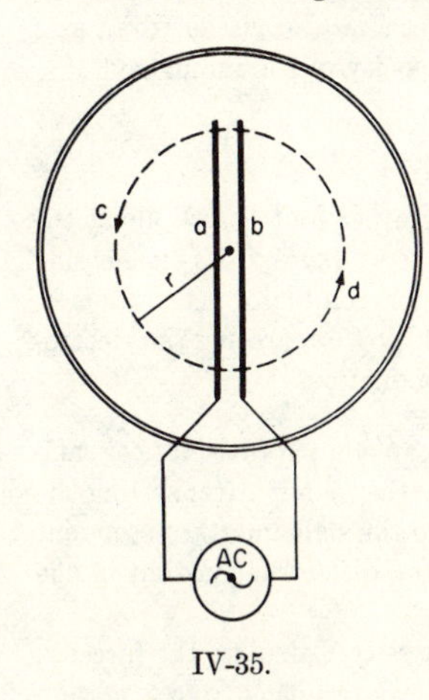

IV-35.

43. In a betatron, an increasing magnetic field accelerates electrons and keeps them in an orbit of fixed radius r_0. Show that we must have $(dB_0/dt) = \frac{1}{2}(dB_{av}/dt)$, where B_0 is the field at the orbit, r_0, and B_{av} is the average field within the orbit, if the electrons are to remain in the orbit of constant radius. Suppose that the field increases linearly with the time. Now if the electrons are injected into the field with an initial energy V_0 by a sort of accelerating "gun" at a time t_0 seconds after the field has zero B, find the value of t_0 such that the electron will circulate in the orbit of radius r_0.

44. (a) Solve for the torque on a square wire loop of side a and carrying current i, when the normal to the plane of the loop makes angle θ with a uniform magnetic field B. Is there any net force on the loop?

(b) Recalculate for a circular loop of radius r.

(c) Show that one can, in general, express the force on a loop as $\mathbf{T} = \mathbf{m} \times \mathbf{B}$, where $\mathbf{m}$ is the magnetic moment of the loop of area A, $\mathbf{m} = iA\mathbf{n}/c$.

(d) Note the difference in the calculation resulting from the use of Gaussian vs. MKS units.

45. (a) Two current-bearing wires run parallel to one another. Calculate the force (magnitude and direction) between the wires as a function of current and separation. Are any electrostatic forces involved?

(b) Consider two parallel electron beams, wherein the electrons move with velocity v_0, in the same direction. Calculate the force between the beams. Are any electrostatic forces involved?

(c) Consider the force determined by an observer moving with velocity v parallel to the wires or beam. What force will the observer measure in the two respective cases? What fraction is electrostatic and what magnetic?

46. Consider a coordinate system wherein the xy-plane is horizontal. An electromagnet produces a uniform field, B, parallel to the x-axis. Within this field, hold a non-conducting rod of cross section A, length l, and mass m parallel to the y-axis. Now drop the rod so that it falls parallel to the z-axis while retaining its horizontal position.

(a) Find the voltage difference from one end of the rod to the other.

Ans. $V = (d\varphi/dt) = Blv = Blgt.$

(b) Let the rod in part (a) have resistivity ρ and connect its two ends by a wire of resistance R. We suppose that the magnetic field ends at the limit of the rod and that the connecting wire is free of the field. Determine the viscous drag as a function of the velocity and calculate the terminal velocity of the falling rod.

Ans. $F = B^2l^2v/(R + \rho l/A).$

$v_{\text{terminal}} = mg(R + \rho l/A)/B^2l^2.$

47. A cylindrical capacitor, of length L has a charge per unit length along the inner conductor whose radius is a. Calculate the field as a function of r, neglecting end effects. Calculate the capacitance when the outer conductor has radius b. Verify that the energy contained in a capacitor is $Q^2/2C$, by integrating the electrostatic energy density over the volume between the conductors.

48. A long slab of dielectric material slides into the gap of a parallel-plate capacitor. The thickness of the slab just equals the spacing of the plates. After setting the slab part way in, release it. Discuss the forces acting on the slab and its subsequent motion under the following circumstances. Let κ be the dielectric constant of the material.

(a) Suppose that the capacitor plates carry charge Q. Calculate the force on the slab. Hint: Consider the difference in the field energy resulting from a motion dx into the gap. Regard any loss of field energy as corresponding to a gain of kinetic energy of a free slab or as mechanical work done on the slab. Will the slab tend to move into or out of the capacitor gap?

(b) Now suppose that the capacitor is connected to a battery of voltage V_0. Again calculate the force. How does it compare with the result of part (a) for an insulated capacitor? Do the forces tend to move the slab into or out of the capacitor? Hint: Use the energy method, as before, noting that $(\partial E/\partial x) = F$. Be sure to include, however, all sources of energy, including the work done by the battery.)

49. Consider a small, flat disk of dielectric, (something like a hockey puck) whose thickness is equal to the spacing of a parallel-plate capacitor. The diameter of the plates is large compared with either the diameter or thickness of the dielectric disk.

Let the capacitor be insulated and carry a charge Q. Now suppose that we propel the puck horizontally into the gap from an initial position well outside the field of the capacitor. Let v_0 be the initial velocity, m the mass, and κ the dielectric constant of the disk. Calculate the velocity of the disk when it is completely within the space between the capacitor plates. Note that no force acts on the disk, once it is entirely within the capacitor, because the field energy is then a constant. What part of the field, therefore, actively does the attracting? Do we consider this part of the field in our energy calculations?

50. Consider the following statement. "The vector field **D** depends solely on real charges whereas the vector field **E** depends on induced polarization charges as well." Is this statement valid for (a) a plane parallel-plate capacitor and dielectric? (b) a uniform dielectric sphere with charges at the center? (c) a charge adjacent to a dielectric block, as in Fig. 4? Hint: Does **D** remain constant, if we remove the dielectric in the three above cases? (d) What additional condition is necessary to make the above statement generally valid? (e) Modify the statement, to make it generally valid, merely by substituting other electrostatic parameters for "the vector field **D**" and "the vector field **E**."

51. The sun's radiation (solar constant) at the earth's distance from the sun is .94 g cal/cm^2 min by Abbot's measures. What will be the pressure exerted by this radiation on a unit area of the ocean (considered to be a perfect reflector) or of the ground (considered to be a perfect absorber). Use the result of the previous problem to calculate the total radiation pressure on the earth, considered as a perfectly reflecting sphere. Compare this figure with the gravitational force of the sun.

$$\text{One g cal} = 4.183 \times 10^7 \text{ ergs} = 4.183 \text{ joules};$$

$$\text{mass of earth} = 5.98 \times 10^{27} \text{ g};$$

$$\text{mass of sun} = 1.91 \times 10^{33} \text{ g};$$

$$G = 6.67 \times 10^{-8} \text{ dyne cm}^2 \text{ g}^{-2}.$$

The total mass of a body varies as the cube of a radius, whereas the effective cross section for radiation pressure varies as the square of the radius. At what radius, then, can we expect radiation pressure to equal the gravitational attraction? Hold the density of the body constant.

52. A plane electromagnetic wave and a free, charged particle interact to transfer both momentum and energy to the particle. Since we can show that the ratio of the rates of energy transfer and momentum transfer is c, we may reasonably assume that the wave itself possesses energy and momentum in that ratio, as given in equation (21.6). To demonstrate that the ratio is c, consider a plane electromagnetic wave traveling in the x-direction, with the **E** and **H** vectors parallel respectively to the z- and y-axes. A free electron lies at the origin.

(a) What is the force acting on the electron? Write the differential equation of motion.

(b) By taking the dot product of the force and electron velocity determine the rate of energy transfer, dW/dt.

Ans. $dW/dt = q\mathbf{E} \cdot \mathbf{v} = qE_z v_z$.

(c) From the equation of motion written in (a) calculate the average rate of momentum transfer.

Ans. $\overline{F_x} = \overline{dp_x/dt} = qH_y v_z/c$.

Note that the force associated with H_y and v_x averages to zero.

(d) Show that $dW/dt = (d/dt)(pc)$.

53. From the solar constant (see problem 51) calculate, by Einstein's equation, the mass that must turn into energy every second to account for the sun's output of light and heat.

Ans. 4.2×10^{12} g/sec.

54. Let ϕ be the flux in erg/cm^2 sec, of radiation at normal incidence. Show that radiation incident at angle θ upon a totally reflecting plane will produce a pressure, normal to the plane, of magnitude $(2\phi/c)\cos^2\theta$, with a component $(2\phi/c)\cos^3\theta$ along the direction of the original beam. Calculate the total radiation pressure upon a reflecting sphere of radius R, produced by a plane radiation incident from one side.

Ans. $\pi R^2\phi/c$.

55. (a) Describe radiation pressure of a plane wave on a sheet of absorbing material, qualitatively in terms of the interaction between currents in the sheet and the **H** of the incident wave.

(b) Calculate the radiation pressure on a perfect conductor. Hint: Calculate in terms of a surface current density, related to **H** through the appropriate boundary conditions.

56. Consider a 60-watt bulb as a point source of radiation. What force will this radiation exert on a perfectly reflecting radiometer vane of area 1 cm^2, located 1 meter from the bulb? If the vane is blackened, what will be the force?

57. Section (22) gives the boundary conditions on the electric and magnetic vectors in the absence of charges or currents at the interface. By an analogous procedure, derive the boundary conditions for:

(a) The normal component of **D** in the presence of a surface charge density σ at the interface. Will a volume charge density in the regions bordering on the interface affect the boundary conditions? Why?

(b) The tangential component of **H** in the presence of a surface current of density **j** per unit length flowing along the interface perpendicular to the component of **H**. Will volume currents affect these boundary conditions? Why?

(c) State briefly why the normal component of **B** and the tangential component of **E** must always be continuous at the interface, regardless of surface or volume phenomena.

(d) Consider a piece of magnetic material wherein a field **B** exists. Scoop out two small cylindrical cavities in the material, one long and needle-shaped and the other flat like a coin. Let the axes of the two cylinders be parallel to **B**. Calculate the value of **B** at the center of each cavity, assuming that a small cavity does not appreciably distort the field of the material. Hint: Use the boundary conditions of (a), (b), and (c). One should note that this result is consistent with the idea that the resultant field arises from (a) the initial field **H**, and (b) an induced field on the circular faces of the two cylinders. In the needle-shaped cavity (b) is negligible because the faces are small and far removed from the point of measure. In the coin-shaped cavity, the induced field cannot be neglected. By analogy with electrostatic polarization show that: $\mathbf{B} = \mathbf{H} + 4\pi\mathbf{M}$, where $\mathbf{M}$ is the intensity of magnetization.

58. We can define $\mathbf{D}$ and $\mathbf{E}$ in terms of measurements of the field $\mathbf{E}_c$ in needle-shaped and coin-shaped cavities in a dielectric, analogous to the previous problem. Derive the cavity definitions of $\mathbf{D}$ and $\mathbf{E}$ by relating $\mathbf{E}_c$ in the cavity to the uniform field in the dielectric, using the boundary conditions on $\mathbf{D}$ and $\mathbf{E}$.

59. Place a conductor of arbitrary shape in the field of a point charge.

(a) Is the field pattern in space affected?

Now imagine that we can "freeze" the surface charge distribution on the metal in space (or reproduce it on an insulator) and remove the metal.

(b) Is the field pattern different from that referred to in part (a)?

(c) Hence, which of these two statements represents the facts more accurately? "The presence of the metal affects the field of the point charge" or "The presence of charges, available in the metal, affects the field pattern."

(d) Make a similar statement relative to the effect of a dielectric of arbitrary shape on the field pattern. When we introduce a dielectric medium, which alters the initial charge distribution, does this phenomenon require a totally new parameter κ? Or can we define this parameter in terms of the availability and relative mobility of charges in the dielectric, so that the field alterations again arise from the entrance of new charges into the picture, rather than from some mysterious new property of dielectrics?

(e) Consider the gravitational analogy of the force upon a mass at the surface of the earth. How is it affected by the tides, which in turn depend on the position of the moon?

60. The magnetism of some substances arises from partial alignment of the magnetic moments of the individual atoms under the action of an applied external field. Molecular agitation prevents perfect alignment of the magnetic axes. The number of dipoles having orientation within the element $d\omega$ of solid angle is

$$dn = Ae^{-E/kT}\, d\omega,$$

with $E = \mu H \cos\theta$, where θ is the angle between the dipole of magnetic moment μ and the magnetic field H, k is Boltzmann's constant, T the absolute temperature, and A a numerical constant.

$$d\omega = 2\pi \sin\theta\, d\theta.$$

(a) Calculate the magnetic moment per unit volume,

$$M = \int \mu \cos\theta\, dn.$$

(b) Find the ratio of magnetization at temperature T to that at zero temperature, $T = 0°$ (perfect alignment).

(c) Show that for small fields or high temperatures, when $\mu H/kT \ll 1$, the Curie law for the susceptibility,

$$\chi = \frac{M}{H} = \frac{N\mu^2}{3kT},$$

follows directly.

61. Consider a dielectric sphere of radius a and dielectric constant κ in a uniform field **E**.

(a) Show that the boundary conditions to be satisfied are:

(1) As $r \to \infty$, $V \to -E_0 x$.

(2) $\kappa(\partial V/\partial r)$ must be continuous at $r = R$.

(3) V, $(1/r)(\partial V/\partial \theta)$, and E_{tang} are continuous at $r = R$.

(4) $\nabla^2 V = 0$, everywhere else.

(b) From the results of § (2), try the solution

$$V = \left(-E_0 r + \frac{A}{r^2}\right) \cos \theta,$$

outside the sphere and

$$V = -E_1 r \cos \theta$$

inside, where the field must be uniform. Show that these potentials fit the boundary conditions (a) if we set $E_1 = E_0(1 - A/a^3)$ and $\kappa E_1 = E_0(1 + 2A/a^3)$.

(c) Compute the effective dipole moment of the sphere.

(d) By what factor is the uniform field E_0 reduced within the sphere?

62. Use the Fresnel equations, which relate the reflected and transmitted electric vectors to the incident electric vectors, to verify the conservation of energy by computing the appropriate Poynting vectors for

(a) Normal incidence.

(b) Incidence at any angle, E polarized perpendicular to the plane of incidence.

(c) Incidence at any angle, E polarized in the plane of incidence.

63. Show that

$$\frac{\mu\kappa}{c^2}\frac{\partial^2 \mathbf{E}}{\partial t^2} + \frac{4\pi\mu}{R}\frac{\partial \mathbf{E}}{\partial t} = \nabla^2 \mathbf{E},$$

and

$$\frac{\mu\kappa}{c^2}\frac{\partial^2 \mathbf{H}}{\partial t^2} + \frac{4\pi\mu}{R}\frac{\partial \mathbf{H}}{\partial t} = \nabla^2 \mathbf{H}$$

follow directly from Maxwell's equations.

64. Derive Ohm's law for extended media, $\mathbf{J} = \sigma E$, where $\mathbf{J}$ = current/unit area, E the field, and $\sigma = 1/\rho$ = conductivity = 1/resistivity. Resistivity is a characteristic property of the medium defined by $R = \rho l/A$, where R is the total resistance. Consider a cylinder of area A and length l.

65. Evaluate the inward component of the Poynting vector at the surface of the resistor. The power flow into the resistor, represented by the inward component of the Poynting vector, is dissipated as heat. Prove that the power dissipated as heat by a current, in a medium whose resistance is R, is I^2R.

66. Show that $\psi = \varepsilon(t - r/c)/r$ satisfies the wave equation

$$\nabla^2\psi - \frac{1}{c^2}\frac{\partial^2 \psi}{\partial t^2} = 0.$$

Obtain the retarded-potential solution of the equation

$$\nabla^2\varphi - \frac{1}{c^2}\frac{\partial^2\varphi}{\partial t^2} = -4\pi\rho,$$

where ρ is a charge density.

67. Consider the mass of an electron to be related to the energy of its field by $E = mc^2$.

(a) Calculate the radius of an electron, assuming that its charge is uniformly distributed through the volume of a sphere.

68. Solve for the damping constant in the equation of motion of a bound electron moving under the influence of an impressed sinusoidal field

$$m\frac{d^2x}{dt^2} + a\frac{dx}{dt} + \kappa x = \varepsilon E_0 \sin\omega t.$$

The electron will dissipate energy by radiation.

69. (a) Show that two electric dipoles, oscillating at right angles and 90° out of phase to one another, are equivalent to a single charge rotating in a circle, as far as their radiation is concerned. Hint: Calculate **P**, the electric dipole moment.

(b) Calculate the average power, $\overline{P}$, radiated by such a pair of dipoles.

(c) To keep the charge rotating at constant ω in a circular path, one must supply power. Prove that we are also supplying angular momentum at a rate $\dot{L} = \overline{P}/\omega$.

(d) Since the rotating charge possesses constant ω and constant angular momentum as well, the radiation itself must carry away the angular momentum. Show that the **E** vector, from the dipole, represents a circularly polarized wave. Discuss the possibility of reconverting the electromagnetic energy into mechanical energy with the aid of another dipole, hence recapturing the angular momentum transmitted in the circularly polarized wave.

70. To prove the electromagnetic character of x rays, Barkla, in 1906, projected a beam along the x-axis upon a small block of carbon at the origin. Part of the beam scattered by the first block falls upon a second block located at some point on the y-axis. Observation shows that scattering by the second block is zero in the z-direction. Why does this experiment indicate the electromagnetic character of x rays?

71. A light beam falls on a volume of gas containing N atoms/cm^3. Assume that the atomic electrons scatter independently, calculate the attenuation of the original beam as a function of distance, x, traversed in the gas. Let Z be the number of electrons per atom.

Ans. $I = I_0 e^{-NZ\alpha x}$.

72. Use the Rayleigh formula for molecular scattering, as given in § (32), to explain why the sky is normally blue and why sunsets are red.

73. In an ionized atmosphere, the complex propagation constant, k, in MKS units, follows from the expression

$$k^2 = \frac{\omega^2}{c^2}\left[1 - \frac{N\varepsilon/m\epsilon_0}{(\omega^2 \pm \varepsilon\mu_0 H_0\omega/m)}\right],$$

where N is the number of electrons per unit volume (i.e. per m^3), ε and m the respective charge and mass of the electron, and H_0 the component of the earth's magnetic field along the direction of propagation of the radio wave of circular frequency ω. Rewrite this expression in Gaussian units. Calculate the refractive index and absorption coefficient. Determine the phase and group velocity of the waves.

74. Show that the orbital plane of an atomic electron will precess about the axis of a uniform magnetic field, **B**, with a frequency equal to the Larmor value.

75. Consider an atomic electron to be rotating in a fixed circular orbit.

(a) What force is necessary to keep it in this orbit, when it moves with angular velocity ω_0? (Neglect forces caused by the radiation of an accelerated electron.)

(b) What additional force acts on the electron if we introduce a uniform magnetic field **B**, perpendicular to the orbit plane?

(c) Setting the total force on the electron equal to the product of its mass and centripetal acceleration, employ (a) and (b) to evaluate the new angular frequency, in terms of ω_0, B, and fundamental constants. When $\varepsilon B/2m \ll \omega_0$, show that $\omega = \omega_0 \pm \omega_{\text{Larmor}}$.

76. Refer back to problem (III-13), page 233. Using equation (29.51) or equivalent, calculate the electric moment and total radiation from the oscillating electron on each frequency.

Ans. $P^2 = 4\varepsilon^2 l^2/\pi^4(2k + 1)^4$, where k is an integer.

Show that the total radiation on all frequencies is infinite. Why? Hint: What are the values of $\dot{v}$ and $\ddot{v}$?

PART V

Relativity

1. The special theory of relativity. Newton's laws of motion, which constitute the framework of ordinary mechanics, do not depend on the existence of an absolute frame of reference. We have used the words "stationary axes," in our previous discussions, to specify coordinates at rest with respect to some particular system, e.g., the surface of the earth. We have ignored the motion of the earth itself. Strictly speaking, we should not have done so. The axial rotation of the earth and the curvature of the earth's orbit introduce extraneous force fields of a variable nature. We can transform these forces away only by choosing a frame of reference with appropriate motion relative to the original frame. We have referred to coordinate axes of this character as "inertial systems."

Apart from the minor corrections just mentioned, a railroad train moving uniformly along a straight track is an example of an inertial reference frame. Newton's laws imply that any mechanical experiment performed on board such a train will give the same results, *relative to the train*, as it would when performed in the railway station. For example, an observer on the train will record as a straight vertical line the trajectory of a ball dropped from the ceiling. A stationary observer, however, will chart the motion as parabolic. A reciprocal relationship clearly exists. This experiment very clearly implies a relativity principle of a sort. We are required to find the mathematical transformations connecting the phenomena in the two frames.

Let us adopt two such reference systems, S and S′, with rectangular axes parallel and so orientated that the relative motion is in the x-direction. Then the transformation equations of Newtonian relativity are

$$x' = x - ut, \quad y' = y, \quad z' = z, \quad t' = t. \tag{1}$$

where u is the relative velocity of the systems. (1) is sometimes called the Galilean transformation.

At first sight, the last of these equations almost seems superfluous. But a closer examination of the problem reveals a fundamental defect of the Newtonian equations. We note, first of all, that we are privileged to write $t = t'$ only if we are able to set the clocks on S and S′ in absolute synchronism. But here we encounter unforeseen difficulties even when $u = 0$.

Suppose that we, as observers on S, note that the clock on S′ reads noon. We must apply a correction for the light-time between S′ and S. If SS′ $= x'$, we infer that our clock must be advanced by an amount

$$\Delta t = x'/c \tag{2}$$

where c is the velocity of light.

Now S and S′ may both be stations fixed on the surface of the earth. But if the earth is traveling through space with a velocity component u in the direction SS′, we should have used $c + u$ for the effective velocity of light, in which case the correction becomes

$$\Delta t' = x(c + u). \tag{3}$$

If, in our attempts to set our clocks synchronously, we employ (2) where we should have employed (3), as the light-time correction, the error in our clock setting is

$$\varepsilon = \Delta t - \Delta t' \sim ux'/c, \tag{4}$$

if $u \ll c$.

At this point the experimental physicist of the nineteenth century would ask, "Why employ (2)? Why not determine experimentally the effective velocity of the light signal and then, by (3), set your clocks in exact synchronism?" The man who was accustomed to set his watch by the chimes of Big Ben, correcting his observations for the finite speed of sound, could similarly attain a better correction if he used an anemometer to gauge the speed of the wind. He could thus ascertain how much the wind would retard or advance the arrival of the sound wave.

The principle enunciated in the previous paragraph is in full accord with Newtonian mechanics. There is no apparent a priori reason why light should be exempt from the transformation equations. Physicists confidently expected to detect, for example, the changes in the velocity of light at different seasons of the year. The Michelson interferometer is, in effect, an "ether anemometer." The numerous attempts to discover and measure an ether "wind" or "drift" have, however, come to naught. The velocity of light appears to be constant no matter how the observer is moving. We are faced with the paradox that $c + u = c$. It also appears that we can no longer avoid the error of (4). We must discard or drastically revise the concept of simultaneity and, at the same time, we must alter the transformation equations (1).

Einstein pointed the way out of the dilemma. We are first to accept as universally true the experimental result that the velocity of light is constant, as measured from any inertial frame of reference. And then we must modify our concept of time and space in accord with the primary postulate.

Let S′ move with velocity u relative to S in the direction of the x-axis. Suppose that the two systems are coincident at time $t = 0$, at which instant a flash of light is sent out. After an interval t, the observer at S will find conditions as in Fig. 1a. The wave front will lie on a sphere of radius ct whose equation is

$$x^2 + y^2 + z^2 = c^2t^2 \tag{5}$$

and S′ will have moved on in the direction of the arrow. It seems paradoxical even to suggest that the form of the wave front as viewed from S′ could be considered by that observer as a sphere *centered at* S′. But that is exactly

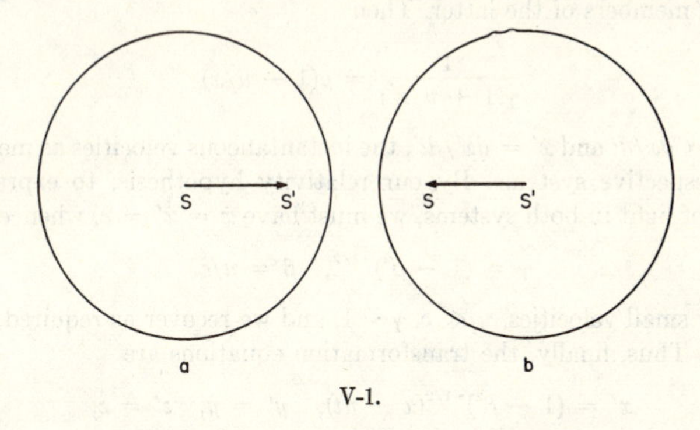

V-1.

what the relativity postulate requires. The velocity of light, as measured in either system must be a constant. There is no essential difference between S and S′ since both are inertial systems. Thus Fig. 1b presents the reciprocal conditions from the viewpoint of observer S′. For him the sphere will be represented by the equation

$$x'^2 + y'^2 + z'^2 = c^2t'^2. \tag{6}$$

We are required to find the transformation equations analogous to (1) that will satisfy (5) and (6) identically. Further the desired set must reduce approximately to (1) when u is small, because the Newtonian equations must then be closely fulfilled, to agree with experiment.

Certainly we cannot possibly satisfy (5) and (6), which imply the existence of one sphere with two different centers, if we retain Euclidean ideas of measurement. We have already seen, however, that the classical concept of simultaneity and therefore of a universal time system runs into difficulties. Perhaps spatial conditions are similarly altered; e.g., the length of a measuring rod may depend on the velocity.

Let us try the following transformation equations in x:

$$x' = \gamma(x - ut), \quad x = \gamma(x' + ut'), \tag{7}$$

where γ is a multiplicative constant to be determined. The principle of relativity demands the reciprocal character of these equations. Further, since the relative velocity along y and z is zero, we take

$$y' = y \quad \text{and} \quad z' = z. \tag{8}$$

The transformation equation for t still remains to be determined.

Take the differentials of (7). Then

$$dx' = \gamma(dx - u\,dt) \quad \text{and} \quad \gamma(dx' + u\,dt') = dx. \tag{9}$$

Divide the left- and right-hand members of the first equation by the corresponding members of the latter. Then

$$\frac{1}{\gamma(1 + u/\dot{x}')} = \mu(1 - u/\dot{x}) \tag{10}$$

where $\dot{x} = dx/dt$ and $\dot{x}' = dx'/dt'$, the instantaneous velocities as measured in the respective systems. By our relativity hypothesis, to express the velocity of light in both systems, we must have $\dot{x} = \dot{x}' = c$, whence

$$\gamma = (1 - \beta^2)^{-1/2}, \quad \beta = u/c. \tag{11}$$

Thus, for small velocities, $u \ll c$, $\gamma \sim 1$, and we recover as required, equations (1). Thus, finally, the transformation equations are

$$\begin{aligned} x' &= (1 - \beta^2)^{-1/2}(x - ut), \quad y' = y, \quad z' = z, \\ t' &= (1 - \beta^2)^{-1/2}(t - \beta x/c). \end{aligned} \tag{12}$$

To obtain the last equation, eliminate x' between the pair of equations (7). The reader will verify that equation (6), transformed by means of (12), reduces to (5). The relativity conditions are thus satisfied. We also write the reciprocal transformations

$$\begin{aligned} x &= (1 - \beta^2)^{-1/2}(x' + ut'), \quad y = y', \quad z = z', \\ t &= (1 - \beta^2)^{-1/2}(t' + \beta x'/c). \end{aligned} \tag{13}$$

Solving (10) and (11), we obtain as the relativity addition theorem for velocities,

$$\dot{x} = \frac{\dot{x}' + u}{1 + \dot{x}'u/c^2}, \tag{14}$$

which replaces the Newtonian theorem $\dot{x} = \dot{x}' + u$.

Let us investigate some of the consequences of these relationships. Consider a rod, of length $l = x_2 - x_1$, at rest in system S. A physicist in S′, at a given instant, say $t' = 0$, as recorded by his own clock, will find the length $l' = x_2' - x_1'$ to be

$$l = (1 - \beta^2)^{-1/2}l' \quad \text{or} \quad l' = (1 - \beta^2)^{1/2}l. \tag{15}$$

The rod thus appears to be shortened along the direction of its relative motion. Note that l' is the length of the rod as recorded by observer S′.

An observer on S will find a rod of length l attached to S′, similarly shortened. The relationships are reciprocal.

This effect is generally known as the Lorentz-Fitzgerald contraction. Fitzgerald suggested that the null result of the Michelson-Morley experiment to detect the relative motion of the earth and the ether might arise from a contraction of this character. Lorentz proved the effect to be consistent with the electron theory of matter. Einstein's contribution was to have deduced it from the principle of relativity.

The independent clocks in the two systems show a somewhat analogous behavior. Note that, for small velocities, we have approximately, from (13),

$$t - t' = \beta x'/c = ux'/c^2. \tag{16}$$

This term must represent the lack of synchronism deduced in (4). We are ordinarily concerned, however, only with time intervals measured in the respective systems. If t_1 and t_2 denote two times measured on a clock stationary in system S and t_1' and t_2' the analogous times recorded on a clock stationary in system S′,

$$t_2 - t_1 = (1 - \beta^2)^{-1/2}(t_2' - t_1'). \tag{17}$$

Thus the interval $(t_2' - t_1')$ as perceived in the unaccented system, will appear longer than the interval $t_2 - t_1$. The clock in S′ will appear to run slow, as checked by an observer on S, and vice versa.

The foregoing transformations have an important geometrical interpretation, given by Minkowski. We have seen that the equations (5) and (6) represent, in their respective coordinate systems, spheres whose radii are expanding uniformly with the time. Let us make the substitutions

$$\tau = ict \quad \text{and} \quad \tau' = ict'; \quad i = \sqrt{-1}. \tag{18}$$

Then the equations take the symmetrical form

$$x^2 + y^2 + z^2 + \tau^2 = 0; \quad x'^2 + y'^2 + z'^2 + \tau'^2 = 0. \tag{19}$$

By analogy with the two-dimensional equation

$$x^2 - c^2t^2 = 0, \tag{20}$$

which represents a conical surface, we see that (19) is the equation of a four-dimensional cone, its axes extending toward the past or future.

In relativity, time and space are no longer considered separately, but are fused into one geometry by means of the transformation (18). In special relativity we regard this four-dimensional geometry as Euclidean. We further assume that the relativistic laws of mechanics follow from extrapolation of the Newtonian laws to four dimensions, assuming invariance of the significant physical quantities, and applying the formulae in a coordinate frame wherein the particle is momentarily at rest. We term this frame the *proper* system of coordinates. We may consider four-dimensional vectors, and four-dimensional tensors analogous to the three-dimensional ones discussed earlier. For example, we have by simple extension of the three-dimensional formulae

$$\begin{aligned} d\mathbf{s} &= \mathbf{i}\,dx + \mathbf{j}\,dy + \mathbf{k}\,dz + \mathbf{1}\,d\tau, \\ ds^2 &= dx^2 + dy^2 + dz^2 + d\tau^2, \end{aligned} \tag{21}$$

where $\mathbf{1}$ is a unit vector along τ. The only restriction on our previous vector formulae is that we must interpret the cross-products, like the axial vector $\mathbf{k} \times \mathbf{1}$, as a tensor of second rank. In the primed system, we have

$$\begin{aligned} d\mathbf{s}' &= \mathbf{i}'\,dx' + \mathbf{j}'\,dy' + \mathbf{k}'\,dz' + \mathbf{1}'\,d\tau', \\ ds'^2 &= dx'^2 + dy'^2 + dz'^2 + d\tau'^2. \end{aligned} \tag{22}$$

Let us now make the assumption that the vector element $d\mathbf{s}$ is invariant to a change of coordinates, i.e.,

$$d\mathbf{s}' = d\mathbf{s} \quad \text{and} \quad ds'^2 = ds^2, \tag{23}$$

which are at least consistent with the fundamental equation (19) and its primed analogue, where $d\mathbf{s}$ shrinks to a point. Equation (23) is equivalent to assuming that (21) and (22) are related by a mere rotation of axes, because the resulting vectors are identical. Further, let us take $dy = dy'$ and $dz = dz'$, in accord with (13), and assume that the primed system is moving along x with constant velocity u with respect to the unprimed. Then we have

$$dx^2 + d\tau^2 = dx'^2 + d\tau'^2, \tag{24}$$

or, by (18),

$$dt^2\left[1 - \frac{1}{c^2}\left(\frac{dx}{dt}\right)^2\right] = dt'^2\left[1 - \frac{1}{c^2}\left(\frac{dx'}{dt'}\right)^2\right]. \tag{25}$$

But if the velocity of a particle in the primed system is zero, $dx'/dt' = 0$, and $dx/dt = u$. Therefore

$$dt(1 - \beta^2)^{1/2} = dt', \tag{26}$$

which is equivalent to transformation equation (17) for the time dilatation. It is interesting to note, though difficult to interpret physically, that through the afore-mentioned rotation, part of the vector $\mathbf{1}\,d\tau$ becomes a true spatial vector $\mathbf{i}'\,dx'$, and vice versa. The fusion of time and space is more than formal or symbolic; it is fundamental to relativity.

We introduce a four-dimensional velocity $d\mathbf{s}/dt$ related to the velocity in the primed system, by

$$\begin{aligned} \frac{d\mathbf{s}}{dt'} = \frac{d\mathbf{s}}{dt}\frac{dt}{dt'} &= \left(\mathbf{i}\frac{dx}{dt} + \mathbf{j}\frac{dy}{dt} + \mathbf{k}\frac{dz}{dt} + \mathbf{1}\frac{d\tau}{dt}\right)\frac{dt}{dt'} \\ &= \frac{\mathbf{i}u + \mathbf{j}v + \mathbf{k}w + \mathbf{1}ic}{(1 - \beta^2)^{1/2}}, \end{aligned} \tag{27}$$

by (26). We have written u, v, and w for the velocities in the unprimed system. Since we have now allowed motion in all coordinates, we must define

$$\beta^2 = v_0^2/c^2, \quad v_0^2 = u^2 + v^2 + w^2. \tag{28}$$

The physical significance of this four-vector of (27) is not immediately obvious, although the first three components are clearly the ordinary velocities, multiplied by $(1 - \beta^2)^{1/2}$. The vector $d\mathbf{s}$, apart from the imaginary character of the time variable, is the separation of two points in time and space. It thus represents the separation between events at two points in four-dimensional space. If dx, dy, and dz are zero, $d\mathbf{s}$ is the interval, measured in τ units, between two events at some point in three-dimensional space. By analogy, we infer that $d\mathbf{s}/dt'$ defines the rate at which events occur in the primed system. The four-dimensional coordinate system represents a space-time frame. A line in this space represents the progress of a point through ordinary space and time. We call it a "world line." A point that is regarded as stationary in space will move parallel to and hence define the direction of the ict axis. From this argument we can visualize how a transformation from S to S′ is in effect a rotation of our coordinate system.

Since the rate of occurrence of events will depend on the field of force, we *assume* that Newton's equation of motion applies to a four- as well as to a three-vector, and write, in accord with (II-2.1), a four-vector force:

$$\mathbf{F}_4 = \frac{d}{dt'}\left(m_0 \frac{d\mathbf{s}}{dt}\right), \tag{29}$$

where m_0 is the mass of the particle in the proper system. If now, we set

$$\mathbf{v}_0 = \mathbf{i}u + \mathbf{j}v + \mathbf{k}w, \tag{30}$$

the spatial part of the force vector becomes, by (27),

$$\mathbf{F} = \frac{d}{dt'}\left[\frac{m_0\mathbf{v}_0}{(1 - \beta^2)^{1/2}}\right] = \frac{1}{(1 - \beta^2)^{1/2}}\frac{d}{dt}\left[\frac{m_0\mathbf{v}_0}{(1 - \beta^2)^{1/2}}\right] \tag{31}$$

At this point we now understand the reason for specifying m_0 as the "rest mass," i.e., the mass measured in a frame in which the particle is at rest. For if we introduce an effective mass

$$m = m_0/(1 - \beta^2)^{1/2}, \tag{32}$$

and define $\mathbf{F}$, in terms of $\mathbf{F}_0$, the ordinary classical force, by

$$\mathbf{F} = \mathbf{F}_0/(1 - \beta^2)^{1/2}, \tag{33}$$

we shall then be able to write our equation of motion in its customary Newtonian form

$$\mathbf{F} = \frac{d}{dt}(m\mathbf{v}_0). \tag{34}$$

By (32), we see that the effective mass varies with the velocity of the particle. For the velocities commonly met with, β^2 is a small quantity, so that both **F** and m approximate closely to the classical value, The force $\mathbf{F}_4$, when defined as in equation (29), is invariant under a Lorentz transformation, and hence fulfills the requirements of relativity. In relativity, the momentum of the particle becomes

$$\mathbf{p} = m\mathbf{v}_0 = m_0\mathbf{v}_0/(1 - \beta^2)^{1/2}. \tag{35}$$

To derive the energy imparted to the particle, as it moves under the action of a field of force, we calculate, as in the classical case, the work done by a particle that moves along a trajectory. To shorten the algebra, we write $\mathbf{v}_0 = \mathbf{i}u$ and $d\mathbf{s} = \mathbf{i}\,dx$. Then $\beta = u/c$. The final results are not affected by this assumption.

$$W = \int_A^B \mathbf{F} \cdot d\mathbf{s} = m_0 \int_A^B \frac{d}{dt}\left[\frac{u}{(1 - \beta^2)^{1/2}}\right] dx$$

$$= m_0 \int_A^B u \frac{d}{dt}\left[\frac{u}{(1 - \beta^2)^{1/2}}\right] dt.$$

Note that we must treat β as a variable. Integrating this equation by parts, we find that

$$W = m_0 \frac{u^2}{(1 - \beta^2)^{1/2}} - m_0 \int_A^B \frac{u}{(1 - \beta^2)^{1/2}}\, du = \left[\frac{m_0 c^2}{(1 - \beta^2)^{1/2}}\right]_{\beta_A}^{\beta_B}, \tag{36}$$

wherein we are to substitute the values of β at the initial and final limits. If $\beta = 0$, initially, we note that the particle is to be assigned a "rest energy" of magnitude

$$W_0 = m_0 c^2. \tag{37}$$

The difference between the initial and final energies, if the particle starts from rest, is the kinetic energy, T, imparted to the system. Its value is

$$T = m_0 c^2 \frac{1}{\sqrt{1 - \beta^2}} - 1 = c^2(m - m_0). \tag{38}$$

All the foregoing equations are important, but (37) is perhaps the most fundamental, because it expresses the equivalence of mass and energy. The widest physical application of this relation is in the field of nuclear physics. For example, if 4×1.0078 ($= 4.0312$) g of hydrogen could be transmuted into helium, the weight of the product would be 4.0000 g, corresponding to a loss of 0.0312 g of mass. By (37), this mass reappears as free energy of the nuclear reaction, of amount

$$E = 0.0312c^2 \sim 2.8 \times 10^{19} \text{ ergs}, \tag{39}$$

which is approximately equal to the energy released by combustion of 100 tons of coal. Nuclear reactions are not as simple as here pictured, but the energies involved are truly enormous. The fission processes of uranium 235 and plutonium are well-known examples. Astronomers generally believe that nuclear reactions furnish the heat radiated by the sun and stars. In relativity the laws of conservation of matter and energy fail individually, but hold collectively through the medium of (37).

Let the symbol $\Box^2$ represent the operator

$$\Box^2 = \frac{\partial^2}{\partial x^2} + \frac{\partial^2}{\partial y^2} + \frac{\partial^2}{\partial z^2} + \frac{\partial^2}{\partial \tau^2}, \tag{40}$$

the analogue of ∇^2 in three dimensions. This operator is equivalent to

$$\Box^2 = \nabla^2 - \frac{1}{c^2}\frac{\partial^2}{\partial t^2}, \tag{41}$$

by (18). Thus the four-dimensional analogue of Laplace's equation,

$$\Box^2\psi = 0 = \nabla^2\psi - \frac{1}{c^2}\frac{\partial^2\psi}{\partial t^2}, \tag{42}$$

where ψ is any potential that is a function of x, y, z, and t, is merely the wave equation. Hence, if ψ represents the "retarded" scalar potential or any component of the "retarded" vector potential, as described in the coordinate system S and ψ', the similar potential in S′, the relativity transformations will enable us to express the wave equation in the symmetrical form,

$$\Box^2\psi' = 0. \tag{43}$$

Further, since we can express **H**, **H**′, **E**, **E**′, etc., in terms of these potentials, we infer that the Maxwellian equations must have the same form for both coordinate systems. Lorentz foresaw the necessity of this result from data of the Michelson-Morley experiment. From this condition alone he was able to deduce the transformation equations (12) and (13), which we generally refer to as the Lorentz transformation.

To demonstrate the invariance of Maxwell's equations, we shall employ the basic methods of tensor analysis, as developed in (II-31). To make our notation consistent with that of tensors, we adopt the following coordinates to represent the system at rest.

$$x^1 = x; \quad x^2 = y; \quad x^3 = z; \quad x^4 = \tau = ict. \tag{44}$$

We write out the pair of equations

$$\begin{aligned} \nabla \times \mathbf{H} - \frac{1}{c}\frac{\partial \mathbf{D}}{\partial t} &= \frac{4\pi}{c}\mathbf{J}, \\ \nabla \cdot \mathbf{D} &= 4\pi\rho, \end{aligned} \tag{45}$$

to give the following set of four linear equations:

$$
\begin{aligned}
0 + \frac{\partial H_3}{\partial x^2} - \frac{\partial H_2}{\partial x^3} - i\frac{\partial D_1}{\partial x^4} &= \frac{4\pi}{c} J_1. \\
-\frac{\partial H_3}{\partial x^2} + 0 + \frac{\partial H_1}{\partial x^3} - i\frac{\partial D_2}{\partial x^4} &= \frac{4\pi}{c} J_2. \\
\frac{\partial H_2}{\partial x^1} - \frac{\partial H_1}{\partial x^2} + 0 - i\frac{\partial D_3}{\partial x^4} &= \frac{4\pi}{c} J_3. \\
i\frac{\partial D_2}{\partial x^1} + i\frac{\partial D_2}{\partial x^2} + i\frac{\partial D_3}{\partial x^3} + 0 &= 4\pi i\rho.
\end{aligned}
\tag{46}
$$

We have multiplied the last equation through by i, to produce a symmetrical array. Now consider the antisymmetric tensor

$$
\mathcal{G}^{ij} = \begin{bmatrix} 0 & H_3 & -H_2 & -iD_1 \\ -H_3 & 0 & H_1 & -iD_2 \\ H_2 & -H_1 & 0 & -iD_3 \\ iD_1 & iD_2 & iD_3 & 0 \end{bmatrix}, \tag{47}
$$

where i denotes the row and j the column, in the usual matrix sense. Also define a four-vector $\mathcal{J}^i$ as

$$
\mathcal{J}^i = \frac{4\pi}{c}\begin{bmatrix} J_1 \\ J_2 \\ J_3 \\ ic\rho \end{bmatrix}. \tag{48}
$$

With these definitions, the single equation

$$
\frac{\partial}{\partial x^j}\mathcal{G}^{ij} = \mathcal{G}^{ij}_{\cdot\,,j} = \mathcal{J}^i \tag{49}
$$

represents the pair of Maxwell equations, (45).

We invoke the summation convention over the repeated index, j. That the introduction of the four-vector $\mathcal{J}^i$ is in no sense artificial follows directly from the process of taking its divergence. If $\mathcal{J}^i$ is a vector in four-dimensional space,

$$
\frac{\partial}{\partial x^i}\mathcal{J}^i = \mathcal{J}^i_{,i}, \tag{50}
$$

must be a scalar invariant. That this is so follows from the equation of continuity (II-34.8),

$$
\mathcal{J}^i_{,i} = \nabla\cdot\mathbf{J} + \frac{\partial\rho}{\partial t} = 0. \tag{51}
$$

We have written $\mathcal{G}^{ij}$ in the notation of a contravariant tensor, despite the fact that the distinction between the two types of indices vanishes for the Cartesian system of special relativity.

In terms of the tensor, we express the displacement vector as

$$\mathbf{D} = i\mathbf{e}_i\mathcal{G}^{i4} = -i\mathbf{e}_i\mathcal{G}^{4i}, \tag{52}$$

but the expression for $\mathbf{H}$ is somewhat more complicated. We can write it in the form

$$\begin{aligned}\mathbf{H} &= \mathbf{e}^1\mathcal{G}^{23} + \mathbf{e}^2\mathcal{G}^{31} + \mathbf{e}^3\mathcal{G}^{12} \\ &= -\mathbf{e}^1\mathcal{G}^{32} - \mathbf{e}^2\mathcal{G}^{13} - \mathbf{e}^3\mathcal{G}^{21}\end{aligned} \tag{53}$$

but the notation indicates that we should consider $\mathbf{H}$ as an antisymmetric tensor rather than as an axial vector. Hence we write

$$\begin{aligned}\mathbf{H} &= \mathbf{e}_1\mathbf{e}_2\mathcal{G}^{12} + \mathbf{e}_2\mathbf{e}_3\mathcal{G}^{23} + \mathbf{e}_3\mathbf{e}_1\mathcal{G}^{31} + \mathbf{e}_2\mathbf{e}_1\mathcal{G}^{21} + \mathbf{e}_3\mathbf{e}_2\mathcal{G}^{32} + \mathbf{e}_1\mathbf{e}_3\mathcal{G}^{13} \\ &= \mathbf{e}_i\mathbf{e}_j\mathcal{G}^{ij}.\end{aligned} \tag{54}$$

If we place a cross between each pair of unit vectors in (54) and carry out the multiplication so indicated, we shall obtain a final result that is twice that of (53).

To these equations we must add the defining relationships

$$\mathbf{B} = \mu\mathbf{H}, \qquad \mathbf{E} = \kappa^{-1}\mathbf{D}, \tag{55}$$

and the other pair of Maxwell equations

$$\begin{gathered}\nabla \times \mathbf{E} + \frac{1}{c}\frac{\partial \mathbf{B}}{\partial t} = 0. \\ \nabla \cdot \mathbf{B} = 0.\end{gathered} \tag{56}$$

First, we note that (55) implies the existence of a second tensor related to the first in some special way. Let us write the tensor in the following form and later justify our procedure.

$$\mathfrak{F}_{rs} = \begin{bmatrix} 0 & B_3 & -B_2 & -iE_1 \\ -B_3 & 0 & B_1 & -iE_2 \\ B_2 & -B_1 & 0 & -iE_3 \\ iE_1 & iE_2 & iE_3 & 0 \end{bmatrix}. \tag{57}$$

This form of the tensor implies a congruence between (51) and (47) in the sense that electric and magnetic quantities occupy similar positions. Then our fundamental vector quantities are

$$\mathbf{E} = i\mathbf{e}^r\mathfrak{F}_{r4} = -i\mathbf{e}^s\mathfrak{F}_{4s} \tag{58}$$

and

$$\mathbf{B} = \mathbf{e}^r\mathbf{e}^s\mathfrak{F}_{rs}, \qquad r, s = 1, 2, \text{ or } 3. \tag{59}$$

Let
$$\frac{\partial}{\partial x^j}\mathfrak{F}_{rs} = \mathfrak{F}_{rs,j}. \tag{60}$$

Our full set of 4 equations is contained in the following abbreviated tensor form:

$$\mathfrak{F}_{rs,j} + \mathfrak{F}_{sj,r} + \mathfrak{F}_{jr,s} = 0, \tag{61}$$

where $r \neq s \neq j$ and $r, s, j = 1, 2, 3,$ or 4.

We must now define the transformation connecting $\mathcal{G}^{ij}$ and $\mathfrak{F}_{rs}$. First of all,

$$g_{ir} = \delta_{ir}, \tag{62}$$

because the coordinate system of special relativity is Euclidean. Hence we can lower the index on $\mathcal{G}^{ij}$ to form the covariant equivalent $\mathcal{G}_{ij}$, whose components are identical with those of $\mathcal{G}^{ij}$.

Now define the mixed tensor

$$\eta^i_r = \begin{bmatrix} \eta^1_1 & \eta^1_2 & \eta^1_3 & 0 \\ \eta^2_1 & \eta^2_2 & \eta^2_3 & 0 \\ \eta^3_1 & \eta^3_2 & \eta^3_3 & 0 \\ 0 & 0 & 0 & \eta^4_4 \end{bmatrix} \tag{63}$$

and carry out the transformation

$$\mathfrak{F}_{rs} = \eta^i_r \eta^j_s \mathcal{G}_{ij}. \tag{64}$$

This equation allows for the fact that both μ and κ may be tensors. Carrying out the indicated multiplication, we find that

$$(\kappa^{-1})^i_r = \eta^4_4 \eta^i_r, \tag{65}$$

and μ^i_r is the minor of det η that results from the suppression of the ith row and rth column of the tensor array. Hence

$$\begin{aligned} \mu^1_1 &= \eta^2_2\eta^3_3 - \eta^2_3\eta^3_2. \\ \mu^1_2 &= \eta^2_3\eta^3_1 - \eta^2_1\eta^3_3. \\ \mu^2_1 &= \eta^1_3\eta^3_2 - \eta^1_2\eta^3_3. \end{aligned} \tag{66}$$

When the medium is isotropic,

$$\eta^1_1 = \eta^2_2 = \eta^3_3 = \mu^{1/2}, \qquad \eta^i_j = 0, \quad i \neq j, \tag{67}$$

while
$$\eta^4_4 = \mu^{1/2}/\kappa\mu. \tag{68}$$

We note, in passing that (κ^{-1}) and (μ) are symmetric tensors only when the axes defined by η^i_r are orthogonal.

The mere fact that we have been able to express both electromagnetic quantities and the Maxwell equations in the form of four-dimensional vectors and tensors is in itself a guarantee of their invariance to coordinate transformations. However, we find it useful to determine the precise laws of the transformation.

Consider, therefore, a system in motion along the z axis with velocity v. This amounts to a change of coordinates in the usual tensor sense. We are to evaluate the coefficient γ^i_j, for contravariant vectors.

$$\bar{x}^i = \gamma^i_j x^j, \tag{69}$$

or

$$\begin{aligned} \bar{x}^1 &= x^1; \qquad \bar{x}^3 = \gamma^3_3 x^3 + \gamma^3_4 x^4. \\ \bar{x}^2 &= x^2; \qquad \bar{x}^4 = \gamma^4_3 x^3 + \gamma^4_4 x^4. \end{aligned} \tag{70}$$

With β defined as in (11), we have

$$\bar{x}^3 = \gamma^3_3(x^3 - vt) = \gamma^3_3(x^3 + i\beta x^4). \tag{71}$$

Orthogonality requires that

$$\gamma^3_3\gamma^4_3 + \gamma^3_4\gamma^4_4 = 0,$$

whereas

$$(\gamma^3_3)^2 + (\gamma^3_4)^2 = (\gamma^4_3)^2 + (\gamma^4_4)^2 = 1. \tag{72}$$

Solve these equations, as follows:

$$\begin{aligned} \gamma^3_4 &= i\beta\gamma^3_3 = [1 - (\gamma^3_3)^2]^{1/2}. \\ -\beta^2(\gamma^3_3)^2 &= 1 - (\gamma^3_3)^2. \end{aligned} \tag{73}$$

Thus

$$\begin{gathered} \gamma^3_3 = 1/\sqrt{1-\beta^2}; \quad \gamma^3_4 = i\beta/\sqrt{1-\beta^2}; \quad \gamma^4_3 = -i\beta/\sqrt{1-\beta^2}; \\ \gamma^4_4 = i/\sqrt{1-\beta^2}; \quad \gamma^1_1 = \gamma^2_2 = 1. \end{gathered} \tag{74}$$

Other than these, the values of

$$\gamma^i_j = 0, \qquad i \neq j. \tag{75}$$

Now if we have $\mathfrak{F}_{lm}$ or $\mathfrak{G}^{ij}$, in a stationary coordinate system, we have, for a moving system

$$\begin{aligned} \bar{\mathfrak{G}}^{rs} &= \gamma^r_i\gamma^s_j\mathfrak{G}^{ij}. \\ \bar{\mathfrak{F}}_{rs} &= \gamma^l_r\gamma^m_s\mathfrak{F}_{lm}. \end{aligned} \tag{76}$$

These transformations give

$$\bar{\mathfrak{G}}^{\mu\nu} = \begin{bmatrix} 0 & \bar{H}_3 & -\bar{H}_2 & -i\bar{D}_1 \\ -\bar{H}_3 & 0 & \bar{H}_1 & -i\bar{D}_2 \\ \bar{H}_2 & -\bar{H}_1 & 0 & -i\bar{D}_3 \\ i\bar{D}_1 & i\bar{D}_2 & i\bar{D}_3 & 0 \end{bmatrix}, \tag{77}$$

where

$$\overline{H}_3 = H_3; \quad \overline{H}_2 = \frac{H_2 - \beta D_1}{\sqrt{1-\beta^2}}; \quad \overline{H}_1 = \frac{H_1 + \beta D_2}{\sqrt{1-\beta^2}}.$$

$$\overline{D}_1 = \frac{D_1 - \beta H_2}{\sqrt{1-\beta^2}}; \quad \overline{D}_2 = \frac{D_2 + \beta H_1}{\sqrt{1-\beta^2}}; \quad \overline{D}_3 = D_3. \tag{78}$$

Likewise

$$\overline{\mathfrak{F}}^{rs} = \begin{bmatrix} 0 & \overline{B}_3 & -\overline{B}_2 & -i\overline{E}_1 \\ \overline{B}_3 & 0 & \overline{B}_1 & -i\overline{E}_2 \\ \overline{B}_2 & -\overline{B}_1 & 0 & -i\overline{E}_3 \\ i\overline{E}_1 & i\overline{E}_2 & i\overline{E}_3 & 0 \end{bmatrix}, \tag{79}$$

where

$$\overline{B}_3 = B_3; \quad \overline{B}_2 = \frac{B_2 - \beta E_1}{\sqrt{1-\beta^2}}; \quad \overline{B}_1 = \frac{B_1 + \beta E_2}{\sqrt{1-\beta^2}}.$$

$$\overline{E}_1 = \frac{E_1 - \beta B_2}{\sqrt{1-\beta^2}}; \quad \overline{E}_2 = \frac{E_2 + \beta B_1}{\sqrt{1-\beta^2}}; \quad \overline{E}_3 = E_3. \tag{80}$$

The components of the four-vector, $\mathfrak{J}$, obey the transformation

$$\overline{\mathfrak{J}}^r = \gamma^r_{\cdot i}\mathfrak{J}^i, \tag{81}$$

so that

$$\overline{J}_1 = J_1; \quad \overline{J}_2 = J_2; \quad \overline{J}_3 = \frac{J_3 - v\rho}{\sqrt{1-\beta^2}};$$

$$\overline{J}_4 = ic\overline{\rho} = i\,\frac{c\rho - \beta J_3}{\sqrt{1-\beta^2}}. \tag{82}$$

The form of all the equations, therefore, is unaltered by the relative motion, though the magnitudes assigned to the various quantities **H**, **P**, **E**, **D**, **J** and ρ change.

The Maxwell equations, when written in the above manner, transform as tensors. For example, (49) becomes

$$\overline{\mathfrak{G}}^{rs}_{\cdot\cdot,s} = \gamma^r_{\cdot i}\mathfrak{G}^{ij}_{\cdot\cdot,j}. \tag{83}$$

Similar relationships hold for the other equations (50) and (61).

Numerous consequences of scientific importance follow directly from these equations. As a primary example, consider the case where a stationary charge "sees" only the electric field E_1, with all other components of E and B equal to zero. A charge moving with velocity v, parallel to the z-axis, will under the same circumstances experience the force components

$$\overline{B}_2 = -\frac{\beta E_1}{\sqrt{1-\beta^2}} \quad \text{and} \quad \overline{E}_1 = \frac{E_1}{\sqrt{1-\beta^2}}. \tag{84}$$

In other words, the charge moving in what we originally defined as a purely electric field is now subject to magnetic forces as well. By the same token, we can show that purely magnetic forces may appear as electric forces to a charge in motion.

The primary conclusion to draw from this analysis is that electromagnetic forces possess no unique character by whose means one can determine absolute motion. Electric vectors and magnetic vectors do not possess individual properties of conservation. The two sets of parameters are interrelated in a manner that depends on the motion of the observer.

Any observer can determine all physical parameters in a manner consistent with the assumption that his own velocity is zero. If any other observer in uniform relative motion reports to him a simultaneous set of physical measurements, the foregoing equations will enable him to reconcile both sets of observations. However, the same equations will enable observer number two to regard his own velocity as zero and attribute the difference between the two sets as due to relative motion of the first observer. Thus no experiment can possibly enable one to establish a system of absolute coordinates. All motion is relative. Hence the theory of relativity.

As an alternative representation, in terms of scalar and vector potentials, define a four-dimensional potential Φ whose components, Φ^i, are

$$\Phi^1 = A^1; \quad \Phi^2 = A^2; \quad \Phi^3 = A^3; \quad \Phi^4 = i\phi. \tag{85}$$

Then form the tensor Ψ through multiplication of Φ by the tensor

$$\zeta^i_i = \begin{bmatrix} 1 & 0 & 0 & 0 \\ 0 & 1 & 0 & 0 \\ 0 & 0 & 1 & 0 \\ 0 & 0 & 0 & \overline{\kappa\mu} \end{bmatrix} = \mu(\eta^i_r)^{-1}(\eta^r_i)^{-1}, \tag{86}$$

where $(\eta^i_r)^{-1}$ is the reciprocal of the matrix (63), with the condition of (67) and (68), for the medium, which we here take to be isotropic. Then

$$\Psi^i = \zeta^i_i \Phi^i. \tag{87}$$

The divergence of Ψ, given by the covariant derivative,

$$\Psi^i_{,i} = 0, \tag{88}$$

represents the equation (IV-26.14),

$$\nabla \cdot \mathbf{A} + \frac{\kappa\mu}{c}\frac{\partial\phi}{\partial t} = 0. \tag{89}$$

Similarly,

$$\Psi^i_{,kk} = -\mu^{1/2}\eta^i_i \mathfrak{J}^i \tag{90}$$

represents both the vector equation (IV-26.16),

$$\nabla^2\mathbf{A} - \frac{\mu\kappa}{c^2}\frac{\partial^2\mathbf{A}}{\partial t^2} = -\frac{4\pi\mu\mathbf{J}}{c}, \tag{91}$$

and the scalar equation (IV-26.17),

$$\nabla^2\phi - \frac{\mu\kappa}{c^2}\frac{\partial^2\phi}{\partial t^2} = -\frac{4\pi\rho}{\kappa}. \tag{92}$$

Then we find that we can represent the components of our basic tensor $\mathfrak{F}_{ij}$ in terms of the Φ_i, whose components are identical with those of Φ^i:

$$\mathfrak{F}_{ij} = \frac{\partial}{\partial x^j}\Phi_i - \frac{\partial}{\partial x^i}\Phi_j = \Phi_{i,j} - \Phi_{j,i}. \tag{93}$$

These equations are the equivalents of equations (IV-26.11) and (IV-26.5),

$$\mathbf{E} = -\nabla\phi - \frac{1}{c}\frac{\partial \mathbf{A}}{\partial t} \tag{94}$$

and $$\mathbf{B} = \nabla \times \mathbf{A}. \tag{95}$$

The vector Φ is also invariant to a Lorentz transformation, so that equations (85), (93), and (55) completely define the electromagnetic vector field.

Now we shall investigate the physical problem of a beam of radiation moving in the xz-plane at angle α with respect to the z-axis, in free space. The stationary observer will measure the x and z components of electric field as

$$E_1 = E\cos\alpha; \qquad E_3 = E\sin\alpha, \tag{96}$$

and the magnetic component, parallel to the y-axis, as

$$H_2 = E, \tag{97}$$

where E is the intensity of the electric vector. In the moving system, we have, similarly

$$\bar{E}_1 = \bar{E}\cos\bar{\alpha}; \quad \bar{E}_3 = \bar{E}\sin\bar{\alpha}; \quad \bar{H}_2 = \bar{E}. \tag{98}$$

But by (79) and (80),

$$\begin{gathered}\bar{E}_1 = \frac{E_1 - \beta H_2}{\sqrt{1-\beta^2}} = \frac{E_1 - \beta E}{\sqrt{1-\beta^2}}; \quad \bar{E}_3 = E_3. \\ \bar{H}_2 = \bar{E} = \frac{H_2 - \beta H_1}{\sqrt{1-\beta^2}} = \frac{E - \beta E_1}{\sqrt{1-\beta^2}}.\end{gathered} \tag{99}$$

Hence

$$\begin{gathered}\cos\bar{\alpha} = \frac{\bar{E}_1}{\bar{E}} = \frac{E_1 - \beta E}{E - \beta E_1} = \frac{\cos\alpha - \beta}{1 - \beta\cos\alpha}. \\ \sin\bar{\alpha} = \frac{\sin\alpha}{1 - \beta\cos\alpha}(1-\beta^2)^{1/2}.\end{gathered} \tag{100}$$

The apparent direction of the incident beam thus depends upon the velocity of the moving system. This equation is the relativistic expression for what physicists commonly term the "aberration" or wandering of light, first discovered by Bradley. The earth's orbital motion causes the position of stars to traverse an ellipse or circle, a result of the changing direction of the motion.

As a final calculation, we determine the effect that relative motion has upon the wavelength, λ, of a light beam, a phenomenon generally called the "Doppler effect." As in the previous problem, let the beam be incident at angle α with respect to the z-axis. Thus, in the inertial system, λ has the component $\lambda \sin \alpha$ along x and $\lambda \cos \alpha$ along the z-axis. Similar expressions $\bar{\lambda} \sin \bar{\alpha}$ and $\bar{\lambda} \cos \bar{\alpha}$ hold for the moving system. The projection along x must be constant in both systems, because the motion is along z not x, by hypothesis. Hence

$$\bar{\lambda} \sin \bar{\alpha} = \lambda \sin \alpha, \tag{101}$$

or

$$\lambda = \bar{\lambda} \frac{\sin \bar{\alpha}}{\sin \alpha} = \bar{\lambda} \frac{\sqrt{1 - \beta^2}}{1 - \beta \cos \alpha}, \quad \text{by (100).} \tag{102}$$

The frequency

$$\nu = \frac{c}{\lambda} = \bar{\nu} \frac{1 - \beta \cos \alpha}{\sqrt{1 - \beta^2}}. \tag{103}$$

Hence, if we observe a frequency $\bar{\nu}$, of light originating on a system receding from the earth with velocity u, the observed frequency will be diminished according to the above formula. The light will thus suffer a shift to the red. If the light beam contains lines of known frequency, the observer may judge the relative velocity of himself and the source by measuring the amount of the Doppler shift. When $\beta \ll 1$, we have

$$\nu = \bar{\nu}\left[1 - \frac{u}{c} \cos \alpha + \frac{1}{2}\left(\frac{u}{c}\right)^2 - \ldots\right]. \tag{104}$$

The first two terms of the parenthesis give the old classical expression, which holds sufficiently well as long as $\beta \ll 1$.

2. General relativity. In special relativity we regarded the four-dimensional space-time manifold as Euclidean. The electromagnetic field equations take an invariant form under such circumstances, as we have seen. However, we must introduce gravitational fields by some sort of ad hoc hypothesis. Furthermore we are led to assume that gravitational fields, like electric ones, are propagated with the finite velocity c. Instead of representing the field $\mathbf{F}$ as the three components of the spatial potential

V, we introduce a four-dimensional gravitational vector $\mathbf{F}$, whose components are

$$\mathbf{F} = -\Box V. \tag{1}$$

Then, in place of Poisson's equation, we have

$$\Box \cdot \mathbf{F} = \Box^2 V = 4\pi G\rho, \tag{2}$$

an equation that implies the existence of gravitational waves.*

In general relativity we attempt to avoid all assumptions concerning external fields of force by introduction of a curved metric. We consider that such a procedure may be possible from the basic equation (II-31.186) for a geodesic:

$$\frac{d^2q^\epsilon}{ds^2} + \Gamma_{\mu\nu}^{\ \ \epsilon} \frac{dq^\mu}{ds}\frac{dq^\nu}{ds} = 0. \tag{3}$$

We adopt the convention of using Greek letters for the tensor summation indices, in this relativistic treatment.

Compare equation (3) with the law of gravitation:

$$\frac{d^2\mathbf{r}}{dt^2} + \frac{GM}{r^3}\mathbf{r} = 0. \tag{4}$$

In (3), introduction of Cartesian coordinates immediately gave the equation of a straight line because the Γ's vanish. If we can introduce a non-Euclidean space, whose non-vanishing Christoffel symbols can reproduce the force term $GM\mathbf{r}/r^3$, we can say that a body moving in a geodesic in warped space traverses the same orbit that it would ordinarily follow, in Euclidean space under the law of gravitational attraction.

To visualize the physical meaning of this new viewpoint, consider again the picture of the navigator (page 131) sailing in a triangular course around an island. Since the triangle is a spherical one, on the curved surface of the earth, the sum of the three angles must exceed 180° by an amount E, the so-called "spherical excess."

Now, if the navigator is unaware that the earth's surface is curved, he will be puzzled, at first, by the fact that he must make each angle of his triangle exceed 60° in order to return to the starting point. And then a simple explanation may occur to him: "The island is exerting an attraction upon the ship!" This "attraction" draws the ship toward the island during the time that the ship traverses a given leg of the triangle, so that the angular excess of the initial course just balances out.

The navigator, with this clue, can then investigate the nature of the force field associated with the island. The larger the triangle, the greater the excess. Hence he concludes that the attractive force varies as the area

*What we have here called $\Box^2$, a four-dimensional operator, many writers express as $\Box$. They have no symbol analogous to $\Box$ the four-dimensional ∇.

of the triangle, i.e., the force increases (approximately) as the square of the distance. The navigator can prepare an entire set of sailing directions, based on the hypothesis of force fields and a plane geometry. And yet we know that introduction of a curved metric or Riemannian space could completely eliminate the fictitious field of force.

We shall consider further the navigator's problem. Meanwhile, can we not learn a lesson from the above analogy? In our solar system we describe the motion of planets about the sun in terms of a pure Euclidean space and force fields. Instead, may not the mere presence of the great solar mass (and other masses as well) warp the surrounding space or rather, space-time continuum? Can we not regard orbital trajectories as geodesics in four-dimensional curved geometry?

We can draw some conclusions about the character of this non-Euclidean space. It should not possess a curvature like that of the earth's surface, as in the analogy, because then the equivalent gravitational force would increase with distance. Instead we must expect the curvature to decrease with distance from the sun; as a rough picture we may visualize a space shaped something like a rapidly flaring horn of an old-fashioned phonograph. And, unlike the analogy, we must attribute the curvature directly to the presence of the "island." But let us try to state these results in quantitative form.

To summarize the significant tensor formulae, we shall define the curvature in terms of the Riemann-Christoffel tensor (II-31.172):

$$B_{\nu\mu\sigma\cdot}^{\cdot\cdot\cdot\epsilon} = D_\nu \Gamma_{\mu\sigma}^{\epsilon} - D_\mu \Gamma_{\nu\sigma}^{\epsilon} + \Gamma_{\mu\sigma}^{\rho} \Gamma_{\nu\rho}^{\epsilon} - \Gamma_{\nu\sigma}^{\rho} \Gamma_{\mu\rho}^{\epsilon}, \tag{5}$$

where

$$D_\mu = \frac{\partial}{\partial q^\mu} \tag{6}$$

and

$$\Gamma_{\mu\nu}^{\sigma} = \Gamma_{\nu\mu}^{\sigma} = \frac{1}{2} g^{\sigma\epsilon}(D_\mu g_{\nu\epsilon} + D_\nu g_{\epsilon\mu} - D_\epsilon g_{\mu\nu}), \tag{7}$$

with the $g_{\mu\nu}$ given by

$$(ds)^2 = g_{\mu\nu}\, dq^\mu\, dq^\nu. \tag{8}$$

The sixteen components of $g_{\mu\nu}$, in four dimensions, form the symmetric metric tensor. We define the reciprocal tensor, $g^{\sigma\nu}$, by

$$g_{\mu\nu} g^{\sigma\nu} = \delta_\mu^\sigma. \tag{9}$$

The foregoing equations, in their general sense, contain far too many variables for us even to attempt a solution. We must find some way of restricting the problem.

If we concern ourselves with motion of planets in the neighborhood of the sun, we readily see that spherical coordinates, r, θ, ϕ, and t are the natural coordinates for the problem. We expect, therefore, to express our line element as

$$(ds)^2 = (dr)^2 + r^2(d\theta)^2 + r^2 \sin^2 \theta(d\phi)^2 - c^2(dt)^2, \tag{10}$$

with

$$\begin{gathered} g_{11} = 1; \quad g_{22} = r^2; \quad g_{33} = r^2 \sin^2 \theta; \\ g_{44} = -c^2; \quad g_{\mu\nu} = 0, \quad \mu \neq \nu. \end{gathered} \tag{11}$$

This system, however, is basically Euclidean, because the curvature tensor vanishes. We readily adapt it to the problem of special relativity already discussed. But if we are to introduce force fields as described above, we require a more complicated system of $g_{\mu\nu}$'s. The spherical symmetry inherent in the problem leads us to maintain g_{22} and g_{33} as in (11). However, we shall suppose that g_{11} and g_{44} are functions of r, which, respectively, must go to unity and $-c^2$ for large values of r.

At this point we introduce some minor changes of notation, unimportant mathematically or physically but desirable in order to make the notation agree with that in common usage. Instead of (10), we write

$$(ds)^2 = -e^{\xi}(dr)^2 - r^2(d\theta)^2 - r^2 \sin^2 \theta(d\phi)^2 + e^{\zeta}(dt)^2. \tag{12}$$

This change of sign makes $(ds)^2$ positive and ds real, because

$$\left(\frac{ds}{dt}\right)^2 = c^2 - v^2 \quad \text{and} \quad v \leq c. \tag{13}$$

In equation (12) we have arbitrarily written

$$g_{11} = -e^{\xi}, \quad g_{44} = e^{\zeta}, \quad g_{22} = -r^2, \quad g_{33} = -r^2 \sin^2 \theta, \tag{14}$$

a substitution that simplifies our mathematics and involves no loss of generality, since we regard both ξ and ζ as functions of r, still to be determined.

We now calculate the significant Γ's. The contravariant metric tensor possesses the components

$$\begin{gathered} g^{11} = -e^{-\xi}; \quad g^{22} = -\frac{1}{r^2}; \quad g^{33} = -\frac{1}{r^2} \sin^2 \theta; \quad g^{44} = e^{-\zeta}; \\ g^{\mu\nu} = 0, \quad \mu \neq \nu. \end{gathered} \tag{15}$$

$$\left.\begin{aligned} \Gamma^1_{11} &= \frac{1}{2} \frac{\partial}{\partial r} \ln g_{11} = \frac{1}{2} \xi'; \quad \Gamma^2_{12} = \Gamma^2_{21} = \Gamma^3_{13} = \Gamma^1_{31} = \frac{1}{r}, \\ \Gamma^4_{14} &= \Gamma^4_{41} = \frac{1}{2} \frac{\partial}{\partial r} \ln g_{44} = \frac{1}{2} \zeta'; \quad \Gamma^1_{22} = \frac{r}{g_{11}} = -re^{-\xi}, \\ \Gamma^3_{23} &= \Gamma^3_{32} = \cot \theta; \quad \Gamma^1_{33} = \frac{r \sin^2 \theta}{g_{11}} = -re^{-\xi} \sin^2 \theta, \\ \Gamma^2_{33} &= -\sin \theta \cos \theta; \quad \Gamma^1_{44} = -\frac{1}{2g_{11}} \frac{\partial}{\partial r} g_{44} = \frac{1}{2} e^{\zeta - \xi} \zeta', \end{aligned}\right\} \tag{16}$$

where the primes signify derivatives with respect to r. All other values of Γ vanish.

We consider substituting these symbols into $B^{\cdots\epsilon}_{\nu\mu\sigma}$, which task is not impossible, because only 20 of the 256 components that this tensor possesses are linearly independent. However even this number is excessively large. Rather than operate with the complete tensor, which provides us with far too many equations, we follow the precedent set by Einstein and consider the contracted tensor

$$G_{\mu\nu} = B^{\cdots\sigma}_{\nu\mu\sigma}. \tag{17}$$

with $\epsilon = \sigma$. Then

$$G_{\mu\nu} = D_\nu\Gamma^\sigma_{\mu\sigma} - D_\sigma\Gamma^\sigma_{\mu\nu} + \Gamma^\rho_{\mu\sigma}\Gamma^\sigma_{\rho\nu} - \Gamma^\rho_{\mu\nu}\Gamma^\sigma_{\rho\sigma}. \tag{18}$$

The Γ's with duplicate indices take the form

$$\Gamma^\sigma_{\mu\sigma} = D_\mu \ln \sqrt{-g}, \tag{19}$$

where

$$g = \det |g_{\mu\nu}| = -e^{\xi+\zeta}r^4 \sin^2\theta. \tag{20}$$

Hence

$$G_{\mu\nu} = D_\mu D_\nu \ln \sqrt{-g} - \Gamma^\rho_{\mu\nu}D_\rho \ln \sqrt{-g} - D_\rho\Gamma^\rho_{\mu\nu} + \Gamma^\rho_{\mu\sigma}\Gamma^\sigma_{\rho\nu}. \tag{21}$$

Write out the expressions in detail. For example,

$$\begin{aligned} G_{11} &= \left(\frac{\partial^2}{\partial r}\ln\sqrt{-g}\right) - \left(\Gamma^1_{11}\frac{\partial}{\partial r}\ln\sqrt{-g}\right) - \left(\frac{\partial}{\partial r}\Gamma^1_{11}\right) \\ &\qquad + (\Gamma^1_{11}\Gamma^1_{11} + \Gamma^2_{12}\Gamma^2_{21} + \Gamma^3_{13}\Gamma^3_{31} + \Gamma^4_{14}\Gamma^4_{41}) \\ &= \frac{\zeta''}{2} - \frac{\zeta'\xi'}{4} + \frac{(\zeta')^2}{4} - \frac{\xi'}{r}. \end{aligned} \tag{22}$$

Similarly,

$$G_{22} = e^{-\xi}\left[1 + \frac{r(\zeta' - \xi')}{2} - e^\xi\right], \tag{23}$$

$$G_{33} = e^{-\xi}\left[1 + \frac{r(\zeta' - \xi')}{2} - e^\xi\right]\sin^2\theta, \tag{24}$$

$$G_{44} = -e^{\zeta-\xi}\left[\frac{\zeta''}{2} - \frac{\zeta'\xi'}{4} + \frac{(\zeta')^2}{4} + \frac{\zeta'}{r}\right], \tag{25}$$

$$G_{\mu\nu} = 0, \quad \mu \neq \nu, \tag{26}$$

so that we have four equations from which to determine ξ and ζ. Of these, (23) and (24) are identical, as far as functional dependence on r is concerned.

Up to this point, our discussion has been mathematical rather than physical, except for the fact that we have allowed the spherical symmetry of the problem to influence our choice of the metric tensor. But here mathematics fails us. We have determined what mathematical structure the tensor $G_{\mu\nu}$ must possess. We have found, in other words, the mathematical form of the left-hand side of the equation, but we do not know what the right-hand side should be. Therefore, as out-and-out hypothesis, let us make the simplest assumption open to us, viz., that

$$G_{\mu\nu} = 0 \tag{27}$$

for all sixteen components. This equation is Einstein's alternative expression for the law of planetary motions. We can say, if we wish, that it is Einstein's equation for the law of gravitation. The three basic equations then become

$$\frac{\zeta''}{2} - \frac{\zeta'\xi'}{4} - \frac{(\zeta')^2}{4} - \frac{\xi'}{r} = 0, \tag{28}$$

$$1 + \frac{r(\zeta' - \xi')}{2} - e^{\xi} = 0, \tag{29}$$

$$\frac{\zeta''}{2} - \frac{\zeta'\xi'}{4} - \frac{(\zeta')^2}{4} + \frac{\zeta'}{r} = 0. \tag{30}$$

Subtracting (28) from (30) we get

$$\zeta' = -\xi' \quad \text{or} \quad \zeta = -\xi + \text{CONST.} \tag{31}$$

Substituting this value into (29), we obtain the differential equation

$$r\xi' = -e^{\xi} + 1. \tag{32}$$

Now set

$$e^{\xi} = \frac{1}{\gamma}; \quad \xi = -\ln\gamma; \quad \xi' = -\frac{\gamma'}{\gamma};$$

$$\frac{\gamma'}{1-\gamma} = \frac{1}{r}; \quad -\ln(1-\gamma) = \ln r + \text{const}; \quad \gamma = 1 - \kappa/r, \tag{33}$$

where κ is a constant to be determined. Note that this solution makes γ and therefore e^{ξ} go to unity as $r \to \infty$, a condition we previously expressed so that g_{11} will be -1 at infinity. The quantity, g_{44}, must approach c^2 as $r \to \infty$, a relationship that sets the constant in (31). Or

$$e^{\zeta} = \gamma c^2. \tag{34}$$

These relations satisfy all three of the basic equations. Hence we now take the line element

$$(ds)^2 = -\gamma^{-1}(dr)^2 - r^2(d\theta)^2 - r^2\sin^2\theta(d\phi)^2 + \gamma c^2(dt)^2. \tag{35}$$

The metric tensor

$$g_{11} = -\gamma^{-1}; \quad g_{22} = -r^2; \quad g_{33} = -r^2 \sin^2 \theta; \quad g_{44} = \gamma c^2 \tag{36}$$

is thus a particular solution of the basic equation (27).

We are now ready to test our assumption that the equation of a geodesic, in this four-dimensional continuum, represents the path of a particle in the gravitational field of the sun. We still have one constant at our disposal: κ, which we expect to fit by observation, since it expresses or depends on some physical property of the sun itself.

If, in (3) we set $\epsilon = 2$ (i.e., $q^2 = \theta$), we have

$$\frac{d^2\theta}{ds^2} + \Gamma^2_{12} \frac{dr}{ds}\frac{d\theta}{ds} + \Gamma^2_{21} \frac{d\theta}{ds}\frac{dr}{ds} + \Gamma^2_{33} \frac{d\phi}{ds}\frac{d\phi}{ds} = 0. \tag{37}$$

Employing (16), we get

$$\frac{d^2\theta}{ds^2} + \frac{2}{r}\frac{dr}{ds}\frac{d\theta}{ds} - \cos\theta \sin\theta \left(\frac{d\phi}{ds}\right)^2 = 0. \tag{38}$$

If we take our initial condition so that the particle moves in the plane $\theta = \pi/2$, this equation is satisfied identically and for all time. With this condition, we get

$$\frac{d^2r}{ds^2} + \frac{1}{2}\xi'\left(\frac{dr}{ds}\right)^2 - re^{-\xi}\left(\frac{d\phi}{ds}\right) + \frac{1}{2}e^{\zeta-\xi}\zeta'\left(\frac{dt}{ds}\right)^2 = 0, \tag{39}$$

$$\frac{d^2\phi}{ds^2} + \frac{2}{r}\frac{dr}{ds}\frac{d\phi}{ds} = 0, \tag{40}$$

$$\frac{d^2t}{ds^2} + \zeta'\frac{dr}{ds}\frac{dt}{ds} = 0. \tag{41}$$

These last two equations are exact differentials and we get the first integrals:

$$r^2\frac{d\phi}{ds} = h; \quad \frac{dt}{ds} = Ke^{-\zeta} = \frac{K}{\gamma c^2}, \tag{42}$$

where h and K are constants of integration. Using these expressions, we may simplify (39) and carry out the indicated integration. As a simpler, but equivalent, alternative, we can substitute into (12), and get, after some algebraic reduction,

$$\left(\frac{dr}{d\phi}\right)^2 + \gamma r^2 + \left(\gamma - \frac{K^2}{c^2}\right)\frac{r^4}{h^2} = 0. \tag{43}$$

Now compare this equation with that previously obtained by classical methods, viz., (II-26.27). Replace the constants c and c' in (26.27) by α and E, respectively. Square both sides and find

$$\left(\frac{dr}{d\phi}\right)^2 + r^2 - \frac{2GM}{\alpha^2}r^3 + \frac{Er^4}{\alpha^2} = 0. \tag{44}$$

With γ defined as in (33), the equation (43) assumes the analogous form

$$\left(\frac{dr}{d\phi}\right)^2 + r^2 - \frac{\kappa}{h^2} r^3 + \left(1 - \frac{K^2}{c^2}\right)\frac{r^4}{h^2} = \kappa r. \tag{45}$$

Now if κ is very small, so that the term on the right-hand side of (45) is negligible or nearly so, we can identify the constants

$$2GM/\alpha^2 = \kappa/h^2 \tag{46}$$

and

$$\frac{1 - K^2/c^2}{h^2} = \frac{E}{\alpha^2}. \tag{47}$$

Of the constants in these equations, we regard G, M, and c as basic. E and α, which measure orbital properties, represent the energy and angular momentum of a unit mass in a given orbit. Thus we have to determine K, h, and κ. We need one more equation. From (II-26.25),

$$r^2 \frac{d\phi}{dt} = \alpha. \tag{48}$$

Hence

$$r^2 \frac{d\phi}{dt} = r^2 \frac{d\phi}{ds}\frac{ds}{dt} = \frac{h\gamma c^2}{K} = \alpha. \tag{49}$$

This relationship is supposed to be one between the constants of the two sets of equations. However γ is not a constant, because it is by definition a function of r. Even so, if κ is very small, so that $\kappa/r \ll 1$ for significant values of r, we can set $\gamma \sim 1$ into (49) and take

$$\alpha \sim \frac{hc^2}{K}. \tag{50}$$

Then, from (46) and (47) we obtain

$$\kappa = \frac{2GMK^2}{c^4} \tag{51}$$

and

$$\frac{K^2}{c^2} = \frac{1}{1 + E/c^2} \tag{52}$$

The total energy of a unit mass is usually small compared with c^2, and of the order v^2/c^2. Hence

$$\frac{K}{c} \simeq 1, \quad \text{and} \quad \kappa \sim \frac{2GM}{c^2}. \tag{53}$$

For the sun, with $G = 6.66 \times 10^{-8}\ g^{-1}\ \text{cm}^3\ \text{sec}^{-2}$; $M = 1.99 \times 10^{33}\ g$; $c = 3 \times 10^{10}\ \text{cm sec}^{-1}$, so that

$$\kappa \sim 2.95 \times 10^5\ \text{cm}. \tag{54}$$

This calculation indicates that the greatest part of the warped space occurs in the first few kilometers of solar radius. Even at the solar surface, where $r = 7 \times 10^{10}$ cm, $\gamma \sim 1 - 4.3 \times 10^{-6}$.

The term on the right-hand side of (45) thus proves to be small, and our neglecting it in the first approximation is justified. Our analysis thus justifies our original contention, that we can express motions of planets in terms of geodesics in a Riemannian space whose curvature we attribute to the presence of the solar mass.

In deciding between this interesting possibility and the original hypothesis of Newton, we find the neglected term on the right-hand side of equation (45) of some assistance.

In equation (45) set $r = 1/u$. Then

$$\left(\frac{du}{d\phi}\right)^2 + u^2 - \frac{\kappa}{h^2}u + \left(1 - \frac{K^2}{c^2}\right)\frac{1}{h^2} = \kappa u^3. \tag{55}$$

Differentiate this equation with respect to ϕ and cancel the factor $(du/d\phi)$ that appears in every term. Then

$$2\frac{d^2u}{d\phi^2} + 2u - \frac{\kappa}{h^2} = 3\kappa u^2. \tag{56}$$

The equation of the ellipse

$$\frac{1}{r} = u = A[1 + \epsilon \cos(\varphi - \varphi_0)] \tag{57}$$

when substituted into (56), gives zero for the left-hand side:

$$-2A\epsilon \cos(\varphi - \varphi_0) + 2A + 2A\epsilon \cos(\varphi - \varphi_0) - \frac{\kappa}{h^2} = 0, \tag{58}$$

if we set

$$2A = \frac{\kappa}{h^2}. \tag{59}$$

In these equations ϵ is the eccentricity.

To allow for the effect of the extra term on the right-hand side of equation (56), we shall consider the quantity ϕ_0 to be a slowly varying function of ϕ. Then

$$\frac{du}{d\phi} = -A\epsilon(1 - \phi_0') \sin(\phi - \phi_0) \tag{60}$$

and

$$\frac{d^2u}{d\phi^2} = -A\epsilon[(1 - \phi_0')^2 \cos(\phi - \phi_0) - \phi_0'' \sin(\phi - \phi_0)] \tag{61}$$

where the primes signify differentiation with respect to ϕ. In this equation we shall neglect the higher order terms, $(\phi_0')^2$ and ϕ_0''. Then, when we substitute this expression back into (56), we get

$$\left[2A - \frac{\kappa}{h^2} - 3\kappa A^2\right]$$

$$+ A\epsilon[4\phi_0' - 6A\kappa - 3A\kappa\epsilon\cos(\phi - \phi_0)]\cos(\varphi - \varphi_0) = 0. \quad (62)$$

This equation requires that each individual bracket equal zero. In fact, the relation

$$A = \frac{\kappa}{2h^2} \quad (63)$$

is an excellent approximation to the first equation. The second approximation gives

$$A = \frac{\kappa}{2h^2}(1 + 3\kappa^2/4h^2). \quad (64)$$

This second bracket in (62) leads to a differential equation in ϕ_0:

$$4\frac{d\phi_0}{d\phi} = 6A\kappa + 3A\kappa\epsilon\cos(\phi - \phi_0). \quad (65)$$

$$\phi_0 = \frac{3}{2}A\kappa\phi + \frac{3}{4}A\kappa\epsilon\sin(\phi - \phi_0) + \delta. \quad (66)$$

On the right-hand side, we can regard ϕ_0 as a constant, because it varies so slowly.

Equation (66) shows that ϕ_0 consists of a fluctuating sine term and a progressive term: $3A\kappa\phi/2$. The effect is that of an ellipse whose major axis slowly rotates at the rate ϕ_0', i.e., an advance of the line of apsides. The secular change is simply

$$\phi_0' = \frac{3\kappa^2}{4h^2} = \frac{3G^2M^2}{c^4h^2}. \quad (67)$$

To reduce these, we use

$$\alpha = \sqrt{\frac{2h^2}{\kappa}GM} = \sqrt{\frac{GM}{A}} = \sqrt{GMp} = \sqrt{GMa(1 - \epsilon^2)} = hc, \quad (68)$$

by (46), (63), (II-26.32), (50), and (53). Therefore

$$\phi_0' = \frac{3GM}{c^2a(1 - \epsilon^2)}, \quad (69)$$

where a is the major semi-axis. The rate at which the line of apsides progresses is

$$\frac{d\phi_0}{dt} = \frac{d\phi_0}{d\phi}\frac{d\phi}{dt} = 3\frac{c}{a}\left(\frac{GM}{ac^2}\right)^{3/2}\frac{1}{(1 - \epsilon^2)}, \quad (70)$$

since, by Kepler's third law, (I-14.9),

$$\left(\frac{d\phi}{dt}\right)^2 = \left(\frac{2\pi}{P}\right)^2 = \frac{GM}{a^3}. \tag{71}$$

For the planet Mercury, $\epsilon = 0.206$, $a = 5.78 \times 10^{12}$ cm, $GM/c^2 = 1.475 \times 10^5$ cm. Hence

$$\frac{d\phi_0}{dt} = 6.68 \times 10^{-14} \text{ rad/sec.} \tag{72}$$

To interpret this number, we note that

$$1 \text{ radian} = 2.06265 \times 10^5 \text{ seconds of arc.}$$

$$100 \text{ years} = 3.15 \times 10^9 \text{ seconds of time.} \tag{73}$$

At this rate, the advance of perihelion is

$$\Delta\phi_0 = 6.68 \times 10^{-14} \times 2.06 \times 10^5 \times 3.15 \times 10^9$$
$$= 43 \text{ seconds of arc per century.} \tag{74}$$

This figure agrees very well with the observed value, and no other acceptable explanation exists for the rotation of the ellipse, other than the Einstein theory of relativity.

Einstein has also pointed out that a light beam should experience a deflection that results from the curvature of the field geometry. A ray of light is a geodesic, with $ds = 0$. Hence, by (42),

$$h \to \infty, \tag{75}$$

and equation (56) takes the form

$$\frac{d^2u}{d\phi^2} + u = \frac{3}{2}\kappa u^2. \tag{76}$$

To the first approximation, when we neglect the right-hand side of this equation,

$$u = \frac{1}{R}\cos\phi; \quad r\cos\phi = R, \tag{77}$$

the equation of a straight line. To get the full solution, substitute this approximate result in the right-hand side of (76), to give

$$\frac{d^2u}{d\phi^2} + u = \frac{3}{2}\frac{\kappa}{R^2}\cos^2\phi. \tag{78}$$

Try the general solution

$$u = A + \frac{1}{R}\cos\phi + B\cos^2\phi. \tag{79}$$

Then

$$-B(2\cos^2\phi - 2\sin^2\phi - \cos^2\phi) - \frac{3}{2}\frac{\kappa}{R^2}\cos^2\phi + A = 0. \qquad (80)$$

Replacing the $\sin^2\phi$ by $1 - \cos^2\phi$, we see that the condition of (80) is satisfied only if

$$A = -2B \text{ and } B = -\kappa/2R^2. \qquad (81)$$

Thus the full solution is

$$\frac{1}{r} = \frac{2GM}{R^2c^2} + \frac{1}{R}\cos\phi_1 - \frac{GM}{R^2c^2}\cos^2\phi_1. \qquad (82)$$

Without the correction terms, the angle corresponding to the value of $r = \infty$ follows from the equation $\cos\phi_1 = 0$, or $\phi_1 = \pi/2$ or $3\pi/2$. The two asymptotes are 180° apart and the ray is a straight line. Now, for the full equation, with $r = \infty$ we get for the first approximation

$$\cos\phi_1 = \frac{-2GM}{Rc^2}. \qquad (83)$$

Substituting this result for $\cos^2\phi$, we get

$$\cos\phi_1 = -\frac{2GM}{Rc^2}\left(1 - \frac{2G^2M^2}{R^2c^4}\right). \qquad (84)$$

The first approximation is amply accurate.

The two solutions for ϕ_1 lie respectively in the second and third quadrants. The values are

$$\sin\left(\frac{\pi}{2} + \frac{\delta}{2}\right) = \sin\left(\frac{3\pi}{2} - \frac{\delta}{2}\right) = -\frac{2GM}{Rc^2}. \qquad (85)$$

The total angle of deflection

$$\left(\frac{3\pi}{2} - \frac{\delta}{2}\right) - \left(\frac{\pi}{2} + \frac{\delta}{2}\right) - \pi = \delta = \frac{4GM}{Rc^2}. \qquad (86)$$

At the surface of the sun,

$$\delta = 8.5 \times 10^{-6} \text{ radians} = 1.75 \text{ seconds of arc}. \qquad (87)$$

A ray of starlight grazing the sun should suffer a deflection of just this amount. The observations designed to measure the deflection must be made at a total solar eclipse and are among the most exacting of scientific experiments. Some uncertainties have unavoidably crept into the data, and the evidence is not absolutely conclusive. However, the measured deflections from many eclipses appear to be about 2 seconds or slightly less. The values thus provide some confirmation of the general theory of relativity.

The third observational test of the theory consists of a displacement of the Fraunhofer lines in the solar spectrum. We assume that each atom, in its own reference frame, emits radiation at a constant frequency. An atom on the sun, at rest relative to the earth, will measure the timelike interval between successive waves as

$$(ds)^2 = \gamma c^2 (dt)^2, \tag{88}$$

from (35), wherein we have set dr, $d\theta$, and $d\phi$ equal to zero, since the atom is motionless, by hypothesis. In this equation, ds is the true invariant; hence the ratio between the time intervals of two similar atoms, one located on the earth and the other on the sun, is

$$\gamma_S (dt)_S^2 = \gamma_E (dt)_E^2, \tag{89}$$

where the subscripts S and E refer to sun and earth, respectively. If dt denotes the time for one cycle of the emitting mechanism, the frequency of the radiation is

$$\nu = 1/(dt). \tag{90}$$

Hence

$$\nu_S/\nu_E = \sqrt{\gamma_S/\gamma_E} \sim \sqrt{1 - 2GM/Rc^2}, \tag{91}$$

by (33) and (53), where R is the solar radius. We can set $\gamma_E = 1$, without loss of generality, not only because M_E/R_E is small, but because we measure ν_E in time units appropriate to our location at the surface of the earth.

For the sun, $GM/c^2R \sim 2.1 \times 10^{-6}$. Hence, at $\lambda = 5000$A (in the green of the solar spectrum), solar wavelengths are shifted to the red by 0.01 A, approximately. Doppler shifts, caused by differential vertical currents in the solar atmosphere, somewhat mask the effect, but the observations appear to be in reasonably good accord with prediction.

The equations

$$G_{\mu\nu} = 0 \tag{92}$$

represent only one possible solution, selected from a large number of alternatives. It applies to regions of space that are empty, insofar as the departure from a non-Euclidean geometry arises from the presence of a large mass of matter in the neighborhood, the sun for example.

When we attempt to extend this theory to encompass the enormous distances within the universe, we cannot assume that the relative emptiness of space means that the geometry of the extended world must be Euclidean. In fact the above type of argument is essentially meaningless. A Riemannian space possesses a natural metric, imposed on it by the existence of a definite radius of curvature. A Euclidean space possesses no such metric. How, then, can the expression "relative emptiness of space," used above, acquire meaning? May not the great extent of space more than offset the low density of material contained within it, and thus automatically induce non-Euclidean regions at great distances?

The nature of the quantity γ implies that any universe containing matter must be finite. If, about some region of space, we can draw a boundary such that

$$\gamma = 1 - 2GM/Rc^2 = 0, \tag{93}$$

light waves would be unable to move from that inner region into some outer region. A light wave approaching such a boundary from within would travel slower and slower.

We have seen that, for the sun,

$$2GM/Rc^2 = 4.3 \times 10^{-6}, \tag{94}$$

so that light escapes readily from its surface, although it is subject to the minor red shift already discussed. For a star in general, whose mass depends on the density,

$$M = 4\pi\rho R^3/3, \tag{95}$$

the condition of (93) specifies that

$$8\pi G\rho R^2/3c^2 = 1. \tag{96}$$

A star with mean density equal to that of our sun, but with a radius increased by the factor $(4.3 \times 10^{-6})^{-1/2}$, or approximately 500 times, would be a closed system. No light could escape from its surface. We should become aware of it, however, because the warp of the surrounding space would make itself felt as a gravitational force field.

Similarly, if we could compress all the mass of the sun into a volume of radius less than 2.95 km, equation (54), light could not escape through this barrier, but the dynamical properties of the solar system would not be altered appreciably. Attempts to explore the characteristics of space just outside the barrier—say at 3 km from the center—would lead to the peculiar situation that the barrier was still an infinite distance away. The paradox arises from the fact that the unit of radial measure (like the unit of time) approaches zero in the vicinity of this region. As we approach the boundary, our measuring rods shrink to zero. To show this, we note that as we hold t, θ, and ϕ constant, $ds \rightarrow \infty$ as $\gamma \rightarrow 0$.

If we try to manufacture a Euclidean universe from a space filled with matter of constant density, ρ, no matter how small, equation (96) shows that there exists some finite value of R, which makes the universe close up on itself. We can even try to estimate this value of R.

Observations indicate that our own galaxy possesses a density of the order of 10^{-24} g/cm^3, about 1 atom per cm^3. This density is perhaps 10^4 times greater than for the universe as a whole, although the exact figure is not known.

Try $\rho = 10^{-28}$, and solve for R. The value is $R = 4 \times 10^{27}$ cm. Since light travels a distance of 9.46×10^{-17} cm/yr, we see that this theory indicates a radius of the order of 5×10^{9} light years, a figure that probably is a maximum since the value assumed for ρ is, very likely, a lower limit.

Our largest telescopes have recorded external galaxies out to distances of 5×10^{8} light years or so. One of the most surprising facts of observation is the increasing red shift of the spectra of the more distant galaxies, an effect that appears to be roughly linear with distance over the range thus far studied.

To formulate the general theory of a finite universe, Einstein replaced his original law (92) by

$$G_{\mu\nu} = \lambda g_{\mu\nu} \tag{97}$$

where λ is a universal constant, so small that his original approximation was excellent, except over distances of the order of the dimensions of the material universe. The non-vanishing components of $G_{\mu\nu}$ are the same as before, viz., those for which $\nu = \mu$.

The form of our equations and the character of the laws of the universe differ considerably according to the various basic assumptions. Einstein considers a world wherein the spatial geometry is spherical and the time geometry linear.

With the aid of equations (14) and (22) to (26), we write out the non-vanishing components of (97), which become

$$\begin{aligned}
G_{11} &= \frac{\zeta''}{2} - \frac{\zeta'\xi'}{4} + \frac{(\zeta')^2}{4} - \frac{\xi'}{r} = -\lambda e^{\xi}. \\
G_{22} &= e^{-\xi}\left[1 + \frac{r(\zeta' - \xi')}{2} - e^{-\xi}\right] = -\lambda r^2 = \frac{G_{33}}{\sin^2\theta}. \\
G_{44} &= -e^{\zeta-\xi}\left[\frac{\zeta''}{2} - \frac{\zeta'\xi'}{4} + \frac{(\zeta')^2}{4} + \frac{\zeta'}{r}\right] = \lambda e^{\zeta}. \\
G_{\mu\nu} &= 0, \quad \mu \neq \nu.
\end{aligned} \tag{98}$$

From G_{11} and G_{44}, we find that

$$e^{-\xi} = \gamma, \quad \zeta' = -\xi' = \gamma'/\gamma, \quad e^{\zeta} = c^2\gamma, \tag{99}$$

as before. These values substituted into G_{22} give

$$\gamma + r\gamma' = 1 - \lambda r^2. \tag{100}$$

This equation leads to the integral

$$\gamma = 1 - \frac{2GM}{c^2 r} - \frac{\lambda r^3}{3}. \tag{101}$$

We have previously noted that the condition, $\gamma = 0$, represents a surface that no light ray can traverse since all time ceases, in effect, when

$\gamma = 0$. Or, in other words, the distance ds, as expressed by (35), measured in terms of a dr, i.e.,

$$\frac{ds}{dr} = \sqrt{-1/\gamma} \rightarrow -i\infty. \tag{102}$$

The two roots of (101) are, approximately

$$r_1 \sim 2GM/c^2 \quad \text{and} \quad r_2 \sim \sqrt{3/\lambda}. \tag{103}$$

Of these two roots r_1 is very small and r_2 very great. A particle of mass M and radius r_1 would behave like the impenetrable solid sphere that we once regarded as the ultimate particle of matter.

We can regard r_2 as the radius of the universe. If we set $r_2 \sim 4 \times 10^{27}$ cm, as previously derived,

$$\lambda \sim 2 \times 10^{-55}. \tag{104}$$

Thus we can now justify our previous contention that the equation, $G_{\mu\nu} = 0$, is a satisfactory approximation in the solar neighborhood.

We are faced with two major relativistic problems. One is related to motion within the solar system and the other to the structure of the entire universe, wherein we regard such condensations as stars, or even galaxies, as only minor irregularities in the otherwise essentially uniform continuum of matter.

The solar-system problem is the one whose solution is most definite and least subject to question. The primary reason for our success here is the fact that the distances are small compared with the dimensions of the universe. We arbitrarily adopted a Euclidean metric, equation (10), for empty space and then computed the character of the Riemannian warp introduced by the presence of the sun.

But by what right could we assume our initial space to be Euclidean? Our reply is that we could have adopted almost any kind of space whatever, so long as its curvature was great enough to include the observed universe, and we could still employ a Euclidean approximation for an empty volume the size of our solar system. However, when we extend our solution farther and farther into space, can we or should we adopt as our starting point, a Euclidean rather than a Riemannian space?

Our analogy of the navigator proved useful, as a means of showing how apparent force fields can arise from the metric alone. It is even more apt when we apply it to the universe as a whole rather than to the solar system, because unless the island itself produced the curvature, its presence is not necessary. The navigator, exploring his full universe, would find that the force field seemed to exist everywhere, for all circuits that he traversed. The analogy might have been improved had we replaced the island by some geyserlike upheaval that altered the shape of the ocean surface in some major fashion.

The navigator finds that the spherical surface of radius R curved in the third dimension follows the law

$$(ds)^2 = R^2(d\theta_1^2 + \sin^2 \theta_1 \, d\theta_2), \tag{105}$$

where θ_1 and θ_2 are angular coordinates. If, about some origin, the navigator proceeds a distance $R\theta_1$, and then draws a circle, the circumference of such a circle will be $2\pi R \sin \theta_1$, not $2\pi R\theta_1$, unless θ_1 is very small. For small distances the result is the same as for Euclidean space. But as we increase θ_1 to $\pi/2$ we shall find some largest circle, an equator. Thereafter the circumference diminishes for still greater distance, to the value zero at the antipodes.

Analogously, a three-dimensional spherical world curved into four dimensions conforms to the law

$$ds^2 = R^2[d\theta_1^2 + \sin^2 \theta_1(d\theta_2^2 + \sin^2 \theta_2 \, d\theta_3)]. \tag{106}$$

If, to this world, we add time as a linear, Cartesian coordinate, we get

$$ds^2 = -R^2[d\theta_1^2 + \sin^2 \theta_1(d\theta_2^2 + \sin^2 \theta_2 \, d\theta_3)] + c^2 \, dt^2, \tag{107}$$

where we have changed the sign to correspond with the convention previously adopted.

As we proceed from the origin, the radial coordinate is $R\theta_1$. As before, we reach, not a greatest circle, but a greatest sphere, for $\theta_1 = \pi/2$. The area of this sphere is $4\pi R^2 \sin^2 \theta_1$, rather than $4\pi R^2\theta_1^2$.

The volume of space is finite and equal to

$$V = \int_0^{\pi} \int_0^{\pi} \int_0^{2\pi} R^3 \, d\theta_1 \, (\sin \theta_1 \, d\theta_2)(\sin \theta_1 \sin \theta_2 \, d\theta_3) = 2\pi^2 R^3. \tag{108}$$

Where Einstein chooses time to be a linear coordinate, de Sitter adopts a more general metric! Analogous to (106), he takes

$$ds^2 = -R^2\{d\theta_1^2 + \sin^2 \theta_1[d\theta_2^2 + \sin^2 \theta_2(d\theta_3^2 + \sin^2 \theta_3 \, d\theta_4^2)]\}. \tag{109}$$

Within this spherical world we transform the coordinates to

$$\cos \theta_1 = \cos \rho \cos ict/R, \tag{110}$$

$$\cot \theta_2 = \cot \rho \sin ict/R, \tag{111}$$

$$\sin \rho = \sin \theta_2 \sin \theta_1, \tag{112}$$

$$\tan ict = \cos \theta_2 \sin \theta_1. \tag{113}$$

Then

$$ds^2 = -R^2 \, d\rho^2 - R^2 \sin^2\rho(d\theta_3^2 + \sin^2\theta_3 \, d\theta_4^2) + c^2 \cos^2\rho \, dt^2. \tag{114}$$

This equation agrees with Einstein's (107) as far as the spatial parameters are concerned. As for time behavior, however, the presence of the $\cos^2 \rho$ factor creates a marked difference. This $\cos^2 \rho$ factor is analogous to the

γ of our previous discussion. Hence for larger distances, i.e., large values of $R\rho$, clocks begin to slow down. At the distance $R\pi/2$, $\cos \pi/2 = 0$, and time utterly ceases. Thus we find an apparent horizon. Spectral lines from distant objects should show a red shift of amount $\cos \rho$, increasing with distance.

We cannot expect to study in detail all the implications of these two worlds. In Einstein's we find a red shift caused by the presence of matter. In de Sitter's we find a red shift produced by the pure geometry of space and time, even in a world devoid of matter. These two points of view represent extreme cases, between which present evidence cannot easily decide. However, de Sitter's curved world appears to be unstable and thus capable of expanding—perhaps carrying the distant galaxies along in this expansion.

These Riemannian geometries carry hope for the future. The relativistic tensor forces of the universe, which display themselves in the form of a warped space, may resemble those that hold the atom together. They may well hold the key to the future as well as to the past of the entire universe.

SELECTED PROBLEMS FOR PART V

1. The index of refraction for x rays is less than 1 for certain crystals. Hence the phase velocity should be greater than c. Does this result conflict with any conclusion or postulate of special relativity?

2. Using the relativistic expressions for the total energy E and momentum p, show that $E^2 = p^2c^2 + (m_0c^2)^2$. Show that the kinetic energy, T, can be expressed as a function of p, that is,

$$T = \sqrt{p^2c^2 - (m_0c^2)^2} - m_0c^2.$$

3. In classical mechanics, $T = p^2/2m$. Show that the relativistic expression reduces to the classical value when $\beta = v/c \ll 1$. Hint: Expand the radical in a power series in $p^2c^2/(m_0c^2)^2$.

4. (a) What will be the kinetic energy of an electron traveling at 99 per cent of the velocity of light?

(b) If we increase the kinetic energy by a factor of 100, what will be the increase in velocity of the electron?

5. Discuss problem IV-42 from the standpoint of special relativity. The differences between the relativistic and classical formulae are large for particles moving with high velocity. Experiment confirms the relativistic relations and thus provides the best single experimental verification of the correctness of special relativity.

Index

S

T

U

V

W

Z

A CATALOGUE OF SELECTED DOVER BOOKS IN ALL FIELDS OF INTEREST

A CATALOGUE OF SELECTED DOVER BOOKS IN ALL FIELDS OF INTEREST

LEATHER TOOLING AND CARVING, Chris H. Groneman. One of few books concentrating on tooling and carving, with complete instructions and grid designs for 39 projects ranging from bookmarks to bags. 148 illustrations. 111pp. 7⅞ x 10.
23061-9 Pa. $2.50

THE CODEX NUTTALL, A PICTURE MANUSCRIPT FROM ANCIENT MEXICO, as first edited by Zelia Nuttall. Only inexpensive edition, in full color, of a pre-Columbian Mexican (Mixtec) book. 88 color plates show kings, gods, heroes, temples, sacrifices. New explanatory, historical introduction by Arthur G. Miller. 96pp. 11⅜ x 8½.
23168-2 Pa. $7.50

AMERICAN PRIMITIVE PAINTING, Jean Lipman. Classic collection of an enduring American tradition. 109 plates, 8 in full color—portraits, landscapes, Biblical and historical scenes, etc., showing family groups, farm life, and so on. 80pp. of lucid text. 8⅜ x 11¼.
22815-0 Pa. $4.00

WILL BRADLEY: HIS GRAPHIC ART, edited by Clarence P. Hornung. Striking collection of work by foremost practitioner of Art Nouveau in America: posters, cover designs, sample pages, advertisements, other illustrations. 97 plates, including 8 in full color and 19 in two colors. 97pp. 9⅜ x 12¼.
20701-3 Pa. $4.00
22120-2 Clothbd. $10.00

THE UNDERGROUND SKETCHBOOK OF JAN FAUST, Jan Faust. 101 bitter, horrifying, black-humorous, penetrating sketches on sex, war, greed, various liberations, etc. Sometimes sexual, but not pornographic. Not for prudish. 101pp. 6½ x 9¼.
22740-5 Pa. $1.50

THE GIBSON GIRL AND HER AMERICA, Charles Dana Gibson. 155 finest drawings of effervescent world of 1900-1910: the Gibson Girl and her loves, amusements, adventures, Mr. Pipp, etc. Selected by E. Gillon; introduction by Henry Pitz. 144pp. 8¼ x 11⅜.
21986-0 Pa. $3.50

STAINED GLASS CRAFT, J.A.F. Divine, G. Blachford. One of the very few books that tell the beginner exactly what he needs to know: planning cuts, making shapes, avoiding design weaknesses, fitting glass, etc. 93 illustrations. 115pp.
22812-6 Pa. $1.50

SLEEPING BEAUTY, illustrated by Arthur Rackham. Perhaps the fullest, most delightful version ever, told by C.S. Evans. Rackham's best work. 49 illustrations. 110pp. 7⅞ x 10¾. 22756-1 Pa. $2.00

THE WONDERFUL WIZARD OF OZ, L. Frank Baum. Facsimile in full color of America's finest children's classic. Introduction by Martin Gardner. 143 illustrations by W.W. Denslow. 267pp. 20691-2 Pa. $3.00

GOOPS AND HOW TO BE THEM, Gelett Burgess. Classic tongue-in-cheek masquerading as etiquette book. 87 verses, 170 cartoons as Goops demonstrate virtues of table manners, neatness, courtesy, more. 88pp. 6½ x 9¼. 22233-0 Pa. $2.00

THE BROWNIES, THEIR BOOK, Palmer Cox. Small as mice, cunning as foxes, exuberant, mischievous, Brownies go to zoo, toy shop, seashore, circus, more. 24 verse adventures. 266 illustrations. 144pp. 6⅝ x 9¼. 21265-3 Pa. $2.50

BILLY WHISKERS: THE AUTOBIOGRAPHY OF A GOAT, Frances Trego Montgomery. Escapades of that rambunctious goat. Favorite from turn of the century America. 24 illustrations. 259pp. 22345-0 Pa. $2.75

THE ROCKET BOOK, Peter Newell. Fritz, janitor's kid, sets off rocket in basement of apartment house; an ingenious hole punched through every page traces course of rocket. 22 duotone drawings, verses. 48pp. 6⅞ x 8⅜. 22044-3 Pa. $1.50

PECK'S BAD BOY AND HIS PA, George W. Peck. Complete double-volume of great American childhood classic. Hennery's ingenious pranks against outraged pomposity of pa and the grocery man. 97 illustrations. Introduction by E.F. Bleiler. 347pp. 20497-9 Pa. $2.50

THE TALE OF PETER RABBIT, Beatrix Potter. The inimitable Peter's terrifying adventure in Mr. McGregor's garden, with all 27 wonderful, full-color Potter illustrations. 55pp. 4¼ x 5½. USO 22827-4 Pa. $1.00

THE TALE OF MRS. TIGGY-WINKLE, Beatrix Potter. Your child will love this story about a very special hedgehog and all 27 wonderful, full-color Potter illustrations. 57pp. 4¼ x 5½. USO 20546-0 Pa. $1.00

THE TALE OF BENJAMIN BUNNY, Beatrix Potter. Peter Rabbit's cousin coaxes him back into Mr. McGregor's garden for a whole new set of adventures. A favorite with children. All 27 full-color illustrations. 59pp. 4¼ x 5½. USO 21102-9 Pa. $1.00

THE MERRY ADVENTURES OF ROBIN HOOD, Howard Pyle. Facsimile of original (1883) edition, finest modern version of English outlaw's adventures. 23 illustrations by Pyle. 296pp. 6½ x 9¼. 22043-5 Pa. $4.00

TWO LITTLE SAVAGES, Ernest Thompson Seton. Adventures of two boys who lived as Indians; explaining Indian ways, woodlore, pioneer methods. 293 illustrations. 286pp. 20985-7 Pa. $3.00

AUSTRIAN COOKING AND BAKING, Gretel Beer. Authentic thick soups, wiener schnitzel, veal goulash, more, plus dumplings, puff pastries, nut cakes, sacher tortes, other great Austrian desserts. 224pp. USO 23220-4 Pa. $2.50

CHEESES OF THE WORLD, U.S.D.A. Dictionary of cheeses containing descriptions of over 400 varieties of cheese from common Cheddar to exotic Surati. Up to two pages are given to important cheeses like Camembert, Cottage, Edam, etc. 151pp. 22831-2 Pa. $1.50

TRITTON'S GUIDE TO BETTER WINE AND BEER MAKING FOR BEGINNERS, S.M. Tritton. All you need to know to make family-sized quantities of over 100 types of grape, fruit, herb, vegetable wines; plus beers, mead, cider, more. 11 illustrations. 157pp. USO 22528-3 Pa. $2.25

DECORATIVE LABELS FOR HOME CANNING, PRESERVING, AND OTHER HOUSEHOLD AND GIFT USES, Theodore Menten. 128 gummed, perforated labels, beautifully printed in 2 colors. 12 versions in traditional, Art Nouveau, Art Deco styles. Adhere to metal, glass, wood, most plastics. 24pp. 8¼ x 11. 23219-0 Pa. $2.00

FIVE ACRES AND INDEPENDENCE, Maurice G. Kains. Great back-to-the-land classic explains basics of self-sufficient farming: economics, plants, crops, animals, orchards, soils, land selection, host of other necessary things. Do not confuse with skimpy faddist literature; Kains was one of America's greatest agriculturalists. 95 illustrations. 397pp. 20974-1 Pa. $3.00

GROWING VEGETABLES IN THE HOME GARDEN, U.S. Dept. of Agriculture. Basic information on site, soil conditions, selection of vegetables, planting, cultivation, gathering. Up-to-date, concise, authoritative. Covers 60 vegetables. 30 illustrations. 123pp. 23167-4 Pa. $1.35

FRUITS FOR THE HOME GARDEN, Dr. U.P. Hedrick. A chapter covering each type of garden fruit, advice on plant care, soils, grafting, pruning, sprays, transplanting, and much more! Very full. 53 illustrations. 175pp. 22944-0 Pa. $2.50

GARDENING ON SANDY SOIL IN NORTH TEMPERATE AREAS, Christine Kelway. Is your soil too light, too sandy? Improve your soil, select plants that survive under such conditions. Both vegetables and flowers. 42 photos. 148pp. USO 23199-2 Pa. $2.50

THE FRAGRANT GARDEN: A BOOK ABOUT SWEET SCENTED FLOWERS AND LEAVES, Louise Beebe Wilder. Fullest, best book on growing plants for their fragrances. Descriptions of hundreds of plants, both well-known and overlooked. 407pp. 23071-6 Pa. $4.00

EASY GARDENING WITH DROUGHT-RESISTANT PLANTS, Arno and Irene Nehrling. Authoritative guide to gardening with plants that require a minimum of water: seashore, desert, and rock gardens; house plants; annuals and perennials; much more. 190 illustrations. 320pp. 23230-1 Pa. $3.50

DECORATIVE ALPHABETS AND INITIALS, edited by Alexander Nesbitt. 91 complete alphabets (medieval to modern), 3924 decorative initials, including Victorian novelty and Art Nouveau. 192pp. 7¾ x 10¾. 20544-4 Pa. $4.00

CALLIGRAPHY, Arthur Baker. Over 100 original alphabets from the hand of our greatest living calligrapher: simple, bold, fine-line, richly ornamented, etc. —all strikingly original and different, a fusion of many influences and styles. 155pp. 11⅜ x 8¼. 22895-9 Pa. $4.50

MONOGRAMS AND ALPHABETIC DEVICES, edited by Hayward and Blanche Cirker. Over 2500 combinations, names, crests in very varied styles: script engraving, ornate Victorian, simple Roman, and many others. 226pp. 8⅛ x 11. 22330-2 Pa. $5.00

THE BOOK OF SIGNS, Rudolf Koch. Famed German type designer renders 493 symbols: religious, alchemical, imperial, runes, property marks, etc. Timeless. 104pp. 6⅛ x 9¼. 20162-7 Pa. $1.75

200 DECORATIVE TITLE PAGES, edited by Alexander Nesbitt. 1478 to late 1920's. Baskerville, Dürer, Beardsley, W. Morris, Pyle, many others in most varied techniques. For posters, programs, other uses. 222pp. 8⅜ x 11¼. 21264-5 Pa. $5.00

DICTIONARY OF AMERICAN PORTRAITS, edited by Hayward and Blanche Cirker. 4000 important Americans, earliest times to 1905, mostly in clear line. Politicians, writers, soldiers, scientists, inventors, industrialists, Indians, Blacks, women, outlaws, etc. Identificatory information. 756pp. 9¼ x 12¾. 21823-6 Clothbd. $30.00

ART FORMS IN NATURE, Ernst Haeckel. Multitude of strangely beautiful natural forms: Radiolaria, Foraminifera, jellyfishes, fungi, turtles, bats, etc. All 100 plates of the 19th century evolutionist's Kunstformen der Natur (1904). 100pp. 9⅜ x 12¼. 22987-4 Pa. $4.00

DECOUPAGE: THE BIG PICTURE SOURCEBOOK, Eleanor Rawlings. Make hundreds of beautiful objects, over 550 florals, animals, letters, shells, period costumes, frames, etc. selected by foremost practitioner. Printed on one side of page. 8 color plates. Instructions. 176pp. 9³⁄₁₆ x 12¼. 23182-8 Pa. $5.00

AMERICAN FOLK DECORATION, Jean Lipman, Eve Meulendyke. Thorough coverage of all aspects of wood, tin, leather, paper, cloth decoration — scapes, humans, trees, flowers, geometrics — and how to make them. Full instructions. 233 illustrations, 5 in color. 163pp. 8⅜ x 11¼. 22217-9 Pa. $3.95

WHITTLING AND WOODCARVING, E.J. Tangerman. Best book on market; clear, full. If you can cut a potato, you can carve toys, puzzles, chains, caricatures, masks, patterns, frames, decorate surfaces, etc. Also covers serious wood sculpture. Over 200 photos. 293pp. 20965-2 Pa. $3.00

EAST O' THE SUN AND WEST O' THE MOON, George W. Dasent. Considered the best of all translations of these Norwegian folk tales, this collection has been enjoyed by generations of children (and folklorists too). Includes True and Untrue, Why the Sea is Salt, East O' the Sun and West O' the Moon, Why the Bear is Stumpy-Tailed, Boots and the Troll, The Cock and the Hen, Rich Peter the Pedlar, and 52 more. The only edition with all 59 tales. 77 illustrations by Erik Werenskiold and Theodor Kittelsen. xv + 418pp. 22521-6 Paperbound **$4.00**

GOOPS AND HOW TO BE THEM, Gelett Burgess. Classic of tongue-in-cheek humor, masquerading as etiquette book. 87 verses, twice as many cartoons, show mischievous Goops as they demonstrate to children virtues of table manners, neatness, courtesy, etc. Favorite for generations. viii + 88pp. 6½ x 9¼.
22233-0 Paperbound **$2.00**

ALICE'S ADVENTURES UNDER GROUND, Lewis Carroll. The first version, quite different from the final *Alice in Wonderland,* printed out by Carroll himself with his own illustrations. Complete facsimile of the "million dollar" manuscript Carroll gave to Alice Liddell in 1864. Introduction by Martin Gardner. viii + 96pp. Title and dedication pages in color. 21482-6 Paperbound **$1.50**

THE BROWNIES, THEIR BOOK, Palmer Cox. Small as mice, cunning as foxes, exuberant and full of mischief, the Brownies go to the zoo, toy shop, seashore, circus, etc., in 24 verse adventures and 266 illustrations. Long a favorite, since their first appearance in St. Nicholas Magazine. xi + 144pp. 6⅝ x 9¼.
21265-3 Paperbound **$2.50**

SONGS OF CHILDHOOD, Walter De La Mare. Published (under the pseudonym Walter Ramal) when De La Mare was only 29, this charming collection has long been a favorite children's book. A facsimile of the first edition in paper, the 47 poems capture the simplicity of the nursery rhyme and the ballad, including such lyrics as I Met Eve, Tartary, The Silver Penny. vii + 106pp. (USO) 21972-0 Paperbound **$2.00**

THE COMPLETE NONSENSE OF EDWARD LEAR, Edward Lear. The finest 19th-century humorist-cartoonist in full: all nonsense limericks, zany alphabets, Owl and Pussycat, songs, nonsense botany, and more than 500 illustrations by Lear himself. Edited by Holbrook Jackson. xxix + 287pp. (USO) 20167-8 Paperbound **$3.00**

BILLY WHISKERS: THE AUTOBIOGRAPHY OF A GOAT, Frances Trego Montgomery. A favorite of children since the early 20th century, here are the escapades of that rambunctious, irresistible and mischievous goat—Billy Whiskers. Much in the spirit of *Peck's Bad Boy,* this is a book that children never tire of reading or hearing. All the original familiar illustrations by W. H. Fry are included: 6 color plates, 18 black and white drawings. 159pp. 22345-0 Paperbound **$2.75**

MOTHER GOOSE MELODIES. Faithful republication of the fabulously rare Munroe and Francis "copyright 1833" Boston edition—the most important Mother Goose collection, usually referred to as the "original." Familiar rhymes plus many rare ones, with wonderful old woodcut illustrations. Edited by E. F. Bleiler. 128pp. 4½ x 6⅜. 22577-1 Paperbound **$1.50**

HOUDINI ON MAGIC, Harold Houdini. Edited by Walter Gibson, Morris N. Young. How he escaped; exposés of fake spiritualists; instructions for eye-catching tricks; other fascinating material by and about greatest magician. 155 illustrations. 280pp. 20384-0 Pa. $2.75

HANDBOOK OF THE NUTRITIONAL CONTENTS OF FOOD, U.S. Dept. of Agriculture. Largest, most detailed source of food nutrition information ever prepared. Two mammoth tables: one measuring nutrients in 100 grams of edible portion; the other, in edible portion of 1 pound as purchased. Originally titled Composition of Foods. 190pp. 9 x 12. 21342-0 Pa. $4.00

COMPLETE GUIDE TO HOME CANNING, PRESERVING AND FREEZING, U.S. Dept. of Agriculture. Seven basic manuals with full instructions for jams and jellies; pickles and relishes; canning fruits, vegetables, meat; freezing anything. Really good recipes, exact instructions for optimal results. Save a fortune in food. 156 illustrations. 214pp. 6 1/8 x 9 1/4. 22911-4 Pa. $2.50

THE BREAD TRAY, Louis P. De Gouy. Nearly every bread the cook could buy or make: bread sticks of Italy, fruit breads of Greece, glazed rolls of Vienna, everything from corn pone to croissants. Over 500 recipes altogether. including buns, rolls, muffins, scones, and more. 463pp. 23000-7 Pa. $3.50

CREATIVE HAMBURGER COOKERY, Louis P. De Gouy. 182 unusual recipes for casseroles, meat loaves and hamburgers that turn inexpensive ground meat into memorable main dishes: Arizona chili burgers, burger tamale pie, burger stew, burger corn loaf, burger wine loaf, and more. 120pp. 23001-5 Pa. $1.75

LONG ISLAND SEAFOOD COOKBOOK, J. George Frederick and Jean Joyce. Probably the best American seafood cookbook. Hundreds of recipes. 40 gourmet sauces, 123 recipes using oysters alone! All varieties of fish and seafood amply represented. 324pp. 22677-8 Pa. $3.50

THE EPICUREAN: A COMPLETE TREATISE OF ANALYTICAL AND PRACTICAL STUDIES IN THE CULINARY ART, Charles Ranhofer. Great modern classic. 3,500 recipes from master chef of Delmonico's, turn-of-the-century America's best restaurant. Also explained, many techniques known only to professional chefs. 775 illustrations. 1183pp. 6 5/8 x 10. 22680-8 Clothbd. $22.50

THE AMERICAN WINE COOK BOOK, Ted Hatch. Over 700 recipes: old favorites livened up with wine plus many more: Czech fish soup, quince soup, sauce Perigueux, shrimp shortcake, filets Stroganoff, cordon bleu goulash, jambonneau, wine fruit cake, more. 314pp. 22796-0 Pa. $2.50

DELICIOUS VEGETARIAN COOKING, Ivan Baker. Close to 500 delicious and varied recipes: soups, main course dishes (pea, bean, lentil, çheese, vegetable, pasta, and egg dishes), savories, stews, whole-wheat breads and cakes, more. 168pp. USO 22834-7 Pa. $1.75

VISUAL ILLUSIONS: THEIR CAUSES, CHARACTERISTICS, AND APPLICATIONS, Matthew Luckiesh. Thorough description and discussion of optical illusion, geometric and perspective, particularly; size and shape distortions, illusions of color, of motion; natural illusions; use of illusion in art and magic, industry, etc. Most useful today with op art, also for classical art. Scores of effects illustrated. Introduction by William H. Ittleson. 100 illustrations. xxi + 252pp.
21530-X Paperbound $2.50

A HANDBOOK OF ANATOMY FOR ART STUDENTS, Arthur Thomson. Thorough, virtually exhaustive coverage of skeletal structure, musculature, etc. Full text, supplemented by anatomical diagrams and drawings and by photographs of undraped figures. Unique in its comparison of male and female forms, pointing out differences of contour, texture, form. 211 figures, 40 drawings, 86 photographs. xx + 459pp. 5⅜ x 8⅜. 21163-0 Paperbound $5.00

150 MASTERPIECES OF DRAWING, Selected by Anthony Toney. Full page reproductions of drawings from the early 16th to the end of the 18th century, all beautifully reproduced: Rembrandt, Michelangelo, Dürer, Fragonard, Urs, Graf, Wouwerman, many others. First-rate browsing book, model book for artists. xviii + 150pp. 8⅜ x 11¼. 21032-4 Paperbound $4.00

THE LATER WORK OF AUBREY BEARDSLEY, Aubrey Beardsley. Exotic, erotic, ironic masterpieces in full maturity: Comedy Ballet, Venus and Tannhauser, Pierrot, Lysistrata, Rape of the Lock, Savoy material, Ali Baba, Volpone, etc. This material revolutionized the art world, and is still powerful, fresh, brilliant. With *The Early Work,* all Beardsley's finest work. 174 plates, 2 in color. xiv + 176pp. 8⅛ x 11.
21817-1 Paperbound $4.00

DRAWINGS OF REMBRANDT, Rembrandt van Rijn. Complete reproduction of fabulously rare edition by Lippmann and Hofstede de Groot, completely reedited, updated, improved by Prof. Seymour Slive, Fogg Museum. Portraits, Biblical sketches, landscapes, Oriental types, nudes, episodes from classical mythology—All Rembrandt's fertile genius. Also selection of drawings by his pupils and followers. "Stunning volumes," *Saturday Review.* 550 illustrations. lxxviii + 552pp. 9⅛ x 12¼. 21485-0, 21486-9 Two volumes, Paperbound $12.00

THE DISASTERS OF WAR, Francisco Goya. One of the masterpieces of Western civilization—83 etchings that record Goya's shattering, bitter reaction to the Napoleonic war that swept through Spain after the insurrection of 1808 and to war in general. Reprint of the first edition, with three additional plates from Boston's Museum of Fine Arts. All plates facsimile size. Introduction by Philip Hofer, Fogg Museum. v + 97pp. 9⅜ x 8¼. 21872-4 Paperbound $3.00

GRAPHIC WORKS OF ODILON REDON. Largest collection of Redon's graphic works ever assembled: 172 lithographs, 28 etchings and engravings, 9 drawings. These include some of his most famous works. All the plates from *Odilon Redon: oeuvre graphique complet,* plus additional plates. New introduction and caption translations by Alfred Werner. 209 illustrations. xxvii + 209pp. 9⅛ x 12¼.
21966-8 Paperbound $6.00

MODERN CHESS STRATEGY, Ludek Pachman. The use of the queen, the active king, exchanges, pawn play, the center, weak squares, etc. Section on rook alone worth price of the book. Stress on the moderns. Often considered the most important book on strategy. 314pp. 20290-9 Pa. $3.50

CHESS STRATEGY, Edward Lasker. One of half-dozen great theoretical works in chess, shows principles of action above and beyond moves. Acclaimed by Capablanca, Keres, etc. 282pp. USO 20528-2 Pa. $3.00

CHESS PRAXIS, THE PRAXIS OF MY SYSTEM, Aron Nimzovich. Founder of hypermodern chess explains his profound, influential theories that have dominated much of 20th century chess. 109 illustrative games. 369pp. 20296-8 Pa. $3.50

HOW TO PLAY THE CHESS OPENINGS, Eugene Znosko-Borovsky. Clear, profound examinations of just what each opening is intended to do and how opponent can counter. Many sample games, questions and answers. 147pp. 22795-2 Pa. $2.00

THE ART OF CHESS COMBINATION, Eugene Znosko-Borovsky. Modern explanation of principles, varieties, techniques and ideas behind them, illustrated with many examples from great players. 212pp. 20583-5 Pa. $2.50

COMBINATIONS: THE HEART OF CHESS, Irving Chernev. Step-by-step explanation of intricacies of combinative play. 356 combinations by Tarrasch, Botvinnik, Keres, Steinitz, Anderssen, Morphy, Marshall, Capablanca, others, all annotated. 245 pp. 21744-2 Pa. $3.00

HOW TO PLAY CHESS ENDINGS, Eugene Znosko-Borovsky. Thorough instruction manual by fine teacher analyzes each piece individually; many common endgame situations. Examines games by Steinitz, Alekhine, Lasker, others. Emphasis on understanding. 288pp. 21170-3 Pa. $2.75

MORPHY'S GAMES OF CHESS, Philip W. Sergeant. Romantic history, 54 games of greatest player of all time against Anderssen, Bird, Paulsen, Harrwitz; 52 games at odds; 52 blindfold; 100 consultation, informal, other games. Analyses by Anderssen, Steinitz, Morphy himself. 352pp. 20386-7 Pa. $4.00

500 MASTER GAMES OF CHESS, S. Tartakower, J. du Mont. Vast collection of great chess games from 1798-1938, with much material nowhere else readily available. Fully annotated, arranged by opening for easier study. 665pp. 23208-5 Pa. $6.00

THE SOVIET SCHOOL OF CHESS, Alexander Kotov and M. Yudovich. Authoritative work on modern Russian chess. History, conceptual background. 128 fully annotated games (most unavailable elsewhere) by Botvinnik, Keres, Smyslov, Tal, Petrosian, Spassky, more. 390pp. 20026-4 Pa. $3.95

WONDERS AND CURIOSITIES OF CHESS, Irving Chernev. A lifetime's accumulation of such wonders and curiosities as the longest won game, shortest game, chess problem with mate in 1220 moves, and much more unusual material —356 items in all, over 160 complete games. 146 diagrams. 203pp. 23007-4 Pa. $3.50

MANUAL OF THE TREES OF NORTH AMERICA, Charles S. Sargent. The basic survey of every native tree and tree-like shrub, 717 species in all. Extremely full descriptions, information on habitat, growth, locales, economics, etc. Necessary to every serious tree lover. Over 100 finding keys. 783 illustrations. Total of 986pp.
20277-1, 20278-X Pa., Two vol. set $9.00

BIRDS OF THE NEW YORK AREA, John Bull. Indispensable guide to more than 400 species within a hundred-mile radius of Manhattan. Information on range, status, breeding, migration, distribution trends, etc. Foreword by Roger Tory Peterson. 17 drawings; maps. 540pp. 23222-0 Pa. $6.00

THE SEA-BEACH AT EBB-TIDE, Augusta Foote Arnold. Identify hundreds of marine plants and animals: algae, seaweeds, squids, crabs, corals, etc. Descriptions cover food, life cycle, size, shape, habitat. Over 600 drawings. 490pp.
21949-6 Pa. $5.00

THE MOTH BOOK, William J. Holland. Identify more than 2,000 moths of North America. General information, precise species descriptions. 623 illustrations plus 48 color plates show almost all species, full size. 1968 edition. Still the basic book. Total of 551pp. 6½ x 9¼. 21948-8 Pa. $6.00

AN INTRODUCTION TO THE REPTILES AND AMPHIBIANS OF THE UNITED STATES, Percy A. Morris. All lizards, crocodiles, turtles, snakes, toads, frogs; life history, identification, habits, suitability as pets, etc. Non-technical, but sound and broad. 130 photos. 253pp. 22982-3 Pa. $3.00

OLD NEW YORK IN EARLY PHOTOGRAPHS, edited by Mary Black. Your only chance to see New York City as it was 1853-1906, through 196 wonderful photographs from N.Y. Historical Society. Great Blizzard, Lincoln's funeral procession, great buildings. 228pp. 9 x 12. 22907-6 Pa. $6.00

THE AMERICAN REVOLUTION, A PICTURE SOURCEBOOK, John Grafton. Wonderful Bicentennial picture source, with 411 illustrations (contemporary and 19th century) showing battles, personalities, maps, events, flags, posters, soldier's life, ships, etc. all captioned and explained. A wonderful browsing book, supplement to other historical reading. 160pp. 9 x 12. 23226-3 Pa. $4.00

PERSONAL NARRATIVE OF A PILGRIMAGE TO AL-MADINAH AND MECCAH, Richard Burton. Great travel classic by remarkably colorful personality. Burton, disguised as a Moroccan, visited sacred shrines of Islam, narrowly escaping death. Wonderful observations of Islamic life, customs, personalities. 47 illustrations. Total of 959pp. 21217-3, 21218-1 Pa., Two vol. set $10.00

INCIDENTS OF TRAVEL IN CENTRAL AMERICA, CHIAPAS, AND YUCATAN, John L. Stephens. Almost single-handed discovery of Maya culture; exploration of ruined cities, monuments, temples; customs of Indians. 115 drawings. 892pp.
22404-X, 22405-8 Pa., Two vol. set $8.00

AGAINST THE GRAIN (A REBOURS), Joris K. Huysmans. Filled with weird images, evidences of a bizarre imagination, exotic experiments with hallucinatory drugs, rich tastes and smells and the diversions of its sybarite hero Duc Jean des Esseintes, this classic novel pushed 19th-century literary decadence to its limits. Full unabridged edition. Do not confuse this with abridged editions generally sold. Introduction by Havelock Ellis. xlix + 206pp. 22190-3 Paperbound **$2.50**

VARIORUM SHAKESPEARE: HAMLET. Edited by Horace H. Furness; a landmark of American scholarship. Exhaustive footnotes and appendices treat all doubtful words and phrases, as well as suggested critical emendations throughout the play's history. First volume contains editor's own text, collated with all Quartos and Folios. Second volume contains full first Quarto, translations of Shakespeare's sources (Belleforest, and Saxo Grammaticus), Der Bestrafte Brudermord, and many essays on critical and historical points of interest by major authorities of past and present. Includes details of staging and costuming over the years. By far the best edition available for serious students of Shakespeare. Total of xx + 905pp.
21004-9, 21005-7, 2 volumes, Paperbound **$11.00**

A LIFE OF WILLIAM SHAKESPEARE, Sir Sidney Lee. This is the standard life of Shakespeare, summarizing everything known about Shakespeare and his plays. Incredibly rich in material, broad in coverage, clear and judicious, it has served thousands as the best introduction to Shakespeare. 1931 edition. 9 plates. xxix + 792pp. 21967-4 Paperbound $4.50

MASTERS OF THE DRAMA, John Gassner. Most comprehensive history of the drama in print, covering every tradition from Greeks to modern Europe and America, including India, Far East, etc. Covers more than 800 dramatists, 2000 plays, with biographical material, plot summaries, theatre history, criticism, etc. "Best of its kind in English," *New Republic*. 77 illustrations. xxii + 890pp.
20100-7 Clothbound **$10.00**

THE EVOLUTION OF THE ENGLISH LANGUAGE, George McKnight. The growth of English, from the 14th century to the present. Unusual, non-technical account presents basic information in very interesting form: sound shifts, change in grammar and syntax, vocabulary growth, similar topics. Abundantly illustrated with quotations. Formerly *Modern English in the Making*. xii + 590pp.
21932-1 Paperbound **$4.00**

AN ETYMOLOGICAL DICTIONARY OF MODERN ENGLISH, Ernest Weekley. Fullest, richest work of its sort, by foremost British lexicographer. Detailed word histories, including many colloquial and archaic words; extensive quotations. Do not confuse this with the Concise Etymological Dictionary, which is much abridged. Total of xxvii + 830pp. 6½ x 9¼.
21873-2, 21874-0 Two volumes, Paperbound **$10.00**

FLATLAND: A ROMANCE OF MANY DIMENSIONS, E. A. Abbott. Classic of science-fiction explores ramifications of life in a two-dimensional world, and what happens when a three-dimensional being intrudes. Amusing reading, but also useful as introduction to thought about hyperspace. Introduction by Banesh Hoffmann. 16 illustrations. xx + 103pp. 20001-9 Paperbound **$1.50**

JEWISH GREETING CARDS, Ed Sibbett, Jr. 16 cards to cut and color. Three say "Happy Chanukah," one "Happy New Year," others have no message, show stars of David, Torahs, wine cups, other traditional themes. 16 envelopes. 8¼ x 11. 23225-5 Pa. $2.00

AUBREY BEARDSLEY GREETING CARD BOOK, Aubrey Beardsley. Edited by Theodore Menten. 16 elegant yet inexpensive greeting cards let you combine your own sentiments with subtle Art Nouveau lines. 16 different Aubrey Beardsley designs that you can color or not, as you wish. 16 envelopes. 64pp. 8¼ x 11. 23173-9 Pa. $2.00

RECREATIONS IN THE THEORY OF NUMBERS, Albert Beiler. Number theory, an inexhaustible source of puzzles, recreations, for beginners and advanced. Divisors, perfect numbers. scales of notation, etc. 349pp. 21096-0 Pa. $4.00

AMUSEMENTS IN MATHEMATICS, Henry E. Dudeney. One of largest puzzle collections, based on algebra, arithmetic, permutations, probability, plane figure dissection, properties of numbers, by one of world's foremost puzzlists. Solutions. 450 illustrations. 258pp. 20473-1 Pa. $3.00

MATHEMATICS, MAGIC AND MYSTERY, Martin Gardner. Puzzle editor for Scientific American explains math behind: card tricks, stage mind reading, coin and match tricks, counting out games, geometric dissections. Probability, sets, theory of numbers, clearly explained. Plus more than 400 tricks, guaranteed to work. 135 illustrations. 176pp. 20335-2 Pa. $2.00

BEST MATHEMATICAL PUZZLES OF SAM LOYD, edited by Martin Gardner. Bizarre, original, whimsical puzzles by America's greatest puzzler. From fabulously rare Cyclopedia, including famous 14-15 puzzles, the Horse of a Different Color, 115 more. Elementary math. 150 illustrations. 167pp. 20498-7 Pa. $2.50

MATHEMATICAL PUZZLES FOR BEGINNERS AND ENTHUSIASTS, Geoffrey Mott-Smith. 189 puzzles from easy to difficult involving arithmetic, logic, algebra, properties of digits, probability. Explanation of math behind puzzles. 135 illustrations. 248pp. 20198-8 Pa. $2.75

BIG BOOK OF MAZES AND LABYRINTHS, Walter Shepherd. Classical, solid, and ripple mazes; short path and avoidance labyrinths; more — 50 mazes and labyrinths in all. 12 other figures. Full solutions. 112pp. 8⅛ x 11. 22951-3 Pa. $2.00

COIN GAMES AND PUZZLES, Maxey Brooke. 60 puzzles, games and stunts — from Japan, Korea, Africa and the ancient world, by Dudeney and the other great puzzlers, as well as Maxey Brooke's own creations. Full solutions. 67 illustrations. 94pp. 22893-2 Pa. $1.50

HAND SHADOWS TO BE THROWN UPON THE WALL, Henry Bursill. Wonderful Victorian novelty tells how to make flying birds, dog, goose, deer, and 14 others. 32pp. 6½ x 9¼. 21779-5 Pa. $1.25

HOW TO SOLVE CHESS PROBLEMS, Kenneth S. Howard. Practical suggestions on problem solving for very beginners. 58 two-move problems, 46 3-movers, 8 4-movers for practice, plus hints. 171pp. 20748-X Pa. $2.00

A GUIDE TO FAIRY CHESS, Anthony Dickins. 3-D chess, 4-D chess, chess on a cylindrical board, reflecting pieces that bounce off edges, cooperative chess, retrograde chess, maximummers, much more. Most based on work of great Dawson. Full handbook, 100 problems. 66pp. 7⅞ x 10¾. 22687-5 Pa. $2.00

WIN AT BACKGAMMON, Millard Hopper. Best opening moves, running game, blocking game, back game, tables of odds, etc. Hopper makes the game clear enough for anyone to play, and win. 43 diagrams. 111pp. 22894-0 Pa. $1.50

BIDDING A BRIDGE HAND, Terence Reese. Master player "thinks out loud" the binding of 75 hands that defy point count systems. Organized by bidding problem—no-fit situations, overbidding, underbidding, cueing your defense, etc. 254pp. EBE 22830-4 Pa. $3.00

THE PRECISION BIDDING SYSTEM IN BRIDGE, C.C. Wei, edited by Alan Truscott. Inventor of precision bidding presents average hands and hands from actual play, including games from 1969 Bermuda Bowl where system emerged. 114 exercises. 116pp. 21171-1 Pa. $1.75

LEARN MAGIC, Henry Hay. 20 simple, easy-to-follow lessons on magic for the new magician: illusions, card tricks, silks, sleights of hand, coin manipulations, escapes, and more —all with a minimum amount of equipment. Final chapter explains the great stage illusions. 92 illustrations. 285pp. 21238-6 Pa. $2.95

THE NEW MAGICIAN'S MANUAL, Walter B. Gibson. Step-by-step instructions and clear illustrations guide the novice in mastering 36 tricks; much equipment supplied on 16 pages of cut-out materials. 36 additional tricks. 64 illustrations. 159pp. 6⅝ x 10. 23113-5 Pa. $3.00

PROFESSIONAL MAGIC FOR AMATEURS, Walter B. Gibson. 50 easy, effective tricks used by professionals —cards, string, tumblers, handkerchiefs, mental magic, etc. 63 illustrations. 223pp. 23012-0 Pa. $2.50

CARD MANIPULATIONS, Jean Hugard. Very rich collection of manipulations; has taught thousands of fine magicians tricks that are really workable, eye-catching. Easily followed, serious work. Over 200 illustrations. 163pp. 20539-8 Pa. $2.00

ABBOTT'S ENCYCLOPEDIA OF ROPE TRICKS FOR MAGICIANS, Stewart James. Complete reference book for amateur and professional magicians containing more than 150 tricks involving knots, penetrations, cut and restored rope, etc. 510 illustrations. Reprint of 3rd edition. 400pp. 23206-9 Pa. $3.50

THE SECRETS OF HOUDINI, J.C. Cannell. Classic study of Houdini's incredible magic, exposing closely-kept professional secrets and revealing, in general terms, the whole art of stage magic. 67 illustrations. 279pp. 22913-0 Pa. $2.50

CATALOGUE OF DOVER BOOKS

THE ART DECO STYLE, ed. by Theodore Menten. Furniture, jewelry, metalwork, ceramics, fabrics, lighting fixtures, interior decors, exteriors, graphics from pure French sources. Best sampling around. Over 400 photographs. 183pp. 8 3/8 x 11 1/4.
22824-X Pa. $4.00

THE GENTLEMAN AND CABINET MAKER'S DIRECTOR, Thomas Chippendale. Full reprint, 1762 style book, most influential of all time; chairs, tables, sofas, mirrors, cabinets, etc. 200 plates, plus 24 photographs of surviving pieces. 249pp. 9 7/8 x 12 3/4. 21601-2 Pa. $6.00

PINE FURNITURE OF EARLY NEW ENGLAND, Russell H. Kettell. Basic book. Thorough historical text, plus 200 illustrations of boxes, highboys, candlesticks, desks, etc. 477pp. 7 7/8 x 10 3/4. 20145-7 Clothbd. $12.50

ORIENTAL RUGS, ANTIQUE AND MODERN, Walter A. Hawley. Persia, Turkey, Caucasus, Central Asia, China, other traditions. Best general survey of all aspects: styles and periods, manufacture, uses, symbols and their interpretation, and identification. 96 illustrations, 11 in color. 320pp. 6 1/8 x 9 1/4.
22366-3 Pa. $5.00

DECORATIVE ANTIQUE IRONWORK, Henry R. d'Allemagne. Photographs of 4500 iron artifacts from world's finest collection, Rouen. Hinges, locks, candelabra, weapons, lighting devices, clocks, tools, from Roman times to mid-19th century. Nothing else comparable to it. 420pp. 9 x 12. 22082-6 Pa. $8.50

THE COMPLETE BOOK OF DOLL MAKING AND COLLECTING, Catherine Christopher. Instructions, patterns for dozens of dolls, from rag doll on up to elaborate, historically accurate figures. Mould faces, sew clothing, make doll houses, etc. Also collecting information. Many illustrations. 288pp. 6 x 9. 22066-4 Pa. $3.00

ANTIQUE PAPER DOLLS: 1915-1920, edited by Arnold Arnold. 7 antique cut-out dolls and 24 costumes from 1915-1920, selected by Arnold Arnold from his collection of rare children's books and entertainments, all in full color. 32pp. 9 1/4 x 12 1/4. 23176-3 Pa. $2.00

ANTIQUE PAPER DOLLS: THE EDWARDIAN ERA, Epinal. Full-color reproductions of two historic series of paper dolls that show clothing styles in 1908 and at the beginning of the First World War. 8 two-sided, stand-up dolls and 32 complete, two-sided costumes. Full instructions for assembling included. 32pp. 9 1/4 x 12 1/4.
23175-5 Pa. $2.00

A HISTORY OF COSTUME, Carl Köhler, Emma von Sichardt. Egypt, Babylon, Greece up through 19th century Europe; based on surviving pieces, art works, etc. Full text and 595 illustrations, including many clear, measured patterns for reproducing historic costume. Practical. 464pp. 21030-8 Pa. $4.00

EARLY AMERICAN LOCOMOTIVES, John H. White, Jr. Finest locomotive engravings from late 19th century: historical (1804-1874), main-line (after 1870), special, foreign, etc. 147 plates. 200pp. 11 3/8 x 8 1/4. 22772-3 Pa. $3.50

THE JOURNAL OF HENRY D. THOREAU, edited by Bradford Torrey, F.H. Allen. Complete reprinting of 14 volumes, 1837-1861, over two million words; the sourcebooks for Walden, etc. Definitive. All original sketches, plus 75 photographs. Introduction by Walter Harding. Total of 1804pp. 8½ x 12¼.
20312-3, 20313-1 Clothbd., Two vol. set $50.00

MASTERS OF THE DRAMA, John Gassner. Most comprehensive history of the drama, every tradition from Greeks to modern Europe and America, including Orient. Covers 800 dramatists, 2000 plays; biography, plot summaries, criticism, theatre history, etc. 77 illustrations. 890pp. 20100-7 Clothbd. $10.00

GHOST AND HORROR STORIES OF AMBROSE BIERCE, Ambrose Bierce. 23 modern horror stories: The Eyes of the Panther, The Damned Thing, etc., plus the dream-essay Visions of the Night. Edited by E.F. Bleiler. 199pp. 20767-6 Pa. $2.00

BEST GHOST STORIES, Algernon Blackwood. 13 great stories by foremost British 20th century supernaturalist. The Willows, The Wendigo, Ancient Sorceries, others. Edited by E.F. Bleiler. 366pp. USO 22977-7 Pa. $3.00

THE BEST TALES OF HOFFMANN, E.T.A. Hoffmann. 10 of Hoffmann's most important stories, in modern re-editings of standard translations: Nutcracker and the King of Mice, The Golden Flowerpot, etc. 7 illustrations by Hoffmann. Edited by E.F. Bleiler. 458pp. 21793-0 Pa. $3.95

BEST GHOST STORIES OF J.S. LEFANU, J. Sheridan LeFanu. 16 stories by greatest Victorian master: Green Tea, Carmilla, Haunted Baronet, The Familiar, etc. Mostly unavailable elsewhere. Edited by E.F. Bleiler. 8 illustrations. 467pp.
20415-4 Pa. $4.00

SUPERNATURAL HORROR IN LITERATURE, H.P. Lovecraft. Great modern American supernaturalist brilliantly surveys history of genre to 1930's, summarizing, evaluating scores of books. Necessary for every student, lover of form. Introduction by E.F. Bleiler. 111pp. 20105-8 Pa. $1.50

THREE GOTHIC NOVELS, ed. by E.F. Bleiler. Full texts Castle of Otranto, Walpole; Vathek, Beckford; The Vampyre, Polidori; Fragment of a Novel, Lord Byron. 331pp. 21232-7 Pa. $3.00

SEVEN SCIENCE FICTION NOVELS, H.G. Wells. Full novels. First Men in the Moon, Island of Dr. Moreau, War of the Worlds, Food of the Gods, Invisible Man, Time Machine, In the Days of the Comet. A basic science-fiction library. 1015pp.
USO 20264-X Clothbd. $6.00

LADY AUDLEY'S SECRET, Mary E. Braddon. Great Victorian mystery classic, beautifully plotted, suspenseful; praised by Thackeray, Boucher, Starrett, others. What happened to beautiful, vicious Lady Audley's husband? Introduction by Norman Donaldson. 286pp. 23011-2 Pa. $3.00

THE RED FAIRY BOOK, Andrew Lang. Lang's color fairy books have long been children's favorites. This volume includes Rapunzel, Jack and the Bean-stalk and 35 other stories, familiar and unfamiliar. 4 plates, 93 illustrations x + 367pp.
21673-X Paperbound **$3.00**

THE BLUE FAIRY BOOK, Andrew Lang. Lang's tales come from all countries and all times. Here are 37 tales from Grimm, the Arabian Nights, Greek Mythology, and other fascinating sources. 8 plates, 130 illustrations. xi + 390pp.
21437-0 Paperbound **$3.50**

HOUSEHOLD STORIES BY THE BROTHERS GRIMM. Classic English-language edition of the well-known tales — Rumpelstiltskin, Snow White, Hansel and Gretel, The Twelve Brothers, Faithful John, Rapunzel, Tom Thumb (52 stories in all). Translated into simple, straightforward English by Lucy Crane. Ornamented with headpieces, vignettes, elaborate decorative initials and a dozen full-page illustrations by Walter Crane. x + 269pp. 21080-4 Paperbound **$3.00**

THE MERRY ADVENTURES OF ROBIN HOOD, Howard Pyle. The finest modern versions of the traditional ballads and tales about the great English outlaw. Howard Pyle's complete prose version, with every word, every illustration of the first edition. Do not confuse this facsimile of the original (1883) with modern editions that change text or illustrations. 23 plates plus many page decorations. xxii + 296pp.
22043-5 Paperbound **$4.00**

THE STORY OF KING ARTHUR AND HIS KNIGHTS, Howard Pyle. The finest children's version of the life of King Arthur; brilliantly retold by Pyle, with 48 of his most imaginative illustrations. xviii + 313pp. $6\frac{1}{8}$ x $9\frac{1}{4}$.
21445-1 Paperbound **$3.50**

THE WONDERFUL WIZARD OF OZ, L. Frank Baum. America's finest children's book in facsimile of first edition with all Denslow illustrations in full color. The edition a child should have. Introduction by Martin Gardner. 23 color plates, scores of drawings. iv + 267pp. 20691-2 Paperbound **$3.00**

THE MARVELOUS LAND OF OZ, L. Frank Baum. The second Oz book, every bit as imaginative as the Wizard. The hero is a boy named Tip, but the Scarecrow and the Tin Woodman are back, as is the Oz magic. 16 color plates, 120 drawings by John R. Neill. 287pp. 20692-0 Paperbound **$3.00**

THE MAGICAL MONARCH OF MO, L. Frank Baum. Remarkable adventures in a land even stranger than Oz. The best of Baum's books not in the Oz series. 15 color plates and dozens of drawings by Frank Verbeck. xviii + 237pp.
21892-9 Paperbound **$2.95**

THE BAD CHILD'S BOOK OF BEASTS, MORE BEASTS FOR WORSE CHILDREN, A MORAL ALPHABET, Hilaire Belloc. Three complete humor classics in one volume. Be kind to the frog, and do not call him names . . . and 28 other whimsical animals. Familiar favorites and some not so well known. Illustrated by Basil Blackwell. 156pp. (USO) 20749-8 Paperbound **$2.00**

150 MASTERPIECES OF DRAWING, edited by Anthony Toney. 150 plates, early 15th century to end of 18th century; Rembrandt, Michelangelo, Dürer, Fragonard, Watteau, Wouwerman, many others. 150pp. 8⅜ x 11¼. 21032-4 Pa. $4.00

THE GOLDEN AGE OF THE POSTER, Hayward and Blanche Cirker. 70 extraordinary posters in full colors, from Maîtres de l'Affiche, Mucha, Lautrec, Bradley, Cheret, Beardsley, many others. 9⅜ x 12¼. 22753-7 Pa. $4.95
21718-3 Clothbd. $7.95

SIMPLICISSIMUS, selection, translations and text by Stanley Appelbaum. 180 satirical drawings, 16 in full color, from the famous German weekly magazine in the years 1896 to 1926. 24 artists included: Grosz, Kley, Pascin, Kubin, Kollwitz, plus Heine, Thöny, Bruno Paul, others. 172pp. 8½ x 12¼. 23098-8 Pa. $5.00
23099-6 Clothbd. $10.00

THE EARLY WORK OF AUBREY BEARDSLEY, Aubrey Beardsley. 157 plates, 2 in color: Manon Lescaut, Madame Bovary, Morte d'Arthur, Salome, other. Introduction by H. Marillier. 175pp. 8½ x 11. 21816-3 Pa. $4.00

THE LATER WORK OF AUBREY BEARDSLEY, Aubrey Beardsley. Exotic masterpieces of full maturity: Venus and Tannhäuser, Lysistrata, Rape of the Lock, Volpone, Savoy material, etc. 174 plates, 2 in color. 176pp. 8½ x 11. 21817-1 Pa. $4.00

DRAWINGS OF WILLIAM BLAKE, William Blake. 92 plates from Book of Job, Divine Comedy, Paradise Lost, visionary heads, mythological figures, Laocoön, etc. Selection, introduction, commentary by Sir Geoffrey Keynes. 178pp. 8½ x 11.
22303-5 Pa. $3.50

LONDON: A PILGRIMAGE, Gustave Doré, Blanchard Jerrold. Squalor, riches, misery, beauty of mid-Victorian metropolis; 55 wonderful plates, 125 other illustrations, full social, cultural text by Jerrold. 191pp. of text. 8⅛ x 11.
22306-X Pa. $5.00

THE COMPLETE WOODCUTS OF ALBRECHT DÜRER, edited by Dr. W. Kurth. 346 in all: Old Testament, St. Jerome, Passion, Life of Virgin, Apocalypse, many others. Introduction by Campbell Dodgson. 285pp. 8½ x 12¼. 21097-9 Pa. $6.00

THE DISASTERS OF WAR, Francisco Goya. 83 etchings record horrors of Napoleonic wars in Spain and war in general. Reprint of 1st edition, plus 3 additional plates. Introduction by Philip Hofer. 97pp. 9⅜ x 8¼. 21872-4 Pa. $3.00

ENGRAVINGS OF HOGARTH, William Hogarth. 101 of Hogarth's greatest works: Rake's Progress, Harlot's Progress, Illustrations for Hudibras, Midnight Modern Conversation, Before and After, Beer Street and Gin Lane, many more. Full commentary. 256pp. 11 x 14. 22479-1 Pa. $7.00
23023-6 Clothbd. $13.50

PRIMITIVE ART, Franz Boas. Great anthropologist on ceramics, textiles, wood, stone, metal, etc.; patterns, technology, symbols, styles. All areas, but fullest on Northwest Coast Indians. 350 illustrations. 378pp. 20025-6 Pa. $3.75